AF840237

VOLUME ONE HUNDRED AND FORTY SEVEN

ADVANCES IN CANCER RESEARCH

Receptor Tyrosine Kinases

VOLUME ONE HUNDRED AND FORTY SEVEN

ADVANCES IN CANCER RESEARCH

Receptor Tyrosine Kinases

Edited by

RAKESH KUMAR

Cancer Research Program, Rajiv Gandhi Centre for Biotechnology, Trivandrum, Kerala, India; Department of Medicine, Division of Hematology & Oncology, Rutgers New Jersey Medical School, Newark, NJ; Department of Human and Molecular Genetics, Virginia Commonwealth University, School of Medicine, Richmond, VA, United States

PAUL B. FISHER

Department of Human and Molecular Genetics, Virginia Commonwealth University, School of Medicine; VCU Institute of Molecular Medicine; VCU Massey Cancer Center, Virginia Commonwealth University, School of Medicine, Richmond, VA, United States

ELSEVIER

ACADEMIC PRESS
An imprint of Elsevier

Academic Press is an imprint of Elsevier
50 Hampshire Street, 5th Floor, Cambridge, MA 02139, United States
525 B Street, Suite 1650, San Diego, CA 92101, United States
The Boulevard, Langford Lane, Kidlington, Oxford OX5 1GB, United Kingdom
125 London Wall, London, EC2Y 5AS, United Kingdom

First edition 2020

Copyright © 2020 Elsevier Inc. All rights reserved.

No part of this publication may be reproduced or transmitted in any form or by any means, electronic or mechanical, including photocopying, recording, or any information storage and retrieval system, without permission in writing from the publisher. Details on how to seek permission, further information about the Publisher's permissions policies and our arrangements with organizations such as the Copyright Clearance Center and the Copyright Licensing Agency, can be found at our website: www.elsevier.com/permissions.

This book and the individual contributions contained in it are protected under copyright by the Publisher (other than as may be noted herein).

Notices
Knowledge and best practice in this field are constantly changing. As new research and experience broaden our understanding, changes in research methods, professional practices, or medical treatment may become necessary.

Practitioners and researchers must always rely on their own experience and knowledge in evaluating and using any information, methods, compounds, or experiments described herein. In using such information or methods they should be mindful of their own safety and the safety of others, including parties for whom they have a professional responsibility.

To the fullest extent of the law, neither the Publisher nor the authors, contributors, or editors, assume any liability for any injury and/or damage to persons or property as a matter of products liability, negligence or otherwise, or from any use or operation of any methods, products, instructions, or ideas contained in the material herein.

ISBN: 978-0-12-820142-8
ISSN: 0065-230X

For information on all Academic Press publications
visit our website at https://www.elsevier.com/books-and-journals

Publisher: Zoe Kruze
Editorial Project Manager: Leticia M. Lima
Production Project Manager: Vijayaraj Purushothaman
Cover Designer: Miles Hitchen

Typeset by SPi Global, India

ELSEVIER Book Aid International
Working together to grow libraries in developing countries
www.elsevier.com • www.bookaid.org

Contents

Contributors

Wadih Arap
Rutgers Cancer Institute of New Jersey; Division of Hematology/Oncology, Department of Medicine, Rutgers New Jersey Medical School, Newark, NJ, United States

Marcia R. Campbell
Department of Medicine, Helen Diller Family Comprehensive Cancer Center, University of California, San Francisco, CA, United States

Neal Chawla
Department of Medicine, Advocate Illinois Masonic Medical Center, Chicago, IL, United States

Mei-Kuang Chen
Department of Molecular and Cellular Oncology, The University of Texas MD Anderson Cancer Center; The University of Texas MD Anderson Cancer Center UTHealth Graduate School of Biomedical Sciences, Houston, TX, United States

Louis W.C. Chow
UNIMED Medical Institute; Organisation for Oncology and Translational Research, Hong Kong, China

Luis Marques da Costa
Medical Oncology Department, Hospital de Santa Maria, Centro Hospitalar Universitário Lisboa Norte; Instituto de Medicina Molecular, Faculdade de Medicina, Universidade de Lisboa, Lisbon, Portugal

Swadesh K. Das
Department of Human and Molecular Genetics; VCU Institute of Molecular Medicine; VCU Massey Cancer Center, Virginia Commonwealth University, School of Medicine, Richmond, VA, United States

Deodate Davis
Rutgers Cancer Institute of New Jersey; Division of Cancer Biology, Department of Radiation Oncology, Rutgers New Jersey Medical School, Newark, NJ, United States

Luni Emdad
Department of Human and Molecular Genetics; VCU Institute of Molecular Medicine; VCU Massey Cancer Center, Virginia Commonwealth University, School of Medicine, Richmond, VA, United States

Lynsey M. Fettig
Masonic Cancer Center, University of Minnesota, Minneapolis, MN, United States

Paul B. Fisher
Department of Human and Molecular Genetics; VCU Institute of Molecular Medicine; VCU Massey Cancer Center, Virginia Commonwealth University, School of Medicine, Richmond, VA, United States

Jeremy Fricke
Department of Medical Oncology and Therapeutics Research, City of Hope National Medical Center, Duarte, CA, United States

Bijesh George
Cancer Research Program, Rajiv Gandhi Centre for Biotechnology, Trivandrum, Kerala, India

Soma Ghosh
Department of Biological Regulation, Weizmann Institute of Science, Rehovot, Israel

Jennifer L. Hsu
Department of Molecular and Cellular Oncology, The University of Texas MD Anderson Cancer Center, Houston, TX, United States

Mien-Chie Hung
Department of Molecular and Cellular Oncology, The University of Texas MD Anderson Cancer Center, Houston, TX, United States; Graduate Institute of Biomedical Sciences, Research Center for Cancer Biology, and Center for Molecular Medicine, China Medical University, Taichung, Taiwan

Rakesh Kumar
Cancer Research Program, Rajiv Gandhi Centre for Biotechnology, Trivandrum, Kerala, India; Department of Medicine, Division of Hematology & Oncology, Rutgers New Jersey Medical School, Newark, NJ; Department of Human and Molecular Genetics, Virginia Commonwealth University, School of Medicine, Richmond, VA, United States

Erich F. Lie
Organisation for Oncology and Translational Research, Hong Kong, China

Raeva Malik
George Washington University Hospital, Washington, DC, United States

Isa Mambetsariev
Department of Medical Oncology and Therapeutics Research, City of Hope National Medical Center, Duarte, CA, United States

Ilaria Marrocco
Department of Biological Regulation, Weizmann Institute of Science, Rehovot, Israel

Cecília Melo-Alvim
Medical Oncology Department, Hospital de Santa Maria, Centro Hospitalar Universitário Lisboa Norte, Lisbon, Portugal

Mark M. Moasser
Department of Medicine, Helen Diller Family Comprehensive Cancer Center, University of California, San Francisco, CA, United States

William J. Muller
Rosalind and Morris Goodman Cancer Research Center; Department of Biochemistry; Faculty of Medicine, McGill University, Montreal, QC, Canada

Revathy Nadhan
Cancer Research Program, Rajiv Gandhi Centre for Biotechnology, Trivandrum, Kerala, India

Arin Nam
Department of Medical Oncology and Therapeutics Research, City of Hope National Medical Center, Duarte, CA, United States

Renata Pasqualini
Rutgers Cancer Institute of New Jersey; Division of Cancer Biology, Department of Radiation Oncology, Rutgers New Jersey Medical School, Newark, NJ, United States

Aswathy Mary Paul
Cancer Research Program, Rajiv Gandhi Centre for Biotechnology, Trivandrum, Kerala, India

Rebecca Pharaon
Department of Medical Oncology and Therapeutics Research, City of Hope National Medical Center, Duarte, CA, United States

M. Radhakrishna Pillai
Cancer Research Program, Rajiv Gandhi Centre for Biotechnology, Trivandrum, Kerala, India

Leonor Ribeiro
Medical Oncology Department, Hospital de Santa Maria, Centro Hospitalar Universitário Lisboa Norte, Lisbon, Portugal

Ravi Salgia
Department of Medical Oncology and Therapeutics Research, City of Hope National Medical Center, Duarte, CA, United States

Alexandra M. Simond
Rosalind and Morris Goodman Cancer Research Center; Department of Biochemistry, McGill University, Montreal, QC, Canada

Priya Srinivas
Cancer Research Program, Rajiv Gandhi Centre for Biotechnology, Trivandrum, Kerala, India

Fernanda I. Staquicini
Rutgers Cancer Institute of New Jersey; Division of Cancer Biology, Department of Radiation Oncology, Rutgers New Jersey Medical School, Newark, NJ, United States

Sarmistha Talukdar
Department of Human and Molecular Genetics; VCU Institute of Molecular Medicine, Virginia Commonwealth University, School of Medicine, Richmond, VA, United States

Fenny H.F. Tang
Rutgers Cancer Institute of New Jersey; Division of Cancer Biology, Department of Radiation Oncology, Rutgers New Jersey Medical School, Newark, NJ, United States

Masakazu Toi
Organisation for Oncology and Translational Research, Hong Kong, China; Department of Breast Surgery, Graduate School of Medicine, Kyoto University, Kyoto, Japan

Nandini Verma
Advanced Centre for Treatment, Research and Education in Cancer, Mumbai, India

Yosef Yarden
Department of Biological Regulation, Weizmann Institute of Science, Rehovot, Israel

Douglas Yee
Masonic Cancer Center, University of Minnesota, Minneapolis, MN, United States

Preface

Cellular homeostasis demands efficient transduction of signals from dynamic changes in the extracellular milieu, allowing cells to respond appropriately for the maintenance of metabolic equilibrium. Malfunction in the underlying biological system and its components that control this fundamental cellular process can lead to a variety of diseases, including, excessive or hyperproliferation in cells leading to cancer. Cell surface receptor tyrosine kinases (RTKs) play a seminal role in capturing and transducing signals from the extracellular milieu in both healthy and cancerous cells. Among RTKs, the human epidermal growth factor receptor (HER) family has emerged as a prototype providing mechanistic insights and helping to expand our understanding of RTK biology and the role of RTKs in human cancer development and progression. This information provides a new appreciation of HER biology that is relevant to additional RTK family members and establishes a framework for developing new targeted cancer therapeutics exploiting RTK-mediated pathways in cancer cells.

RTKs continue to be the focal point of targeted cancer therapies and this concerted effort has culminated in the development of new anticancer drugs. Considering the excellent progress that has been made in understanding the mode of action of RTKs and their current and future impact on cancer therapy, it was considered timely to develop a dedicated thematic issue of *Advances in Cancer Research* (ACR) focused on RTKs. To achieve this objective Rakesh Kumar and Paul B. Fisher (Co-Editors of this thematic issue of ACR) have recruited well-recognized international experts with a rich history of original contributions to the RTK field and its clinical utility over the preceding decades. In the present thematic collection, leaders in the areas of one or more RTK family members provide a historical background and updated summary of the most impactful advances of these molecule(s) in cancer research, and how the scientific community has exploited the core mechanistic principles of action to develop novel and effective targeted cancer therapeutics.

We launch this special volume with the pioneering work of Yosef Yarden. Yarden and colleagues (Gosh and Marrocco) discuss a series of guidepost advances in the action of RTKs and their role in cancer progression, the function of distinct gene-specific structural features of these receptors in ligand-triggered receptor activation, and why RTKs have emerged as molecules of interest for targeted cancer therapeutics. In the same vein, Hung and colleagues (Chen and Hsu) discuss their discovery of HER2, the significance of RTK signaling in distinct intracellular compartments, and how noncanonical signaling could influence the process of cancer progression as well as offer superior strategies for therapeutic intervention in human cancer. Kumar and Moasser and colleagues (George, Campbell, Verma, Paul, Melo-Alvim, Ribeiro, Pillai and Costa) provide a brief account of discoveries and defining moments in the history of the HER family since its inception. They also discuss early advances in the action of heregulin and developing anti-HER blocking antibodies and how these early advances catalyzed the uncovering of new aspects of HER biology in cancer research and treatment with a particular focus on HER3 oncobiology. Kumar and Moasser also highlight the significance of HER3 as one of the shared effectors of acquired resistance to a range of diverse cancer drugs and present a postulated model for consideration. This review article also shares some of the on-going limitations of the current RTK-directed therapies and a few examples of anticipated advances in the field in the coming decade. Fisher and colleagues (Talukdar, Emdad, and Das) emphasize the role of EGFR and its naturally occurring variants as driving forces in the process of tumorigenesis through supporting the functions of cancer stem cells, including stemness, metabolism, immunomodulatory-activity, dormancy, and therapy-resistance. The authors also link the EGFR—cancer stem cell pathway to acquired therapeutic resistance to EGFR inhibitors, and how novel combinatorial therapeutic approaches are proving to be useful to achieve superior clinical responses. Muller and colleague (Simond) overview the utility of genetically engineered mouse models for *in vivo* modeling of HER family members to better comprehend the initiation, progression, and metastasis of human cancer. The article describes the use of HER2 mouse models to illustrate the value and power of whole animal models in shaping our understanding of HER signaling and vetting therapeutic approaches. Yee and colleague (Fettig) discuss exciting advances in the areas of regulation and/or trans-regulation of the HER family members and insulin-like growth factor-1 receptor (IGF1R) in cancer cells and how dual targeting of two hyper-activated RTKs can contribute to a superior therapeutic response in cancer patients. Salgia and colleagues (Malik, Mambetsariev, Fricke, Chawla, Nam, and Pharaon) summarize the history, genetics, and clinical utility of the MET receptor and its variant in the context of mechanism of resistance to targeted therapy. Arap, Pasqualini, and Staquicini and colleagues

(Tang and Davis) describe the significance of the Eph receptors and ephrin ligands in a range of physiologic processes and proliferative diseases, and how the *S*election of *P*hage-displayed *A*ccessible *R*ecombinant *T*argeted *A*ntibodies (SPARTA) platform is being used to screen novel antibody libraries against known tumor targets. Pillai and colleagues (Nadhan and Srinivas) encapsulate advances in the field of RTK-based therapies in head and neck cancers and discuss the problem of acquired therapeutic resistance. And, finally, Chow and Toi and colleague (Lie) highlight investigations on the efficacy and safety of HER2-directed therapies and combination cancer treatment strategies for a greater benefit of cancer patients.

We believe the current thematic issue of ACR which focuses on RTKs will serve as a timely and indispensable reference source for faculty, researchers and students alike, and will also be of significant value to the readership of ACR and Cancer Medicine, at-large.

RAKESH KUMAR
PAUL B. FISHER

CHAPTER ONE

Roles for receptor tyrosine kinases in tumor progression and implications for cancer treatment

Soma Ghosh, Ilaria Marrocco, Yosef Yarden*
Department of Biological Regulation, Weizmann Institute of Science, Rehovot, Israel
*Corresponding author: e-mail address: yosef.yarden@weizmann.ac.il

Contents

Advances in Cancer Research, Volume 147
ISSN 0065-230X
https://doi.org/10.1016/bs.acr.2020.04.002
© 2020 Elsevier Inc.
All rights reserved.

Abstract

Growth factors and their receptor tyrosine kinases (RTKs), a group of transmembrane molecules harboring cytoplasm-facing tyrosine-specific kinase functions, play essential roles in migration of multipotent cell populations and rapid proliferation of stem cells' descendants, transit amplifying cells, during embryogenesis and tissue repair. These intrinsic functions are aberrantly harnessed when cancer cells undergo intertwined phases of cell migration and proliferation during cancer progression. For example, by means of clonal expansion growth factors fixate the rarely occurring driver mutations, which initiate tumors. Likewise, autocrine and stromal growth factors propel angiogenesis and penetration into the newly sprouted vessels, which enable seeding micrometastases at distant organs. We review genetic and other mechanisms that preempt ligand-mediated activation of RTKs, thereby supporting sustained cancer progression. The widespread occurrence of aberrant RTKs and downstream signaling pathways in cancer, identifies molecular targets suitable for pharmacological intervention. We list all clinically approved cancer drugs that specifically intercept oncogenic RTKs. These are mainly tyrosine kinase inhibitors and monoclonal antibodies, which can inhibit cancer but inevitably become progressively less effective due to adaptive rewiring processes or emergence of new mutations, processes we overview. Similarly important are patient treatments making use of radiation, chemotherapeutic agents and immune checkpoint inhibitors. The many interfaces linking RTK-targeted therapies and these systemic or local regimens are described in details because of the great promise offered by combining pharmacological modalities.

1. Preface: An introduction to growth factors and the family of receptor tyrosine kinases (RTKs)

Growth factors are relatively small and compact molecules that play key roles in cell-to-cell communication, including control of cellular growth, proliferation, differentiation, survival and migration. These processes are propelled by binding of growth factor molecules to their specific receptors, which relay, by means of phosphorylation events, biochemical signals to intracellular organelles, such as mitochondria and the nucleus. The majority of growth factor receptors belong to the receptor tyrosine kinase (RTK) family (Blume-Jensen & Hunter, 2001; Yarden & Ullrich, 1988). The human genome encodes for 58 RTKs, which fall into

20 subfamilies, as defined by genetic phylogeny (Lemmon & Schlessinger, 2010). The growth factors and the respective RTKs, along with the intracellular signals generated following receptor activation, play crucial roles in cell fate determination, hence they are highly conserved in evolution, from the nematode *C. elegans* to humans (Pires-daSilva & Sommer, 2003). Mutations in RTKs or proteins that facilitate their downstream signaling have been implicated in the onset and progression of a wide-range of pathological states, such as diabetes, inflammation, bone disorders, atherosclerosis, angiogenesis, and various malignancies (Blume-Jensen & Hunter, 2001). Thus, an increasing number of RTK-targeted therapies are currently emerging. This chapter describes the processes whereby RTKs are activated under normal physiological conditions, and discusses several mechanisms involving aberrant activation of RTKs in human cancers, as well as the implications for selection of anti-cancer therapies.

2. Classification and activation of RTKs

2.1 Classification of RTKs

It has been shown that among 90 identified genes encoding proteins with tyrosine kinase activity, 58 encode receptors, and these might be classified into 20 classes (Table 1). Of these subfamilies, EGFR/ERBB (class I), the receptor for insulin (class II), for PDGF (Class III), for FGF (class IV), for VEGF (class V) and HGF (MET, class VI) are strongly associated with malignant diseases (Grassot, Mouchiroud, & Perriere, 2003). All RTKs share a common structure characterized by a single transmembrane domain, a highly glycosylated N-terminal extracellular domain containing a large number of disulfide bonds, and an intracellular tyrosine kinase domain flanked by a hydrophilic carboxyl-terminal tail. The extracellular domain is involved in ligand recognition and in dimerization of monomeric receptors. The extracellular domains of RTKs harbor a variety of structural motifs (e.g., immunoglobulin-like domains, regions rich in leucines or cysteines, and fibronectin type III domains), which define the respective class of RTKs and dictate recognition of specific growth factor ligands. The cytoplasmic domains of RTKs are less variables, since they harbor the evolutionary conserved tyrosine kinase region involved in catalysis of ATP-dependent phosphorylation of receptor chains. The bi-lobular tyrosine kinase domain includes an activation loop, the orientation and phosphorylation of which determine the state of activation. The tyrosine kinase domain is flanked at the amino terminus by a juxtamembrane region composed of 40–80 amino

Table 1 Sub-families of receptor tyrosine kinases and their roles in cancer.

Class	RTK family	Members	Ligands	Extracellular motives	Relevance to cancer
I	ERBB	EGFR, ERBB2, ERBB3, ERBB4	EGF, TGF-a, HB-EGF, AREG, BTC, EPG, EPR, NRG1–4	2 Cysteine-rich domains	*EGFR* mutation is a driver of tumorigenesis and biomarker of resistance to oncologic treatments (Witsch, Sela, & Yarden, 2010). Usually overexpressed in cancer cells, and also related to PD-L1 overexpression (Akbay et al., 2013)
II	Insulin	INSR, IGFR, IRR	IGF-1, IGF-2, Insulin	2 Chains α and β, with one cysteine-rich and 2 FNIII domains	Extensively assessed for its biological activity, and its contribution to tumor progression. Involved in regulation of ATP-binding cassette transporter proteins and interactions with the extracellular matrix (Yuan, Yin, Tao, Wang, & Gao, 2018)
III	PDGF	PDGFRα, PDGFRβ, CSF1R, KIT, FLT3L/FLK2	PDGF-AA, PDGF- AB, PDGF-BB, PDGF-CC, PDGF-DD, CSF-1, IL-34, SCF	5 Ig-like domains	Closely associated with tumor development, cell proliferation, metastasis, chemotherapy resistance, angiogenesis (Farooqi & Siddik, 2015), immunosuppressive tumor microenvironment and apoptosis inhibition (García-Aranda & Redondo, 2019)
IV	VEGF	VEGFR1, VEGFR2, VEGFR3	VEGF A-D, PLGF1–2	7 Ig-like domains	Primarily promotes tumor angiogenesis through stimulating the proliferation and survival of endothelial cells and also by increasing the permeability of vessels (Yang, Yan, & Liu, 2018)
V	FGF	FGFR1, FGFR2, FGFR3, FGFR4	FGF 1–25	3 Ig-like domains, 1 acidic box	Involved in carcinogenesis (Helsten et al., 2016) through cell signaling deregulation, angiogenesis and resistance to cancer therapies (Ghedini, Ronca, Presta, & Giacomini, 2018).

VI	PTK7	PTK7/ CCK4	No ligands identified	7 Ig-like domains	Involved in tumorigenesis and metastasis (Golubkov et al., 2014)
VII	TRK	TRKA, TRKB, TRKC	NGF, NTF-4, NTF-4, BDNF	2 Ig-like domains, rich leucin domains	*NTRK* gene fusions are oncogenic drivers of different types of cancers, which are involved in cell differentiation and survival (Cocco, Scaltriti, & Drilon, 2018). Twenty-three percent of cancer patients presenting *NTRK* fusions express PD-L1 (Gatalica, Xiu, Swensen, & Vranic, 2019)
VIII	MET	MET, RON	HGF	4 Ig-like domains, 1 transmembrane α chain linked with one extracellular β chain	*MET* is mainly activated by receptor overexpression, co-expression with HGF, or genomic up-regulation (Mo & Liu, 2017)
IX	EPH	EPHA1 to 8, EPHB1 to 6	EphrinB1–3, ephrinA1-A5	1 Ig-like, 1 cysteine-rich and 2 FNIII-like domains	These kinases are dependence receptors, playing dual functions as both oncogenes and tumor suppressors depending on the presence of their Ephrin ligands (Andres & Ziemiecki, 2003)
X	AXL	AXL, MER, TYRO3	Gas6, Pros-1	2 Ig-line, 2 FNIII-like domains	Involved in tumor cell proliferation, survival, stem cell phenotype, metastasis, and resistance to cancer therapy. Also, contributes to immunosuppressive and pro-tumorigenic phenotypes (Rankin & Giaccia, 2016)
XI	TIE	TIE, TEK	Ang1–4	2 Ig-like, 1 EGF, and 3 FNIII-like domains	It is recognized as an initial step in tumor progression and angiogenesis. Increased expression of Ang-2 correlates with poor prognosis of patients with different malignancies (Hong et al., 2017)

Continued

Table 1 Sub-families of receptor tyrosine kinases and their roles in cancer.—cont'd

Class	RTK family	Members	Ligands	Extracellular motives	Relevance to cancer
XII	RYK	RYK	Wnt3a, Wnt5a	1 Transmembrane β chain linked with one extracellular α chain	RYK is overexpressed in some types of cancer and correlates to worse survival (Trivier & Ganesan, 2002). RYK activates the WNT/β-catenin pathway essential for cancer stem cell maintenance
XIII	DDR	DDR1, DDR2	Collagens	1 Discoidin-like domain	Regulate EMT, cell migration, invasion and survival, apoptosis inhibition and chemoresistance (Gadiya & Chakraborty, 2018)
XIV	RET	RET	GDNF, Neurturin, artemin, persephin	1 Cadherin-like domain, 1 cysteine-rich domain	Frequently mutated in different malignancies leading to cell survival and apoptosis inhibition (Bagheri-Yarmand et al., 2015)
XV	ROS	ROS	Unidentified	6 FNIII-like domains	Aberrant expression of ROS (fusion isoform) is an important driver of different types of cancers (El-Deeb, Yoo, & Lee, 2011)
XVI	LTK	LTK, ALK	PTN, MK, ALKALS, FAM150A, FAM150B	1 Cysteine-rich domain	Genomic fusions affecting ALK gene are common in cancer and cause ALK constitutive activation, leading to uncontrolled cell proliferation (García-Aranda & Redondo, 2019)

XVII	ROR	ROR1, ROR2	Wnt5A	1 Ig-domain, 1 cysteine-rich domain, one kringle-like domains and Fz like domain	As non-canonical Wnt signaling mediator, this protein kinase has a dual role as a tumor suppressor or activator depending on tumor type or stage (García-Aranda & Redondo, 2019)
XVIII	MUSK	MUSK	ARGIN	4 Ig-like, 1 cysteine-rich domains and Fz like domain	Activates cell proliferation, apoptosis, differentiation and tumorigenesis (Cheusova et al., 2006)
XIX	LMR	AATYK1, AATYK2, AATYK3	Unidentified	A short extracellular domain	Frequently mutated in colorectal cancer and prostate cancer and is associated with the initiation and progression of the cancer phenotype (Rossi, Ionov, Bakin, & Cowell, 2005)
XX	STYK1	SuRTK106, STYK1	Lacks ECD for binding specific ligands	A short receptor chain with a short extracellular domain	STYK aberrant expression is found in different malignancies, including colorectal cancer (Abbaspour Babaei, Kamalidehghan, Saleem, Huri, & Ahmadipour, 2016)

Listed are all receptors of the super family, the structural motifs that characterize each sub-family and some of the main functions in cancer progression. The abbreviations used are: ALK: anaplastic lymphoma kinase; CCK: colon carcinoma kinase; CSF-1R: colony-stimulating factor 1 receptor; DDR: discoidin domain receptor; ECD, extracellular domain; EGFR, epidermal growth factor receptor; EphR: ephrin receptor; FGFR: fibroblast growth factor receptor; FLT3: FMS related tyrosine kinase 3; FNIII, fibronectin type III; Fz, frizzled; HGFR: hepatocyte growth factor receptor; Ig, immunoglobulin; IGFR: insulin growth factor receptor; IRR: insulin receptor-related receptor; InsR: insulin receptor; LMR: lemur receptor kinases; LTK: leukocyte tyrosine kinase; MuSK: muscle-specific kinase; MET: mesenchymal-epithelial transition factor; NTRK: neurotrophin receptor kinase; PDGFR: platelet-derived growth factor receptor; PTK7: protein tyrosine kinase-like 7; RET: rearranged during transfection; RON: receptour d'origine nantais; RO-R: receptor tyrosine kinase-like orphan receptors; ROS: reactive oxygen species receptors; RYK: receptor-related to tyrosine kinases; STYK: serine/threonine/tyrosine kinase; TRK: tropomyosin receptor kinase; TAM: TYRO3-, AXL-, MER-TK receptors; TIE: tyrosine kinase receptor in endothelial cells; VEGFR: vascular endothelial growth factor receptor.

acids. The number of intracellular tyrosine residues and their distribution widely vary among RTKs. Thus, PDGFRβ carries 27 tyrosine residues, of which 19 can be phosphorylated, whereas TRKA harbors only 11 tyrosine, of which 6 may undergo phosphorylation (Bradshaw, Chalkley, Biarc, & Burlingame, 2013; Grassot et al., 2003). These tyrosine residues stabilize the "open" conformation of the activation loop, which allows binding of ATP molecules and peptidergic substrates (Hubbard & Till, 2000).

2.2 Mechanisms underlying RTK dimerization and activation

Binding of a growth factor molecule to the extracellular region of the cognate RTK activates the receptor by means of ligand-induced receptor dimerization (Yarden & Schlessinger, 1987a, 1987b). One receptor in the dimer/oligomer then phosphorylates one or more tyrosines in a neighboring receptor, which results in *trans*-autophosphorylation of each kinase domain and releases *cis*-autoinhibition (Lemmon & Schlessinger, 2010). Thereafter, the phosphorylated receptor serves as a focal site for assembly (and activation) of intracellular downstream signaling proteins, which propagate critical signals into the cell. In general, four molecular modes permit RTK dimerization and consequent activation of the tyrosine kinase domain (Fig. 1). In the first mode, receptor dimerization is solely "ligand mediated." For example, TRKA, the RTK that binds with the nerve growth factor (NGF/neurotrophin) forms dimers without any direct contact between the extracellular regions of the two receptors (Wehrman et al., 2007; Wiesmann, Ultsch, Bass, & de Vos, 1999). In the second mode, dimerization is completely receptor mediated, without any physical interaction between two activating ligands, as in the case of the ERBB family members (EGFR, HER2/ERBB2, HER3/ERBB3, and HER4/ERBB4) (Zhang, Gureasko, Shen, Cole, & Kuriyan, 2006). According to the third setting, ligand homodimers bind to two receptor molecules, which then interact with each other. This is the case for KIT, the stem cell factor (SCF) receptor (Yuzawa et al., 2007). In the fourth mode, receptors employ a combination of bivalent ligand binding, direct receptor-receptor contacts, and an accessory molecule. For example, the receptors for the fibroblasts growth factors (FGFs) use heparin or heparan sulfate as accessory molecules in this mode (Schlessinger et al., 2000; Stauber, DiGabriele, & Hendrickson, 2000).

2.3 The example of EGFR

The founding class of the RTK family, which includes EGFR, HER2 and two receptors for neuregulins (NRGs), ERBB3/HER3 and ERBB4/HER4, are

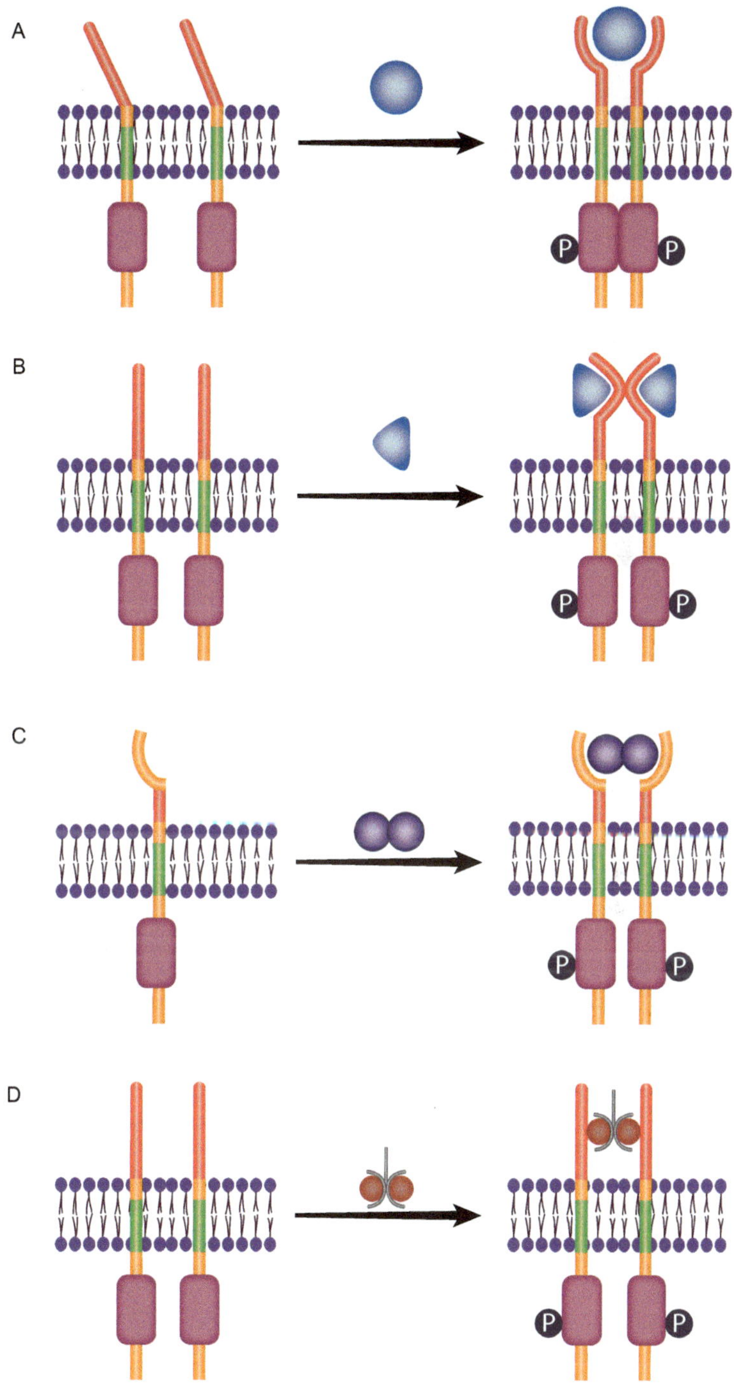

Fig. 1 See legend on next page.

folded such that dimerization is prevented (Burgess et al., 2003; Cho & Leahy, 2002). Only on ligand binding does the conformation of the extracellular domain change to expose a dimerization loop in a way that allows receptor dimerization. What follows is the juxtaposition of the tyrosine kinase domains in a head-to-tail configuration that removes and trans-phosphorylates an inhibitory carboxyl-terminal tail (Jura et al., 2009; Zhang et al., 2006). Conceivably, binding of EGF to its receptor increases the proximity of the other monomer of the dimer, hence permits autophosphorylation and trans-phosphorylation. In addition, receptor dimerization induces rotation of the N- and C-lobes of the kinase domain, as well as the major axis of EGFR. In the majority of RTKs, the activation loop moves to enable ATP binding and autophosphorylation of tyrosine residues located on the opposite receptor chain. Thus, transphosphorylation of key tyrosine residues located in the activation loop stabilizes the "open" conformation, which makes it possible to access the C-lobe and activates the catalytic function (Hubbard, 2002; Segaliny, Tellez-Gabriel, Heymann, & Heymann, 2015). Phosphorylation of the carboxyl tail leads to the recruitment of downstream signaling proteins, to initiate dimer-specific signaling and transcription cascades.

2.4 Activation of intracellular signaling pathways

Phosphorylation within the catalytic domain of most RTKs often increases the activity of the kinase domain, whereas phosphorylation within non-catalytic domains creates various anchoring sites for cytoplasmic proteins involved in intracellular signal transduction. These tyrosines are located on the juxtamembrane segment, the carboxyl-terminal domain, or within

Fig. 1 Variation on the theme of RTK dimerization. Shown are 4 variants of ligand-induced dimerization of RTKs. Note that the ligand may directly contribute to the dimer interface, for example when the ligand itself forms dimers. Alternatively, the ligand may only induce a conformational change that exposes a receptor's domain that establishes a new receptor-receptor interface. (A) The nerve growth factor (NGF) induces dimerization of TRKA without any direct contact between the extracellular regions of the two receptors. (B) Ligand binding to EGFR induces a conformational change in the extracellular domain, which exposes a dimerization arm and permits receptor-mediated dimerization. (C) The stem cell factor (SCF) receptor, KIT, forms dimers by means of a dimeric ligand, as well as a receptor-receptor interface established by one of the five immunoglobulin-like domains of KIT. (D) Dimerization of the receptors for the fibroblast growth factors (FGFs) employs a specific domain of the receptor, along with ligands that are bound to heparin or heparan sulfate proteoglycans. *Adapted from Lemmon, M. A., & Schlessinger, J. (2010). Cell signaling by receptor tyrosine kinases.* Cell, 141*(7), 1117–1134. doi:10.1016/j.cell.2010.06.011.*

insert regions found within some kinase domains (e.g., in PDGFRs). The phosphorylated residues create the cores of binding sites for SH2 (Src homology 2) or PTB (phosphotyrosine-binding) domains of a variety of signaling proteins. SH2/PTB domain containing proteins can be recruited directly to the receptor, or indirectly through docking proteins. Thus, docking proteins function as "assembly platforms" to recruit additional signaling molecules (Anderson et al., 1990; Brummer, Schmitz-Peiffer, & Daly, 2010). This mechanism allows simultaneous binding, activation and phosphorylation of numerous cytoplasmic proteins, which relay the signal towards various intracellular pathways. Some of the proteins recruited to activated RTKs harbor intrinsic enzymatic activity, such as in SRC (a cytoplasmic tyrosine-specific protein kinase) or phospholipase C gamma (PLCγ), a lipid degrading enzyme. Alternatively, they may serve as adaptor proteins that recruit other enzymes, such as growth factor receptor binding protein 2 (Grb2), which instigates the mitogen-activated protein kinase (MAPK) pathway. In summary, the presence of several phosphotyrosines on each RTK and the involvement of various docking proteins confer upon activated RTKs the ability to recruit and regulate a wide-range of linear signaling cascades, including the RAS-to-ERK, PI3K-to-AKT and JAK-to-STAT pathways.

2.5 Mechanisms controlling RTKs

The stimulation and fine tuning of RTKs are typically controlled by feedback loops, including receptor down-regulation, which restricts further RTK signaling. This type of molecular machinery exists prior to receptor activation, hence limits signal propagation through attenuation of RTK autophosphorylation or through promoting receptor ubiquitination, endocytosis and degradation. Many more control loops require synthesis of new RNAs and proteins. For example, ligand-activated RTKs, such as EGFR, set in motion multiple transcriptional and other control loops that strongly regulated EGFR signaling (Amit, Citri, et al., 2007; Avraham et al., 2010). Such events might be activated in-trans by other signaling pathways, such as the activated glucocorticoid receptor (Gelfo et al., 2019). This type of regulation is out of the scope of our review.

2.5.1 *Inhibition of RTK autophosphorylation* (see Fig. 2)

Protein phosphatases are known to inhibit the functions of RTK by causing dephosphorylation and subsequently blocking downstream signaling. Several studies have shown that these phosphatases specifically dephosphorylate certain subsets of RTK's phosphorylated sites, indicating a certain

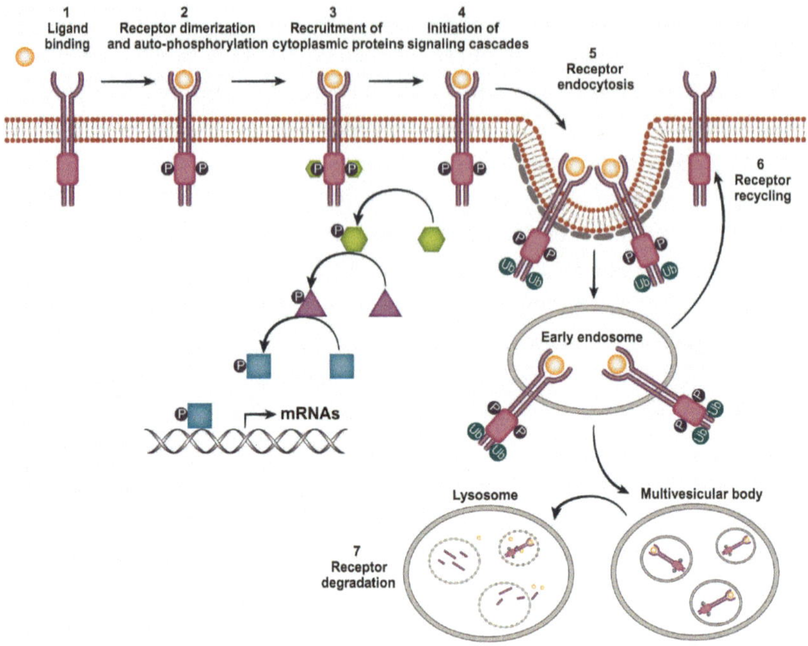

Fig. 2 Mechanisms controlling RTK signaling. The canonical RTK signaling process begins with receptor binding to its ligand (step 1). Thereafter, the occupied receptor undergoes dimerization, followed by transphosphorylation (2). The formed phosphotyrosine moieties serve as docking sites for cytoplasmic proteins, either enzymes or adaptors containing SRC homology-2 (SH2) or phosphotyrosine-binding domains (3). This step initiates signaling cascades, such as serial phosphorylation events, affecting membrane, cytoplasmic and nuclear proteins, including transcription factors that regulate specific gene expression programs (4). RTK endocytosis initiates at clathrin-coated pits and other domains of the plasma membrane (step 5). This step is facilitated by physical recruitment of E3 ubiquitin ligases, such as CBL family proteins, which covalently tag RTKs with mono- or di-ubiquitins. At the early endosome, the ubiquitinated RTK may undergo de-ubiquitination, by means of specific deubiquitylating enzymes (DUBs), which route internalized receptors back to the plasma membrane (6). CBL remains associated with the activated RTKs to ensure ubiquitin-dependent sorting into the inner vesicles of the multi-vesicular body, which are destined for lysosomal degradation (step 7).

degree of selectivity (Kovalenko et al., 2000; Ostman, Hellberg, & Bohmer, 2006). The phosphatases are typically divided into two classes, protein Ser/Thr phosphatases (PSPs) and protein tyrosine phosphatases (PTPs) (Li, Wilmanns, Thornton, & Kohn, 2013). The cysteine-based PTP superfamily includes approximately 100 members (Tonks, 2006), which are grouped into classical PTPs and dual specificity phosphatases (DUSPs).

Classical PTPs play critical roles in tyrosine kinase signaling (Neel & Tonks, 1997), whereas DUSPs can dephosphorylate both Tyr and Ser/Thr residues. Studies have shown that protein tyrosine phosphatase 1B (PTP1B; also known as PTPN1), which resides in the endoplasmic reticulum (ER), can regulate EGFR signaling. Fibroblasts from PTP1B-deficient mice showed an increase and sustained EGFR phosphorylation after growth factor treatment (Haj, Markova, Klaman, Bohmer, & Neel, 2003). It has also been demonstrated that PTP1B can regulate receptor endosomal trafficking. Other studies have demonstrated that PTP1B dephosphorylates EGFR just prior to its internalization into multi-vesicular bodies, by dephosphorylating signal transducing adapter molecule 2 (STAM2), a component of a large protein complex called ESCRT-0 (Neben, Lo, Jura, & Klein, 2019; Stuible & Tremblay, 2010). Another enzyme, known as receptor-like protein tyrosine phosphatase-κ (RPTPκ), dephosphorylates EGFR in human keratinocytes (Xu, Tan, Grachtchouk, Voorhees, & Fisher, 2005). Interestingly, RPTPκ levels increase in keratinocytes as they reach confluence and establish cell contact-mediated growth inhibition. Levels of density-enhanced phosphatase (DEP1; also known as RPTPη), a ubiquitously expressed RPTP, also increase in response to growth inhibition. This receptor-like phosphatase decreases EGFR activation and regulates receptor recycling (Tarcic et al., 2009). Thus, Tyr dephosphorylation regulates crucial sorting event, namely transport from early to late endosomes and the overall balance between receptor recycling and degradation.

2.5.2 Ligand-induced receptor ubiquitination and degradation

Tyrosine autophosphorylation of ligand-activated RTKs, such as EGFR, initiates recruitment of positively-acting signaling proteins, as well as negative regulators, such as CBL and GRB2. This event usually coincides with budding of clathrin-coated vesicles enriched in the internalized, activated RTK (Goh, Huang, Kim, Gygi, & Sorkin, 2010; Waterman & Yarden, 2001). Next, the endocytosed receptors are transferred to early endosomes and, later, to an acidic pre-lysosomal compartment called the multi-vesicular body (MVB). Any RTKs remaining in the late endosome and MVB are covalently modified by several ubiquitin ligases, such as the CBL and NEDD4 families of E3 ubiquitin-protein ligases. Several studies have linked RTK ubiquitination to receptor endocytosis (Marmor & Yarden, 2004; Mosesson et al., 2003). A number of endocytic regulatory proteins have been demonstrated to interact with ubiquitin and coordinate the trafficking of ubiquitinated RTKs from endosomes to lysosomes. Examples include

HRS, EPS15, STAM, Epsin and TSG101 (Amit et al., 2004; Dikic & Giordano, 2003; Polo & Di Fiore, 2006). The E3 ubiquitin ligase NEDD4–2 is an enzyme that binds specifically to the C-terminal portion of the TRKA receptor (Arevalo et al., 2006) and enhances ubiquitination and down-regulation of TRKA. In contrast, several other activated receptors, such as EGFR, the platelet-derived growth factor receptor (PDGFR), RET and MET are ubiquitinated upon interaction with CBL family members (Peschard et al., 2001). CBL can directly associate with activated receptors by binding of its PTB domain to a receptor's phosphotyrosine. It can also interact with the SH3 domain of Grb2, an adaptor protein known to associate with phosphorylated receptors and link RTKs to the activation of the RAS pathway (Waterman, Alroy, Strano, Seger, & Yarden, 1999). Thus, in mammalian cells GRB2 can indirectly recruit CBL to EGFR, MET and RET receptors (Jiang, Huang, Marusyk, & Sorkin, 2003; Scott, Eketjall, Aineskog, & Ibanez, 2005). Interestingly, the kinetic properties and the magnitude of the signaling response of the RTK may be regulated by the location of the activated receptor along the endocytic pathway (Burke, Schooler, & Wiley, 2001). Therefore, any alterations that uncouple RTKs from E3 ligase-mediated ubiquitination, internalization and down-regulation are tightly associated with cancer (Mellman & Yarden, 2013).

3. Oncogenic activation of receptor tyrosine kinases

Under normal physiologic conditions, signaling through RTKs is a major mechanism of intercellular communication and homeostasis. RTKs bind to their polypeptide ligand growth factors, which are responsible for key processes such as cellular growth, differentiation, metabolism and motility (Hubbard & Miller, 2007). The functions of RTKs are tightly regulated in normal cells. However, they may acquire transforming potential due to mutations, overexpression, chromosomal rearrangements and autocrine/paracrine stimulation (Du & Lovly, 2018; Lemmon & Schlessinger, 2010).

3.1 Activation by mutations

An important mechanism leading to uncontrolled tyrosine kinase activation involves mutagenesis. In many cases, RTK's mutations are considered "driver mutations," which have selective growth advantages and can propel cancer initiation and progression. These driver mutations also provide opportunities for targeted treatments, since cancer cells might become "addicted" to their driver mutations (Weinstein & Joe, 2006). Recent sequencing efforts of a wide variety of tumors have identified mutations in numerous RTKs, collected in the COSMIC (Catalog of Somatic Mutations in Cancer) database (Forbes et al., 2010). Recurring examples involves EGFR, HER2, KIT and the FGF-receptor family (Yu et al., 2013). Somatic EGFR mutations serve as excellent examples to illustrate the mutational spectrum of RTKs. In fact, EGFR was the first cell surface receptor to be recognized as an oncogene (Downward et al., 1984; Ullrich et al., 1984). Over 30% of breast cancers, 60% of non-small cell lung cancers, and 40% of glioblastomas either overexpress or contain activating mutations in EGFR family members (Harari & Yarden, 2000; Sharma, Bell, Settleman, & Haber, 2007). The entire EGFR kinase domain is encoded by exons 18–24, but oncogenic mutations predominantly cluster in exons 18–21, which are adjacent to the ATP-binding pocket (Lynch et al., 2004; Paez et al., 2004; Pao et al., 2004; Wang et al., 2011). Most of these mutations are missense substitutions such as G719A/S and L858R, or deletions like E746 to A750 (removal of amino acids Glutamine-Leucine-Arginine-Glutamine-Alanine; ELREA), which are associated with sensitivity to tyrosine kinase inhibitors (Marchetti et al., 2005; Masood, Kancha, & Subramanian, 2019; Shigematsu et al., 2005; Teng et al., 2011). These mutations result in constitutive activation of the kinase domain (Red Brewer et al., 2013). Many clinical trials have now shown that patients harboring an activating mutation are uniquely sensitive to treatment with EGFR-specific tyrosine kinase inhibitors (TKIs) (Janne et al., 2015; Maemondo et al., 2010; Mitsudomi et al., 2010; Mok et al., 2009; Stinchcombe, 2016; Tan, Gilligan, & Pacey, 2015). Transforming mutations may be found also in the extracellular domain (ECD), transmembrane domain (TMD) and juxtamembrane domain (JMD) of RTKs. Three missense mutations within the EGFR's ectodomain (R108G/K, G598 V, and A289D/T/V) have previously been reported in glioblastoma multiforme (GBM) (Arjona et al., 2005; Binder et al., 2018; Cerami et al., 2012). Structural studies proposed mutation-induced increases in ligand binding affinity. Thus, the overall impact of these mutations could be anticipated to be over-activation of EGFR-driven pathways, leading to increased cell proliferation and invasion (Bessman, Bagchi, Ferguson, & Lemmon, 2014; Talasila et al., 2013). In contrast to lung cancer patients with EGFR TKD mutations, GBM patients with EGFR ECD aberrations have shown disappointing clinical outcomes when treated with the EGFR TKIs, erlotinib and gefitinib. Recently, it has been observed that the ECD of EGFR is methylated, suggesting that posttranslational protein methylation events directed at the extracellular dimer interface provide another mechanism to regulate the

EGFR activation state by modulating receptor dimerization. Critically, these methylation events abrogate response to conformation-specific therapeutic antibodies such as cetuximab (Epstein & Buck, 2015; Liao et al., 2015). Point mutations in the FGFR3's ECD (specifically, S249C) were reported in carcinomas of the uterine cervix. These mutations create unpaired cysteine residues that result in the formation of inappropriate intramolecular or intermolecular disulfide bonds between the extracellular domains of mutant FGFR3 monomers, with ligand-independent dimerization and constitutive activation of the receptor complex (Wu et al., 2000). Mutations within the ECD of other RTKs have also been reported, including RET (in thyroid cancer) (Tallini & Asa, 2001) and KIT (in gastrointestinal stromal tumors; GIST) (Heinrich et al., 2003). The transmembrane domain (TMD) of HER2 harboring the V659E mutation may act as a driver mutation in non-small cell lung cancer (NSCLC) (Serra et al., 2013). These mutations disrupt specific protein-protein and protein-lipid interactions within the HER2 TMD, which are essential for proper receptor dimerization (Bocharov et al., 2016). HER2 V659E exhibits in vitro sensitivity to two TKIs, lapatinib (Serra et al., 2013) and afatinib (Yamamoto et al., 2018), indicating that TMD mutations could serve as actionable therapeutic targets. Finally, mutations within the JMD release autoinhibitory juxtamembrane interactions and subsequently hyperactivate the respective RTKs, such as KIT (V560G) and PDGFRA (V561D) in GIST (Heinrich et al., 2003). Therefore, mutations within the ECD, TMD and JM of RTKs adopt alternative activating mechanisms compared to mutations within the catalytic domain. Interestingly, it has been reported that patients with GIST harboring mutations within the ECD, TMD, and/or JMD exhibit different responses to imatinib, a competitive inhibitor of KIT and PDGFRA (Buchdunger et al., 2000), as compared to patients harboring TKD mutations (Heinrich et al., 2003).

3.2 Overexpression of RTKs

Overexpression of RTKs has been detected in a variety of human cancers. For instance, although EGFR is overexpressed in several cancers, amplification of the EGFR-encoding gene appears to be restricted to glioblastomas, where it occurs in ~35% of cases, and leads to overexpression of both wild-type and mutated forms of the receptor in GBM (Libermann et al., 1985; Lopez-Gines et al., 2010). Overexpression of EGFR has been reported also in lung (Selvaggi et al., 2004), esophageal (Hanawa et al., 2006) and thyroid cancer (Rodriguez-Antona et al., 2010). Likewise, MET is overexpressed in lung (Xu et al., 2010) and gastric cancer (Ha et al., 2013), and HER2/ErbB2 is overexpressed in breast and in ovarian tumors (Slamon et al., 1987; Slamon et al., 1989), as well as in bladder and lung cancer (Selvaggi et al., 2004). While gene amplification is the major mechanism that leads to overexpression of RTKs, additional mechanisms include transcriptional/translational enhancement (Ludes-Meyers et al., 1996), infection by oncogenic viruses (Hanawa et al., 2006), aberrant regulatory mechanisms, such as loss of phosphatases (Sun et al., 2011), or other negative regulators (Maiti et al., 2013; Mudduluru et al., 2011). Regardless of the exact underlying mechanism, RTK overexpression has generally been associated with poor patient outcomes. For example, not only HER2 levels, but also the abundance of EGFR and HER3 has been associated with poor prognosis of patients with breast cancer. Importantly, RTK amplification might serve as a strategy enabling tumor cells to escape pharmacological stress. For example, MET and HER2 amplification frequently associates with emergence of drug resistance in patients with EGFR-mutated lung cancer (Yu et al., 2013).

3.3 Chromosomal rearrangements

Since the first discovery of a fused kinase-encoding gene, *BCR-ABL*, which is encoded by the so-called "Philadelphia chromosome" of chronic myeloid leukemia (CML), a number of other chromosomal rearrangements and fusion kinases have been identified in various tumors (Stephens et al., 2011). RET, ALK, ROS1 and TRK rearrangements have been well studied in human cancers. The increased signaling, which occurs as a result of this rearrangement, can be therapeutically targeted by specific tyrosine kinase inhibitors (TKIs), resulting in clinically relevant responses. In spite of the diversity of tyrosine kinase fusions, a common feature has emerged: the 3′ partner encodes an intact TKD and the 5′ partner encodes a dimerization domain. Functionally, the dimerization domain of the fused protein mediates dimerization of the kinase even without ligand binding, thereby activating the TKD (Du & Lovly, 2018). Four kinase rearrangements, RET, ROS1, ALK and TRK, have been identified in NSCLC, with 1–7% frequency (Shaw, Hsu, Awad, & Engelman, 2013). Similarly, RET fusions were found in more than one-third of papillary thyroid cancer (Drilon, Hu, Lai, & Tan, 2018; Khotskaya et al., 2017). Congruently, inhibiting RTK fusion proteins has proven to be an effective therapeutic strategy, including blocking ALK fusions in anaplastic large cell lymphoma (Iragavarapu et al., 2015), inflammatory myofibroblastic tumors (IMT) (Butrynski et al., 2010) and in lung cancer (Kwak et al., 2010), RET in lung and thyroid cancer (Kloos et al., 2009), ROS1 in GBM (Kiehna et al., 2017), lung cancer, and IMT, EGFR in lung cancer (Konduri et al., 2016), and NTRK in IMT, lung, kidney, colon and other malignancies (Drilon et al., 2017; Du & Lovly, 2018).

3.4 Autocrine RTK activation

Unlike the paracrine (heterotypic) mode of action of growth factors, which dominates physiological processes like embryogenesis, inflammation and wound healing, many cancer cells acquire the ability to synthesize growth factors to which they are responsive (Sporn & Todaro, 1980). Abnormal autocrine growth factor secretion contributes to constitutive RTK signaling and results in sustained receptor activation. This is best exemplified by aberrant stimulation of EGFR and the insulin-like growth factor-I receptor (IGF-IR) (Derynck et al., 1987; Kaleko, Rutter, & Miller, 1990; Salomon, Brandt, Ciardiello, & Normanno, 1995). This mechanism of activation occurs when the relevant RTK is aberrantly expressed, or overexpressed, in the presence of its cognate ligand, or when overexpression of the ligand occurs in the presence of the respective receptor. For example, in many solid tumors elevated levels of both a receptor and its ligand are frequently observed (Kaleko et al., 1990). Constitutive autocrine activation might lead to clonal expansion and tumor progression. Accordingly, autocrine activation of various RTKs has been characterized in diverse tumors, including the autocrine loops TGFα-EGFR (Ciardiello & Tortora, 2001), HGF-MET (Kentsis et al., 2012) and SCF-KIT (Esposito et al., 2002; Krystal, Hines, & Organ, 1996). RTK's autocrine loop may act synergistically with other autocrine pathways to drive tumor development. Interestingly, the growth advantage conferred by the SCF-KIT loop partially synergizes with two other autocrine loops, IGF-l and bombesin secretion, to drive development of small cell lung cancer (SCLC) (Krystal et al., 1996). Lastly, autocrine pathways could act as rational targets for cancer therapy (Ciardiello & Tortora, 2001). For example, ligand-receptor autocrine loops render EGFR-mutant lung cancer cells less sensitive to EGFR-specific kinase inhibition (Fujimoto et al., 2005).

4. Supportive functions of RTKs during tumor progression

4.1 Roles for RTKs in tumor initiation

It is well recognized that replication stress acts as a major mechanism underlying the initiation and progression of many cancers (Taylor & Lindsay, 2016). It is also widely accepted that sustained cell proliferation can lead to DNA replication stress and genome instability. Conceivably, the stem cells of a tissue, due to their slow rate of proliferation, are relatively protected from DNA replication stress and genome instability. In contrast, their rapidly proliferating descendants, called transit amplifying cells (TACs), suffer from replication stress, hence accumulate mutations (Schneider & Yarden, 2016). Thus, by means of their ability to instigate repeated transitions of growth-arrested cells from the G0 phase of the cell cycle to the G1 and S-phases, growth factors and RTKs can play active roles in the earliest events in tumor formation. RTKs similarly play critical roles in the subsequent clonal expansion processes, which select the fittest clones and fixate the underlying founder mutations. This chain of events enables growth factors like EGF, NRGs and HGF to expand the pool of initiated cells susceptible to additional mutations and to the branching evolution, which establishes the clonal diversity responsible for intra-tumoral heterogeneity (Yap, Gerlinger, Futreal, Pusztai, & Swanton, 2012).

4.2 RTKs control pre-malignant precursors of cancer

Several lines of evidence propose that RTKs, such as EGFR, are involved in the initiation and/or maintenance of tissue precursors of cancer. These structures, called intraepithelial neoplasia or carcinoma in situ, are morphologically distinct lesions surrounded by an integral basement membrane. For example, pancreatic intraepithelial neoplasia (PanIN), which is the predominant precancerous lesion of pancreatic ductal adenocarcinoma, is frequently detected in the normal and inflamed pancreas, but only a few of PanINs eventually progress. In line with involvement of EGFR in the progression of low-grade PanIN into high-grade carcinoma, it has been reported that intraductal infusion of EGFR inhibitors significantly decreased the incidence of high-grade PanINs in a mouse model (He et al., 2015). Similar structures are frequently detected in breast (ductal carcinoma in situ, DCIS) and in prostate cancer (prostate intraepithelial neoplasia; PIN (Epstein, 2009)). Importantly, HER2, a close relative of EGFR and an enhancer of EGF and NRG signals, is overexpressed in 40% of DCIS and up to 70% of the high-grade group (Meijnen, Peterse, Antonini, Rutgers, & van de Vijver, 2008), but HER2 overexpression in breast cancer is limited to 15–20% of all mammary tumors. This may hint that RTKs are essential relatively early in tumor progression, but when the intraepithelial precursors accumulate more mutations, the contribution of RTKs to cancer progression is diminished.

4.3 Regulation of angiogenesis and vasculogenesis by RTKs

Microscopic and asymptomatic cancerous lesions (i.e., dormant tumors) are highly prevalent in humans, and they may remain occult for prolonged periods of time (months to years). The switch to become fast-growing angiogenic tumors relates to the evolvement of networks of blood and lymph vessels, which supply sugars and amino acids, as well as remove toxic metabolites. Alarmingly, cancer therapy, including high dose chemotherapy, might induce mobilization of endothelial progenitor cells (EPCs) to the viable rim of tumors, thereby support cancer progression (Shaked et al., 2006). Independently, animal models mimicking exit from dormancy have proposed that the angiogenic switch associates with elevated RNA levels of receptors for EGF and for the insulin-like growth factor type 1 (IGF-1) (Almog et al., 2009). Indeed, RTKs control both angiogenesis, the growth of new capillaries from preexisting blood vessels, and vasculogenesis, which is characterized by de novo formation of a primitive vascular network by bone marrow-derived EPCs (Asahara et al., 1997; Carmeliet & Jain, 2000). Growth factors like the vascular endothelial growth factors (VEGFs), angiopoietins, FGFs and TGF-beta play important roles in both vasculogenesis and/or angiogenesis. Accordingly, blocking the receptors for the respective growth factors and intercepting the ligands, before they activate the cognate RTKs, are widely recognized as major strategies that inhibit the pro-tumorigenic effects of angiogenesis and vasculogenesis (Saharinen, Eklund, & Alitalo, 2017). For example, bevacizumab, a human mAb, targets VEGF has been approved, in combination with chemotherapeutic drugs, for the treatment of colorectal cancer and certain types of lung, kidney and brain cancer. Similarly, several TKIs that target VEGFRs, along with additional receptors, have been approved for clinical use (e.g., sunitinib and sorafenib).

4.4 Growth factors and RTKs controlling intravasation and invasive growth

Recruitment of blood vessels that densely interweave primary tumors might accelerate cancer progression and colonization of distant organs. However, an essential preceding step requires dissolution of basement membranes. This step, as well as the subsequent processes that propel cancer progression, such as invasion into the vascular or lymphatic systems (intravasation), departure from the bloodstream (extravasation), and seeding distant organs, require ligand-induced activation of specific RTKs (Yilmaz & Christofori, 2009). The origin of RTK ligands might be the tumor microenvironment (e.g., tumor-associated fibroblasts and macrophages) or autocrine loops. Interestingly, when under pharmacological stress, tumors might secrete growth factors and cytokines, which stimulate the AKT survival pathway. For example, targeted therapy of melanoma using BRAF kinase inhibitors induces a complex network of secreted signals in drug-stressed cells (Obenauf et al., 2015). This therapy-induced secretome stimulates the outgrowth, dissemination and metastasis of drug-resistant cancer cell clones, and supports the survival of drug-sensitive cancer cells. An ordered sequence of events, collectively known as epithelial-mesenchymal transition (EMT), converts tightly packed polarized epithelial cells into motile individual cells. Along with TGF-beta, Notch and the WNT pathway, RTKs like HER3, EGFR, MET and FGF-receptors act as major mediators of EMT. The role of AXL, an RTK that binds with GAS6, has been intrinsically linked not only to EMT but also to anoikis resistance, invasion and metastasis, in different cancers (Gonzalez & Medici, 2014). In ovarian cancer, AXL is primarily expressed in metastases and advanced-stage tumors, but it is only lowly expressed in the normal ovarian epithelium (Kariolis et al., 2017). Furthermore, inhibition of AXL in human metastatic ovarian tumor cells is sufficient to prevent the initiation of metastatic disease in model animals, likely due to decreased invasion and matrix metalloproteinase activity. Similarly, studies performed in vitro propose that TGF-beta, FGFs and the hepatocyte growth factor enhance the invasive potential of cancer cells by up-regulating matrix metalloproteinases like MMP-2 and MMP-9, along with down-regulating several protease inhibitors (Stetler-Stevenson, 1999; Werb, 1997). EMT entails several molecular switches, involving newly synthesized mRNAs and microRNAs, as well as posttranslational regulation (e.g., protein ubiquitination and degradation), which alter cell-to-matrix and cell-to-cell adhesion, such as the loss of the epithelial E-cadherin and gain of the mesenchymal N-cadherin (Nieman, Prudoff, Johnson, & Wheelock, 1999). Both EGF and TGF-beta elevate expression of ZEB-2 and Twist, which are potent E-cadherin repressors. Alternatively, mutational inactivation, transcriptional repression, ubiquitination and endocytosis of E-cadherin can downregulate this adhesion molecule in tumors (Fujita et al., 2002). Another switch involves tensins, actin-binding molecules that link the cytoskeleton to integrin-based adhesion sites: EGF downregulates tensin-3 expression, and concomitantly upregulates CTEN, a tensin family member that lacks an actin-binding domain. Hence, CTEN displaces tensin-3 from the cytoplasmic tail of integrin beta1, which instigates disassembly of the actin cytoskeleton and enables cell migration (Katz et al., 2007).

4.5 RTKs involved in extravasation and metastasis

The multistep metastatic cascade represents a Darwinian process that selects individual tumor's clones, from a genetically and epigenetically heterogeneous tumor cell population, on the basis of the ability to reach and survive at specific tissue microenvironments (Obenauf & Massague, 2015). This process is highly inefficient; only a few cells succeed at reestablishing a secondary tumor at a distant location. To survive in the circulation, cancer cells adhere to platelets or to neutrophil extracellular traps (NETs), which are pathogen-trapping structures generated by expulsion of the neutrophil's DNA. For instance, metastatic breast cancer cells can induce neutrophils to form metastasis-supporting NETs, such that digesting NETs with DNase I can reduce metastasis (Park et al., 2016). The efficacy of metastasis, as well as its organ-specific patterns, are influenced by many growth factors, RTKs and downstream signaling pathways, especially the RAF-to-ERK and PI3K-to-AKT axes. For example, breast cancer cells that preferentially infiltrate the brain express the cyclooxygenase COX2, the alpha 2,6-sialyltransferase ST6GALNAC5, and HBEGF, a ligand of EGFR, which function as mediators of cancer cell passage through the blood-brain barrier and regulate organ-specific metastatic interactions (Bos et al., 2009). Paget's "seed and soil" hypothesis predicts that survival of disseminated metastatic cells requires colonization of supportive niches within an organ. Moreover, during the final colonization phase of the metastatic cascade, cancer cells might actively modify the niche, such as the bone microenvironment. For instance, in osteoblastic lesions of prostate cancer metastasis, tumor cells stimulate bone matrix deposition and displace the bone marrow (Logothetis & Lin, 2005). Several autocrine factors of prostate cancer cells promote osteoblastic bone metastasis. They include FGFs, IGFs, HGF, VEGF and PDGFs, along with other soluble mediators, such as endothelin 1 and bone morphogenetic proteins (BMPs) (Weilbaecher, Guise, & McCauley, 2011). In line with these findings, inhibition of MET and VEGFR2 using a dual kinase inhibitor, cabozantinib, was shown to reduce prostate cancer growth in bone, with evidence for suppressing osteoblastic activity (Lee et al., 2018). In addition to autocrine regulation, paracrine interactions involving macrophages, fibroblasts and endothelial cells might accelerate intravasation and metastasis. For example, breast cancer cells secrete the colony-stimulating factor 1 (CSF-1), which binds to an RTK of the PDGFR family and attracts macrophages to tumors. CSF-1 produced by carcinoma cells promotes the expression of EGF-like ligands by macrophages and enhances the expression of CSF-1 by carcinoma cells, thereby generating a positive feedback loop.

In conclusion, along the metastasis cascade, recurring paracrine mechanisms mediated by stromal cells appear to complement essential autocrine loops of cancer cells.

5. Response and resistance to anti-RTK treatments

5.1 An introduction to cancer therapy (see Table 2)

Despite the enormous progress made in cancer treatment in the past two decades, the incidence and morbidity due to cancer are expected to increase in the coming years (NIH, 2018). The major approaches to treat cancer include surgery, chemotherapy, radiation therapy, targeted therapy and immune therapy, as well as combinations of pharmacological and other strategies. Surgery is the preferable treatment in case the cancer has not spread. For instance, it is the primary option for patients with non-metastatic melanoma. Alternatively, it can be used to reduce the tumor mass before cytotoxic treatment. In some other cases, if the mass is too large, surgery can be employed after cycles of chemo- or radiotherapy. For example, 49% of women with breast cancer (stage I or II) undergo surgery followed by radiotherapy, while 56% of women diagnosed with stage III breast cancer are treated with mastectomy followed by chemotherapy (Society, 2019). Although chemotherapy has remained the most widely used regimen in oncology, in the past 20 years, targeted therapy has been extensively employed in the treatment of specific types of cancer. Targeted therapy comprises two classes of drugs: protein kinase inhibitors (PKIs) and monoclonal antibodies (mAbs), both classes target molecules that drive or support tumor growth (Hanahan & Weinberg, 2011). Many RTKs are commonly mutated or upregulated in cancer, thus promoting cell growth and proliferation. Importantly, most RTK molecules are druggable, meaning that relatively effective drugs can be designed and implemented. Examples of mutated RTKs that might be targeted using specific drugs include EGFR, KIT, MET, RET and HER2, which have been implicated in the development of gastrointestinal tumors, thyroid, lung and breast cancer (Amit, Wides, & Yarden, 2007). Similarly, cytoplasmic kinases that are frequently mutated in tumors are often targetable. Thus, mutations in the PI3K family have been implicated in 30–50% of all human cancers, including breast, ovarian, endometrial and colorectal cancer (Samuels et al., 2004; Samuels & Waldman, 2010). Similarly, BRAF mutations are typically found in 40–50% of all melanoma patients and in smaller fractions of patients with thyroid, colorectal and lung cancer (Dankner, Rose, Rajkumar, Siegel, & Watson, 2018). The majority of PKIs bind the ATP-binding pocket in the

Table 2 Kinase inhibitors and antibodies approved for application in oncology (November 2019).

Target receptor	Name (Trade name)	Class	Year	Company	Indication
HER2	Trastuzumab (Herceptin)	mAb	1998	Genentech	Metastatic breast cancer, gastric or gastroesophageal junction adenocarcinoma
BCR-ABL, KIT, PDGFR	Imatinib (Gleevec)	TKI	2001	Novartis	CML, GIST, ALL
EGFR	Gefitinib (Iressa)	TKI	2003	AstraZeneca	NSCLC
EGFR	Erlotinib (Tarceva)	TKI	2004	Genentech OSI	NSCLC, Pancreatic cancer
EGFR	Cetuximab (Erbitux)	mAb	2004	Imclone BMS	Metastatic CRC, head and neck cancer
VEGF-A	Bevacizumab (Avastin)	mAb	2004	Genentech	Metastatic breast cancer, NSCLC, CRC, glioblastoma, RCC, cervical, ovarian and fallopian tube cancer
RAF, VEGFR2, PDGFRB,	Sorafenib (Nexavar)	TKI	2005	Onyx, Bayer	RCC, HCC, thyroid carcinoma
EGFR	Panitumumab (Vectibix)	mAb	2006	Amgen	Metastatic CRC
BCR-ABL, SRC, KIT, PDGFR	Dasatinib (Sprycel)	TKI	2006	Bristol-Meyers Squib	CML, ALL
KIT, FLT3, PDGFR	Sunitinib (Sutent)	TKI	2006	Pfizer	GIST, RCC, pancreatic neuroendocrine tumors
EGFR, HER2	Lapatinib (Tykerb)	TKI	2007	GSK	Breast cancer
BCR-ABL, PDGFR	Nilotinib (Tasigna)	TKI	2007	Novartis	CML
PDGFR, KIT	Pazopanib (Votrient)	TKI	2009	GlaxoSmithKline	RCC, soft tissue sarcoma
EML4–ALK, ROS1, MET	Crizotinib (Xalkori)	TKI	2011	Pfizer	NSCLC

EGFR, VEGFR, RET	Vandetanib (Caprelsa)	TKI	2011	AstraZeneca	MTC
HER2	Pertuzumab (Perjeta)	mAb	2012	Genentech	Metastatic breast cancer
VEGFR	Axitinib (Inlyta)	TKI	2012	Pfizer	RCC
PDGFR, TIE2, VEGFR, KIT	Regorafenib (Stivarga)	TKI	2012	Bayer	CRC, GIST, HCC
RET, VEGFR, MET, TRKB, TIE2	Cabozantinib (Cabometyx)	TKI	2012	Exelixis	MTC, RCC, HCC
BCR-ABL, VEGFR, PDGFR, FGFR, EphR, Src, KIT, RET, Tie2, Flt3	Ponatinib (Iclusig)	TKI	2012	Ariad	CML, Ph chromosome positive ALL
EGFR	Afatinib (Gilotrif)	TKI	2013	Boehringer Ingelheim	NSCLC
HER2	Ado-Trastuzumab emtansine (Kadcyla)	mAb	2013	Genentech	Metastatic breast cancer
VEGFR2	Ramucirumab (Cyramza)	mAb	2014	Eli Lilly	Gastric and breast cancers, HCC, NSCLC, CRC
VEGFR, FGFR, PDGFR	Nintedanib (Ofev)	TKI	2014	Boehringer Ingelheim	Idiopathic pulmonary fibrosis
VEGFRs, FGFRs, PDGFR, KIT, RET	Lenvatinib (Lenvima)	TKI	2015	Eisai	Thyroid cancer, RCC, HCC, endometrial carcinoma
EGFR	Osimertinib (Tagrisso)	TKI	2015	AstraZeneca	NSCLC
EGFR	Necitumumab (Portrazza)	mAb	2015	Eli Lilly	NSCLC
PDGFR	Olaratumab (Lartruvo)	mAb	2016	Eli Lilly	Soft tissue sarcoma
HER2	Neratinib (Nerlynx)	TKI	2017	Puma Biotech	Breast cancer
EGFR	Dacomitinib (Vizimpro)	TKI	2018	Pfizer	NSCLC

All currently approved anti-RTK drugs are listed according to the target receptors. The year of first approval and the manufacturer of each drug are indicated, along with the clinical indications. Note that Nintedanib (Ofev) has been approved for fibrosis. The abbreviations used are as follows: ALL: acute lymphocytic leukemia; AML: acute myeloid leukemia; CML: chronic myeloid leukemia; CRC: colorectal carcinoma; EML4: echinoderm microtubule-associated protein-like 4; GIST: gastrointestinal stromal tumors; HCC: hepatocellular carcinoma; mAb: monoclonal antibody; MTC: medullary thyroid cancer; NSCLC: non-small cell lung carcinoma; RCC: renal cell carcinoma; TKI: tyrosine kinase inhibitor.

kinase domain of the protein kinase, while only a small number bind to allosteric sites, thus they are not ATP competitors. The pioneer PKI, imatinib, was approved in 2001, for the treatment of chronic myeloid leukemia (CML). However, since then, more than 30 new PKIs have been approved for clinical applications, including five EGFR tyrosine kinase inhibitors (TKIs), erlotinib, gefitinib, afatinib, dacomitinib, and osimertinib. Another way to specifically target RTKs is by using monoclonal antibodies (mAbs). mAbs can activate the immune system, through ADCC (antibody-dependent cytotoxicity) or CDC (complement-dependent cytotoxicity), or they can directly bind RTKs on the cell membrane to induce their internalization and degradation, thus shut down survival pathways in the cancer cell (Marrocco, Romaniello, & Yarden, 2019). Rituximab, targeting CD20 on B cells, was the first approved antibody (in 1997) for the treatment of non-Hodgkin lymphoma (NHL). This was followed by the approval of the first antibody targeting an RTK: in 1998, the anti-HER2 antibody trastuzumab was approved for the treatment of HER2-positive metastatic breast cancer. Since then, several other mAbs have been approved, for instance the anti-EGFR antibody cetuximab, and the anti-VEGF antibody bevacizumab (Marrocco et al., 2019).

5.2 Roles for RTKs in resistance to cytotoxic treatments

Due to their ability to inhibit actively replicating cells, chemotherapy and radiotherapy are potentially able to defeat all types of cancers. Unfortunately, these scenarios are limited owing to common onset of resistance, or due to extensive adverse effects. All main classes of cytotoxic drugs, including alkylating agents, antimetabolites and platinum compounds, target both cancer cells and normal cells, while they undergo rapid proliferation. In other words, because chemotherapy and radiotherapy affect rapidly proliferating cancer cells, other rapidly proliferating cells will also be targeted, including blood-forming cells in the bone marrow, reproductive system, hair follicles, along with epithelial cells lining the digestive tract, causing mild to severe side effects, such as fatigue, hair loss, anemia, diarrhea or constipation. Together with surgery and chemotherapy, radiation therapy remains one of the most common treatments used in oncology. Currently, about 50% of cancer patients receive radiotherapy, alone or in association with other treatments (Allen, Her, & Jaffray, 2017). Radiation causes cancer cell death by inducing reactive oxygen species (ROS), DNA damage and stress response (Kim et al., 2019). As for chemotherapy, also in the case of radiation therapy, success rates remain low due to emergence of resistance, or due to extensive side effects, caused by the non-selective mode of action of this treatment (Marrocco et al., 2019).

5.2.1 Roles for RTKs and growth factors in resistance to chemotherapy

Patient resistance to treatment represents a major cause of oncology-related mortality (Housman et al., 2014). Resistance might be intrinsic, if it is present before exposure to the drug, or acquired (emergent), if the sensitivity to the drug is diminished after initial or prolonged exposure to the drug. Resistance to chemotherapy is mainly due to an increased rate of drug efflux, alterations in drug metabolism, or activation of repair systems. An increase in drug efflux leads to decreased intracellular levels of the drug, causing therapy failure. One of the most studied groups of transporters responsible for this mechanism is the ABC (ATP-binding cassette) family, especially P-glycoprotein (P-gp/MDR1 or ABCB1). ABC transporters are responsible for energy-dependent efflux of substrates (Sauna, Kim, & Ambudkar, 2007). P-gp mediates efflux of many cytotoxic drugs, such as etoposide, paclitaxel and olaparib (Lagas et al., 2010; Vaidyanathan et al., 2016). Notably, in prostate cancer cells, EGFR elevates P-gp/ABCB1 expression by stimulating the AKT pathway, thus conferring resistance to docetaxel (Hour et al., 2015). Importantly, multidrug resistance-associated protein 1 (MRP1 or ABCC1), which is upregulated in many types of cancer, including lung, breast and prostate (Munoz, Henderson, Haber, & Norris, 2007), is responsible for the efflux of other anti-cancer drugs, like anthracyclines, epipodophyllotoxins, vinca alkaloids, camptothecins, methotrexate and mitoxantrone (Yin & Zhang, 2011). The third ABC transporter with a role in resistance to cancer treatments is named ABCG2, or breast cancer resistance protein. ABCG2 is responsible for the efflux of drugs like epipodophyllotoxin, camptothecins and anthracyclines. This transporter is also able to transport several PKIs, like imatinib and gefitinib. ABCG2, is overexpressed not only in breast cancer; it was found to be overexpressed in other types of cancer, such as lung cancer and leukemia (Mao & Unadkat, 2015). Overexpression of ABC transporters in cancer is associated with multidrug resistance (MDR) (Robey et al., 2018). MDR is responsible for resistance to a broad spectrum of drugs, which are not necessarily chemically and functionally related. Collectively, MDR has been associated with failure of chemotherapeutic treatments in 90% of cases (Gong et al., 2012). Another resistant mechanism to chemotherapeutic drugs involves the activation of reparatory systems. Cells treated with DNA-damaging agents, such as 5-fluorouracil and platinum compounds, can acquire resistance to these treatments by activating the DNA damage response (DDR). Resistance to cisplatin has been associated with activation of nucleotide excision repair and homologous recombination (Bonanno, Favaretto, & Rosell, 2014; Selvakumaran, Pisarcik, Bao, Yeung, & Hamilton, 2003). Drug metabolism can also decrease the efficacy of anti-cancer drugs. Normally, drug metabolism entails the addition of hydrophilic groups to lipophilic drugs, in order to enhance excretion from the body. While this is a physiological process that maintains low levels of drugs, in some cases it might induce fast excretion, thus reducing treatment efficacy (Alfarouk et al., 2015). Resistance to chemotherapy can also be due to alterations in the target of the drug. For instance, in metastatic colorectal patients, resistance to 5-FU, which interferes with DNA synthesis by inhibiting the enzyme thymidylate synthetase, is accompanied by amplification of the thymidylate synthetase gene (Wang et al., 2004; Watson et al., 2010). At the molecular level, resistance to antiproliferative and death signals often involves transcription factors like the retinoblastoma protein (pRb) and E2Fs, which control gene expression essential for cell proliferation. Resistance to apoptosis can also be acquired through the PI3K-to-AKT axis, which transduces antiapoptotic signals. This antiapoptotic survival pathway likely underlies the extensive involvement of growth factors like IGF-1, as well as several EGF-like ligands, in evasion from apoptosis. Hence, it is not surprising that blocking growth factor-mediated evasion from cell death bears clinical implications. For example, early studies indicated that anti-EGFR antibodies can sensitize tumors to chemotherapy (Aboud-Pirak et al., 1988). This finding has been translated to therapies that combine with chemotherapy certain anti-RTK drugs, like cetuximab and trastuzumab (Citri & Yarden, 2006).

5.2.2 Roles for RTKs and their ligands in resistance to radiotherapy

Radio-resistance represents a major issue in the success of radiation therapy in oncology. Radiation therapy causes cell death by inducing stress and DNA damage, but some cells may survive, by activating DNA repair, unfolded protein response (UPR), or autophagy, thereby leading to tumor regrowth and recurrence (Kim et al., 2019). The main cell death pathway triggered by radiation is apoptosis. Primarily, radiation therapy induces DNA double strand breaks, which cause activation of PARP (polyADP-ribose polymerase). Indeed, it has been shown in several animal models, that PARP is involved in resistance to radiation (Donawho et al., 2007; Khan et al., 2010; Senra et al., 2011). The lipid second messenger ceramide is a potent mediator of apoptosis induced by radiation, and defective ceramide signaling in tumors has been associated with radioresistance (Chmura et al., 1997; Michael, Lavin, & Watters, 1997). Another mechanism responsible for cancer relapse involves cancer stem cells (CSCs) or cancer initiating cells. These rather rare cells have been linked to resistance to cytotoxic treatments in many cancer indications, such as leukemia, glioblastoma, and pancreatic cancer

(Eramo et al., 2006; Hermann et al., 2007; Viale et al., 2009). CSCs comprise a slowly dividing population of cells, which are capable of self-renewal and differentiation (Frank, Schatton, & Frank, 2010). These cells are normally found in a quiescent state, hence are intrinsically resistant to cytotoxic treatment (Casals, Gusta, Cobaleda-Siles, Garcia-Sanz, & Puntes, 2017). EGFR seems to be the main RTK involved in mechanisms of radiation resistance (Harari, 2007). For example, by means of accelerating cell proliferation and avoiding cell death, EGFR signaling increases the number of cancer stem cells between irradiation sessions. One supportive line of evidence stems from post hoc analyses showing that tumors with high EGFR expression have significantly lower local tumor control probabilities (Bentzen et al., 2005). Additional RTKs, such as MET and IGF-1R, likely play important roles in tumor radioresistance. Hence, there is clearly the potential for these RTKs to be used as biomarkers to predict response to radiotherapy treatment. Moreover, phase 3 randomized trials showed that adding an anti-EGFR mAb, cetuximab, to primary radiotherapy increased overall survival in patients with locoregionally advanced squamous cell carcinoma of the head and neck, which led to the approval of this treatment (Bonner et al., 2006).

5.3 Roles for RTKs and their ligands in resistance to molecular targeted therapies (see Fig. 3)

5.3.1 *Resistance to kinase inhibitors: Secondary mutations, autocrine and bypass mechanisms*

The most common mechanisms of resistance to kinase inhibitors consist of acquisition of new secondary mutations within the kinase domain, selection of preexisting sub-clones, already harboring mutations in the kinase domain (Hata et al., 2016), up-regulation of other kinases and activation of bypass signaling pathways and phenotypical changes, such as EMT (epithelial-mesenchymal transition) (Carvalho, Levi-Schaffer, Sela, & Yarden, 2016; Garraway & Janne, 2012; Mancini & Yarden, 2016). The onset of resistance limits the long-term success rate of PKI-based cancer therapy. For example, imatinib, with up to 80% response rate, has changed the clinical outcome of chronic myelogenous leukemia (CML) (O'Brien et al., 2003). However, 15–25% of patients show intrinsic resistance to imatinib, and 7–15% become resistant after an initial phase in which they respond to the therapy (Barouch-Bentov & Sauer, 2011). Resistance to imatinib is mainly due to mutations in the kinase domain and, secondarily, to amplification of the *ABL-BCR* genomic locus (Shah et al., 2004). The gatekeeper mutation T315I in ABL, which leads to the substitution of the hydrophilic and small amino acid

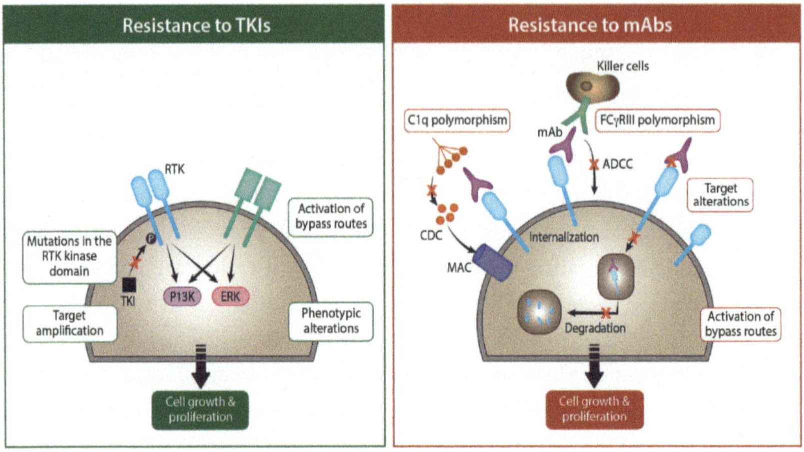

Fig. 3 Mechanisms of resistance to RTK-targeted therapy. The left panel shows the following, most common resistance mechanisms to TKIs: Appearance of a secondary mutation within the kinase domain of the already mutated receptor, activation of bypass survival mechanisms, such as alternative routes resulting in the activation of pathways like PI3K/AKT or RAS/ERK, which sustain cell growth and proliferation. Other resistance mechanisms to TKIs include amplification of the target receptor, mutations in signaling downstream molecules (e.g., RAS), and phenotypic alterations, such as EMT. The right panel presents resistance mechanisms to anti-RTK mAbs. Examples include polymorphic forms of CD16, the FCγRIII involved in antibody-dependent cellular cytotoxicity (ADCC), or the complement component C1q, which might impair complement-dependent cytotoxicity (CDC). Other mechanisms of resistance include activation of bypass RTKs and alterations in the target RTK (e.g., expression of a truncated form of the receptor), which might affect the mAb binding leading, for example, to inhibition of internalization/degradation of the mAb/receptor complex.

threonine with a hydrophobic and bulky amino acid, such as isoleucine, inhibits the binding of the TKI to the kinase because of steric hindrance (Zhang, Yang, & Gray, 2009). These attributes are remarkably similar to the emergence of resistance of NSCLC to EGFR-specific PKIs (Maicol Mancini & Yarden, 2016). EGFR-targeting first-generation PKIs, erlotinib and gefitinib, have revolutionized the treatment of NSCLC patients with activating mutations in EGFR. However, despite the initial good response, patients become resistant to the drugs: in about 60% of cases the emerging resistance is due to the gatekeeper mutation T790M, which dramatically decreases the affinity of PKIs for the ATP-binding pocket of the receptor (Oxnard et al., 2011; Pao et al., 2005). Other resistance mechanisms include MET amplification (Bean et al., 2007), HER2 amplification (Takezawa et al., 2012) and activation of other RTKs, such as AXL, IGF1R and FGFR (Guix et al., 2008; Ware et al., 2013; Zhang et al., 2012). To overcome T790M resistance, the T790M-specific EGFR-PKI called osimertinib was developed, and approved in 2015, after demonstrating high response rates in clinical trials (Janne et al., 2015; Mok et al., 2017). Unfortunately, resistance to osimertinib emerges in treated patients: a tertiary mutation, C797S, which prevents binding of osimertinib to the receptor (Thress et al., 2015), activation of other RTKs, such as HER2 or MET (Planchard et al., 2015), or aberrant expression of NRAS and KRAS (Eberlein et al., 2015) underlay tolerance to osimertinib. Importantly, more recently osimertinib has also been approved for treatment-naïve patients (first-line treatment setting), who do not present the T790M mutation. This approval was based on the FLAURA study, which showed improvement in progression-free survival of patients when osimertinib was compared to first-generation PKIs (Soria et al., 2018). Once again, resistance to first-line osimertinib involves emergence of the C797S mutation, along with HER2 or MET amplification. Similarly, resistance to another PKI, crizotinib, which targets ALK, a rearranged RTK, occur through secondary mutations or variation in copy number, as well as EGFR activation and amplification of other RTKs (e.g., KIT and IGF-1R). Lapatinib, an ATP-competitive inhibitor of HER2 and EGFR (Rusnak et al., 2001), is used in combination with chemotherapy to treat patients with HER2-positive breast cancer. However, many patients acquire resistance after an initial phase of response (Geyer et al., 2006; Johnston et al., 2008). Resistance to lapatinib seems to be mainly dependent on activation of bypass survival pathways, such as increased expression of the HER3 ligand neuregulin 1 (NRG1) and activation of the EGFR-HER3-PI3K pathway (Wilson et al., 2012; Xia et al., 2013). Additional mechanisms might involve loss of PTEN and constitutive activation of AKT (Eichhorn et al., 2008; Gayle, Arnold, O'Regan, & Nahta, 2012; Hegde et al., 2007). In conclusion, several small molecule inhibitors of RTKs are initially effective in patients, but secondary mutations, along with emergence of compensatory signaling pathways frequently confer resistance to drugs.

5.3.2 Resistance to anti-RTK antibodies

Since the approval of rituximab (an antibody against CD20) in 1997 for the treatment of NHL, some 30 monoclonal antibodies have been approved in cancer therapy, and the number of mAbs in clinical trials is increasing (Marrocco et al., 2019). Despite the recognized potential of this type of therapy, when used as monotherapy, mAbs often show limited efficacy. Therapy failure might relate to rapid onset of resistance and the observation that, unlike PKIs, mAbs are typically weak inducers of cell death. Resistance to monoclonal antibodies can be due to alterations in the target.

For example, loss of CD20 has been described as a resistance mechanism to rituximab (Pavanello, Zucca, & Ghielmini, 2017). Breast cancer patients that express the truncated form of HER2, p95HER2, are often resistant to the anti-HER2 antibody trastuzumab, which is currently used to treat patients with *HER2*-amplified breast and gastric cancer (Scaltriti et al., 2007). Masking the binding site of the antibody on the target is another resistance mechanism: it has been shown that mucin 4, a cell surface glycoprotein, engages the very same site that is recognized by trastuzumab (Nagy et al., 2005). The mechanisms of action of mAbs frequently involve activation of complement- and antibody-dependent cytotoxicity (CDC and ADCC). It has been shown that impairment in ADCC or CDC can be responsible for resistance to mAbs. Other common mechanisms of resistance comprise emergence of mutations in molecules that act downstream to the target RTK, or compensatory activation of yet other signaling pathways. For instance, colorectal cancer patients with KRAS mutations do not respond to anti-EGFR antibodies, cetuximab or panitumumab (Lievre et al., 2006). Similarly, mutations in PIK3CA or loss of PTEN expression correlate with resistance to trastuzumab in patients with breast cancer (Berns et al., 2007). Likewise, overexpression of IGF1R or MET have been associated with resistance to trastuzumab (Lu, Zi, Zhao, Mascarenhas, & Pollak, 2001; Shattuck, Miller, Carraway, & Sweeney, 2008). Treatment of in vitro and in vivo models of NSCLC with an anti-EGFR antibody, caused up-regulation of other ERbB family members, HER2 and HER3 (Mancini et al., 2015). In conclusion, in line with different mechanisms of action, resistance to mAbs and resistance to PKIs remarkably differ; secondary mutations frequently drive tolerance to PKIs, but this mechanism has rarely been implicated in refractoriness to mAbs.

5.4 Combination therapies targeting RTKs

As we describe below, targeting RTKs using either kinase inhibitors or mAbs, is a valid option in many types of cancer, but inevitably, slowly emerging resistance drives treatment failure. Like in infectious diseases, combining drugs that engage different mechanisms of action might overcome resistance. This approach has been used in cancer: combining rituximab with CHOP (cyclophosphamide, doxorubicin, vincristine and prednisone) increases overall survival and progression-free survival of patients with Non-Hodgkin lymphoma, as compared to chemotherapy alone (Kahl, 2008). Antibodies against EGFR and HER2 have also been approved in combination with chemotherapy.

Combining the anti-HER2 antibody trastuzumab with either paclitaxel (or cyclophosphamide plus anthracyclines) prolonged time to disease progression in comparison to chemotherapy alone, in patients with HER2-positive metastatic breast cancer (Slamon et al., 2001). The dual HER2-EGFR kinase inhibitor lapatinib, has been approved in combination with chemotherapy in HER2-positive metastatic breast cancer patients. In similarity, lapatinib was combined with capecitabine, in HER2-positive breast cancer patients who had progressed after treatment with an anthracycline and trastuzumab; the median time to progression was 8.4 months, compared to 4.4 months in the group treated with capecitabine as monotherapy (Geyer et al., 2006). In the same line, the anti-EGFR antibody cetuximab is currently indicated for head-and-neck cancer in combination with radiotherapy, because the combination of drugs reduced mortality without increasing toxicity of radiotherapy (Bonner et al., 2006). Combinations of different chemotherapy regimens with either the anti-VEGF antibody bevacizumab, or the anti-EGFR antibodies, cetuximab and panitumumab, are standard treatments in RAS-wild-type metastatic colorectal cancer, showing 50–60% response rates (Matos et al., 2019). Another approach by combining chemotherapy and mAbs is to use ADC (antibody-drug conjugate). Ado-trastuzumab emantisine (a conjugate of trastuzumab and the anti-microtubule compound DM1, a derivative of maytansine) was approved in 2013 for the treatment of HER2-positive metastatic breast cancer, based on the results of a clinical study that compared the antibody-drug conjugate to lapatinib plus capecitabine (Verma et al., 2012). Combinations of kinase inhibitors targeting RTKs and mAbs have also been tested in cancer. Combining bevacizumab with the EGFR-TKI erlotinib in EGFR-mutated NSCLC showed improvement in progression-free survival, leading to the approval of this combination in Europe in 2016 (Herbst et al., 2007; Rosell et al., 2017). Dual blockade of HER2, either with a kinase inhibitor plus a mAb or with two mAbs, has been tested in metastatic HER2 positive breast cancer. When lapatinib was combined with trastuzumab in HER2-overexpressing breast cancer cell lines, synergistic effects were observed (Xia et al., 2002). When this combination was tested in several clinical trials, the efficacy of the combination was confirmed but the side effects led to relatively high toxicity (Ahn & Vogel, 2012). More successful was the combination of two anti-HER2 antibodies, pertuzumab and trastuzumab. In line with animal studies that tested combinations of mAbs recognizing non-overlapping epitopes of HER2 (Ben-Kasus, Schechter, Lavi, Yarden, & Sela, 2009), these two antibodies recognize different epitopes within the extracellular part of the receptor, hence showed cooperative anti-tumor effects both in vitro and in vivo (Nahta, Hung, & Esteva, 2004; Scheuer et al., 2009). Based on the results of different clinical trials, the combination of trastuzumab and pertuzumab, together with chemotherapy, was approved as adjuvant treatment, for patients with HER2-positive breast cancer who are at high risk of developing recurrence (Gianni et al., 2014; von Minckwitz et al., 2017). In closing, many additional combinations of drugs targeting RTKs are now in preclinical or clinical studies, suggesting that this field will keep expanding in the coming years.

6. Relevance of RTKs to anti-cancer treatments using immune checkpoint inhibitors

6.1 A primer to immune checkpoint inhibitors

Cancer immunotherapy has become a staple of modern oncology, since the first attempts were described more than 30 years ago. These approaches are based on targeting cancer cells by utilizing the patient's own immune system, thereby mitigating many of the side effects. One way by which tumor cells achieve survival and progression requires evasion of immune checkpoints. Normally, the checkpoints monitor autoimmunity and mitigate collateral tissue damage by modulating costimulatory and inhibitory signaling (Pardoll, 2012). However, during tumorigenesis, aberrant checkpoint functions may result in the activation of inhibitory mechanisms, thereby preventing T lymphocytes and other immune cells from recognizing and eliminating cancer cells (Dong et al., 2002; Latchman et al., 2001; Pardoll, 2012). The major immune checkpoint receptors that have gained attention in the past few years, and served as primary targets in cancer immunotherapy, are cytotoxic T lymphocyte-associated antigen 4 (CTLA-4), programmed cell death protein 1 (PD-1), programmed cell death ligand 1 (PD-L1) and lymphocyte activation gene 3 (LAG-3) (Pardoll, 2012; Smyth, Ngiow, Ribas, & Teng, 2016). In recent years, immunotherapies based on the use of antibodies against PD-1 or PD-L1 have been useful in cancer treatment, because they can prolong patient survival, in different cancer indications, compared to standard therapy. However, a significant percentage of treated patients do not respond, or they eventually evolve resistance. In general, response rates associated with monotherapy checkpoint inhibitors rarely exceed 40%, ranging between 20% and 38% (Emens et al., 2017), due to significant fractions of partial responders (Bai et al., 2017), as well as relapses (Gauci et al., 2019). Importantly, resistance to anti-PD-1/PD-L1 therapy occurs in up to 70% of patients (O'Donnell, Long, Scolyer, Teng, & Smyth, 2017), with up to 60% of patients presenting primary resistance. Interestingly, expression levels of PD-L1 in the tumor and nearby tissue environment might alone serve as a predictor of patient response (Herbst et al., 2014). Nevertheless, some clinical trials reported that patient responses may not be affected by PD-L1 expression or histology (Brahmer et al., 2015; Rittmeyer et al., 2017). In conclusion, there is still a need to investigate and develop new biomarker panels, along with new treatment strategies able to predict and improve therapeutic efficacy.

6.2 The currently approved immune checkpoint inhibitors

Although cancer cells differ from their normal counterparts, and they are recognized as foreign by the immune system, tumors are rarely rejected spontaneously, owing to their ability to maintain an immunosuppressive microenvironment. Antibodies targeting checkpoint inhibitors can revert this by inducing T cell-mediated anti-tumor responses (Koury et al., 2018). Below we describe the major antibodies that target PD-1 (pembrolizumab and nivolumab), PD-L1 (atezolizumab; avelumab and durvalumab) and CTLA-4 (ipilimumab and tremelimumab).

6.2.1 Anti-CTLA-4 treatments

The first immune checkpoint receptor to be clinically targeted has been cytotoxic T lymphocyte-associated antigen 4 (CTLA-4) (Perkins et al., 1996). CTLA-4 is an inhibitory immune checkpoint receptor expressed on activated and regulatory T cells and it binds to B7 family ligands (CD80 and CD86) on antigen-presenting cells (Le Mercier, Lines, & Noelle, 2015; Perkins et al., 1996). CTLA-4 functions to downregulate T cell proliferation by outcompeting CD28, a costimulatory receptor (Teft, Kirchhof, & Madrenas, 2006). Anti-CTLA-4 antibodies potentiate anti-tumor effects by mitigating the natural brakes on T cells (D'Errico, Machado, & Sainz, 2017; Hodi et al., 2010). Ipilimumab, a monoclonal anti-CTLA-4 antibody, was the first checkpoint inhibitor to demonstrate an improved overall survival rate in patients with previously treated metastatic melanoma (Delyon et al., 2013; Hodi et al., 2010). The most common immune-related toxicity associated with ipilimumab administration at any grade was diarrhea, which was reported by 31% of patients. The durability of ipilimumab responses was verified by longitudinal follow-up studies of patients treated with ipilimumab (Prieto et al., 2012).

6.2.2 Anti-PD-1/PD-L1 treatments

Except for cells of the macrophage lineage, normal human tissues do not express PD-L1. In contrast, this molecule is abundant in human carcinomas of lung, ovary and colon. In addition to tumor cells, PD-L1 is highly expressed in the tumor microenvironment (Topalian, Drake, & Pardoll, 2015). In accordance, many tumor-specific T cells express the complementary PD-1 receptor (Nowicki, Hu-Lieskovan, & Ribas, 2018). Clinical trials that used anti-PD-1 monoclonal antibody therapies, nivolumab and pembrolizumab, or anti-PD-L1 antibodies, such as atezolizumab, avelumab and durvalumab, have established the usefulness of targeting the PD-1/PD-L1 axis in cancer immunology. Currently, PD-1/PD-L1 checkpoint blockade therapy is becoming part of the standard therapy for multiple malignancies, including melanoma, non-small-cell lung cancer (NSCLC), small cell lung cancer (SCLC), renal cell carcinoma (RCC), classical Hodgkin lymphoma (cHL), head and neck squamous cell carcinoma (HNSCC), colorectal cancer (CRC), hepatocellular carcinoma (HCC), primary mediastinal large B-cell lymphoma (PMLBCL) and bladder cancer.

6.3 Regulation of PD-L1 expression by RTKs

PD-L1 expression is driven by both the adaptive immune system (through IFN-γ receptor signaling) and oncogenic mechanisms, such as activation of MYC and AKT (Parsa et al., 2007). Both INF-γ receptors 1 and 2 participate in the regulation of PD-L1 expression through the JAK/STAT signaling pathway. Likewise, lipopolysaccharide (LPS) also leads to increased PD-L1 expression by means of activating toll-like receptor (TLR) 4 and subsequent induction of NF-κB, which results in elevated secretion of type I interferons (Lu, Yeh, & Ohashi, 2008). More relevant to the topic of this review, EGFR mutations, as well as stimulation with EGF, induce PD-L1 expression in lung cancer, and, accordingly, EGFR-specific PKIs can inhibit transcription of PD-L1 (Akbay et al., 2013; Chen et al., 2015; Concha-Benavente et al., 2016). Although it has been reported that PD-L1 expression downstream to EGFR activation is mediated by the AKT-mTOR pathway (Lastwika et al., 2016), another study demonstrated regulation via ERK stimulation (Shohat, Modan, & Rosenfeld, 1989). This inconsistency may be explained by tissue-specific reactions, or compensatory rewiring of signaling pathways (Sun & Bernards, 2014). A number of transcription factors known to be regulated by RTKs have been identified as activators of PD-L1 transcription. In addition to MYC, which is activated by multiple RTKs, both HIF-1α and HIF-2α can physically interact with the hypoxia response element (HRE) in the promoter region of *PD-L1* (Barsoum, Smallwood, Siemens, & Graham, 2014; Messai et al., 2016). Another downstream effector of RTKs, STAT3, can directly act on PD-L1's promoter to increase expression in human lymphoma and HNSCC cells (Atsaves et al., 2017; Bu et al., 2017). Interestingly, oncogenic RAS signaling can upregulate PD-L1 through a different mechanism, which stabilizes PD-L1's transcripts and explains the innately immunoresistant phenotype of *RAS* mutant cancers (Coelho et al., 2017).

6.4 Predictive value of PD-L1 levels and relations to RTKs

The above described lines of evidence propose that blocking PD-L1 would be beneficial for patients with EGFR mutations, and furthermore, that combinations of PD-1/PD-L1 blockers and RTK-targeting drugs might improve clinical responses. In line with these predictions, comparison of the effects of pembrolizumab, an anti-PD-1 mAb, with those of chemotherapy (docetaxel) in a group of *EGFR* mutation–positive patients with at least 50% of tumor cells expressing PD-L1, reported that the overall survival time was significantly longer with pembrolizumab treatment than with docetaxel therapy (Herbst et al., 2016). Furthermore, a combination of nivolumab and erlotinib prolonged overall survival in a group of 21 patients with lung cancer, who became resistant to erlotinib (Gettinger et al., 2018). Notably, PD-L1 expression, which is strongly activated by EGFR and downstream pathways, such as RAS and AKT, has undergone extensive assessment with respect to its value as a predictive biomarker. Although it was demonstrated, in several cancer types, that patients with PD-L1-positive tumors have better clinical responses as compared to PD-L1-negative subgroups (Borghaei et al., 2015; Jacquelot et al., 2017; Sun, Mezzadra, & Schumacher, 2018; Weber et al., 2017), exceptions have been reported and controversies remain as to the predictive value of this biomarker (Carbognin et al., 2015). Tumor mutational burden (TMB) is another potential biomarker, as is the number tumor-infiltrating lymphocytes (TILs) (Carbone et al., 2017; Schatton, Scolyer, Thompson, & Mihm, 2014). However, which biomarker is preferred and relationships between RTKs, TMB and TILs remain matters for future investigation.

7. Perspectives

The early animal studies performed in embryonic chicks and mice by Rita Levi Montalcini and Stanley Cohen, who respectively discovered the first growth factors, NGF (nerve growth factor) and EGF (epidermal growth factor), opened previously untraveled routes and yielded unexpected medical breakthroughs. Later studies, by Axel Ullrich and other investigators, surprisingly uncovered a common denominator of NGF, EGF and many additional growth factors: all bind with and stimulate transmembrane proteins, the RTKs, sharing a tyrosine-specific kinase domain. The discovery of RTKs and their highly similar mechanisms of downstream signaling pathways, which mainly culminate in transcription programs, greatly influenced the way we now understand development and physiology. The impact on cancer research and medical oncology has been similarly strong. As befits their pivotal functions, permanently active aberrant forms of RTKs, or their downstream effector molecules, have been observed in the last two decades in multiple cancer types. The next logical step entailed attempts to block active forms of RTKs in tumors, which instigated an avalanche of novel anti-cancer drugs and laid the cornerstone for personalized cancer medicine. Nowadays, drugs intercepting RTK signaling outnumber other drug families in oncology, but new challenges are emerging: secondary resistance to the new drugs severely limits drug applications. Similarly, optimal ways to combine anti-RTK drugs with other modalities, such as cytotoxic and immunological agents, are needed in order to keep the remarkably rapid path of discovery and fast translation to clinical practice.

References

Abbaspour Babaei, M., Kamalidehghan, B., Saleem, M., Huri, H. Z., & Ahmadipour, F. (2016). Receptor tyrosine kinase (c-kit) inhibitors: A potential therapeutic target in cancer cells. *Drug Design, Development and Therapy*, *10*, 2443–2459. https://doi.org/10.2147/dddt.S89114.

Aboud-Pirak, E., Hurwitz, E., Pirak, M. E., Bellot, F., Schlessinger, J., & Sela, M. (1988). Efficacy of antibodies to epidermal growth factor receptor against kb carcinoma in vitro and in nude mice. *Journal of the National Cancer Institute*, *80*(20), 1605–1611.

Ahn, E. R., & Vogel, C. L. (2012). Dual her2-targeted approaches in her2-positive breast cancer. *Breast Cancer Research and Treatment*, *131*(2), 371–383. https://doi.org/10.1007/s10549-011-1781-y.

Akbay, E. A., Koyama, S., Carretero, J., Altabef, A., Tchaicha, J. H., Christensen, C. L., et al. (2013). Activation of the pd-1 pathway contributes to immune escape in egfr-driven lung tumors. *Cancer Discovery*, *3*(12), 1355–1363. https://doi.org/10.1158/2159-8290.Cd-13-0310.

Alfarouk, K. O., Stock, C. M., Taylor, S., Walsh, M., Muddathir, A. K., Verduzco, D., et al. (2015). Resistance to cancer chemotherapy: Failure in drug response from adme to p-gp. *Cancer Cell International*, *15*, 71. https://doi.org/10.1186/s12935-015-0221-1.

Allen, C., Her, S., & Jaffray, D. A. (2017). Radiotherapy for cancer: Present and future. *Advanced Drug Delivery Reviews*, *109*, 1–2. https://doi.org/10.1016/j.addr.2017.01.004.

Almog, N., Ma, L., Raychowdhury, R., Schwager, C., Erber, R., Short, S., et al. (2009). Transcriptional switch of dormant tumors to fast-growing angiogenic phenotype. *Cancer Research*, *69*(3), 836–844. https://doi.org/10.1158/0008-5472.CAN-08-2590.

American Cancer Society. (2019). *Cancer treatment and survivorship facts & figures 2019–2021.* Retrieved from https://www.cancer.org/content/dam/cancer-org/research/cancer-facts-and-statistics/cancer-treatment-and-survivorship-facts-and-figures/cancer-treatment-and-survivorship-facts-and-figures-2019-2021.pdf.

Amit, I., Citri, A., Shay, T., Lu, Y., Katz, M., Zhang, F., et al. (2007). A module of negative feedback regulators defines growth factor signaling. *Nature Genetics*, *39*(4), 503–512. https://doi.org/10.1038/ng1987.

Amit, I., Wides, R., & Yarden, Y. (2007). Evolvable signaling networks of receptor tyrosine kinases: Relevance of robustness to malignancy and to cancer therapy. *Molecular Systems Biology*, *3*, 151. https://doi.org/10.1038/msb4100195.

Amit, I., Yakir, L., Katz, M., Zwang, Y., Marmor, M. D., Citri, A., et al. (2004). Tal, a tsg101-specific e3 ubiquitin ligase, regulates receptor endocytosis and retrovirus budding. *Genes & Development*, *18*(14), 1737–1752. https://doi.org/10.1101/gad.294904. 18/14/1737 (pii).

Anderson, D., Koch, C. A., Grey, L., Ellis, C., Moran, M. F., & Pawson, T. (1990). Binding of sh2 domains of phospholipase c gamma 1, gap, and src to activated growth factor receptors. *Science*, *250*(4983), 979–982.

Andres, A. C., & Ziemiecki, A. (2003). Eph and ephrin signaling in mammary gland morphogenesis and cancer. *Journal of Mammary Gland Biology and Neoplasia*, *8*(4), 475–485. https://doi.org/10.1023/b:Jomg.0000017433.83226.22.

Arevalo, J. C., Waite, J., Rajagopal, R., Beyna, M., Chen, Z. Y., Lee, F. S., et al. (2006). Cell survival through trk neurotrophin receptors is differentially regulated by ubiquitination. *Neuron*, *50*(4), 549–559. https://doi.org/10.1016/j.neuron.2006.03.044.

Arjona, D., Bello, M. J., Alonso, M. E., Aminoso, C., Isla, A., De Campos, J. M., et al. (2005). Molecular analysis of the egfr gene in astrocytic gliomas: Mrna expression, quantitative-pcr analysis of non-homogeneous gene amplification and DNA sequence alterations. *Neuropathology and Applied Neurobiology*, *31*(4), 384–394.

Asahara, T., Murohara, T., Sullivan, A., Silver, M., van der Zee, R., Li, T., et al. (1997). Isolation of putative progenitor endothelial cells for angiogenesis. *Science*, *275*(5302), 964–967.

Atsaves, V., Tsesmetzis, N., Chioureas, D., Kis, L., Leventaki, V., Drakos, E., et al. (2017). Pd-l1 is commonly expressed and transcriptionally regulated by stat3 and myc in alk-negative anaplastic large cell lymphoma. *Leukemia*, *31*(7), 1633–1637. https://doi.org/10.1038/leu.2017.103.

Avraham, R., Sas-Chen, A., Manor, O., Steinfeld, I., Shalgi, R., Tarcic, G., et al. (2010). Egf decreases the abundance of micrornas that restrain oncogenic transcription factors. *Science Signaling*, *3*(124), ra43. https://doi.org/10.1126/scisignal.2000876.

Bagheri-Yarmand, R., Sinha, K. M., Gururaj, A. E., Ahmed, Z., Rizvi, Y. Q., Huang, S. C., et al. (2015). A novel dual kinase function of the ret proto-oncogene negatively regulates activating transcription factor 4-mediated apoptosis. *The Journal of Biological Chemistry*, *290*(18), 11749–11761. https://doi.org/10.1074/jbc.M114.619833.

Bai, J., Gao, Z., Li, X., Dong, L., Han, W., & Nie, J. (2017). Regulation of pd-1/pd-l1 pathway and resistance to pd-1/pd-l1 blockade. *Oncotarget*, *8*(66), 110693–110707. https://doi.org/10.18632/oncotarget.22690.

Barouch-Bentov, R., & Sauer, K. (2011). Mechanisms of drug resistance in kinases. *Expert Opinion on Investigational Drugs*, *20*(2), 153–208. https://doi.org/10.1517/13543784.2011.546344.

Barsoum, I. B., Smallwood, C. A., Siemens, D. R., & Graham, C. H. (2014). A mechanism of hypoxia-mediated escape from adaptive immunity in cancer cells. *Cancer Research*, *74*(3), 665–674. https://doi.org/10.1158/0008-5472.CAN-13-0992.

Bean, J., Brennan, C., Shih, J.-Y., Riely, G., Viale, A., Wang, L., et al. (2007). *MET* amplification occurs with or without *T790M* mutations in *EGFR* mutant lung tumors with acquired resistance to gefitinib or erlotinib. *Proceedings of the National Academy of Sciences, 104*(52), 20932–20937. https://doi.org/10.1073/pnas.0710370104.

Ben-Kasus, T., Schechter, B., Lavi, S., Yarden, Y., & Sela, M. (2009). Persistent elimination of erbb-2/her2-overexpressing tumors using combinations of monoclonal antibodies: Relevance of receptor endocytosis. *Proceedings of the National Academy of Sciences of the United States of America, 106*(9), 3294–3299. https://doi.org/10.1073/pnas.0812059106. 0812059106 [pii].

Bentzen, S. M., Atasoy, B. M., Daley, F. M., Dische, S., Richman, P. I., Saunders, M. I., et al. (2005). Epidermal growth factor receptor expression in pretreatment biopsies from head and neck squamous cell carcinoma as a predictive factor for a benefit from accelerated radiation therapy in a randomized controlled trial. *Journal of Clinical Oncology, 23*(24), 5560–5567. https://doi.org/10.1200/JCO.2005.06.411.

Berns, K., Horlings, H. M., Hennessy, B. T., Madiredjo, M., Hijmans, E. M., Beelen, K., et al. (2007). A functional genetic approach identifies the pi3k pathway as a major determinant of trastuzumab resistance in breast cancer. *Cancer Cell, 12*(4), 395–402.

Bessman, N. J., Bagchi, A., Ferguson, K. M., & Lemmon, M. A. (2014). Complex relationship between ligand binding and dimerization in the epidermal growth factor receptor. *Cell Reports, 9*(4), 1306–1317. https://doi.org/10.1016/j.celrep.2014.10.010.

Binder, Z. A., Thorne, A. H., Bakas, S., Wileyto, E. P., Bilello, M., Akbari, H., et al. (2018). Epidermal growth factor receptor extracellular domain mutations in glioblastoma present opportunities for clinical imaging and therapeutic development. *Cancer Cell, 34*(1), 163–177. e167. https://doi.org/10.1016/j.ccell.2018.06.006.

Blume-Jensen, P., & Hunter, T. (2001). Oncogenic kinase signalling. *Nature, 411*(6835), 355–365. https://doi.org/10.1038/35077225.

Bocharov, E. V., Lesovoy, D. M., Pavlov, K. V., Pustovalova, Y. E., Bocharova, O. V., & Arseniev, A. S. (2016). Alternative packing of egfr transmembrane domain suggests that protein-lipid interactions underlie signal conduction across membrane. *Biochimica et Biophysica Acta, 1858*(6), 1254–1261. https://doi.org/10.1016/j.bbamem.2016.02.023.

Bonanno, L., Favaretto, A., & Rosell, R. (2014). Platinum drugs and DNA repair mechanisms in lung cancer. *Anticancer Research, 34*(1), 493–501.

Bonner, J. A., Harari, P. M., Giralt, J., Azarnia, N., Shin, D. M., Cohen, R. B., et al. (2006). Radiotherapy plus cetuximab for squamous-cell carcinoma of the head and neck. *The New England Journal of Medicine, 354*(6), 567–578.

Borghaei, H., Paz-Ares, L., Horn, L., Spigel, D. R., Steins, M., Ready, N. E., et al. (2015). Nivolumab versus docetaxel in advanced nonsquamous non-small-cell lung cancer. *The New England Journal of Medicine, 373*(17), 1627–1639. https://doi.org/10.1056/NEJMoa1507643.

Bos, P. D., Zhang, X. H., Nadal, C., Shu, W., Gomis, R. R., Nguyen, D. X., et al. (2009). Genes that mediate breast cancer metastasis to the brain. *Nature, 459*(7249), 1005–1009. https://doi.org/10.1038/nature08021.

Bradshaw, R. A., Chalkley, R. J., Biarc, J., & Burlingame, A. L. (2013). Receptor tyrosine kinase signaling mechanisms: Devolving trka responses with phosphoproteomics. *Advances in Biological Regulation, 53*(1), 87–96. https://doi.org/10.1016/j.jbior.2012.10.006.

Brahmer, J., Reckamp, K. L., Baas, P., Crino, L., Eberhardt, W. E., Poddubskaya, E., et al. (2015). Nivolumab versus docetaxel in advanced squamous-cell non-small-cell lung cancer. *The New England Journal of Medicine, 373*(2), 123–135. https://doi.org/10.1056/NEJMoa1504627.

Brummer, T., Schmitz-Peiffer, C., & Daly, R. J. (2010). Docking proteins. *The FEBS Journal, 277*(21), 4356–4369. https://doi.org/10.1111/j.1742-4658.2010.07865.x.

Bu, L. L., Yu, G. T., Wu, L., Mao, L., Deng, W. W., Liu, J. F., et al. (2017). Stat3 induces immunosuppression by upregulating pd-1/pd-l1 in hnscc. *Journal of Dental Research, 96*(9), 1027–1034. https://doi.org/10.1177/0022034517712435.

Buchdunger, E., Cioffi, C. L., Law, N., Stover, D., Ohno-Jones, S., Druker, B. J., et al. (2000). Abl protein-tyrosine kinase inhibitor sti571 inhibits in vitro signal transduction mediated by c-kit and platelet-derived growth factor receptors. *The Journal of Pharmacology and Experimental Therapeutics*, *295*(1), 139–145.

Burgess, A. W., Cho, H. S., Eigenbrot, C., Ferguson, K. M., Garrett, T. P., Leahy, D. J., et al. (2003). An open-and-shut case? Recent insights into the activation of egf/erbb receptors. *Molecular Cell*, *12*(3), 541–552.

Burke, P., Schooler, K., & Wiley, H. S. (2001). Regulation of epidermal growth factor receptor signaling by endocytosis and intracellular trafficking. *Molecular Biology of the Cell*, *12*(6), 1897–1910.

Butrynski, J. E., D'Adamo, D. R., Hornick, J. L., Dal Cin, P., Antonescu, C. R., Jhanwar, S. C., et al. (2010). Crizotinib in alk-rearranged inflammatory myofibroblastic tumor. *The New England Journal of Medicine*, *363*(18), 1727–1733. https://doi.org/10.1056/NEJMoa1007056.

Carbognin, L., Pilotto, S., Milella, M., Vaccaro, V., Brunelli, M., Calio, A., et al. (2015). Differential activity of nivolumab, pembrolizumab and mpdl3280a according to the tumor expression of programmed death-ligand-1 (pd-l1): Sensitivity analysis of trials in melanoma, lung and genitourinary cancers. *PLoS One*, *10*(6). e0130142https://doi.org/10.1371/journal.pone.0130142.

Carbone, D. P., Reck, M., Paz-Ares, L., Creelan, B., Horn, L., Steins, M., et al. (2017). First-line nivolumab in stage iv or recurrent non-small-cell lung cancer. *The New England Journal of Medicine*, *376*(25), 2415–2426. https://doi.org/10.1056/NEJMoa1613493.

Carmeliet, P., & Jain, R. K. (2000). Angiogenesis in cancer and other diseases. *Nature*, *407*(6801), 249–257. https://doi.org/10.1038/35025220.

Carvalho, S., Levi-Schaffer, F., Sela, M., & Yarden, Y. (2016). Immunotherapy of cancer: From monoclonal to oligoclonal cocktails of anti-cancer antibodies: Luphar review 18. *British Journal of Pharmacology*, *173*(9), 1407–1424.

Casals, E., Gusta, M. F., Cobaleda-Siles, M., Garcia-Sanz, A., & Puntes, V. F. (2017). Cancer resistance to treatment and antiresistance tools offered by multimodal multifunctional nanoparticles. *Cancer Nanotechnology*, *8*(1), 7. https://doi.org/10.1186/s12645-017-0030-4.

Cerami, E., Gao, J., Dogrusoz, U., Gross, B. E., Sumer, S. O., Aksoy, B. A., et al. (2012). The cbio cancer genomics portal: An open platform for exploring multidimensional cancer genomics data. *Cancer Discovery*, *2*(5), 401–404. https://doi.org/10.1158/2159-8290.CD-12-0095.

Chen, N., Fang, W., Zhan, J., Hong, S., Tang, Y., Kang, S., et al. (2015). Upregulation of pd-l1 by egfr activation mediates the immune escape in egfr-driven nsclc: Implication for optional immune targeted therapy for nsclc patients with egfr mutation *Journal of Thoracic Oncology*, *10*(6), 910–923. https://doi.org/10.1097/JTO.0000000000000500.

Cheusova, T., Khan, M. A., Schubert, S. W., Gavin, A. C., Buchou, T., Jacob, G., et al. (2006). Casein kinase 2-dependent serine phosphorylation of musk regulates acetylcholine receptor aggregation at the neuromuscular junction. *Genes & Development*, *20*(13), 1800–1816. https://doi.org/10.1101/gad.375206.

Chmura, S. J., Nodzenski, E., Beckett, M. A., Kufe, D. W., Quintans, J., & Weichselbaum, R. R. (1997). Loss of ceramide production confers resistance to radiation-induced apoptosis. *Cancer Research*, *57*(7), 1270–1275.

Cho, H. S., & Leahy, D. J. (2002). Structure of the extracellular region of her3 reveals an interdomain tether. *Science*, *297*(5585), 1330–1333.

Ciardiello, F., & Tortora, G. (2001). A novel approach in the treatment of cancer: Targeting the epidermal growth factor receptor. *Clinical Cancer Research*, 7(10), 2958–2970.

Citri, A., & Yarden, Y. (2006). Egf-erbb signalling: Towards the systems level. *Nature Reviews. Molecular Cell Biology*, 7(7), 505–516.

Cocco, E., Scaltriti, M., & Drilon, A. (2018). Ntrk fusion-positive cancers and trk inhibitor therapy. *Nature Reviews Clinical Oncology*, *15*(12), 731–747. https://doi.org/10.1038/s41571-018-0113-0.

Coelho, M. A., de Carne Trecesson, S., Rana, S., Zecchin, D., Moore, C., Molina-Arcas, M., et al. (2017). Oncogenic ras signaling promotes tumor immunoresistance by stabilizing pd-l1 mrna. *Immunity*, *47*(6), 1083–1099 e1086. https://doi.org/10.1016/j.immuni.2017.11.016.

Concha-Benavente, F., Srivastava, R. M., Trivedi, S., Lei, Y., Chandran, U., Seethala, R. R., et al. (2016). Identification of the cell-intrinsic and -extrinsic pathways downstream of egfr and ifngamma that induce pd-l1 expression in head and neck cancer. *Cancer Research*, *76*(5), 1031–1043. https://doi.org/10.1158/0008-5472.CAN-15-2001.

Dankner, M., Rose, A. A. N., Rajkumar, S., Siegel, P. M., & Watson, I. R. (2018). Classifying braf alterations in cancer: New rational therapeutic strategies for actionable mutations. *Oncogene*, *37*(24), 3183–3199. https://doi.org/10.1038/s41388-018-0171-x.

Delyon, J., Mateus, C., Lefeuvre, D., Lanoy, E., Zitvogel, L., Chaput, N., et al. (2013). Experience in daily practice with ipilimumab for the treatment of patients with metastatic melanoma: An early increase in lymphocyte and eosinophil counts is associated with improved survival. *Annals of Oncology*, *24*(6), 1697–1703. https://doi.org/10.1093/annonc/mdt027.

D'Errico, G., Machado, H. L., & Sainz, B., Jr. (2017). A current perspective on cancer immune therapy: Step-by-step approach to constructing the magic bullet. *Clinical and Translational Medicine*, *6*(1), 3. https://doi.org/10.1186/s40169-016-0130-5.

Derynck, R., Goeddel, D. V., Ullrich, A., Gutterman, J. U., Williams, R. D., Bringman, T. S., et al. (1987). Synthesis of messenger rnas for transforming growth factors alpha and beta and the epidermal growth factor receptor by human tumors. *Cancer Research*, *47*(3), 707–712.

Dikic, I., & Giordano, S. (2003). Negative receptor signalling. *Current Opinion in Cell Biology*, *15*(2), 128–135.

Donawho, C. K., Luo, Y., Luo, Y., Penning, T. D., Bauch, J. L., Bouska, J. J., et al. (2007). Abt-888, an orally active poly(adp-ribose) polymerase inhibitor that potentiates DNA-damaging agents in preclinical tumor models. *Clinical Cancer Research*, *13*(9), 2728–2737. https://doi.org/10.1158/1078-0432.Ccr-06-3039.

Dong, H., Strome, S. E., Salomao, D. R., Tamura, H., Hirano, F., Flies, D. B., et al. (2002). Tumor-associated b7-h1 promotes t-cell apoptosis: A potential mechanism of immune evasion. *Nature Medicine*, *8*(8), 793–800. https://doi.org/10.1038/nm730.

Downward, J., Yarden, Y., Mayes, E., Scrace, G., Totty, N., Stockwell, P., et al. (1984). Close similarity of epidermal growth factor receptor and v-erb-b oncogene protein sequences. *Nature*, *307*(5951), 521–527.

Drilon, A., Hu, Z. I., Lai, G. G. Y., & Tan, D. S. W. (2018). Targeting ret-driven cancers: Lessons from evolving preclinical and clinical landscapes. *Nature Reviews. Clinical Oncology*, *15*(3), 151–167. https://doi.org/10.1038/nrclinonc.2017.175.

Drilon, A., Siena, S., Ou, S. I., Patel, M., Ahn, M. J., Lee, J., et al. (2017). Safety and antitumor activity of the multitargeted pan-trk, ros1, and alk inhibitor entrectinib: Combined results from two phase i trials (alka-372-001 and startrk-1). *Cancer Discovery*, 7(4), 400–409. https://doi.org/10.1158/2159-8290.CD-16-1237.

Du, Z., & Lovly, C. M. (2018). Mechanisms of receptor tyrosine kinase activation in cancer. *Molecular Cancer*, *17*(1), 58. https://doi.org/10.1186/s12943-018-0782-4.

Eberlein, C. A., Stetson, D., Markovets, A. A., Al-Kadhimi, K. J., Lai, Z., Fisher, P. R., et al. (2015). Acquired resistance to the mutant-selective egfr inhibitor azd9291 is associated with increased dependence on ras signaling in preclinical models. *Cancer Research*, *75*(12), 2489–2500. https://doi.org/10.1158/0008-5472.CAN-14-3167.

Eichhorn, P. J., Gili, M., Scaltriti, M., Serra, V., Guzman, M., Nijkamp, W., et al. (2008). Phosphatidylinositol 3-kinase hyperactivation results in lapatinib resistance that is reversed by the mtor/phosphatidylinositol 3-kinase inhibitor nvp-bez235. *Cancer Research, 68*(22), 9221–9230. https://doi.org/10.1158/0008-5472.CAN-08-1740. 68/22/9221 [pii].

El-Deeb, I. M., Yoo, K. H., & Lee, S. H. (2011). Ros receptor tyrosine kinase: A new potential target for anticancer drugs. *Medicinal Research Reviews, 31*(5), 794–818. https://doi.org/10.1002/med.20206.

Emens, L. A., Ascierto, P. A., Darcy, P. K., Demaria, S., Eggermont, A. M. M., Redmond, W. L., et al. (2017). Cancer immunotherapy: Opportunities and challenges in the rapidly evolving clinical landscape. *European Journal of Cancer, 81*, 116–129. https://doi.org/10.1016/j.ejca.2017.01.035.

Epstein, J. I. (2009). Precursor lesions to prostatic adenocarcinoma. *Virchows Archiv, 454*(1), 1–16. https://doi.org/10.1007/s00428-008-0707-5.

Epstein, D. M., & Buck, E. (2015). Old dog, new tricks: Extracellular domain arginine methylation regulates egfr function. *The Journal of Clinical Investigation, 125*(12), 4320–4322. https://doi.org/10.1172/JCI85001.

Eramo, A., Ricci-Vitiani, L., Zeuner, A., Pallini, R., Lotti, F., Sette, G., et al. (2006). Chemotherapy resistance of glioblastoma stem cells. *Cell Death and Differentiation, 13*(7), 1238–1241. https://doi.org/10.1038/sj.cdd.4401872.

Esposito, I., Kleeff, J., Bischoff, S. C., Fischer, L., Collecchi, P., Iorio, M., et al. (2002). The stem cell factor-c-kit system and mast cells in human pancreatic cancer. *Laboratory Investigation, 82*(11), 1481–1492.

Farooqi, A. A., & Siddik, Z. H. (2015). Platelet-derived growth factor (pdgf) signalling in cancer: Rapidly emerging signalling landscape. *Cell Biochemistry and Function, 33*(5), 257–265. https://doi.org/10.1002/cbf.3120.

Forbes, S. A., Tang, G., Bindal, N., Bamford, S., Dawson, E., Cole, C., et al. (2010). Cosmic (the catalogue of somatic mutations in cancer): A resource to investigate acquired mutations in human cancer. *Nucleic Acids Research, 38*(Database issue), D652–D657. https://doi.org/10.1093/nar/gkp995.

Frank, N. Y., Schatton, T., & Frank, M. H. (2010). The therapeutic promise of the cancer stem cell concept. *The Journal of Clinical Investigation, 120*(1), 41–50. https://doi.org/10.1172/jci41004.

Fujimoto, N., Wislez, M., Zhang, J., Iwanaga, K., Dackor, J., Hanna, A. E., et al. (2005). High expression of erbb family members and their ligands in lung adenocarcinomas that are sensitive to inhibition of epidermal growth factor receptor. *Cancer Research, 65*(24), 11478–11485. https://doi.org/10.1158/0008-5472.CAN-05-1977.

Fujita, Y., Krause, G., Scheffner, M., Zechner, D., Leddy, H. E., Behrens, J., et al. (2002). Hakai, a c-cbl-like protein, ubiquitinates and induces endocytosis of the e-cadherin complex. *Nature Cell Biology, 4*(3), 222–231. https://doi.org/10.1038/ncb758. ncb758 [pii].

Gadiya, M., & Chakraborty, G. (2018). Signaling by discoidin domain receptor 1 in cancer metastasis. *Cell Adhesion & Migration, 12*(4), 315–323. https://doi.org/10.1080/19336918.2018.1520556.

García-Aranda, M., & Redondo, M. (2019). Targeting receptor kinases in colorectal cancer. *Cancers (Basel), 11*(4), 433. https://doi.org/10.3390/cancers11040433.

Garraway, L. A., & Janne, P. A. (2012). Circumventing cancer drug resistance in the era of personalized medicine. *Cancer Discovery, 2*(3), 214–226. https://doi.org/10.1158/2159-8290.CD-12-0012.

Gatalica, Z., Xiu, J., Swensen, J., & Vranic, S. (2019). Molecular characterization of cancers with ntrk gene fusions. *Modern Pathology, 32*(1), 147–153. https://doi.org/10.1038/s41379-018-0118-3.

Gauci, M. L., Lanoy, E., Champiat, S., Caramella, C., Ammari, S., Aspeslagh, S., et al. (2019). Long-term survival in patients responding to anti-pd-1/pd-l1 therapy and disease outcome upon treatment discontinuation. *Clinical Cancer Research*, *25*(3), 946–956. https://doi.org/10.1158/1078-0432.CCR-18-0793.

Gayle, S. S., Arnold, S. L., O'Regan, R. M., & Nahta, R. (2012). Pharmacologic inhibition of mtor improves lapatinib sensitivity in her2-overexpressing breast cancer cells with primary trastuzumab resistance. *Anti-Cancer Agents in Medicinal Chemistry*, *12*(2), 151–162. https://doi.org/10.2174/187152012799015002.

Gelfo, V., Pontis, F., Mazzeschi, M., Sgarzi, M., Mazzarini, M., Solmi, R., et al. (2019). Glucocorticoid receptor modulates egfr feedback upon acquisition of resistance to monoclonal antibodies. *Journal of Clinical Medicine*, *8*(5), 600. https://doi.org/10.3390/jcm8050600.

Gettinger, S., Hellmann, M. D., Chow, L. Q. M., Borghaei, H., Antonia, S., Brahmer, J. R., et al. (2018). Nivolumab plus erlotinib in patients with egfr-mutant advanced nsclc. *Journal of Thoracic Oncology*, *13*(9), 1363–1372. https://doi.org/10.1016/j.jtho.2018.05.015.

Geyer, C. E., Forster, J., Lindquist, D., Chan, S., Romieu, C. G., Pienkowski, T., et al. (2006). Lapatinib plus capecitabine for her2-positive advanced breast cancer. *The New England Journal of Medicine*, *355*(26), 2733–2743.

Ghedini, G. C., Ronca, R., Presta, M., & Giacomini, A. (2018). Future applications of fgf/fgfr inhibitors in cancer. *Expert Review of Anticancer Therapy*, *18*(9), 861–872. https://doi.org/10.1080/14737140.2018.1491795.

Gianni, L., Eiermann, W., Semiglazov, V., Lluch, A., Tjulandin, S., Zambetti, M., et al. (2014). Neoadjuvant and adjuvant trastuzumab in patients with her2-positive locally advanced breast cancer (noah): Follow-up of a randomised controlled superiority trial with a parallel her2-negative cohort. *The Lancet Oncology*, *15*(6), 640–647. https://doi.org/10.1016/S1470-2045(14)70080-4.

Goh, L. K., Huang, F., Kim, W., Gygi, S., & Sorkin, A. (2010). Multiple mechanisms collectively regulate clathrin-mediated endocytosis of the epidermal growth factor receptor. *The Journal of Cell Biology*, *189*(5), 871–883. https://doi.org/10.1083/jcb.201001008. jcb.201001008 [pii].

Golubkov, V. S., Prigozhina, N. L., Zhang, Y., Stoletov, K., Lewis, J. D., Schwartz, P. E., et al. (2014). Protein-tyrosine pseudokinase 7 (ptk7) directs cancer cell motility and metastasis. *The Journal of Biological Chemistry*, *289*(35), 24238–24249. https://doi.org/10.1074/jbc.M114.574459.

Gong, J., Jaiswal, R., Mathys, J. M., Combes, V., Grau, G. E., & Bebawy, M. (2012). Microparticles and their emerging role in cancer multidrug resistance. *Cancer Treatment Reviews*, *38*(3), 226–234. https://doi.org/10.1016/j.ctrv.2011.06.005.

Gonzalez, D. M., & Medici, D. (2014). Signaling mechanisms of the epithelial-mesenchymal transition. *Science Signaling*, 7(344), re8. https://doi.org/10.1126/scisignal.2005189.

Grassot, J., Mouchiroud, G., & Perriere, G. (2003). Rtkdb: Database of receptor tyrosine kinase. *Nucleic Acids Research*, *31*(1), 353–358. https://doi.org/10.1093/nar/gkg036.

Guix, M., Faber, A. C., Wang, S. E., Olivares, M. G., Song, Y., Qu, S., et al. (2008). Acquired resistance to egfr tyrosine kinase inhibitors in cancer cells is mediated by loss of igf-binding proteins. *The Journal of Clinical Investigation*, *118*(7), 2609–2619. https://doi.org/10.1172/JCI34588.

Ha, S. Y., Lee, J., Kang, S. Y., Do, I. G., Ahn, S., Park, J. O., et al. (2013). Met overexpression assessed by new interpretation method predicts gene amplification and poor survival in advanced gastric carcinomas. *Modern Pathology*, *26*(12), 1632–1641. https://doi.org/10.1038/modpathol.2013.108.

Haj, F. G., Markova, B., Klaman, L. D., Bohmer, F. D., & Neel, B. G. (2003). Regulation of receptor tyrosine kinase signaling by protein tyrosine phosphatase-1b. *The Journal of*

Biological Chemistry, 278(2), 739–744. https://doi.org/10.1074/jbc.M210194200 [M210194200 [pii].

Hanahan, D., & Weinberg, R. A. (2011). Hallmarks of cancer: The next generation. *Cell, 144*(5), 646–674. https://doi.org/10.1016/j.cell.2011.02.013.

Hanawa, M., Suzuki, S., Dobashi, Y., Yamane, T., Kono, K., Enomoto, N., et al. (2006). Egfr protein overexpression and gene amplification in squamous cell carcinomas of the esophagus. *International Journal of Cancer, 118*(5), 1173–1180. https://doi.org/10.1002/ijc.21454.

Harari, P. M. (2007). Stepwise progress in epidermal growth factor receptor/radiation studies for head and neck cancer. *International Journal of Radiation Oncology, Biology, Physics, 69*(2 Suppl), S25–S27. https://doi.org/10.1016/j.ijrobp.2007.04.087, S0360-3016(07)00995-9 [pii].

Harari, D., & Yarden, Y. (2000). Molecular mechanisms underlying erbb2/her2 action in breast cancer. *Oncogene, 19*(53), 6102–6114. https://doi.org/10.1038/sj.onc.1203973.

Hata, A. N., Niederst, M. J., Archibald, H. L., Gomez-Caraballo, M., Siddiqui, F. M., Mulvey, H. E., et al. (2016). Tumor cells can follow distinct evolutionary paths to become resistant to epidermal growth factor receptor inhibition. *Nature Medicine, 22*(3), 262–269.

He, X., Zhang, H., Xiao, M., Kong, Y., Li, W., & Zhang, H. (2015). Inhibition of progression of panin through antagonizing egfr. *Tumour Biology, 36*(5), 3245–3249. https://doi.org/10.1007/s13277-014-2953-2.

Hegde, P. S., Rusnak, D., Bertiaux, M., Alligood, K., Strum, J., Gagnon, R., et al. (2007). Delineation of molecular mechanisms of sensitivity to lapatinib in breast cancer cell lines using global gene expression profiles. *Molecular Cancer Therapeutics, 6*(5), 1629–1640. https://doi.org/10.1158/1535-7163.MCT-05-0399.

Heinrich, M. C., Corless, C. L., Demetri, G. D., Blanke, C. D., von Mehren, M., Joensuu, H., et al. (2003). Kinase mutations and imatinib response in patients with metastatic gastrointestinal stromal tumor. *Journal of Clinical Oncology, 21*(23), 4342–4349. https://doi.org/10.1200/JCO.2003.04.190.

Helsten, T., Elkin, S., Arthur, E., Tomson, B. N., Carter, J., & Kurzrock, R. (2016). The fgfr landscape in cancer: Analysis of 4,853 tumors by next-generation sequencing. *Clinical Cancer Research, 22*(1), 259–267. https://doi.org/10.1158/1078-0432.Ccr-14-3212.

Herbst, R. S., Baas, P., Kim, D. W., Felip, E., Perez-Gracia, J. L., Han, J. Y., et al. (2016). Pembrolizumab versus docetaxel for previously treated, pd-l1-positive, advanced non-small-cell lung cancer (keynote-010): A randomised controlled trial. *Lancet, 387*(10027), 1540–1550. https://doi.org/10.1016/S0140-6736(15)01281-7.

Herbst, R. S., O'Neill, V. J., Fehrenbacher, L., Belani, C. P., Bonomi, P. D., Hart, L., et al. (2007). Phase ii study of efficacy and safety of bevacizumab in combination with chemotherapy or erlotinib compared with chemotherapy alone for treatment of recurrent or refractory non small-cell lung cancer. *Journal of Clinical Oncology, 25*(30), 4743–4750. https://doi.org/10.1200/jco.2007.12.3026.

Herbst, R. S., Soria, J. C., Kowanetz, M., Fine, G. D., Hamid, O., Gordon, M. S., et al. (2014). Predictive correlates of response to the anti-pd-l1 antibody mpdl3280a in cancer patients. *Nature, 515*(7528), 563–567. https://doi.org/10.1038/nature14011.

Hermann, P. C., Huber, S. L., Herrler, T., Aicher, A., Ellwart, J. W., Guba, M., et al. (2007). Distinct populations of cancer stem cells determine tumor growth and metastatic activity in human pancreatic cancer. *Cell Stem Cell, 1*(3), 313–323. https://doi.org/10.1016/j.stem.2007.06.002.

Hodi, F. S., O'Day, S. J., McDermott, D. F., Weber, R. W., Sosman, J. A., Haanen, J. B., et al. (2010). Improved survival with ipilimumab in patients with metastatic melanoma. *The New England Journal of Medicine, 363*(8), 711–723. https://doi.org/10.1056/NEJMoa1003466.

Hong, S., Jung, H. I., Ahn, T. S., Kim, H. J., Lee, K. T., Baek, M. J., et al. (2017). Expressions and clinical significances of angiopoietin-1, angiopoietin-2, and tie-2 receptor in patients with colorectal cancer. *Annals of Coloproctology, 33*(1), 9–15. https://doi.org/10.3393/ac.2017.33.1.9.

Hour, T. C., Chung, S. D., Kang, W. Y., Lin, Y. C., Chuang, S. J., Huang, A. M., et al. (2015). Egfr mediates docetaxel resistance in human castration-resistant prostate cancer through the akt-dependent expression of abcb1 (mdr1). *Archives of Toxicology, 89*(4), 591–605. https://doi.org/10.1007/s00204-014-1275-x.

Housman, G., Byler, S., Heerboth, S., Lapinska, K., Longacre, M., Snyder, N., et al. (2014). Drug resistance in cancer: An overview. *Cancers (Basel), 6*(3), 1769–1792. https://doi.org/10.3390/cancers6031769.

Hubbard, S. R. (2002). Autoinhibitory mechanisms in receptor tyrosine kinases. *Frontiers in Bioscience, 7*, d330–d340.

Hubbard, S. R., & Miller, W. T. (2007). Receptor tyrosine kinases: Mechanisms of activation and signaling. *Current Opinion in Cell Biology, 19*(2), 117–123.

Hubbard, S. R., & Till, J. H. (2000). Protein tyrosine kinase structure and function. *Annual Review of Biochemistry, 69*, 373–398. https://doi.org/10.1146/annurev.biochem.69.1.373.

Iragavarapu, C., Mustafa, M., Akinleye, A., Furqan, M., Mittal, V., Cang, S., et al. (2015). Novel alk inhibitors in clinical use and development. *Journal of Hematology & Oncology, 8*, 17. https://doi.org/10.1186/s13045-015-0122-8.

Jacquelot, N., Roberti, M. P., Enot, D. P., Rusakiewicz, S., Ternes, N., Jegou, S., et al. (2017). Predictors of responses to immune checkpoint blockade in advanced melanoma. *Nature Communications, 8*(1), 592. https://doi.org/10.1038/s41467-017-00608-2.

Janne, P. A., Yang, J. C., Kim, D. W., Planchard, D., Ohe, Y., Ramalingam, S. S., et al. (2015). Azd9291 in egfr inhibitor-resistant non-small-cell lung cancer. *The New England Journal of Medicine, 372*(18), 1689–1699. https://doi.org/10.1056/NEJMoa1411817.

Jiang, X., Huang, F., Marusyk, A., & Sorkin, A. (2003). Grb2 regulates internalization of egf receptors through clathrin-coated pits. *Molecular Biology of the Cell, 14*(3), 858–870.

Johnston, S., Trudeau, M., Kaufman, B., Boussen, H., Blackwell, K., LoRusso, P., et al. (2008). Phase ii study of predictive biomarker profiles for response targeting human epidermal growth factor receptor 2 (her-2) in advanced inflammatory breast cancer with lapatinib monotherapy. *Journal of Clinical Oncology, 26*(7), 1066–1072.

Jura, N., Endres, N. F., Engel, K., Deindl, S., Das, R., Lamers, M. H., et al. (2009). Mechanism for activation of the egf receptor catalytic domain by the juxtamembrane segment. *Cell, 137*(7), 1293–1307. https://doi.org/10.1016/j.cell.2009.04.025. S0092-8674(09)00450-4 [pii].

Kahl, B. (2008). Chemotherapy combinations with monoclonal antibodies in non-hodgkin's lymphoma. *Seminars in Hematology, 45*(2), 90–94. https://doi.org/10.1053/j.seminhematol.2008.02.003.

Kaleko, M., Rutter, W. J., & Miller, A. D. (1990). Overexpression of the human insulinlike growth factor i receptor promotes ligand-dependent neoplastic transformation. *Molecular and Cellular Biology, 10*(2), 464–473. https://doi.org/10.1128/mcb.10.2.464.

Kariolis, M. S., Miao, Y. R., Diep, A., Nash, S. E., Olcina, M. M., Jiang, D., et al. (2017). Inhibition of the gas6/axl pathway augments the efficacy of chemotherapies. *The Journal of Clinical Investigation, 127*(1), 183–198. https://doi.org/10.1172/JCI85610.

Katz, M., Amit, I., Citri, A., Shay, T., Carvalho, S., Lavi, S., et al. (2007). A reciprocal tensin-3-cten switch mediates egf-driven mammary cell migration. *Nature Cell Biology, 9*(8), 961–969. https://doi.org/10.1038/ncb1622. ncb1622 [pii].

Kentsis, A., Reed, C., Rice, K. L., Sanda, T., Rodig, S. J., Tholouli, E., et al. (2012). Autocrine activation of the met receptor tyrosine kinase in acute myeloid leukemia. *Nature Medicine, 18*(7), 1118–1122. https://doi.org/10.1038/nm.2819.

Khan, K., Araki, K., Wang, D., Li, G., Li, X., Zhang, J., et al. (2010). Head and neck cancer radiosensitization by the novel poly(adp-ribose) polymerase inhibitor gpi-15427. *Head & Neck*, *32*(3), 381–391. https://doi.org/10.1002/hed.21195.

Khotskaya, Y. B., Holla, V. R., Farago, A. F., Mills Shaw, K. R., Meric-Bernstam, F., & Hong, D. S. (2017). Targeting trk family proteins in cancer. *Pharmacology & Therapeutics*, *173*, 58–66. https://doi.org/10.1016/j.pharmthera.2017.02.006.

Kiehna, E. N., Arnush, M. R., Tamrazi, B., Cotter, J. A., Hawes, D., Robison, N. J., et al. (2017). Novel gopc(fig)-ros1 fusion in a pediatric high-grade glioma survivor. *Journal of Neurosurgery. Pediatrics*, *20*(1), 51–55. https://doi.org/10.3171/2017.2.PEDS16679.

Kim, W., Lee, S., Seo, D., Kim, D., Kim, K., Kim, E., et al. (2019). Cellular stress responses in radiotherapy. *Cell*, *8*(9), 1105. https://doi.org/10.3390/cells8091105.

Kloos, R. T., Ringel, M. D., Knopp, M. V., Hall, N. C., King, M., Stevens, R., et al. (2009). Phase ii trial of sorafenib in metastatic thyroid cancer. *Journal of Clinical Oncology*, *27*(10), 1675–1684. https://doi.org/10.1200/JCO.2008.18.2717.

Konduri, K., Gallant, J. N., Chae, Y. K., Giles, F. J., Gitlitz, B. J., Gowen, K., et al. (2016). Egfr fusions as novel therapeutic targets in lung cancer. *Cancer Discovery*, *6*(6), 601–611. https://doi.org/10.1158/2159-8290.CD-16-0075.

Koury, J., Lucero, M., Cato, C., Chang, L., Geiger, J., Henry, D., et al. (2018). Immunotherapies: Exploiting the immune system for cancer treatment. *Journal of Immunology Research*, *2018*, 9585614. https://doi.org/10.1155/2018/9585614.

Kovalenko, M., Denner, K., Sandstrom, J., Persson, C., Gross, S., Jandt, E., et al. (2000). Site-selective dephosphorylation of the platelet-derived growth factor beta-receptor by the receptor-like protein-tyrosine phosphatase dep-1. *The Journal of Biological Chemistry*, *275*(21), 16219–16226.

Krystal, G. W., Hines, S. J., & Organ, C. P. (1996). Autocrine growth of small cell lung cancer mediated by coexpression of c-kit and stem cell factor. *Cancer Research*, *56*(2), 370–376.

Kwak, E. L., Bang, Y. J., Camidge, D. R., Shaw, A. T., Solomon, B., Maki, R. G., et al. (2010). Anaplastic lymphoma kinase inhibition in non-small-cell lung cancer. *The New England Journal of Medicine*, *363*(18), 1693–1703. https://doi.org/10.1056/NEJMoa1006448.

Lagas, J. S., Fan, L., Wagenaar, E., Vlaming, M. L., van Tellingen, O., Beijnen, J. H., et al. (2010). P-glycoprotein (p-gp/abcb1), abcc2, and abcc3 determine the pharmacokinetics of etoposide. *Clinical Cancer Research*, *16*(1), 130–140. https://doi.org/10.1158/1078-0432.Ccr-09-1321.

Lastwika, K. J., Wilson, W., 3rd, Li, Q. K., Norris, J., Xu, H., Ghazarian, S. R., et al. (2016). Control of pd-l1 expression by oncogenic activation of the akt-mtor pathway in non-small cell lung cancer. *Cancer Research*, *76*(2), 227–238. https://doi.org/10.1158/0008-5472.CAN-14-3362.

Latchman, Y., Wood, C. R., Chernova, T., Chaudhary, D., Borde, M., Chernova, I., et al. (2001). Pd-l2 is a second ligand for pd-1 and inhibits t cell activation. *Nature Immunology*, *2*(3), 261–268. https://doi.org/10.1038/85330.

Le Mercier, I., Lines, J. L., & Noelle, R. J. (2015). Beyond ctla-4 and pd-1, the generation z of negative checkpoint regulators. *Frontiers in Immunology*, *6*, 418. https://doi.org/10.3389/fimmu.2015.00418.

Lee, C., Whang, Y. M., Campbell, P., Mulcrone, P. L., Elefteriou, F., Cho, S. W., et al. (2018). Dual targeting c-met and vegfr2 in osteoblasts suppresses growth and osteolysis of prostate cancer bone metastasis. *Cancer Letters*, *414*, 205–213. https://doi.org/10.1016/j.canlet.2017.11.016.

Lemmon, M. A., & Schlessinger, J. (2010). Cell signaling by receptor tyrosine kinases. *Cell*, *141*(7), 1117–1134. https://doi.org/10.1016/j.cell.2010.06.011.

Li, X., Wilmanns, M., Thornton, J., & Kohn, M. (2013). Elucidating human phosphatase-substrate networks. *Science Signaling*, *6*(275), rs10. https://doi.org/10.1126/scisignal.2003203.

Liao, H. W., Hsu, J. M., Xia, W., Wang, H. L., Wang, Y. N., Chang, W. C., et al. (2015). Prmt1-mediated methylation of the egf receptor regulates signaling and cetuximab response. *The Journal of Clinical Investigation*, *125*(12), 4529–4543. https://doi.org/10.1172/JCI82826.

Libermann, T. A., Nusbaum, H. R., Razon, N., Kris, R., Lax, I., Soreq, H., et al. (1985). Amplification, enhanced expression and possible rearrangement of egf receptor gene in primary human brain tumours of glial origin. *Nature*, *313*(5998), 144–147.

Lievre, A., Bachet, J. B., Le Corre, D., Boige, V., Landi, B., Emile, J. F., et al. (2006). Kras mutation status is predictive of response to cetuximab therapy in colorectal cancer. *Cancer Research*, *66*(8), 3992–3995. https://doi.org/10.1158/0008-5472.CAN-06-0191. 66/8/3992 [pii].

Logothetis, C. J., & Lin, S. H. (2005). Osteoblasts in prostate cancer metastasis to bone. *Nature Reviews. Cancer*, *5*(1), 21–28. https://doi.org/10.1038/nrc1528.

Lopez-Gines, C., Gil-Benso, R., Ferrer-Luna, R., Benito, R., Serna, E., Gonzalez-Darder, J., et al. (2010). New pattern of egfr amplification in glioblastoma and the relationship of gene copy number with gene expression profile. *Modern Pathology*, *23*(6), 856–865. https://doi.org/10.1038/modpathol.2010.62.

Lu, Y. C., Yeh, W. C., & Ohashi, P. S. (2008). Lps/tlr4 signal transduction pathway. *Cytokine*, *42*(2), 145–151. https://doi.org/10.1016/j.cyto.2008.01.006.

Lu, Y., Zi, X., Zhao, Y., Mascarenhas, D., & Pollak, M. (2001). Insulin-like growth factor-i receptor signaling and resistance to trastuzumab (herceptin). *Journal of the National Cancer Institute*, *93*(24), 1852–1857.

Ludes-Meyers, J. H., Subler, M. A., Shivakumar, C. V., Munoz, R. M., Jiang, P., Bigger, J. E., et al. (1996). Transcriptional activation of the human epidermal growth factor receptor promoter by human p53. *Molecular and Cellular Biology*, *16*(11), 6009–6019.

Lynch, T. J., Bell, D. W., Sordella, R., Gurubhagavatula, S., Okimoto, R. A., Brannigan, B. W., et al. (2004). Activating mutations in the epidermal growth factor receptor underlying responsiveness of non-small-cell lung cancer to gefitinib. *The New England Journal of Medicine*, *350*(21), 2129–2139.

Maemondo, M., Inoue, A., Kobayashi, K., Sugawara, S., Oizumi, S., Isobe, H., et al. (2010). Gefitinib or chemotherapy for non-small-cell lung cancer with mutated egfr. *The New England Journal of Medicine*, *362*(25), 2380–2388. https://doi.org/10.1056/NEJMoa0909530.

Maiti, G. P., Mondal, P., Mukherjee, N., Ghosh, A., Ghosh, S., Dey, S., et al. (2013). Overexpression of egfr in head and neck squamous cell carcinoma is associated with inactivation of sh3gl2 and cdc25a genes. *PLoS One*, *8*(5). e63440https://doi.org/10.1371/journal.pone.0063440.

Mancini, M., Gaborit, N., Lindzen, M., Salame, T. M., Dall'Ora, M., Sevilla-Sharon, M., et al. (2015). Combining three antibodies nullifies feedback-mediated resistance to erlotinib in lung cancer. *Science Signaling*, *8*(379), ra53. https://doi.org/10.1126/scisignal.aaa0725.

Mancini, M., & Yarden, Y. (2016). Mutational and network level mechanisms underlying resistance to anti-cancer kinase inhibitors. *Seminars in Cell & Developmental Biology*, *50*, 164–176.

Mao, Q., & Unadkat, J. D. (2015). Role of the breast cancer resistance protein (bcrp/abcg2) in drug transport–an update. *The AAPS Journal*, *17*(1), 65–82. https://doi.org/10.1208/s12248-014-9668-6.

Marchetti, A., Martella, C., Felicioni, L., Barassi, F., Salvatore, S., Chella, A., et al. (2005). Egfr mutations in non-small-cell lung cancer: Analysis of a large series of cases and development of a rapid and sensitive method for diagnostic screening with potential implications on pharmacologic treatment. *Journal of Clinical Oncology*, *23*(4), 857–865. https://doi.org/10.1200/JCO.2005.08.043.

Marmor, M. D., & Yarden, Y. (2004). Role of protein ubiquitylation in regulating endocytosis of receptor tyrosine kinases. *Oncogene, 23*(11), 2057–2070. https://doi.org/10.1038/sj.onc.1207390. 1207390 [pii].

Marrocco, I., Romaniello, D., & Yarden, Y. (2019). Cancer immunotherapy: The dawn of antibody cocktails. *Methods in Molecular Biology, 1904*, 11–51. https://doi.org/10.1007/978-1-4939-8958-4_2.

Masood, A., Kancha, R. K., & Subramanian, J. (2019). Epidermal growth factor receptor (egfr) tyrosine kinase inhibitors in non-small cell lung cancer harboring uncommon egfr mutations: Focus on afatinib. *Seminars in Oncology, 46*(3), 271–283. https://doi.org/10.1053/j.seminoncol.2019.08.004.

Matos, I., Noguerido, A., Ros, J., Mulet, N., Argiles, G., Elez, E., et al. (2019). Triple-drug chemotherapy regimens in combination with an anti-egfr agent in metastatic colorectal cancer—Prospects from phase ii clinical trials. *Expert Opinion on Investigational Drugs, 28*(5), 463–471. https://doi.org/10.1080/13543784.2019.1599860.

Meijnen, P., Peterse, J. L., Antonini, N., Rutgers, E. J., & van de Vijver, M. J. (2008). Immunohistochemical categorisation of ductal carcinoma in situ of the breast. *British Journal of Cancer, 98*(1), 137–142. https://doi.org/10.1038/sj.bjc.6604112. 6604112 [pii].

Mellman, I., & Yarden, Y. (2013). Endocytosis and cancer. *Cold Spring Harbor Perspectives in Biology, 5*(12). https://doi.org/10.1101/cshperspect.a016949.

Messai, Y., Gad, S., Noman, M. Z., Le Teuff, G., Couve, S., Janji, B., et al. (2016). Renal cell carcinoma programmed death-ligand 1, a new direct target of hypoxia-inducible factor-2 alpha, is regulated by von hippel-Lindau gene mutation status. *European Urology, 70*(4), 623–632. https://doi.org/10.1016/j.eururo.2015.11.029.

Michael, J. M., Lavin, M. F., & Watters, D. J. (1997). Resistance to radiation-induced apoptosis in burkitt's lymphoma cells is associated with defective ceramide signaling. *Cancer Research, 57*(16), 3600–3605.

Mitsudomi, T., Morita, S., Yatabe, Y., Negoro, S., Okamoto, I., Tsurutani, J., et al. (2010). Gefitinib versus cisplatin plus docetaxel in patients with non-small-cell lung cancer harbouring mutations of the epidermal growth factor receptor (wjtog3405): An open label, randomised phase 3 trial. *The Lancet Oncology, 11*(2), 121–128. https://doi.org/10.1016/S1470-2045(09)70364-X.

Mo, H. N., & Liu, P. (2017). Targeting met in cancer therapy. *Chronic Diseases and Translational Medicine, 3*(3), 148–153. https://doi.org/10.1016/j.cdtm.2017.06.002.

Mok, T. S., Wu, Y. L., Ahn, M. J., Garassino, M. C., Kim, H. R., Ramalingam, S. S., et al. (2017). Osimertinib or platinum-pemetrexed in egfr t790m-positive lung cancer. *The New England Journal of Medicine, 376*(7), 629–640. https://doi.org/10.1056/NEJMoa1612674.

Mok, T. S., Wu, Y. L., Thongprasert, S., Yang, C. H., Chu, D. T., Saijo, N., et al. (2009) Gefitinib or carboplatin-paclitaxel in pulmonary adenocarcinoma. *The New England Journal of Medicine, 361*(10), 947–957.

Mosesson, Y., Shtiegman, K., Katz, M., Zwang, Y., Vereb, G., Szollosi, J., et al. (2003). Endocytosis of receptor tyrosine kinases is driven by monoubiquitylation, not polyubiquitylation. *The Journal of Biological Chemistry, 278*(24), 21323–21326.

Mudduluru, G., Ceppi, P., Kumarswamy, R., Scagliotti, G. V., Papotti, M., & Allgayer, H. (2011). Regulation of axl receptor tyrosine kinase expression by mir-34a and mir-199a/b in solid cancer. *Oncogene, 30*(25), 2888–2899. https://doi.org/10.1038/onc.2011.13.

Munoz, M., Henderson, M., Haber, M., & Norris, M. (2007). Role of the mrp1/abcc1 multidrug transporter protein in cancer. *IUBMB Life, 59*(12), 752–757. https://doi.org/10.1080/15216540701736285.

Nagy, P., Friedlander, E., Tanner, M., Kapanen, A. I., Carraway, K. L., Isola, J., et al. (2005). Decreased accessibility and lack of activation of erbb2 in jimt-1, a herceptin-resistant, muc4-expressing breast cancer cell line. *Cancer Research, 65*(2), 473–482. 65/2/473 pii.

Nahta, R., Hung, M. C., & Esteva, F. J. (2004). The her-2-targeting antibodies trastuzumab and pertuzumab synergistically inhibit the survival of breast cancer cells. *Cancer Research*, *64*(7), 2343–2346.

Neben, C. L., Lo, M., Jura, N., & Klein, O. D. (2019). Feedback regulation of rtk signaling in development. *Developmental Biology*, *447*(1), 71–89. https://doi.org/10.1016/j.ydbio.2017.10.017.

Neel, B. G., & Tonks, N. K. (1997). Protein tyrosine phosphatases in signal transduction. *Current Opinion in Cell Biology*, *9*(2), 193–204. https://doi.org/10.1016/s0955-0674(97)80063-4.

Nieman, M. T., Prudoff, R. S., Johnson, K. R., & Wheelock, M. J. (1999). N-cadherin promotes motility in human breast cancer cells regardless of their e-cadherin expression. *The Journal of Cell Biology*, *147*(3), 631–644.

NIH. (2018). Cancer statistics. Retrieved from https://www.cancer.gov/about-cancer/understanding/statistics

Nowicki, T. S., Hu-Lieskovan, S., & Ribas, A. (2018). Mechanisms of resistance to pd-1 and pd-l1 blockade. *Cancer Journal*, *24*(1), 47–53. https://doi.org/10.1097/PPO.0000000000000303.

Obenauf, A. C., & Massague, J. (2015). Surviving at a distance: Organ-specific metastasis. *Trends in Cancer*, *1*(1), 76–91. https://doi.org/10.1016/j.trecan.2015.07.009.

Obenauf, A. C., Zou, Y., Ji, A. L., Vanharanta, S., Shu, W., Shi, H., et al. (2015). Therapy-induced tumour secretomes promote resistance and tumour progression. *Nature*, *520*(7547), 368–372. https://doi.org/10.1038/nature14336.

O'Brien, S. G., Guilhot, F., Larson, R. A., Gathmann, I., Baccarani, M., Cervantes, F., et al. (2003). Imatinib compared with interferon and low-dose cytarabine for newly diagnosed chronic-phase chronic myeloid leukemia. *The New England Journal of Medicine*, *348*(11), 994–1004. https://doi.org/10.1056/NEJMoa022457.

O'Donnell, J. S., Long, G. V., Scolyer, R. A., Teng, M. W., & Smyth, M. J. (2017). Resistance to pd1/pdl1 checkpoint inhibition. *Cancer Treatment Reviews*, *52*, 71–81. https://doi.org/10.1016/j.ctrv.2016.11.007.

Ostman, A., Hellberg, C., & Bohmer, F. D. (2006). Protein-tyrosine phosphatases and cancer. *Nature Reviews. Cancer*, *6*(4), 307–320.

Oxnard, G. R., Arcila, M. E., Chmielecki, J., Ladanyi, M., Miller, V. A., & Pao, W. (2011). New strategies in overcoming acquired resistance to epidermal growth factor receptor tyrosine kinase inhibitors in lung cancer. *Clinical Cancer Research*, *17*(17), 5530–5537. https://doi.org/10.1158/1078-0432.CCR-10-2571.

Paez, J. G., Janne, P. A., Lee, J. C., Tracy, S., Greulich, H., Gabriel, S., et al. (2004). Egfr mutations in lung cancer: Correlation with clinical response to gefitinib therapy. *Science*, *304*(5676), 1497–1500.

Pao, W., Miller, V. A., Politi, K. A., Riely, G. J., Somwar, R., Zakowski, M. F., et al. (2005). Acquired resistance of lung adenocarcinomas to gefitinib or erlotinib is associated with a second mutation in the egfr kinase domain. *PLoS Medicine*, *2*(3)e73.

Pao, W., Miller, V., Zakowski, M., Doherty, J., Politi, K., Sarkaria, I., et al. (2004). Egf receptor gene mutations are common in lung cancers from "never smokers" and are associated with sensitivity of tumors to gefitinib and erlotinib. *Proceedings of the National Academy of Sciences of the United States of America*, *101*(36), 13306–13311.

Pardoll, D. M. (2012). The blockade of immune checkpoints in cancer immunotherapy. *Nature Reviews. Cancer*, *12*(4), 252–264. https://doi.org/10.1038/nrc3239.

Park, J., Wysocki, R. W., Amoozgar, Z., Maiorino, L., Fein, M. R., Jorns, J., et al. (2016). Cancer cells induce metastasis-supporting neutrophil extracellular DNA traps. *Science Translational Medicine*, *8*(361). https://doi.org/10.1126/scitranslmed.aag1711. 361ra138.

Parsa, A. T., Waldron, J. S., Panner, A., Crane, C. A., Parney, I. F., Barry, J. J., et al. (2007). Loss of tumor suppressor pten function increases b7-h1 expression and immunoresistance in glioma. *Nature Medicine*, *13*(1), 84–88.

Pavanello, F., Zucca, E., & Ghielmini, M. (2017). Rituximab: 13 open questions after 20years of clinical use. *Cancer Treatment Reviews*, *53*, 38–46. https://doi.org/10.1016/j.ctrv.2016.11.015.

Perkins, D., Wang, Z., Donovan, C., He, H., Mark, D., Guan, G., et al. (1996). Regulation of ctla-4 expression during t cell activation. *Journal of Immunology*, *156*(11), 4154–4159.

Peschard, P., Fournier, T. M., Lamorte, L., Naujokas, M. A., Band, H., Langdon, W. Y., et al. (2001). Mutation of the c-cbl tkb domain binding site on the met receptor tyrosine kinase converts it into a transforming protein. *Molecular Cell*, *8*(5), 995–1004. https://doi.org/10.1016/S1097-2765(01)00378-1. S1097-2765(01)00378-1 [pii].

Pires-daSilva, A., & Sommer, R. J. (2003). The evolution of signalling pathways in animal development. *Nature Reviews. Genetics*, *4*(1), 39–49. https://doi.org/10.1038/nrg977.

Planchard, D., Loriot, Y., Andre, F., Gobert, A., Auger, N., Lacroix, L., et al. (2015). Egfr-independent mechanisms of acquired resistance to azd9291 in egfr t790m-positive nsclc patients. *Annals of Oncology*, *26*(10), 2073–2078. https://doi.org/10.1093/annonc/mdv319.

Polo, S., & Di Fiore, P. P. (2006). Endocytosis conducts the cell signaling orchestra. *Cell*, *124*(5), 897–900. https://doi.org/10.1016/j.cell.2006.02.025. S0092-8674(06)00242-X [pii].

Prieto, P. A., Yang, J. C., Sherry, R. M., Hughes, M. S., Kammula, U. S., White, D. E., et al. (2012). Ctla-4 blockade with ipilimumab: Long-term follow-up of 177 patients with metastatic melanoma. *Clinical Cancer Research*, *18*(7), 2039–2047. https://doi.org/10.1158/1078-0432.CCR-11-1823.

Rankin, E. B., & Giaccia, A. J. (2016). The receptor tyrosine kinase axl in cancer progression. *Cancers (Basel)*, *8*(11), 103. https://doi.org/10.3390/cancers8110103.

Red Brewer, M., Yun, C. H., Lai, D., Lemmon, M. A., Eck, M. J., & Pao, W. (2013). Mechanism for activation of mutated epidermal growth factor receptors in lung cancer. *Proceedings of the National Academy of Sciences of the United States of America*, *110*(38), E3595–E3604. https://doi.org/10.1073/pnas.1220050110.

Rittmeyer, A., Barlesi, F., Waterkamp, D., Park, K., Ciardiello, F., von Pawel, J., et al. (2017). Atezolizumab versus docetaxel in patients with previously treated non-small-cell lung cancer (oak): A phase 3, open-label, multicentre randomised controlled trial. *Lancet*, *389*(10066), 255–265. https://doi.org/10.1016/S0140-6736(16)32517-X.

Robey, R. W., Pluchino, K. M., Hall, M. D., Fojo, A. T., Bates, S. E., & Gottesman, M. M. (2018). Revisiting the role of abc transporters in multidrug-resistant cancer. *Nature Reviews. Cancer*, *18*(7), 452–464. https://doi.org/10.1038/s41568-018-0005-8.

Rodriguez-Antona, C., Pallares, J., Montero-Conde, C., Inglada-Perez, L., Castelblanco, E., Landa, I., et al. (2010). Overexpression and activation of egfr and vegfr2 in medullary thyroid carcinomas is related to metastasis. *Endocrine-Related Cancer*, *17*(1), 7 16. https://doi.org/10.1677/ERC-08-0304.

Rosell, R., Dafni, U., Felip, E., Curioni-Fontecedro, A., Gautschi, O., Peters, S., et al. (2017). Erlotinib and bevacizumab in patients with advanced non-small-cell lung cancer and activating egfr mutations (belief): An international, multicentre, single-arm, phase 2 trial. *The Lancet Respiratory Medicine*, *5*(5), 435–444. https://doi.org/10.1016/s2213-2600(17)30129-7.

Rossi, M. R., Ionov, Y., Bakin, A. V., & Cowell, J. K. (2005). Truncating mutations in the acvr2 gene attenuates activin signaling in prostate cancer cells. *Cancer Genetics and Cytogenetics*, *163*(2), 123–129. https://doi.org/10.1016/j.cancergencyto.2005.05.007.

Rusnak, D. W., Affleck, K., Cockerill, S. G., Stubberfield, C., Harris, R., Page, M., et al. (2001). The characterization of novel, dual erbb-2/egfr, tyrosine kinase inhibitors: Potential therapy for cancer. *Cancer Research*, *61*(19), 7196–7203.

Saharinen, P., Eklund, L., & Alitalo, K. (2017). Therapeutic targeting of the angiopoietin-tie pathway. *Nature Reviews. Drug Discovery*, *16*(9), 635–661. https://doi.org/10.1038/nrd.2016.278.

Salomon, D. S., Brandt, R., Ciardiello, F., & Normanno, N. (1995). Epidermal growth factor-related peptides and their receptors in human malignancies. *Critical Reviews in Oncology/Hematology*, *19*(3), 183–232.

Samuels, Y., & Waldman, T. (2010). Oncogenic mutations of pik3ca in human cancers. *Current Topics in Microbiology and Immunology*, *347*, 21–41. https://doi.org/10.1007/82_2010_68.

Samuels, Y., Wang, Z., Bardelli, A., Silliman, N., Ptak, J., Szabo, S., et al. (2004). High frequency of mutations of the pik3ca gene in human cancers. *Science*, *304*(5670), 554. https://doi.org/10.1126/science.1096502. 1096502 [pii].

Sauna, Z. E., Kim, I. W., & Ambudkar, S. V. (2007). Genomics and the mechanism of p-glycoprotein (abcb1). *Journal of Bioenergetics and Biomembranes*, *39*(5–6), 481–487. https://doi.org/10.1007/s10863-007-9115-9.

Scaltriti, M., Rojo, F., Ocana, A., Anido, J., Guzman, M., Cortes, J., et al. (2007). Expression of p95her2, a truncated form of the her2 receptor, and response to anti-her2 therapies in breast cancer. *Journal of the National Cancer Institute*, *99*(8), 628–638.

Schatton, T., Scolyer, R. A., Thompson, J. F., & Mihm, M. C., Jr. (2014). Tumor-infiltrating lymphocytes and their significance in melanoma prognosis. *Methods in Molecular Biology*, *1102*, 287–324. https://doi.org/10.1007/978-1-62703-727-3_16.

Scheuer, W., Friess, T., Burtscher, H., Bossenmaier, B., Endl, J., & Hasmann, M. (2009). Strongly enhanced antitumor activity of trastuzumab and pertuzumab combination treatment on her2-positive human xenograft tumor models. *Cancer Research*, 69(24), 9330–9336. https://doi.org/10.1158/0008-5472.CAN-08-4597. 0008-5472.CAN-08-4597 [pii].

Schlessinger, J., Plotnikov, A. N., Ibrahimi, O. A., Eliseenkova, A. V., Yeh, B. K., Yayon, A., et al. (2000). Crystal structure of a ternary fgf-fgfr-heparin complex reveals a dual role for heparin in fgfr binding and dimerization. *Molecular Cell*, *6*(3), 743–750. https://doi.org/10.1016/s1097-2765(00)00073-3.

Schneider, M. R., & Yarden, Y. (2016). The egfr-her2 module: A stem cell approach to understanding a prime target and driver of solid tumors. *Oncogene*, *35*(23), 2949–2960.

Scott, R. P., Eketjall, S., Aineskog, H., & Ibanez, C. F. (2005). Distinct turnover of alternatively spliced isoforms of the ret kinase receptor mediated by differential recruitment of the cbl ubiquitin ligase. *The Journal of Biological Chemistry*, *280*(14), 13442–13449. https://doi.org/10.1074/jbc.M500507200.

Segaliny, A. I., Tellez-Gabriel, M., Heymann, M. F., & Heymann, D. (2015). Receptor tyrosine kinases: Characterisation, mechanism of action and therapeutic interests for bone cancers. *Journal of Bone Oncology*, *4*(1), 1–12. https://doi.org/10.1016/j.jbo.2015.01.001.

Selvaggi, G., Novello, S., Torri, V., Leonardo, E., De Giuli, P., Borasio, P., et al. (2004). Epidermal growth factor receptor overexpression correlates with a poor prognosis in completely resected non-small-cell lung cancer. *Annals of Oncology*, *15*(1), 28–32. https://doi.org/10.1093/annonc/mdh011.

Selvakumaran, M., Pisarcik, D. A., Bao, R., Yeung, A. T., & Hamilton, T. C. (2003). Enhanced cisplatin cytotoxicity by disturbing the nucleotide excision repair pathway in ovarian cancer cell lines. *Cancer Research*, *63*(6), 1311–1316.

Senra, J. M., Telfer, B. A., Cherry, K. E., McCrudden, C. M., Hirst, D. G., O'Connor, M. J., et al. (2011). Inhibition of parp-1 by olaparib (azd2281) increases the radiosensitivity of a lung tumor xenograft. *Molecular Cancer Therapeutics*, *10*(10), 1949–1958. https://doi.org/10.1158/1535-7163.Mct-11-0278.

Serra, V., Vivancos, A., Puente, X. S., Felip, E., Silberschmidt, D., Caratu, G., et al. (2013). Clinical response to a lapatinib-based therapy for a li-fraumeni syndrome patient with a novel her2v659e mutation. *Cancer Discovery*, *3*(11), 1238–1244. https://doi.org/10.1158/2159-8290.CD-13-0132.

Shah, N. P., Tran, C., Lee, F. Y., Chen, P., Norris, D., & Sawyers, C. L. (2004). Overriding imatinib resistance with a novel abl kinase inhibitor. *Science*, *305*(5682), 399–401. https://doi.org/10.1126/science.1099480.

Shaked, Y., Ciarrocchi, A., Franco, M., Lee, C. R., Man, S., Cheung, A. M., et al. (2006). Therapy-induced acute recruitment of circulating endothelial progenitor cells to tumors. *Science*, *313*(5794), 1785–1787. https://doi.org/10.1126/science.1127592. 313/5794/1785 [pii].

Sharma, S. V., Bell, D. W., Settleman, J., & Haber, D. A. (2007). Epidermal growth factor receptor mutations in lung cancer. *Nature Reviews. Cancer*, 7(3), 169–181.

Shattuck, D. L., Miller, J. K., Carraway, K. L., 3rd, & Sweeney, C. (2008). Met receptor contributes to trastuzumab resistance of her2-overexpressing breast cancer cells. *Cancer Research*, *68*(5), 1471–1477. https://doi.org/10.1158/0008-5472.CAN-07-5962. 68/5/1471 [pii].

Shaw, A. T., Hsu, P. P., Awad, M. M., & Engelman, J. A. (2013). Tyrosine kinase gene rearrangements in epithelial malignancies. *Nature Reviews. Cancer*, *13*(11), 772–787. https://doi.org/10.1038/nrc3612.

Shigematsu, H., Lin, L., Takahashi, T., Nomura, M., Suzuki, M., Wistuba, I. I., et al. (2005). Clinical and biological features associated with epidermal growth factor receptor gene mutations in lung cancers. *Journal of the National Cancer Institute*, *97*(5), 339–346.

Shohat, J., Modan, M., & Rosenfeld, J. B. (1989). Bopindolol: Long-term effects on blood pressure, renal function, plasma renin activity, and plasma aldosterone in mild-to-moderate essential hypertension. *Fundamental & Clinical Pharmacology*, *3*(1), 53–58. https://doi.org/10.1111/j.1472-8206.1989.tb00030.x.

Slamon, D. J., Clark, G. M., Wong, S. G., Levin, W. J., Ullrich, A., & McGuire, W. L. (1987). Human breast cancer: Correlation of relapse and survival with amplification of the her-2/neu oncogene. *Science*, *235*(4785), 177–182.

Slamon, D. J., Godolphin, W., Jones, L. A., Holt, J. A., Wong, S. G., Keith, D. E., et al. (1989). Studies of the her-2/neu proto-oncogene in human breast and ovarian cancer. *Science*, *244*(4905), 707–712.

Slamon, D. J., Leyland-Jones, B., Shak, S., Fuchs, H., Paton, V., Bajamonde, A., et al. (2001). Use of chemotherapy plus a monoclonal antibody against her2 for metastatic breast cancer that overexpresses her2. *The New England Journal of Medicine*, *344*(11), 783–792. https://doi.org/10.1056/NEJM200103153441101.

Smyth, M. J., Ngiow, S. F., Ribas, A., & Teng, M. W. (2016). Combination cancer immunotherapies tailored to the tumour microenvironment. *Nature Reviews. Clinical Oncology*, *13*(3), 143–158. https://doi.org/10.1038/nrclinonc.2015.209.

Soria, J. C., Ohe, Y., Vansteenkiste, J., Reungwetwattana, T., Chewaskulyong, B., Lee, K. H., et al. (2018). Osimertinib in untreated egfr-mutated advanced non-small-cell lung cancer. *The New England Journal of Medicine*, *378*(2), 113–125. https://doi.org/10.1056/NEJMoa1713137.

Sporn, M. B., & Todaro, G. J. (1980). Autocrine secretion and malignant transformation of cells. *The New England Journal of Medicine*, *303*(15), 878–880.

Stauber, D. J., DiGabriele, A. D., & Hendrickson, W. A. (2000). Structural interactions of fibroblast growth factor receptor with its ligands. *Proceedings of the National Academy of Sciences of the United States of America*, *97*(1), 49–54. https://doi.org/10.1073/pnas.97.1.49.

Stephens, P. J., Greenman, C. D., Fu, B., Yang, F., Bignell, G. R., Mudie, L. J., et al. (2011). Massive genomic rearrangement acquired in a single catastrophic event during cancer development. *Cell*, *144*(1), 27–40. https://doi.org/10.1016/j.cell.2010.11.055.

Stetler-Stevenson, W. G. (1999). Matrix metalloproteinases in angiogenesis: A moving target for therapeutic intervention. *The Journal of Clinical Investigation*, *103*(9), 1237–1241.

Stinchcombe, T. E. (2016). Azd9291 in epidermal growth factor receptor inhibitor-resistant non-small-cell lung cancer. *Translational Lung Cancer Research*, *5*(1), 92–94. https://doi.org/10.3978/j.issn.2218-6751.2015.07.19.

Stuible, M., & Tremblay, M. L. (2010). In control at the er: Ptp1b and the down-regulation of rtks by dephosphorylation and endocytosis. *Trends in Cell Biology*, *20*(11), 672–679. https://doi.org/10.1016/j.tcb.2010.08.013.

Sun, T., Aceto, N., Meerbrey, K. L., Kessler, J. D., Zhou, C., Migliaccio, I., et al. (2011). Activation of multiple proto-oncogenic tyrosine kinases in breast cancer via loss of the ptpn12 phosphatase. *Cell*, *144*(5), 703–718. https://doi.org/10.1016/j.cell.2011.02.003.

Sun, C., & Bernards, R. (2014). Feedback and redundancy in receptor tyrosine kinase signaling: Relevance to cancer therapies. *Trends in Biochemical Sciences*, *39*(10), 465–474. https://doi.org/10.1016/j.tibs.2014.08.010.

Sun, C., Mezzadra, R., & Schumacher, T. N. (2018). Regulation and function of the pd-l1 checkpoint. *Immunity*, *48*(3), 434–452. https://doi.org/10.1016/j.immuni.2018.03.014.

Takezawa, K., Pirazzoli, V., Arcila, M. E., Nebhan, C. A., Song, X., de Stanchina, E., et al. (2012). Her2 amplification: A potential mechanism of acquired resistance to egfr inhibition in egfr-mutant lung cancers that lack the second-site egfrt790m mutation. *Cancer Discovery*, *2*(10), 922–933. https://doi.org/10.1158/2159-8290.Cd-12-0108.

Talasila, K. M., Soentgerath, A., Euskirchen, P., Rosland, G. V., Wang, J., Huszthy, P. C., et al. (2013). Egfr wild-type amplification and activation promote invasion and development of glioblastoma independent of angiogenesis. *Acta Neuropathologica*, *125*(5), 683–698. https://doi.org/10.1007/s00401-013-1101-1.

Tallini, G., & Asa, S. L. (2001). Ret oncogene activation in papillary thyroid carcinoma. *Advances in Anatomic Pathology*, *8*(6), 345–354.

Tan, C. S., Gilligan, D., & Pacey, S. (2015). Treatment approaches for egfr-inhibitor-resistant patients with non-small-cell lung cancer. *The Lancet Oncology*, *16*(9), e447–e459. https://doi.org/10.1016/S1470-2045(15)00246-6.

Tarcic, G., Boguslavsky, S. K., Wakim, J., Kiuchi, T., Liu, A., Reinitz, F., et al. (2009). An unbiased screen identifies dep-1 tumor suppressor as a phosphatase controlling egfr endocytosis. *Current Biology*, *19*(21), 1788–1798. https://doi.org/10.1016/j.cub.2009.09.048. S0960-9822(09)01761-8 [pii].

Taylor, E. M., & Lindsay, H. D. (2016). DNA replication stress and cancer: Cause or cure? *Future Oncology*, *12*(2), 221–237. https://doi.org/10.2217/fon.15.292.

Teft, W. A., Kirchhof, M. G., & Madrenas, J. (2006). A molecular perspective of ctla-4 function. *Annual Review of Immunology*, *24*, 65–97. https://doi.org/10.1146/annurev.immunol.24.021605.090535.

Teng, Y. H., Tan, W. J., Thike, A. A., Cheok, P. Y., Tse, G. M., Wong, N. S., et al. (2011). Mutations in the epidermal growth factor receptor (egfr) gene in triple negative breast cancer: Possible implications for targeted therapy. *Breast Cancer Research*, *13*(2), R35.

Thress, K. S., Paweletz, C. P., Felip, E., Cho, B. C., Stetson, D., Dougherty, B., et al. (2015). Acquired egfr c797s mutation mediates resistance to azd9291 in non-small cell lung cancer harboring egfr t790m. *Nature Medicine*, *21*(6), 560–562. https://doi.org/10.1038/nm.3854.

Tonks, N. K. (2006). Protein tyrosine phosphatases: From genes, to function, to disease. *Nature Reviews. Molecular Cell Biology*, 7(11), 833–846.

Topalian, S. L., Drake, C. G., & Pardoll, D. M. (2015). Immune checkpoint blockade: A common denominator approach to cancer therapy. *Cancer Cell*, *27*(4), 450–461.

Trivier, E., & Ganesan, T. S. (2002). Ryk, a catalytically inactive receptor tyrosine kinase, associates with ephb2 and ephb3 but does not interact with af-6. *The Journal of Biological Chemistry*, *277*(25), 23037–23043. https://doi.org/10.1074/jbc.M202486200.

Ullrich, A., Coussens, L., Hayflick, J. S., Dull, T. J., Gray, A., Tam, A. W., et al. (1984). Human epidermal growth factor receptor cdna sequence and aberrant expression of the amplified gene in a431 epidermoid carcinoma cells. *Nature*, *309*(5967), 418–425.

Vaidyanathan, A., Sawers, L., Gannon, A. L., Chakravarty, P., Scott, A. L., Bray, S. E., et al. (2016). Abcb1 (mdr1) induction defines a common resistance mechanism in paclitaxel- and olaparib-resistant ovarian cancer cells. *British Journal of Cancer, 115*(4), 431–441. https://doi.org/10.1038/bjc.2016.203.

Verma, S., Miles, D., Gianni, L., Krop, I. E., Welslau, M., Baselga, J., et al. (2012). Trastuzumab emtansine for her2-positive advanced breast cancer. *The New England Journal of Medicine, 367*(19), 1783–1791. https://doi.org/10.1056/NEJMoa1209124.

Viale, A., De Franco, F., Orleth, A., Cambiaghi, V., Giuliani, V., Bossi, D., et al. (2009). Cell-cycle restriction limits DNA damage and maintains self-renewal of leukaemia stem cells. *Nature, 457*(7225), 51–56. https://doi.org/10.1038/nature07618.

von Minckwitz, G., Procter, M., de Azambuja, E., Zardavas, D., Benyunes, M., Viale, G., et al. (2017). Adjuvant pertuzumab and trastuzumab in early her2-positive breast cancer. *The New England Journal of Medicine, 377*(2), 122–131. https://doi.org/10.1056/NEJMoa1703643.

Wang, T. L., Diaz, L. A., Jr., Romans, K., Bardelli, A., Saha, S., Galizia, G., et al. (2004). Digital karyotyping identifies thymidylate synthase amplification as a mechanism of resistance to 5-fluorouracil in metastatic colorectal cancer patients. *Proceedings of the National Academy of Sciences of the United States of America, 101*(9), 3089–3094. https://doi.org/10.1073/pnas.0308716101.

Wang, Z., Longo, P. A., Tarrant, M. K., Kim, K., Head, S., Leahy, D. J., et al. (2011). Mechanistic insights into the activation of oncogenic forms of egf receptor. *Nature Structural & Molecular Biology, 18*(12), 1388–1393. https://doi.org/10.1038/nsmb.2168.

Ware, K. E., Hinz, T. K., Kleczko, E., Singleton, K. R., Marek, L. A., Helfrich, B. A., et al. (2013). A mechanism of resistance to gefitinib mediated by cellular reprogramming and the acquisition of an fgf2-fgfr1 autocrine growth loop. *Oncogene, 2*, e39. https://doi.org/10.1038/oncsis.2013.4.

Waterman, H., Alroy, I., Strano, S., Seger, R., & Yarden, Y. (1999). The c-terminus of the kinase-defective neuregulin receptor erbb-3 confers mitogenic superiority and dictates endocytic routing. *The EMBO Journal, 18*(12), 3348–3358.

Waterman, H., & Yarden, Y. (2001). Molecular mechanisms underlying endocytosis and sorting of erbb receptor tyrosine kinases. *FEBS Letters, 490*(3), 142–152. https://doi.org/10.1016/s0014-5793(01)02117-2. S0014-5793(01)02117-2 [pii].

Watson, R. G., Muhale, F., Thorne, L. B., Yu, J., O'Neil, B. H., Hoskins, J. M., et al. (2010). Amplification of thymidylate synthetase in metastatic colorectal cancer patients pretreated with 5-fluorouracil-based chemotherapy. *European Journal of Cancer, 46*(18), 3358–3364. https://doi.org/10.1016/j.ejca.2010.07.011.

Weber, J., Mandala, M., Del Vecchio, M., Gogas, H. J., Arance, A. M., Cowey, C. L., et al. (2017). Adjuvant nivolumab versus ipilimumab in resected stage iii or iv melanoma. *The New England Journal of Medicine, 377*(19), 1824–1835. https://doi.org/10.1056/NEJMoa1709030.

Wehrman, T., He, X., Raab, B., Dukipatti, A., Blau, H., & Garcia, K. C. (2007). Structural and mechanistic insights into nerve growth factor interactions with the trka and p75 receptors. *Neuron, 53*(1), 25–38. https://doi.org/10.1016/j.neuron.2006.09.034.

Weilbaecher, K. N., Guise, T. A., & McCauley, L. K. (2011). Cancer to bone: A fatal attraction. *Nature Reviews. Cancer, 11*(6), 411–425. https://doi.org/10.1038/nrc3055.

Weinstein, I. B., & Joe, A. K. (2006). Mechanisms of disease: Oncogene addiction–a rationale for molecular targeting in cancer therapy. *Nature Clinical Practice. Oncology, 3*(8), 448–457.

Werb, Z. (1997). Ecm and cell surface proteolysis: Regulating cellular ecology. *Cell, 91*(4), 439–442.

Wiesmann, C., Ultsch, M. H., Bass, S. H., & de Vos, A. M. (1999). Crystal structure of nerve growth factor in complex with the ligand-binding domain of the trka receptor. *Nature, 401*(6749), 184–188. https://doi.org/10.1038/43705.

Wilson, T. R., Fridlyand, J., Yan, Y., Penuel, E., Burton, L., Chan, E., et al. (2012). Widespread potential for growth-factor-driven resistance to anticancer kinase inhibitors. *Nature*, *487*(7408), 505–509. https://doi.org/10.1038/nature11249.

Witsch, E., Sela, M., & Yarden, Y. (2010). Roles for growth factors in cancer progression. *Physiology (Bethesda)*, *25*(2), 85–101. https://doi.org/10.1152/physiol.00045.2009.

Wu, R., Connolly, D., Ngelangel, C., Bosch, F. X., Munoz, N., & Cho, K. R. (2000). Somatic mutations of fibroblast growth factor receptor 3 (fgfr3) are uncommon in carcinomas of the uterine cervix. *Oncogene*, *19*(48), 5543–5546. https://doi.org/10.1038/sj.onc.1203934.

Xia, W., Mullin, R. J., Keith, B. R., Liu, L. H., Ma, H., Rusnak, D. W., et al. (2002). Antitumor activity of gw572016: A dual tyrosine kinase inhibitor blocks egf activation of egfr/erbb2 and downstream erk1/2 and akt pathways. *Oncogene*, *21*(41), 6255–6263. https://doi.org/10.1038/sj.onc.1205794.

Xia, W., Petricoin, E. F., 3rd, Zhao, S., Liu, L., Osada, T., Cheng, Q., et al. (2013). An heregulin-egfr-her3 autocrine signaling axis can mediate acquired lapatinib resistance in her2+ breast cancer models. *Breast Cancer Research*, *15*(5), R85. https://doi.org/10.1186/bcr3480.

Xu, L., Nilsson, M. B., Saintigny, P., Cascone, T., Herynk, M. H., Du, Z., et al. (2010). Epidermal growth factor receptor regulates met levels and invasiveness through hypoxia-inducible factor-1alpha in non-small cell lung cancer cells. *Oncogene*, *29*(18), 2616–2627. https://doi.org/10.1038/onc.2010.16.

Xu, Y., Tan, L. J., Grachtchouk, V., Voorhees, J. J., & Fisher, G. J. (2005). Receptor-type protein-tyrosine phosphatase-kappa regulates epidermal growth factor receptor function. *The Journal of Biological Chemistry*, *280*(52), 42694–42700.

Yamamoto, H., Toyooka, S., Ninomiya, T., Matsumoto, S., Kanai, M., Tomida, S., et al. (2018). Therapeutic potential of afatinib for cancers with erbb2 (her2) transmembrane domain mutations g660d and v659e. *The Oncologist*, *23*(2), 150–154. https://doi.org/10.1634/theoncologist.2017-0345.

Yang, J., Yan, J., & Liu, B. (2018). Targeting vegf/vegfr to modulate antitumor immunity. *Frontiers in Immunology*, *9*, 978. https://doi.org/10.3389/fimmu.2018.00978.

Yap, T. A., Gerlinger, M., Futreal, P. A., Pusztai, L., & Swanton, C. (2012). Intratumor heterogeneity: Seeing the wood for the trees. *Science Translational Medicine*, *4*(127). https://doi.org/10.1126/scitranslmed.3003854. 127ps110.

Yarden, Y., & Schlessinger, J. (1987a). Epidermal growth factor induces rapid, reversible aggregation of the purified epidermal growth factor receptor. *Biochemistry*, *26*(5), 1443–1451.

Yarden, Y., & Schlessinger, J. (1987b). Self-phosphorylation of epidermal growth factor receptor: Evidence for a model of intermolecular allosteric activation. *Biochemistry*, *26*(5), 1434–1442.

Yarden, Y., & Ullrich, A. (1988). Growth factor receptor tyrosine kinases. *Annual Review of Biochemistry*, *57*, 443–478. https://doi.org/10.1146/annurev.bi.57.070188.002303.

Yilmaz, M., & Christofori, G. (2009). Emt, the cytoskeleton, and cancer cell invasion. *Cancer Metastasis Reviews*, *28*(1–2), 15–33. https://doi.org/10.1007/s10555-008-9169-0.

Yin, J., & Zhang, J. (2011). Multidrug resistance-associated protein 1 (mrp1/abcc1) polymorphism: From discovery to clinical application. *Zhong Nan Da Xue Xue Bao. Yi Xue Ban*, *36*(10), 927–938. https://doi.org/10.3969/j.issn.1672-7347.2011.10.002.

Yu, H. A., Arcila, M. E., Rekhtman, N., Sima, C. S., Zakowski, M. F., Pao, W., et al. (2013). Analysis of tumor specimens at the time of acquired resistance to egfr-tki therapy in 155 patients with egfr-mutant lung cancers. *Clinical Cancer Research*, *19*(8), 2240–2247. https://doi.org/10.1158/1078-0432.CCR-12-2246.

Yuan, J., Yin, Z., Tao, K., Wang, G., & Gao, J. (2018). Function of insulin-like growth factor 1 receptor in cancer resistance to chemotherapy. *Oncology Letters*, *15*(1), 41–47. https://doi.org/10.3892/ol.2017.7276.

Yuzawa, S., Opatowsky, Y., Zhang, Z., Mandiyan, V., Lax, I., & Schlessinger, J. (2007). Structural basis for activation of the receptor tyrosine kinase kit by stem cell factor. *Cell*, 130(2), 323–334. https://doi.org/10.1016/j.cell.2007.05.055S0092-8674(07) 00759-3 [pii].

Zhang, X., Gureasko, J., Shen, K., Cole, P. A., & Kuriyan, J. (2006). An allosteric mechanism for activation of the kinase domain of epidermal growth factor receptor. *Cell, 125*(6), 1137–1149.

Zhang, Z., Lee, J. C., Lin, L., Olivas, V., Au, V., LaFramboise, T., et al. (2012). Activation of the axl kinase causes resistance to egfr-targeted therapy in lung cancer. *Nature Genetics, 44*(8), 852–860. https://doi.org/10.1038/ng.2330.

Zhang, J., Yang, P. L., & Gray, N. S. (2009). Targeting cancer with small molecule kinase inhibitors. *Nature Reviews. Cancer, 9*(1), 28–39. https://doi.org/10.1038/nrc2559.

CHAPTER TWO

Nuclear receptor tyrosine kinase transport and functions in cancer

Mei-Kuang Chen[a,b], Jennifer L. Hsu[a], Mien-Chie Hung[a,c,*]
[a]Department of Molecular and Cellular Oncology, The University of Texas MD Anderson Cancer Center, Houston, TX, United States
[b]The University of Texas MD Anderson Cancer Center UTHealth Graduate School of Biomedical Sciences, Houston, TX, United States
[c]Graduate Institute of Biomedical Sciences, Research Center for Cancer Biology, and Center for Molecular Medicine, China Medical University, Taichung, Taiwan
*Corresponding author: e-mail addresses: mhung@cmu.edu.tw; mhung77030@gmail.com

Contents

Advances in Cancer Research, Volume 147
ISSN 0065-230X
https://doi.org/10.1016/bs.acr.2020.04.010
© 2020 Elsevier Inc.
All rights reserved.

Abstract

Signaling functions of plasma membrane-localized receptor tyrosine kinases (RTKs) have been extensively studied after they were first described in the mid-1980s. Plasma membrane RTKs are activated by extracellular ligands and cellular stress stimuli, and regulate cellular responses by activating the downstream effector proteins to initiate a wide range of signaling cascades in the cells. However, increasing evidence indicates that RTKs can also be transported into the intracellular compartments where they phosphorylate traditional effector proteins and non-canonical substrate proteins. In general, internalization that retains the RTK's transmembrane domain begins with endocytosis, and endosomal RTK remains active before being recycled or degraded. Further RTK retrograde transport from endosome-Golgi-ER to the nucleus is primarily dependent on membranes vesicles and relies on the interaction with the COP-I vesicle complex, Sec61 translocon complex, and importin. Internalized RTKs have non-canonical substrates that include transcriptional co-factors and DNA damage response proteins, and many nuclear RTKs harbor oncogenic properties and can enhance cancer progression. Indeed, nuclear-localized RTKs have been shown to positively correlate with cancer recurrence, therapeutic resistance, and poor prognosis of cancer patients. Therefore, understanding the functions of nuclear RTKs and the mechanisms of nuclear RTK transport will further improve our knowledge to evaluate the potential of targeting nuclear RTKs or the proteins involved in their transport as new cancer therapeutic strategies.

1. Introduction

Cell surface receptor tyrosine kinases (RTKs) initiate signaling cascades when activated by their extracellular ligands and have been extensively studied since the 1960s (Schlessinger, 2014). Upon ligand binding, RTKs undergo conformational changes that lead to receptor dimerization and *trans*-auto-phosphorylation on their cytoplasmic tyrosine kinase domains. This auto-phosphorylation greatly elevates the kinase activities of the RTKs, and the phosphorylated tyrosine residues further serve as docking sites for effector proteins (Schlessinger, 2014). The effector proteins then activate a series of canonical RTK signaling pathways that lead to cellular responses, including cell proliferation and metastasis (Brand, Iida, Li, & Wheeler, 2011). An intricate balance of the physiological RTK activities is crucial for maintaining normal homeostasis whereas aberrant RTK activation is known to contribute to cancer development and progression (Du & Lovly, 2018), and promote resistance to chemo- and radio-therapy (Bhattacharya, Shetake, Pandey, & Kumar, 2018; Chen & Hung, 2016).

Although canonical RTK downstream signaling molecules, such as AKT, PI3K, mTOR, RAF, and MEK, are promising therapeutic targets for cancer treatment [Reviewed in (Akinleye, Furqan, Mukhi, Ravella, & Liu, 2013;

Karoulia, Gavathiotis, & Poulikakos, 2017; Nitulescu et al., 2016; Sanchez, Wang, & Cohen, 2018; Tian, Li, & Zhang, 2019)], recent studies indicated that RTKs have additional signaling functions through the recruitment of non-canonical effector proteins in the intracellular compartments, including endosomes, endoplasmic reticulum (ER) and nucleus [Reviewed in (Casaletto & McClatchey, 2012; Chen & Hung, 2015; Miaczynska, 2013; Platta & Stenmark, 2011)]. Moreover, the interactions between internalized RTKs, especially nuclear localized RTKs, and their non-canonical protein substrates have been shown to promote cancer development and resistance to treatment (Chen & Hung, 2016; Shi et al., 2018; Xie et al., 2014; Xie et al., 2019). Thus, furthering our understanding of the non-canonical effector protein substrates of internalized RTKs and how the RTKs are routed from the plasma membrane to the intracellular compartments may uncover new therapeutic targets for cancer treatment. In this chapter, we summarize the routes for intracellular RTK transport, specifically nuclear RTKs and their role in cancer.

2. Nuclear trafficking of RTKs

Nuclear localized RTKs, also known as membrane receptor in the nucleus (MRINs) (Wang & Hung, 2009), were first detected by Kamio et al. in human adrenocortical carcinoma by immunohistochemical (IHC) staining (Kamio, Shigematsu, Sou, Kawai, & Tsuchiyama, 1990). The biological function of MRIN was first reported in 1994 showing that nuclear p185^{neu} (rat HER2 variant) possesses transcriptional transactivation capability in p185^{neu} transfected NIH3T3 and DHFR-G8 mouse cell lines (Xie & Hung, 1994). The transcriptional regulation functions of nuclear HER2 and EGFR were further validated in subsequent studies performed in human cell lines (Elizalde, Cordo Russo, Chervo, & Schillaci, 2016; Lin et al., 2001; Lo et al., 2005; Wang et al., 2004). The internalization of ligand-activated RTKs begins through either clathrin-dependent or -independent endocytosis (Goh & Sorkin, 2013). The full-length receptor's transmembrane domain containing carboxyl-terminal fragments of RTK (CTF) and soluble intracellular domain (ICD) can be transported from the Golgi apparatus to the ER and subsequently nucleus via the retrograde pathway (Chen & Hung, 2015). However, exceptions to the clathrin-mediated endocytosis pathway have been reported for some RTKs. For instance, nuclear transport of fibroblast growth factor receptor 1 (FGFR1) utilizes a vesicle-independent route (Dunham-Ems et al., 2006; Myers, Martins, Ostrowski, & Stachowiak, 2003; Porebska et al., 2018;

Wang et al., 2012) whereas nuclear transport of EGFR occurs via a vesicle-dependent trafficking mechanism known as integral trafficking from the ER to the nuclear envelope transport (INTERNET) (Chen & Hung, 2015; Porebska et al., 2018; Wang & Hung, 2012). Below, we focus on the nuclear transport mechanisms of full-length RTKs containing the transmembrane domain.

2.1 Nuclear transport of full-length RTKs

2.1.1 INTERNET

Nuclear transport of the members of the EGFR family, such as EGFR and HER2, has been the most extensively studied among all MRINs. Here, we used EGFR nuclear transport the primary example of RTK INTERNET trafficking (Fig. 1).

Nuclear accumulation of EGFR family receptors increases considerably in response to both ligands and cellular stress stimuli (Chen & Hung, 2015; Faria, de Andrade, Goes, Rodrigues, & Gomes, 2016; Tan, Lambert, Rapraeger, & Anderson, 2016). Thus, we focused on ligand-induced EGFR translocation mechanisms as t majority of published studies were conducted under epidermal growth factor (EGF) stimulation. Several studies have reported that nuclear EGFR (nEGFR) accumulation is attenuated in cells experiencing inhibition of dynamin, a key protein of membrane fission during clathrin-coated pit formation, suggesting that nEGFR translocation may begin via dynamin-mediated receptor endocytosis (Bazzani, Donnini, Giachetti, Christofori, & Ziche, 2018; De Angelis Campos et al., 2011; Lo et al., 2006). In addition, transient downregulation of clathrin and caveolin impedes EGFR internalization, suggesting that EGFR can utilize either clathrin- or caveolin-mediated endocytosis (Bazzani et al., 2018; De Angelis Campos et al., 2011; Lo et al., 2006). Moreover, upon ligand stimulation, EGFR can be found in dorsal plasma membrane-derived vesicle-generating tubular network independently of clathrin and caveolin expression, indicating that EGFR internalization can also occur by macropinocytosis (Orth, Krueger, Weller, & McNiven, 2006).

EGFR transport from the plasma membrane to intracellular compartments relies primarily on membrane vesicles. By using digitonin and Triton X-100 to distinguish soluble proteins from those localized inside membranous organelles, Wang et al. demonstrated that the N-terminus of EGFR is detected only when membranous organelles are permeabilized whereas the C-terminus of EGFR is masked by antibodies regardless of organelle permeabilization, suggesting that intracellular full-length EGFR

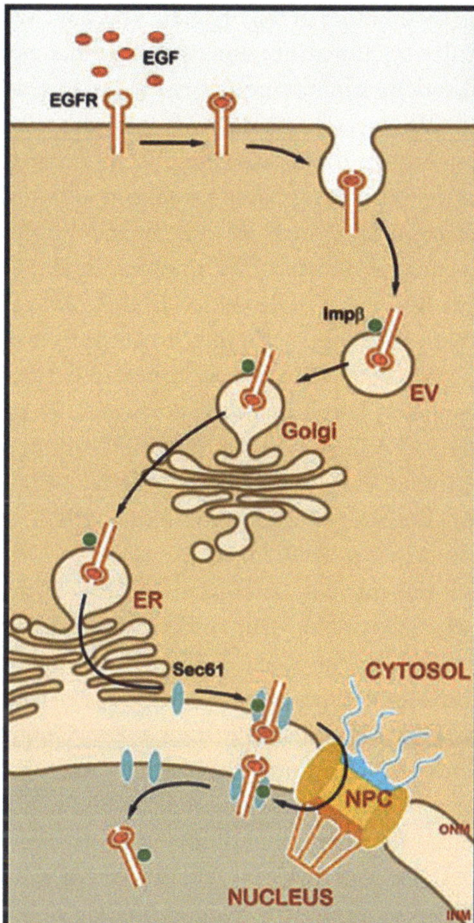

Fig. 1 A proposed nuclear trafficking model of cell surface EGFR. EGF activates cell surface EGFR and triggers EGFR internalization. The EV-located EGFR binds to Impβ and is fused to the Golgi apparatus before retrograde transported to ER and ONM, where it interacts with Sec61. ONM-bound EGFR is then flipped into the INM through the NPC before its release from the INM into the nucleus. The scale of the diagram does not reflect the relative size of the molecules and subcellular organelles. ER, endoplasmic reticulum; NE, nuclear envelope; EV, endocytic vesicle; Impβ, importin β; ONM, outer nuclear membrane; INM, inner nuclear membrane. *This research was originally published in the* Journal of Biological Chemistry. *Wang YN, Yamaguchi H, Huo L, Du Y, Lee HJ, Lee HH, Wang H, Hsu JM, Hung MC. The translocon Sec61beta localized in the inner nuclear membrane transports membrane-embedded EGF receptor to the nucleus.* J Biol Chem. *2010 Dec 3;285(49):38720–9. © the American Society for Biochemistry and Molecular Biology.*

is embedded in the membranes of the organelles (Wang, Wang, et al., 2010). Results from cellular fractionation also supported the notion that EGFR is primarily localized in membranous organelles instead of the soluble cytosolic fractions (Wang et al., 2012). Moving the internalized vesicles across the cell depends on the cargo transported along the cytoskeleton. It has been shown that knocking down proteins mediating vesicle trafficking along the microtubules, such as dynein and COMMD5/HCaRG, decreases EGFR nuclear transport (Campion et al., 2018; Du et al., 2014). A similar phenomenon of decreased EGFR nuclear transport is also observed in cells with disrupted actin fiber organization (Cortes-Reynosa, Robledo, & Salazar, 2009). Moreover, time-lapse tracking shows that EGFR moves from the plasma membrane toward the nucleus at a velocity of up to 2 $\mu m\,s^{-1}$ (Perillo et al., 2015), which is similar to that of cytoskeleton motor protein cargo transport in model organisms (Abraham, Hawley, Hayosh, Webster-Wood, & Akkus, 2018; Wu, Muthaiyan Shanmugam, Bhan, Huang, & Wagner, 2016).

Before entering the nucleus, internalized EGFR is first transported to the Golgi apparatus (Du et al., 2014). Translocation of EGFR to the Golgi requires syntaxin 6 (Du et al., 2014; Zhang et al., 2019), a protein which mediates retrograde endosome-to-*trans* Golgi transport (Laufman, Hong, & Lev, 2011). Moreover, disrupting the Golgi by brefeldin A (BFA) attenuates nuclear transport of EGFR (Wang, Wang, et al., 2010; Zhang et al., 2019). The nuclear-destined EGFR is further transported from the Golgi to the ER. Mutation of the Golgi-to-ER retrograde transport coat protein I (COP-I) complex suppresses nuclear translocation of EGFR (Wang, Wang, et al., 2010), suggesting that EGFR remains in the membranous vesicles during transport.

A putative NLS within EGFR's juxtamembrane domain plays an important role for its interaction with ER-localized importin-β and subsequent nuclear localization (Hsu & Hung, 2007; Lo et al., 2006; Wang et al., 2012). Electron microscopy demonstrated co-localization of EGFR and importin-β in the inner nuclear membrane (Lo et al., 2006; Wang, Yamaguchi, et al., 2010). Although importin-β regulates the permeability of the nuclear pore complex (NPC) (Lowe et al., 2015), localization of EGFR to the inner nuclear membrane largely depends on the interaction between EGFR/importin-β complex and the NPC (Wang, Yamaguchi, et al., 2010). In addition to importin-β, EGFR also interacts and co-localizes with translocon Sec61-β at the inner nuclear membrane (Liao & Carpenter, 2007; Wang, Yamaguchi, et al., 2010). Liao et al. showed that the Sec61-β-interacting EGFRs are mature full-length receptors (Liao & Carpenter, 2007). The transport of EGFR from the plasma membrane to the inner nuclear membrane remains membrane bound, which is then released by translocon Sec61 into the nucleoplasm (Liao & Carpenter, 2007; Wang, Yamaguchi, et al., 2010). Nuclear accumulation of EGFR is a dynamic process and time dependent, can be detected as soon as 1 min after ligand stimulation, and peaks at about 30–45 min before declining at about 1 h post stimulation (Hsu & Hung, 2007; Lin et al., 2001; Lo et al., 2006). The decrease in nEGFR accumulation is consistent with the observed kinetics of the interaction between EGFR and nuclear protein exporter protein exportin-1/CRM1 (Lo et al., 2006). Upon exiting the nucleus, nEGFR signaling is dampened.

2.1.2 INFS

While integrative nuclear FGFR1 signaling (INFS) is a term normally used to specify functions of FGFR1 in nucleus. We used this term to distinguish the transport mechanism of nuclear FGFR1 from INTERNET (Wang et al., 2012; Wang & Hung, 2012) as current literature indicates that FGFR1 is the only receptor which utilizes transport pathways different from INTERNET.

Depending on the isoform of FGFR and the stimuli, internalization of FGFR can occur through either clathrin- or caveolin-dependent endocytosis (Auciello, Cunningham, Tatar, Heath, & Rappoport, 2013; Feng et al., 2012; Haugsten et al., 2011; Haugsten, Sorensen, Brech, Olsnes, & Wesche, 2005). For instance, Haugsten et al. demonstrated in osteosarcoma U2OS cell that nuclear FGFR1 (nFGFR1) and FGFR3 (nFGFR3) accumulation rates differ due to the preference of FGFR1 for clathrin-dependent endocytosis compared with that of FGFR3 for either clathrin-dependent and -independent endocytosis using same concentration of fibroblast growth factor (FGF)-1 stimulation (Haugsten et al., 2011). In HeLa cells, FGFR2 internalization occurs primarily through clathrin-dependent endocytosis in response to FGF and heparin (Auciello et al., 2013) FGFR4 internalization also occurs via clathrin-dependent endocytosis in U2OS cell treated with FGF1 (Haugsten et al., 2016). Interestingly, FGFR1 can also be internalized in a ligand-independent and cell cycle S-phase-dependent manner (Reilly, Mizukoshi, & Maher, 2004). However, inhibitors against clathrin-dependent endocytosis, caveolin-dependent endocytosis, and macropinocytosis have largely failed to block FGFR1 nuclear localization in SWISS 3T3 cell (Reilly et al., 2004), suggesting cell-type dependent mechanisms of FGFR.

Point mutation of FGFR1 at Tyr766 to phenylalanine was reported to inhibit its internalization (Sorokin, Mohammadi, Huang, & Schlessinger, 1994). In addition, blocking FGFR1's kinase activity by small molecule inhibitors effectively inhibited its nuclear accumulation but had little effects on its internalization (Hwang et al., 2017; Reilly et al., 2004), suggesting that internalization occurs independently of its kinase activity. Prior studies indicated that extended-synaptotagmin-2, a protein which mediates ligand-induced FGFR1 conformation change, plays a critical role in initiating receptor internalization (Jean et al., 2010; Tremblay et al., 2015). It appears that the key step to initiate FGFR1 internalization is its conformational change rather than kinase activity.

Even though cytosolic membrane vesicle-associated FGFR1 is detected in cells (Myers et al., 2003) and that importin-β and intact Golgi function are required for FGFR1 nuclear accumulation (Dunham-Ems et al., 2006; Reilly & Maher, 2001), there is evidence to suggest FGFR1 utilizes transport mechanisms that are different from EGFR. The majority of intracellular FGFR1 is located in the soluble cytosolic fraction instead of membranous organelles before nuclear translocation in FGF-2-stimulated cells (Wang et al., 2012). Both membrane-anchored and soluble cytoplasmic FGFR1 exist in the cells; however, cytosolic FGFR1 has higher mobility than the plasma membrane-anchored form, suggesting that majority of nuclear FGFR1 may come from soluble cytoplasmic FGFR1 (Dunham-Ems et al., 2006). In addition, plasma membrane-anchored FGFR1 can downregulate FGFR nuclear accumulation (Bryant, Wylie, & Stow, 2005). Stachowiak et al. proposed that FGFR1 internalization may serve as a signal to trigger the release of intracellular FGFR1 [Reviewed in (Stachowiak, Maher, & Stachowiak, 2007)]. Retrograde transport of internalized FGFR from plasma membrane begins from endocytosis to early endosome. nFGFR accumulation can be blocked by Rab5 mutation or by overexpression of surface membrane adhesive E-cadherin, suggesting that early endosome is involved in FGFR1 and E-cadherin nuclear co-translocation (Bryant et al., 2005). Studies have indicated that a portion of nuclear FGFR1 (nFGFR1) is originated from cytosolic FGFR1, including sources from newly synthesized FGFR1, and Golgi- and ER-localized FGFR1 (Dunham-Ems et al., 2006; Myers et al., 2003; Stachowiak et al., 2007). Destroying Golgi-ER fusion by BFA treatment attenuates both cytosolic soluble FGFR1 generation and nFGFR1 accumulation (Myers et al., 2003).

It is noteworthy that nuclear transportation of FGFR3 may be different from that of FGFR1. FGFR1 is not strictly membrane associated due to the presence of a polar amino acid sequence which interrupts short stretches of hydrophobic amino acids in its transmembrane domain (Myers et al., 2003). Although full-length FGFR3 can translocate into the nucleus, the atypical feature of FGFR1 transmembrane domain does not exist in FGFR3 (Bryant & Stow, 2005). More studies are required to determine the nuclear transport mechanisms of FGFR3 and whether they are similar to that of either EGFR or FGFR1.

2.2 Nuclear transport of RTK fragments

Some RTKs encounter proteolytic cleavage or alternative splicing that generates fragments containing ICD after activation which can serve as positive feedback process to regulate RTK signaling (Volinsky & Kholodenko, 2013). The RTK-ICDs are soluble and can be found in the intracellular compartments, including the nucleus (Carpenter & Pozzi, 2012; Chen & Hung, 2015; Merilahti & Elenius, 2019). To prevent degradation during intracellular transport, RTK-ICDs are stabilized by their interaction with the heat shock protein chaperon complex (Chandarlapaty et al., 2010; Lyu, Wesselschmidt, & Lu, 2009) or by undergoing post-translational modifications, such as SUMOylation (Knittle, Helkkula, Johnson, Sundvall, & Elenius, 2017). Some studies have suggested that nuclear localization of RTK-ICDs may not necessarily require membrane vesicles and that a nuclear localization sequence (NLS) may be needed instead (Bertelsen & Stang, 2014; Hsu & Hung, 2007; Komuro, Nagai, Navin, & Sudol, 2003; Muraoka-Cook et al., 2006; Ni, Murphy, Golde, & Carpenter, 2001; Williams et al., 2004). Current findings suggest that nuclear transport of RTK-ICD binding partners, particularly transcription regulators, may be the primary mode for RTK-ICD translocation into the nucleus although more detailed investigations are needed.

2.3 Summary

From the nuclear trafficking studies of EGFR and FGFR1, we concluded that at least two transportation mechanisms are utilized by nuclear destined full-length RTKs: INTERNET and INFS-related trafficking pathways (Fig. 2). The major difference between these two mechanisms is that nuclear destined receptors utilizing the INTERNET route remain

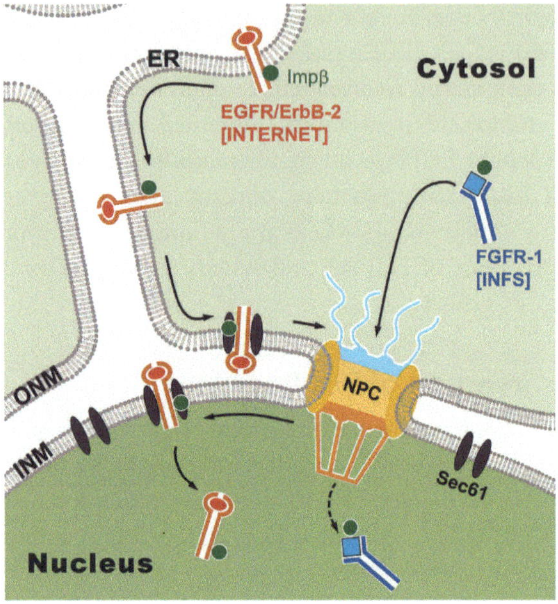

Fig. 2 Schematic of nuclear trafficking of EGFR and FGFR1 through the nuclear pore complex. Cell surface ErbB receptors translocate via INTERNET trafficking from the ER membrane to the INM by interacting with ER membrane-associated importin β. The ErbB receptors then interact with translocon Sec61β and NPC, and are released from the INM to the nucleus. In contrast, the cytosolic FGFR-1 translocates into the nucleus by interacting with importin β and NPC. FGFR-1 is not restricted to the membrane of organelles via the INFS mechanism. Impβ, importin β; ER, endoplasmic reticulum; NPC, nuclear pore complex; ONM, outer nuclear membrane; INM, inner nuclear membrane. *This research was originally published in the* Journal of Biological Chemistry. *Wang YN, Lee HH, Lee HJ, Du Y, Yamaguchi H, Hung MC. Membrane-bound trafficking regulates nuclear transport of integral epidermal growth factor receptor (EGFR) and ErbB-2.* J Biol Chem. *2012 May 11;287(20):16869–79. © the American Society for Biochemistry and Molecular Biology.*

membrane-bound throughout the transport process whereas nuclear FGFR1 transport includes soluble full-length cytosolic RTK (Wang et al., 2012). In this regard, the INTERNET pathway may be the dominant transport mechanism utilized by non-soluble full-length RTKs containing the transmembrane domain.

3. The functions of nuclear RTKs in cancer progression and therapeutic resistance

Human RTKs are grouped into 20 sub-families based on structural homology (Lemmon & Schlessinger, 2010; Pareja, Pines, & Yarden, 2015).

To date, only four RTK families do not have members that translocate into the nucleus, including MuSK, LMR, ALK, and STYK1. Here, we summarize the functions of nuclear RTKs and RTK-ICDs.

3.1 ErbB family

All members in EGFR family, ErbB1/EGFR, ErbB2/HER2, ErbB3 and ErbB4 translocate into the nucleus either as full-length receptor or ICD (Chen & Hung, 2015). Nuclear ErbB receptors and their variants mediate gene transcription and DNA repair through binding with different protein partners, and this in turn upregulates the expression of oncogenes and genomic instability linked to cancer progression. Indeed, the expression of nEGFR, nuclear HER2 (nHER2), and their variants has been shown to positively correlate with poor prognosis and survival in many types of cancers (Hadzisejdic, Mustac, Jonjic, Petkovic, & Grahovac, 2010; Hoshino et al., 2007; Lo et al., 2005; Psyrri et al., 2005; Schillaci et al., 2012; Traynor et al., 2013; Yang, Lin, et al., 2019; Yang, Ren, et al., 2019).

3.1.1 EGFR

Nuclear translocation of EGFR can be triggered by ligand binding, oxidative stress, or EGFR inhibitor gefitinib (Chen & Hung, 2015). Lin et al. first demonstrated that EGFR translocates into nucleus and functions as transcription co-regulator targeting at a specific AT-rich sequence (ATRS) upon ligand stimulation (Lin et al., 2001), and Lo et al. further identified the NLS of EGFR (Hsu & Hung, 2007; Lo et al., 2006; Wang et al., 2012). Due to the lack of DNA binding domain, nEGFR regulates gene expression by interacting with other DNA-binding proteins or transcription factors, such as RNA helicase A, STAT3, STAT5, and E2F1, at the ATRS-containing promoter region of target genes, e.g., *ABCG2*, *iNOS*, *CCND1* (cyclin D1), *c-Fos*, *VEGF*, *PTGS2* (COX-2), *AURKA* (Aurora A), *MMP* (matrix metallopeptidase)-*2*, and *MYBL2* (eB-Myb) (Huang et al., 2011; Huo et al., 2010; Kuo et al., 2014; Lo, Cao, Zhu, & Ali-Osman, 2010; Lo, Hsu, et al., 2005). nEGFR mediates regulation of those genes to enhance cancer cell growth, reduce cell death, and promote cancer cell invasion (Hanada et al., 2006; Hung et al., 2008; Kuo et al., 2014; Lo, Hsu, et al., 2005). Analysis of the ATRS of the promoter, and binding and expression target simulation identified 11 genes that can be transcriptionally upregulated by nEGFR, including *JUN* (c-Jun), *EGR1* (early growth response 1), *JUNB* (transcription factor Jun-B), *IER2* (immediate early response 2), *KLF2* (Kruppel like factor 2), *FOS* (Fos proto-oncogene), *FOSL1* (FOS like 1, AP-1 transcription factor subunit), *RHOB* (Ras homolog family member B), *CCNL1* (cyclin L1), *DUSP2* (dual specificity phosphatase 2), and *DUSP5* (dual specificity phosphatase 5) (Wong, 2018). Kim et al. demonstrated that the inhibition of nuclear EGFR/HER2 complex by the dual EGFR and HER2 inhibitor lapatinib downregulates transcription of *TYMS* which encodes thymidylate synthase, an enzyme essential for DNA replication (Kim et al., 2009). In addition, nEGFR interacts with STAT3 to upregulate gene promoter of *STAT1*, and competition between STAT1 and STAT3 for nEGFR has been reported in multiple breast cancer cell lines (Han, Carpenter, Cao, & Lo, 2013). Given that STAT1 and STAT3 have opposing effects on tumor initiation, with STAT1 as a tumor suppressor and STAT3 as an oncogene (Avalle, Pensa, Regis, Novelli, & Poli, 2012), additional investigations are needed to determine the roles of nEGFR and its interacting proteins in tumorigenesis.

nEGFR also promotes genomic instability and DNA damage repair. EGF-induced nEGFR facilitates DNA synthesis and repair by histone H4 phosphorylation to promote methylation (Chou et al., 2014). Ionizing radiation (IR) stimulates the interaction between nEGFR and DNA repair protein complex, including Ser139 phosphorylated H2A histone family member X (γH2AX), ataxia telangiectasia mutated (ATM), tripartite motif containing 28 (TRIM28/KAP1), promyelocytic leukemia protein (PML), hetero-chromatin binding protein (HP1), and marker of chromatin relaxation histone H3 acetyl-K9 (Dittmann et al., 2011). For the nEGFR-ATM interaction, Lee et al. reported that nEGFR phosphorylates ATM directly to facilitate DNA repair (Lee et al., 2015) and also mediates ATM activity through TIP60, an acetylase critical in ATM activation (Dittmann et al., 2011). Whether nEGFR directly phosphorylates TIP60 to regulate its activity is not known. In response to gefitinib, IR, hydrogen peroxide, or cisplatin treatment, both EGFR and EGFRvIII (a tumor-specific deletion variant of EGFR) can translocate into the nucleus to enhance activity and stability of DNA-dependent protein kinase (DNA-PKcs) in several cancer cell types, and this interaction is specific to cellular stress stimulations (Dittmann et al., 2005; Friedmann et al., 2006; Liccardi, Hartley, & Hochhauser, 2011). In addition, the ribonuclease activity of polynucleotide phosphorylase is inhibited by EGFR-mediated DNA-PKcs activation (Yu et al., 2012). It is worth mentioning that EGFR monoclonal antibody C225 abolishes EGFR nuclear translocation and decreases DNA-PKcs activity (Dittmann et al., 2005). The combination of gefitinib and cisplatin did not further increase the interaction between EGFR and DNA-PKcs compared with gefitinib alone (Friedmann et al., 2006), suggesting that additional sources of DNA damage have no effects on gefitinib-mediated decrease in DNA-PKcs activity. Using artificial stimuli-induced EGFR translocation, nEGFR was shown to phosphorylate proliferating cell nuclear antigen (PCNA) to stabilize the PCNA-chromatin complex and promote cell proliferation and DNA mismatch repair (Ortega et al., 2015; Wang et al., 2006).

Variants of EGFR in nucleus also participate in transcriptional regulations. For instance, EGFRvIII has been shown to interact with STAT3 and STAT5b in glioblastoma nucleus (Fan et al., 2013; Zhang et al., 2019). nuclear EGFRvIII also promotes transcription of cyclo-oxygenase 2 (COX-2) and apoptosis regulator Bcl-X (Bcl-XL), as well as enhances nuclear translocation of pyruvate kinase M2 (PKM2) to enhance tumor cell proliferation (Latha et al., 2013; Lo et al., 2010; Zhang et al., 2019). The EGFR mini-LEEK variant, which lacks the extra cytoplasmic transmembrane and tyrosine kinase domains, is detected in nucleus in response to ER stress (Piccione et al., 2012) and binds to ER stress response element (ERSE) motif on the promoter of genes involved in ER protein folding, including *HSPA5* (heat shock protein family A (Hsp70) member 5), *HSP90B1* (heat shock protein 90 beta family member 1), *HERPUD1* (ubiquitin-like domain member 1), *SDF2L1* (stromal cell derived factor 2 like 1) and *PPIA* (peptidylprolyl isomerase A), and upregulate their protein expression (Piccione et al., 2012).

3.1.2 HER2

Transcriptional transactivation function of $p185^{neu}$ was first discovered in mouse cell lines (Xie & Hung, 1994). Without the well-defined ligand and internalization signals, nHER2 accumulation is facilitated by dimerization with other ErbB receptors, such as EGFR and ErbB3, or by ubiquitination of HER2 (Bertelsen & Stang, 2014). Through recognition of the HER-2-associated DNA sequence, kinase competent nHER2 activates the promoter of *PTGS2* and potentially *PRPK* (TP53 regulating kinase), *MMP-16* (matrix metallopeptidase 16), and *DDX10* (DEAD-box helicase 10) (Wang et al., 2004). Like EGFR, nHER2 also associates with STAT3 to regulate *STAT1*, *CCND1*, and *P21CIP1* (cyclin dependent kinase inhibitor 1A; $p21^{Cip1}$) transcription (Beguelin et al., 2010; Diaz Flaque et al., 2013; Han et al., 2013). Recruited by STAT3, nHER2 also functions as co-activator of microRNA 21 (miR-21) in breast cancer, suggesting an indirect role in promoting breast cancer metastasis (Venturutti et al., 2016). Moreover, nHER2 facilitates cell growth, and interacts with actin and RNA polymerase I at the nucleoli RNA polymerase I active transcription sites to enhance binding of RNA polymerase I to ribosomal DNA to increase ribosomal RNA transcription (Li et al., 2011). In nucleus, HER2 also interacts with atypical histone macroH2A1.2 to upregulate transcription of *HER2* itself and promote cell proliferation (X. Li et al., 2012).

3.1.3 ErbB3

Nuclear ErbB3 (nErbB3) variant enhances transcriptional activity of *CCND1* and *EZR* (Ezrin) promoters while downregulates *HMGB1* (high mobility group box 1) promoter (Adilakshmi, Ness-Myers, Madrid-Aliste, Fiser, & Tapinos, 2011; Andrique et al., 2012; Brand et al., 2013). Both full-length and CTF of ErbB3 have been identified in the nucleus (Koumakpayi et al., 2006; Offterdinger, Schofer, Weipoltshammer, & Grunt, 2002; Reif et al., 2016). Indeed, full-length nErbB3 is positively correlated with enhanced cell proliferation in both breast cancer and prostate cancer cells (Koumakpayi et al., 2006; Reif et al., 2016); however, the nuclear targets of full-length nErbB3 have not been well studied.

3.1.4 ErbB4

Both the full-length (Bueter, Dammann, Zscheppang, Korenbaum, & Dammann, 2006; Thompson et al., 2007) and the γ-secretase-mediated ErbB4-ICD form (Ni et al., 2001; Sardi, Murtie, Koirala, Patten, & Corfas, 2006) of ErbB4 are found in the nucleus. Neuregulin-induced ErbB4-ICD is transported into nucleus via its interaction with TAK1 binding protein 2 (TAB2) and transcriptional regulator protein nuclear receptor corepressor 1 (NCOR1) (Sardi et al., 2006). In mammary cells, nuclear ErbB4-ICD (nErbB4-ICD) associates with Eto2, a transcriptional corepressor in erythrocyte differentiation (Linggi & Carpenter, 2006). nErbB4-ICD interacts with not only transcription co-repressors but also transcription factors, such as estrogen receptor (ERα), hypoxia-inducible factors 1α (HIF-1α) and yes-associated protein (YAP). In breast cancer cells, estrogen-stimulated accumulation of nErbB4-ICD binds to activated ER, and the complex is then recruited to the promoter containing consensus estrogen response element to upregulate expression of estrogen-inducible genes, such as *PGR* (progesterone receptor), *SDF1* (stromal cell-derived factor 1), and *ERBB4* (ErbB4) (Zhu et al., 2006). The ErbB4-ICD also stabilizes HIF-1α, which in turn enhances HIF-1α-mediated gene transcriptions (Paatero et al., 2012). The association of ErbB4-ICD and YAP can activate the Hippo signaling pathway and enhance breast cancer cell mobility (Haskins, Nguyen, & Stern, 2014), In mouse fetal lungs, nErbB4-ICD associates with signal transducer and activator of transcription 5A (STAT5A), YAP, and presenilin-1 to upregulate transcription of genes encoding surfactant protein, which is essential for fetal lung development (Hoeing et al., 2011; Zscheppang, Dork, Schmiedl, Jones, & Dammann, 2011). nErbB4-ICD also interacts with STAT5A to promote transcription of gene encoding β-casein (Long et al., 2003; Williams et al., 2004) in milk production. In addition to its role as transcription regulator, ErbB4-ICD also mediates mouse double minute 2 homolog (Mdm2) phosphorylation to enhance ubiquitination and degradation of Mdm2, which leads to increased level of p53 and p21 to regulate cell cycle (Arasada & Carpenter, 2005; Long et al., 2003; Williams et al., 2004).

3.2 MET family

Nuclear MET and RON RTKs both play a role in transcriptional regulation and DNA damage repair in cancer cells (Table 1). nMET and nRON have been suggested to serve as markers that predict drug resistance and poor prognosis in patients with cancer (Dong et al., 2019; Du et al., 2016; Tey et al., 2017).

Table 1 Functions of nuclear RTKs and their variants.

Nuclear RTK	Nuclear binding partner/target—functional role
DDR1	*Transcription regulation* • Upregulates *collagen IV* transcription (in complex with non-muscle myosin II and β-actin) (Chiusa et al., 2019)
ErbB1/EGFR	*Transcription regulation* • RNA helicase A—mediates interaction between nEGFR and ATRS region on DNA (Huo et al., 2010; Lin et al., 2001) • STAT3—activates promoters of genes including *iNOS, CCND1, c-fos, COX-2, SATA1* and *MMP2* (Han et al., 2013; Kuo et al., 2014; Lo et al., 2010; Lo, Hsu, et al., 2005) • STAT5—upregulates *Aurora A* expression (Hung et al., 2008) • E2F1—upregulates *B-Myb* and *proopiomelanocortin* expression (Araki et al., 2017; Hanada et al., 2006) • Upregulates *ABCG2* (Huang et al., 2011) and *thymidylate synthase* expression (Kim et al., 2009)

Continued

Table 1 Functions of nuclear RTKs and their variants.—Cont'd

Nuclear RTK	Nuclear binding partner/target—functional role
	Cell proliferation and DNA damage repair • EGFRvIII—promotes phosphorylated EGFRvIII/STAT complex to regulate gene expression (Fan et al., 2013) • $PCNA^{Y211}$—stabilizes interaction between PCNA and chromatin (Wang et al., 2006) • Histone $H4^{Y72}$—promotes DNA synthesis and repair (Chou et al., 2014). • TIP60 and ATM^{Y370}—enhances DNA repair activity (Dittmann et al., 2011; Lee et al., 2015). • DNA-PKcs—increases kinase activity and stability of DNA-PKcs (Dittmann et al., 2005; Friedmann et al., 2006; Liccardi et al., 2011).
EGFRvIII	*Transcription regulation* • STAT3—activates *COX-2* transcription (Lo et al., 2010) • STAT5b—activates *Bcl-XL* transcription (Latha et al., 2013) *Cell proliferation and DNA damage repair* • DNA-PKcs—increases kinase activity and stability of DNA-PKcs (Liccardi et al., 2011) • PKM2—enhances cell proliferation through facilitate nuclear localization of PKM2 (Zhang et al., 2019)
EGFR mini-LEEK	*Transcription regulation* • Upregulates proteins involved in response to ER stress, including *HSPA5* and *HSP90B1* (Piccione et al., 2012)
ErbB2/HER2	*Transcription regulation* • HER2-associateD DNA sequence—activates *COX-2* transcription (Wang et al., 2004) • STAT3—upregulates transcription of *STAT1*, *CCND1*, and *P21CIP1* (Beguelin et al., 2010; Diaz Flaque et al., 2013; Han et al., 2013); co-activates miR-21 to promote metastasis (Venturutti et al., 2016) • RNA polymerase I—enhances binding of RNA polymerase I to ribosomal DNA (Li et al., 2011) • Histone macroH2A1.2—upregulates *HER2* transcription (X. Li et al., 2012)
ErbB3-CTF	*Transcription regulation* • Upregulates expression of *CCND1* and *EZR* (Adilakshmi et al., 2011; Andrique et al., 2012) • Downregulates *HMGB1* (Adilakshmi et al., 2011; Ness et al., 2016)

Table 1 Functions of nuclear RTKs and their variants.—Cont'd

Nuclear RTK	Nuclear binding partner/target—functional role
ErbB4-ICD	*Transcription regulation* • TAB2 and NCOR1—decrease transcription of *GFAP* and *S100B* (Sardi et al., 2006) • Eto2—decreases Eto2-mediated gene transcription (Linggi & Carpenter, 2006) • ERα—enhances activity of promoters containing estrogen response element and upregulate transcription of genes such as *PR, SDF1* and *ERBB4* (Rokicki et al., 2010; Zhu et al., 2006) • STAT5A—(associates with YAP) enhances transcription of *surfactant protein* (Hoeing et al., 2011; Zscheppang et al., 2011); promotes transcription of *CSN2* (β-casein) (Long et al., 2003; Williams et al., 2004) • HIF-1α—enhances HIF-1α-mediated gene transcription (Paatero et al., 2012) *Cell cycle regulation* Mdm2—ErbB4-ICD mediates phosphorylation of Mdm2 and leads to Mdm2 degradation (Arasada & Carpenter, 2005)
EphA5	*DNA damage repair* • ATM—regulates cell cycle through p53-mediated pathway (Staquicini et al., 2015)
EphB4	*Transcription regulation* • Regulates transcription of *LEF1* (Mertens-Walker et al., 2015)
FGFR1	*Transcription regulation* • CREB-binding protein and S6 kinase 1—upregulates gene transcriptions including *JUN, CCND1* and *FGF2* (Dunham-Ems et al., 2009; Fang, Stachowiak, Dunham-Ems, Klejbor, & Stachowiak, 2005; Peng et al., 2002; Reilly & Maher, 2001) • ERα—reduces ERα binding to promoters of genes including *TFF1, CCND1, THSB1, CTGF, CCL2* and *EGR3* (Formisano et al., 2017)
FGFR1-ICD	*Transcription regulation* • Downregulates transcription of *EBI3* and *GRINA*; upregulates *KRTAP5–6, PRSS27* and *SFN* gene expression (Chioni & Grose, 2012)
FGFR2	*Transcription regulation* • PR and STAT5—recognizes chromosomal polycomb response element to regulate promoter of genes including *MYC* (Cerliani et al., 2011) • HIF-1α and HIF-2α—attenuate gene transcription mediated by HIF and p300 complex in normoxic conditions (Lee, Shin, Shin, Chun, & Park, 2019)

Continued

Table 1 Functions of nuclear RTKs and their variants.—Cont'd

Nuclear RTK	Nuclear binding partner/target—functional role
InsR	*Transcription regulation* • RNA polymerase II and gene promoters with InsR consensus DNA motif—regulates expression of genes related to insulin-mediated functions (Hancock et al., 2019; Podlecki et al., 1987)
IGF1R	*Transcription regulation* • Chromatin, LEF1, and RNA polymerase II—upregulates transcription of *CCND1, AXIN2, JUN, FAM21,* and *IGF1R* itself (Aleksic et al., 2010, 2018; Sarfstein et al., 2012; Warsito, Sjostrom, Andersson, Larsson, & Sehat, 2012) *DNA damage repair* • $PCNA^{Y60/Y133/Y250}$—promotes PCNA ubiquitination (Waraky et al., 2017)
MET	*Cell proliferation and DNA damage repair* • Transports HGF into the nucleus to induce formation of In_3P (Gomes et al., 2008) • $PARP1^{Y907}$—decreases interaction of PARP1 and PARP inhibitor and enhances DNA repair (Du et al., 2016)
MET-CTF	*Transcription regulation* • NF-κB and TAK1 promoters—enhances cell proliferation and metastasis (Tey et al., 2017)
RON	*Transcription regulation* • Gene promoters—upregulates *CEI, FLJ46072, JUN, MLXIPL, NARG1, NFATC1, PLEKHH3, POLRMT,* and *SSTR1* (Liu et al., 2010) • HIF-1α—upregulates *JUN* expression (Chang et al., 2014) *DNA damage repair* • DNA-PKcs and Ku70—promotes NHEJ (Chang et al., 2016)
PDGFRβ	*Transcription regulation* (potential function) • TATA element–modifying factor 1 (Papadopoulos, Lennartsson, & Heldin, 2018)
PTK7-CTF2	*Cell proliferation and migration* • Enhances cell proliferation and migration in SW480 cells (Na, Shin, Ludwig, & Lee, 2012)
CSF1R	*Transcription regulation* • EGR1 and EGR2 motifs and H3K4me1 in monocyte—downregulate EGR1-mediated gene expression (Bencheikh et al., 2019). • ELK, YY1 and H3K4me3 in macrophage—upregulate *CCND1, JUN and MYC* (Barbetti et al., 2014; Bencheikh et al., 2019)

Table 1 Functions of nuclear RTKs and their variants.—Cont'd

Nuclear RTK	Nuclear binding partner/target—functional role
Kit	*Transcription regulation* • Gene promoters—NFκB inhibitor β (Hsueh et al., 2019)
RET	*Transcription regulation* • ATF4—suppresses ATF4-mediated *NOXA* and *PUMA* transcription (Bagheri-Yarmand et al., 2015)
ROR1	*Transcription regulation* • Upregulate expressions of RDX (radixin), *EZR, SOS2,* and *CALD1* (Tseng et al., 2011)
Ryk-ICD	*Transcription regulation* • Smek1/2—enhances chromatin binding and transcription regulation functions of Ryk-ICD (W. H. Chang et al., 2017) • Downregulate *OLIG2* and upregulate *DLX2* expression to promote neuron differentiation (Lyu, Yamamoto, & Lu, 2008; Zhong et al., 2011) • Suppresses *FOXO3* transcription activity in neuronal cells (Tourette et al., 2014)
Tie2	*DNA damage repair* • Histone $H4^{Y51}$—enhances NHEJ activity (Hossain et al., 2016)
VEGFR2	*Transcription regulation* • Sp1—Upregulates promoter of genes including *VEGFR2* (Domingues, Rino, Demmers, de Lanerolle, & Santos, 2011)

3.2.1 MET

The hepatocyte growth factor receptor (HGFR/MET) is an overexpressed proto-oncogene and a therapeutic target in cancer treatment (Matsumoto, Umitsu, De Silva, Roy, & Bottaro, 2017). Full-length MET nuclear translocation mediated by GRB2-associated binding protein 1 (Gab-1) and importin-β is stimulated by its ligand, hepatocyte growth factor (HGF), in SK-HEP-1 liver cancer cells to initiate inositol 1,4,5-trisphosphate (In_3P) formation through phosphatidylinositol 4,5-bisphosphate hydrolysis induced by nuclear co-transport of HGF and MET (Gomes et al., 2008). The nuclear formation of In_3P then stimulate its receptor to increase calcium ion concentration in the nucleus.

However, in breast cancer cells, translocation of full-length MET into nucleus occurs primarily in response to cellular stress, such as oxidative stress, including hydrogen peroxide, IR and anti-cancer agent doxorubicin, instead of ligand stimulation (Chen et al., 2019). MET contains a putative

NLS which relies on histidine residues to provide positive charges to increase its interaction with importin β (Chaudhary et al., 2014). Immunoprecipitation (IP) analysis revealed that nMET interacts with DNA repair proteins poly [ADP-ribose] polymerase 1 (PARP1) and X-ray repair cross complementing 6 (XRCC6/Ku 70) (Chen et al., 2019; Du et al., 2016). MET can also directly phosphorylate PARP1 at Y907 residue to decrease the interaction between anti-cancer PARP inhibitor and PARP1, which leads to PARP inhibitor resistance in breast cancer (Du et al., 2016). Moreover, under hydrogen peroxide stimulation, MET forms heterodimer with EGFR, and the two receptors co-localize in the nucleus to phosphorylate PARP1 Y907 in liver cancer cells (Dong et al., 2019).

The c-terminal fragment of nMET (nMET-CTF) also exists in cancer cell nucleus (Matteucci, Bendinelli, & Desiderio, 2009; Pozner-Moulis, Pappas, & Rimm, 2006; Xie et al., 2019). nMET-CTF was first reported in A431 epidermoid carcinoma cells and CaCo2 epithelial colorectal adenocarcinoma cells at low cell density (Pozner-Moulis et al., 2006), and its level was later shown to decrease when the cell density increases (Pozner-Moulis et al., 2006). In MDA-MB-231 breast cancer cells, nuclear localization of MET-CTF occurs independently of HGF stimulation and is stabilized by YAP (Matteucci et al., 2009). Exogenous MET-CTF expression revealed co-localization of nMET-CTF with p21, a protein regulates cell cycle progression and maintenance of cancer stem cell (Shamloo & Usluer, 2019), suggesting that MET-CTF may contribute to drug resistance and stemness of cancer cells (Xie et al., 2019). MET-CTF also exhibits trans-activation activity (Matteucci et al., 2009; Tey et al., 2017). Specifically, MET-CTF robustly activates the promoter of *NFKB* (NF-κB) and *MAP3K7* (TAK-1 protein) (Tey et al., 2017). In liver cancer cells, MET-CTF mediates activation of the NF-κB/TAK1 pathway to promote cell proliferation, anchorage-independent growth, and mobility (Tey et al., 2017).

3.2.2 RON

Nuclear transport of RON is largely stimulated by cellular stress (Chang et al., 2014; Liu et al., 2010). It has been reported that serum starvation induces the interaction between nuclear RON-EGFR complex on [?] promoters of *C5orf38* (chromosome 5 open reading frame 38, CEI protein), *JUN, MLXIPL* (MLX interacting protein like), *NAA15* (NMDA receptor regulated 1, NARG1 protein), *NFATC1* (nuclear factor of activated T cells 1), *PLEKHH3* (Pleckstrin homology, MyTH4 and FERM domain containing H3), *POLRMT* (RNA polymerase mitochondrial), and *SSTR1* (somatostatin receptor 1). Although nuclear translocation of the RON-EGFR complex requires docking of importin on to the NLS of RON, nuclear RON (nRON) is not phosphorylated under serum starvation (Liu et al., 2010). Subsequent sequence analysis revealed that nRON contains three Ig-like, plexins, transcription factors (IPT) domains that can interact with DNA, suggesting that nRON may be a transcriptional regulator (Liu et al., 2010). In response to hypoxia, around 50% of RON translocate into the nucleus in an EGFR- and HIF-1α-independent manner (Chang et al., 2014). Nonetheless, nRON can bind to HIF-1α through its kinase domain, and the RON/HIF-1α complex upregulates transcription of *JUN* by binding to the D-box region of the *JUN* promoter (Chang et al., 2014). In addition to acting as co-transcriptional regulator, nRON also interact with DNA-PKcs and Ku 70 proteins under hypoxia to promote non-homologous end joining (NHEJ) DNA repair (Chang et al., 2016).

3.3 FGFR family

3.3.1 FGFR1

Of the transport pathways for the FGFR family members, the INFS pathway is the most extensively studied to date for the FGFR family members. In the nucleoplasm, nFGFR1 has been reported to associate with chromatin by interacting with CREB-binding protein and S6 kinase 1 to further recruit RNA polymerase II for transcriptional regulation of genes, such as *JUN* (Dunham-Ems et al., 2009; Fang et al., 2005; Peng et al., 2002; Reilly & Maher, 2001). Moreover, although the kinase activity of nFGFR1 is not required for its interaction with CREB-binding protein and S6 kinase 1 (Fang et al., 2005; Hu et al., 2004), inhibition of FGFR decreases nFGFR1-mediated gene expression (Dunham-Ems et al., 2009), suggesting that FGFR1 kinase activity is essential for its nuclear function as transcription regulator. nFGFR1 is also present in pancreatic ductal adenocarcinoma tissue specimens, particularly in the stromal fibroblasts and stellate cells where nFGFR1 promotes cell proliferation and invasion (Coleman et al., 2014). In ER-expressing breast cancers, estrogen deprivation-induced nFGFR1 interacts with ERα and reduces the binding of ERα to promoters of its target genes, including *TFF1* (trefoil factor 1), *CCND1, THSB1* (thrombospondin 1), *CCN2* (cellular communication network factor 2, CTGF), *CCL2* (C—C motif chemokine ligand 2) and *EGR3* (early growth response 3) (Formisano et al., 2017).

In addition, following cleavage of FGFR1 by granzyme B to produce a soluble fragment, FGFR1-ICD, which translocates into nucleus to promote breast cancer invasion by downregulating *EBI3* (interleukin-27 subunit beta) and *GRINA* (glutamate ionotropic receptor NMDA type subunit associated protein 1) while upregulating *KRTAP5-6* (keratin associated protein 5-6), *PRSS27* (serine protease 27) and *SFN* (Stratifin) expression (Chioni & Grose, 2012).

3.3.2 FGFR2

In response to medroxyprogesterone acetate and FGF-2 treatment, FGFR2, in complex with PGR and STAT5, translocates into the nucleus in breast cancer cells. The nuclear PGR-FGFR2-STAT5 complex then recognizes chromosomal polycomb response element (PRE) to regulate promoter of genes, such as *Myc* (Cerliani et al., 2011). More recently, FGFR2 was shown to interact with HIF-1α and HIF-2α in prostate cancer cells DU145 and PC3, and the interaction between FGFR2 and HIF proteins is stronger under normoxic than hypoxic conditions to attenuate gene transcription mediated by HIF and p300 complex (Lee et al., 2019). In addition to cancers, nFGFR2 has also been reported to play a role in organ development (Schmahl, Kim, Colvin, Ornitz, & Capel, 2004; Seymour et al., 2012).

3.3.3 FGFR3 and FGFR4

FGFR3 was first reported as a 110-kDa variant form lacking the transmembrane domain but with ligand-binding and kinase domain in the nuclei of breast epithelial cells (Johnston, Cox, Gomm, & Coombes, 1995). Nuclear FGFR3 is also detected in malignant breast cancer (Zammit et al., 2001). In a rat chondrocyte cell line CFK2, overexpression of FGFR3 leads to its co-localization with STAT1 in both the nucleus and cytoplasm, suggesting that FGFR3 may be involved in transcriptional regulation (Li et al., 2010). Interestingly, nuclear FGFR3 is considered as a prognostic marker for pancreatic cancer (Zhou et al., 2015). Moreover, IHC staining for FGFR4 indicated higher levels of nuclear FGFR4 in colorectal adenocarcinomas than in normal colon tissues (Turkington et al., 2014). Nonetheless, more studies are required to delineate the functions of nuclear FGFR3 and FGFR4 in cancer.

3.4 VEGFR family

3.4.1 VEGFR1

Studies in multiple myeloma cells showed that only VEGFR1, but not VEGFR2 or VEGFR3, translocates into the nucleus, and monoclonal antibodies against VEGFR1 leads to suppression of cell growth and motility (Vincent et al., 2005). In breast cancer cells, VEGFR1 co-localizes with lamin A/C located in the nuclear envelope (Lee et al., 2007). Inhibition of endogenous VEGF-activated nuclear-envelope VEGFR1 induces apoptosis in breast cancer cells although VEGF can also promote cell survival in an intracrine manner (Lee et al., 2007). Flow cytometric analysis showed that nuclear-localized VEGFR1 correlates positively with cancer cell malignancy (Tyrsina, Nikulitskiy, Inshakov, Ryabaya, & Tyrsin, 2019). However, in these studies, VEGFR1 appears to mainly localize in the nuclear envelope, and the non-canonical substrate of VEGFR1 has not been determined. The pigment epithelium-derived factor (PEDF) has been reported to induce γ-secretase- and presenilin-mediated VEGFR1-ICD formation, which then accumulates in nucleus (Cai et al., 2011). More studies are required to identify the form of nuclear envelope-localized VEGFR1 and clarify the biological functions of full-length VEGFR1 and VEGFR1-ICD.

3.4.2 VEGFR2

Phosphorylated nuclear VEGFR2 (nVEGFR2) has been reported in the tumor tissues including ovarian granulosa cell tumor, renal tumor, breast carcinomas, colonic carcinomas non-Hodgkin's lymphomas, and acute leukemia (Farkkila et al., 2011; Fox et al., 2004; Santos & Dias, 2004; Stewart et al., 2003). In epithelial cells overexpressing VEGFR2, nVEGFR2 interacts with Sp1 to upregulate the promoter activities of genes, including that of itself, in response to VEGF-induced activation (Domingues et al., 2011). A similar phenomenon occurs in ovarian granulosa tumor cells such that nVEGFR2 signal is positively correlated with VEGF expression and high disease recurrence (Farkkila et al., 2011). VEGF-induced co-transport of VEGFR2, caveolin, and endothelial nitric oxide synthase into the nucleus also suggests that VEGFR2 may play a role as a transcriptional regulator that targets nuclear nitric oxide production (Feng, Venema, Venema, Tsai, & Caldwell, 1999). Moreover, full-length nVEGFR2 translocation, mediated by a crosstalk with VEGFR1 and VEGF-induced PI3K activation, is required to promote epithelial wound healing (Santos et al., 2007). In addition to VEGF, nVEGFR2 accumulation can also be stimulated by hypoxia (Blazquez et al., 2006). Interestingly, mechanical sheer force can also induce nuclear translocation of VEGFR2 in vesicular endothelial cadherin-dependent manner (Shay-Salit et al., 2002).

3.5 Insulin receptor (InsR) family

3.5.1 InsR

Nuclear InsR (nInsR) has been recently shown to serve as prognostic marker for breast cancer patients with high body mass index (Björner et al., 2017). Insulin-induced nuclear transport of InsR depends on clathrin, caveolin, importin-α, and nuclear pore, with potential help from NLS and heat shock protein 70 (Amaya et al., 2014; Hancock et al., 2019; Kesten, Horovitz-Fried, Brutman-Barazani, & Sampson, 2018). InsR interacts with C-terminal phosphorylated RNA polymerase II and HCF-1 transcription co-regulator in the nucleus in human liver cancer cells (Hancock et al., 2019). Chromatin immunoprecipitation (ChIP) analysis further revealed that nInsR interacts with more gene promoters than phosphorylated RNA polymerase II, and that nInsR enhance promoters with the nInsR-preferred consensus DNA motif enhance transcription to upregulate genes with functions related to insulin-mediated lipid metabolism and protein synthesis (Hancock et al., 2019; Podlecki et al., 1987). Although the nInsR-RNA polymerase II interaction and the InsR consensus DNA motif are similar in brain and liver cancer cells, the target genes exhibit tissue specificity (Hancock et al., 2019). nInsR activity is also negatively correlated with nuclear accumulation of apoptotic marker PKCδ, and nInsR initiates nuclear In_3P -dependent calcium signals to regulate cell proliferation (Amaya et al., 2014; Kesten et al., 2018).

3.5.2 Insulin-like growth factor-1 receptor (IGF1R)

Following activation by IGF treatment, IGF1R is phosphorylated and SUMOylated before internalization via clathrin-mediated endocytosis (Aleksic et al., 2010; Sehat et al., 2010). In the nucleus, IGF1R (nIGF1R) interacts with chromatin, transcription factor LEF1, and RNA polymerase II to upregulate transcription of *CCND1*, AXIN2 (axin-related protein), *JUN, WASHC2C* (WASH Complex Subunit 2C), and *IGF1R* itself (Aleksic et al., 2010; Aleksic et al., 2018; Sarfstein et al., 2012; Warsito et al., 2012). It is worth mentioning that nIGF1R-mediated *IGF1R* transcription in breast cancer cells is correlated negatively with ER expression (Sarfstein et al., 2012). Beside its role in transcriptional regulation, nIGF1R also upregulates DNA damage repair by phosphorylating PCNA at Y60, Y133, and Y250 to promote PCNA ubiquitination, which leads cells to escape from G1 arrest (Waraky et al., 2017). Although nIGF1R is enriched in several cancer cell lines and in tumor samples from patients with metastatic colorectal cancer (Aleksic et al., 2010; Aleksic et al., 2018; Codony-Servat et al., 2017), the function of nIGF1R in promoting cell proliferation is not restricted to cancer cells (Solomon-Zemler, Sarfstein, & Werner, 2017), suggesting that more studies are required to evaluate the potential side-effects of targeting IGF1R as a therapeutic strategy.

3.6 Ephrin receptor (Eph) family

The Eph family contains 14 members, including EphA1–8, EphA10, EphB1–4 and EphB6. However, only a few are located in the nucleus. In lung cancer cells, EphA5 translocates into the nucleus in response to irradiation, and its C-terminal domain physically interacts with phosphorylated ATM and associates with chromatin (Staquicini et al., 2015). This EphA5-ATM interaction further regulates cell cycle through a p53-mediated pathway (Staquicini et al., 2015). In prostate cancer cells, nuclear transport of EphB4 depends on two NLS sequences and its interaction with importin-α (Mertens-Walker et al., 2015). ChIP analysis showed that the *LEF1* gene (encoding lymphoid enhancer-binding factor 1) is a target of nuclear EphB4 (Mertens-Walker et al., 2015).

EphA7 and EphA10 are also found in the nucleus in breast cancer cells, and this co-localization of EphA7 and 10 in cell nucleus is interrupted by EphB6 expression (Johnson, Segovia, & Kandpal, 2016). Full-length nuclear EphA1 is detected in rhabdomyosarcoma (LaCombe, 2019), an aggressive form of skeletal muscle cells derived cancer. Nuclear EphA4 has been reported in the nucleus of osteoblastic cells (Kuroda et al., 2008), and nuclear EphB2 in prostate cancer cells treated with demethylation agents (Vardi et al., 2010). In addition to human cancer cells, nuclear EphB1 is present in astrocytes in monkeys without glaucoma (Schmidt, Agapova, Yang, Kaufman, & Hernandez, 2007). Although multiple Eph family members are reported in nucleus, their biological functions and whether they stay as full-length protein remain unknown.

3.7 Receptor tyrosine kinase-like orphan receptor (ROR) family

The ROR family contains two members, ROR1 and ROR2. These two receptors shared 58% amino acid identity and are structurally related (Tseng, Lyu, & Lin, 2010). Nuclear translocation of ROR1 requires a series of KxxK amino acids sequence located in its juxtamembrane domain (Tseng et al., 2010). ROR1 nuclear transport is partially mediated by small GTPase Ran. Expression of a dominant negative Ran mutation increases the levels of cytoplasmic ROR1 while decreasing that of nuclear ROR1 (Tseng et al., 2010). Fusion of ROR1 to two copies of FKBP36v domain and a NLS from c-Myc to enhance its nuclear translocation further showed that nuclear ROR1 upregulates the expression of genes in actin cytoskeleton regulation, including *RDX* (radixin), *EZR*, *SOS2* (son of sevenless homolog 2) and *CALD1* (caldesmon 1) (Tseng et al., 2011). Thus, nuclear ROR1 may play important roles in cytoskeleton remodeling and contribute to cell migration. Similar to ROR1, Tseng et al. demonstrated that the C-terminus of ROR2 is detected within the nucleus (Tseng et al., 2010). Studies on the function of nuclear ROR2 are limited.

3.8 Platelet-derived growth factor receptor (PDGFR) family

PDGFR family contains five members, PDGFRα, PDGFRβ, colony stimulating factor-1 receptor (CSF-1R/FMS), stem cell factor receptor (KIT/SCFR), and fms-like tyrosine kinase 3 (Flt3).

3.8.1 PDGFR

The biological functions of nuclear PDGFRα is not well understood. Nuclear PDGFRα is found in alveolar fibroblasts (Chen, Acciani, Le Cras, Lutzko, & Perl, 2012) and alveolar rhabdomyosarcoma cells expressing high PDGFRα and IGF1R. However, nIGF1R may have regulatory roles in alveolar rhabdomyosarcoma cell as cells with high nIGF1R establish tumors more efficiently (Aslam et al., 2013).

Nuclear PDGFRβ is translocated into the nucleus through a Golgi- and importin-dependent pathway (Papadopoulos et al., 2018). The nuclear accumulated PDGFRβ dramatically increases from 7% to 26% of total PDGFRβ in response to PDGF-BB stimulation in BJhTERT normal human foreskin fibroblasts cells (Papadopoulos et al., 2018). PDGFRβ binds to the Fer non-receptor tyrosine kinase (NRTK) in the cytoplasm for co-transport into the nucleus. PDGFRβ has been shown to regulate the TATA element-modifying factor 1, a substrate protein of Fer (Papadopoulos et al., 2018). Because the TATA element-modifying factor 1 interacts with both BRG-1 and BRM chromatin-remodeling complex proteins (Mori & Kato, 2002), nuclear PDGFRβ may indirectly mediate chromatin remodeling (Papadopoulos et al., 2018).

3.8.2 CSF1R

Chromatin-bound colony stimulating factor 1 receptor (CSF-1R) is present in breast cancer cells, human monocytes, and macrophages (Barbetti et al., 2014; Bencheikh et al., 2019). Subcellular fractionation of breast cancer cells indicated that full-length CSF-1R is enriched in both membrane-bound and chromatin-bound fractions, but not in the soluble nuclear fractions (Barbetti et al., 2014). Although NLS prediction algorithm indicated that CSF-1, a ligand of CSF-1R, but not CSF-1R, contains an NLS, there is evidence to suggest that the NLS in CSF-1 is not required for CSF-1R nuclear accumulation (Barbetti et al., 2014; Bencheikh et al., 2019). ChIP analysis revealed that CSF-1R is recruited to multiple transcription factor binding sites in monocytes, including EGR1 and EGR2 motifs, and downregulates EGR1-mediated gene expression (Bencheikh et al., 2019). In contrast, nuclear CSF-1R (nCSF-1R) in macrophages is located at actively transcribed regions and interacts with ELK and YY1 transcription factors. In monocytes, nCSF-1R preferentially interacts with lysine 4 mono-methylated histone H3 (H3K4me1). However, nCSF-1R mainly interacts with lysine 4 tri-methylated histone H3 (H3K4me3) in macrophages (Bencheikh et al., 2019). Although nCSF-1R-interacting transcription factors have not been identified in breast cancer cells, nCSF-1R also upregulates genes (*CCND1, JUN and c-Myc*) similar that observed in macrophages (Barbetti et al., 2014; Bencheikh et al., 2019).

3.8.3 Kit

Kit is co-localized with the nuclear envelope marker lamin B1 and also localized in the nucleus in gastrointestinal stromal tumors with resistance to small molecule inhibitor imatinib (Hsueh et al., 2019). Analysis of Kit bound DNA sequences indicated that Kit binds to the promoter region of several genes, including *NFKBIB* (NFκB inhibitor β), which can mediate Kit expression (Hsueh et al., 2019). Nuclear localized Kit is also found in normal and neoplastic medullary cells of the adrenal gland (Matsuda et al., 1993) but its functions are not clear.

3.9 Tropomyosin receptor kinase (Trk) family

3.9.1 TrkA

Phosphorylated TrkA utilizes membranous vesicle machinery for transport from the nuclear envelope into the nucleus (Gong et al., 2007). Phosphorylated TrkA-ICD is found constitutively expressed in the nucleus of cultured hepatic stellate cells but is induced by nerve growth factor in rat pheochromocytoma cell line PC12 (Bonacchi et al., 2008). Treatment of PC12 cells with either nerve growth factor or lysophosphatidate increases the nuclear localization of phosphorylated Trk A Y498 along with increased nuclear transport of lysophosphatidate receptor (Moughal, Waters, Sambi, Pyne, & Pyne, 2004). Multiple isoforms of TrkA in both cytoplasm and nucleus have been reported (Luberg, Park, Aleksejeva, & Timmusk, 2015). Although TrkAκ does not possess a conventional NLS, it has been hypothesized that the N-terminal signal sequence of TrkAκ is recognized by importin-7 which facilitates TrkAκ nuclear transport (Luberg et al., 2015). IHC analysis detected nuclear TrkA in melanocytic and serous ovarian carcinoma tumor tissues and multiple cell lines (Davidson et al., 2003; Florenes et al., 2004; Gong et al., 2007). More studies are required to determine the biological functions of nuclear TrkA.

3.9.2 TrkB and TrkC

Amyloid-β peptide treatment has been shown to induce cleavage of full-length TrkB, which increases TrkB-ICD in neuroglioma cells (Jeronimo-Santos et al., 2015). TrkB-ICD is predicted to contain two bipartite NLS for nuclear translocation (Fonseca-Gomes et al., 2019). Moreover, nuclear destined TrkB-ICD requires its kinase activity, and TrkB-ICD induces tyrosine phosphorylation of multiple nuclear and axonal proteins in brain cells from patients with Alzheimer's disease (Fonseca-Gomes et al., 2019). Because the substrate proteins of nuclear TrkB-ICD have not been identified, the biological effects of nuclear TrkB-ICD remain unclear. In addition, IHC staining indicated that TrkC is localized in the nuclei of oocytes in fetus and females (Oron et al., 2011), but the biological function of nuclear TrkC has not yet been studied.

3.10 Tie family

The Tie family contains two members, Tie1 and Tie2. Overexpression of Tie2 was found in human glioma tissues and is positively correlated with chemoresistance and tumor malignancy (Lee et al., 2006; Martin et al., 2009). Using brain tumor stem cells as model, IR-induced expression of Tie2 ligand, ANG1, stimulates nuclear accumulation of Tie2 (Hossain et al., 2016). It is worthwhile to mention that the ANG1-induced nuclear translocation of Tie2 is ligand-receptor specific as ANG2, another ligand of Tie2, fails to stimulate Tie2 nuclear accumulation, and neither ANG1 nor ANG2 can stimulate nuclear accumulation of Tie1 (Hossain et al., 2016). At DNA damage sites, Tie2 is detected in the γH2AX-complex together with other DNA repair proteins, including BRG1, DNA-PKcs, Ku 70, and Ku 80. It is not yet known whether Tie2 phosphorylates these DNA repair proteins. However, Tie2 is reported to phosphorylate histone H4 at Y51 residue, and phosphorylated $H4^{Y51}$ is required by DNA binding protein Abl1 to enhance NHEJ repair (Hossain et al., 2016). These studies point to the role of nuclear Tie2 in facilitating NHEJ of brain tumor stem cells in response to IR.

3.11 Discoidin domain receptor (DDR) family

Nuclear accumulation of DDR1 is induced by its ligand, soluble collagen I, in human kidney cell lines (Chiusa et al., 2019). Nuclear DDR1 (nDDR1) is indeed transported via the retrograde pathway from plasma membrane by surface protein biotinylation (Chiusa et al., 2019). Chiusa and co-worker demonstrated that DDR1 nuclear transport depends on actin-binding non-muscle myosin II and Sec61β using myosin inhibitor blebbistatin (Chiusa et al., 2019). Although DDR1, non-muscle myosin II, and β-actin do not contain an NLS, the authors speculated that nuclear translocation of the complex utilizes NLS from β-actin-interacting protein, cofilin (Chiusa et al., 2019). The nuclear DDR1/non-muscle myosin II/β-actin complex associates with chromatin and upregulates transcription of *COL4A1* (encoding collagen IV) by binding to the promoter region in response to soluble collagen I treatment, which has been suggested to promote kidney fibrosis (Chiusa et al., 2019). Nuclear accumulation of activated DDR2 has been reported in breast cancer mesenchymal stem cells, but whether nuclear DDR2 contribute to tumor growth and mobility is not yet clear (Gonzalez et al., 2017).

3.12 Protein tyrosine kinase 7 (PTK7)

Both full-length PTK7 and its soluble truncated form PTK7-CTF2 are found in the nuclei of SW480 colorectal cancer cells (Na et al., 2012). PTK7-CTF2 is primarily nuclear localized as it has not been detected in cytoplasm of SW480 cells. Expression of PTK7-CTF2 in SW480 cells is sufficient to enhance cell proliferation and migration (Na et al., 2012).

3.13 Receptor-like tyrosine kinase (Ryk)

Nuclear localized Ryk has been studied primarily in neuronal development and neurological diseases instead of cancer. Nuclear transport and stabilization of Ryk-ICD depends on its interaction with chaperon proteins, Smek1, Smek2, and Cdc37 (Chang et al., 2017; Lyu et al., 2009). The Smek1/2 proteins further enhance chromatin binding and transcriptional regulation functions of Ryk-ICD in the nucleus (Chang et al., 2017). Nuclear Ryk-ICD is required during neuron differentiation in response to ligand Wnt3 stimulation (Lyu et al., 2008). A later study demonstrated that Ryk-ICD suppresses FOXO3a transcriptional activity in neuronal cells (Lyu et al., 2008; Tourette et al., 2014; Zhong et al., 2011). Further study showed that Ryk-ICD suppresses FOXO3a transcriptional activity in neuronal cells (Tourette et al., 2014).

3.14 RET

Mutations and overexpression of RET are closely linked to malignant thyroid carcinoma (Santoro, Melillo, Carlomagno, Vecchio, & Fusco, 2004). Bagheri-Yarmand et al. further showed that RET translocates into the nucleus in thyroid medullary carcinoma and NSCLC cell lines (Bagheri-Yarmand et al., 2015). Antibodies against both N- and C-terminus of RET can detect nuclear RET in the cells (Bagheri-Yarmand et al., 2015), suggesting that nuclear RET may be in its full-length form. In silico analysis suggested that nuclear RET binds to activating transcription factor 4 (ATF4) and the promoter of *TP53* and c-*MYC*. The kinase activity of nuclear RET is required to suppress ATF4 activity by enhancing ATF4 degradation, which in turn inhibits ATF4-mediated transcription of *NOXA* (phorbol-12-myristate-13-acetate-induced protein 1) and *PUMA* (p53 upregulated modulator of apoptosis) (Bagheri-Yarmand et al., 2015).

3.15 TAM family

There are three RTKs (Tyro3, Axl and Mer) in the TAM family. Although all three receptors have been detected in the nucleus, the biological functions are not yet clear.

3.15.1 Tyro3

NLS-mediated nuclear localized Tyro3-ICD has been reported in colorectal cancer cells and may be essential to cell survival (Tsai et al., 2018). Nuclear localized Tyro3 is also observed in leiomyosarcoma patient samples and positively correlated with the expression of its ligand, Gas6 (Dantas-Barbosa et al., 2017). These findings suggest that nuclear Tyro3 may contribute to autocrine activation of Tyro3.

3.15.2 Axl

Recent study demonstrated that the NLS is required for nuclear localized Axl-ICD (Lu et al., 2017). However, biological function of nuclear Axl-ICD has not been determined yet. Since Axl inhibition decreases the efficiency of DNA repair (Balaji et al., 2017), further studies on whether nuclear Axl-ICD mediates the activities of DNA repair proteins are warranted.

3.15.3 Mer

Upon activation by its ligand Gas6, a partially glycosylated form of Mer is derived by endoglycosidase cleavage to remove the mannosylated glycans (Migdall-Wilson et al., 2012). Mer contains a conventional monopartite NLS which interacts with importin for nuclear translocation (Migdall-Wilson et al., 2012). Mer can translocate into the nucleus without stimulation by its ligand Gas6, but prolonged Gas6 treatment has been shown to induce soluble nuclear Mer and reduce Mer in chromatin-bound fractions (Migdall-Wilson et al., 2012). Interestingly, chromatin-bound Mer is predominantly a fully glycosylated form whereas soluble nuclear Mer appears to be the partially glycosylated (Migdall-Wilson et al., 2012). It is unclear the functional importance between the two forms of nuclear Mer.

3.16 ROS

ROS expression has been reported to associate with development of tumors, including glioblastoma, non-small cell lung cancer, and oral squamous cell carcinoma (Cheng, Sun, Wang, Yu, & Ding, 2015; El-Deeb, Yoo, & Lee, 2011). Cheng et al. reported the detection of nuclear ROS receptor in a cohort of 31 patients with oral squamous cell carcinoma (Cheng et al., 2015); however, nuclear ROS expression did not show clinical pathological relevance in this cohort. Moreover, as ROS remains an orphan receptor that lacks a defined ligand, more studies are required to better understand the stimuli causing nuclear accumulation of ROS and the nuclear functions of ROS.

4. Conclusion

To date, a majority of the RTK subfamilies (16 out of 20 subfamilies) are found in the nucleus, suggesting that nuclear translocation may be a common trait among members of the RTK family. Although the proportion of nuclear localized RTK is relatively small compared with their cytoplasmic counterparts, studies have consistently demonstrated a positive correlation between nuclear RTK and poor prognosis/drug resistance in patients with cancer (Schillaci et al., 2012; Traynor et al., 2013; Yang, Lin, et al., 2019). Specifically, approximately 2–6% of EGFR translocates into the nucleus in response to ligand stimulation and 10–20% of MET accumulates in the nucleus under hydrogen peroxide treatment (Chen et al., 2019; Wang et al., 2006). As summarized in Table 1, most of the nuclear RTKs function as transcriptional co-regulators whereas some bind directly to the chromatin. Most nuclear RTKs regulate gene transcription by associating with transcription factors, such as STAT3 and STAT5. In response to exogenous cellular stress, nuclear RTKs also enhance DNA damage repair by phosphorylating the core proteins in the repair pathways. In total, the biological functions of nuclear RTK appears to primarily promote cancer progression and drug resistance.

Inhibition of RTK nuclear transport is an emerging strategy to overcome nuclear RTK-mediated resistance to therapeutic agents (Brand et al., 2013; Martins et al., 2011; Qi, Lang, Li, Feng, & Huang, 2017). Although most of the studies regarding the functions of nuclear RTKs utilized cancer cells as model system, there is evidence to support that some nuclear RTK signaling are important in normal physiological regulations, such as embryonic cell development, neuronal cell development, and differentiation (Lyu et al., 2008; Sardi et al., 2006; Tourette et al., 2014). Thus, more detailed comparison of nuclear RTK functions between normal and cancer cells may be critical to evaluating the benefits of therapeutic strategies targeting nuclear RTK accumulation.

Acknowledgments

The authors would like to acknowledge the following grant support: National Institutes of Health T32 Training Grant in Cancer Biology (5T32CA186892), Patel Memorial Breast Cancer Endowment Fund, and Ting Tsung and Wei Fong Chao Research Fund.

References

Abraham, Z., Hawley, E., Hayosh, D., Webster-Wood, V. A., & Akkus, O. (2018). Kinesin and dynein mechanics: Measurement methods and research applications. *Journal of Biomechanical Engineering*, *140*(2), 0208051–02080511. https://doi.org/10.1115/1.4037886.

Adilakshmi, T., Ness-Myers, J., Madrid-Aliste, C., Fiser, A., & Tapinos, N. (2011). A nuclear variant of ErbB3 receptor tyrosine kinase regulates ezrin distribution and Schwann cell myelination. *The Journal of Neuroscience*, *31*(13), 5106–5119. https://doi.org/10.1523/jneurosci.5635-10.2011.

Akinleye, A., Furqan, M., Mukhi, N., Ravella, P., & Liu, D. (2013). MEK and the inhibitors: From bench to bedside. *Journal of Hematology & Oncology*, *6*, 27. https://doi.org/10.1186/1756-8722-6-27.

Aleksic, T., Chitnis, M. M., Perestenko, O. V., Gao, S., Thomas, P. H., Turner, G. D., et al. (2010). Type 1 insulin-like growth factor receptor translocates to the nucleus of human tumor cells. *Cancer Research*, *70*(16), 6412–6419. https://doi.org/10.1158/0008-5472.CAN-10-0052.

Aleksic, T., Gray, N., Wu, X., Rieunier, G., Osher, E., Mills, J., et al. (2018). Nuclear IGF1R interacts with regulatory regions of chromatin to promote RNA polymerase II recruitment and gene expression associated with advanced tumor stage. *Cancer Research*, *78*(13), 3497–3509. https://doi.org/10.1158/0008-5472.CAN-17-3498.

Amaya, M. J., Oliveira, A. G., Guimaraes, E. S., Casteluber, M. C., Carvalho, S. M., Andrade, L. M., et al. (2014). The insulin receptor translocates to the nucleus to regulate cell proliferation in liver. *Hepatology*, *59*(1), 274–283. https://doi.org/10.1002/hep.26609.

Andrique, L., Fauvin, D., El Maassarani, M., Colasson, H., Vannier, B., & Seite, P. (2012). ErbB3(80 kDa), a nuclear variant of the ErbB3 receptor, binds to the Cyclin D1 promoter to activate cell proliferation but is negatively controlled by p14ARF. *Cellular Signalling*, *24*(5), 1074–1085. https://doi.org/10.1016/j.cellsig.2012.01.002.

Araki, T., Liu, X., Kameda, H., Tone, Y., Fukuoka, H., Tone, M., et al. (2017). EGFR induces E2F1-mediated corticotroph tumorigenesis. *Journal of the Endocrine Society*, *1*(2), 127–143. https://doi.org/10.1210/js.2016-1053.

Arasada, R. R., & Carpenter, G. (2005). Secretase-dependent tyrosine phosphorylation of Mdm2 by the ErbB-4 intracellular domain fragment. *The Journal of Biological Chemistry*, *280*(35), 30783–30787. https://doi.org/10.1074/jbc.M506057200.

Aslam, M. I., Hettmer, S., Abraham, J., Latocha, D., Soundararajan, A., Huang, E. T., et al. (2013). Dynamic and nuclear expression of PDGFRalpha and IGF-1R in alveolar Rhabdomyosarcoma. *Molecular Cancer Research*, *11*(11), 1303–1313. https://doi.org/10.1158/1541-7786.MCR-12-0598.

Auciello, G., Cunningham, D. L., Tatar, T., Heath, J. K., & Rappoport, J. Z. (2013). Regulation of fibroblast growth factor receptor signalling and trafficking by Src and Eps8. *Journal of Cell Science*, *126*(Pt. 2), 613–624. https://doi.org/10.1242/jcs.116228.

Avalle, L., Pensa, S., Regis, G., Novelli, F., & Poli, V. (2012). STAT1 and STAT3 in tumorigenesis: A matter of balance. *JAK-STAT*, *1*(2), 65–72. https://doi.org/10.4161/jkst.20045.

Bagheri-Yarmand, R., Sinha, K. M., Gururaj, A. E., Ahmed, Z., Rizvi, Y. Q., Huang, S. C., et al. (2015). A novel dual kinase function of the RET proto-oncogene negatively regulates activating transcription factor 4-mediated apoptosis. *The Journal of Biological Chemistry*, *290*(18), 11749–11761. https://doi.org/10.1074/jbc.M114.619833.

Balaji, K., Vijayaraghavan, S., Diao, L., Tong, P., Fan, Y., Carey, J. P., et al. (2017). AXL inhibition suppresses the DNA damage response and sensitizes cells to PARP inhibition in multiple cancers. *Molecular Cancer Research*, *15*(1), 45–58. https://doi.org/10.1158/1541-7786.MCR-16-0157.

Barbetti, V., Morandi, A., Tusa, I., Digiacomo, G., Riverso, M., Marzi, I., et al. (2014). Chromatin-associated CSF-1R binds to the promoter of proliferation-related genes in breast cancer cells. *Oncogene*, *33*(34), 4359–4364. https://doi.org/10.1038/onc.2013.542.

Bazzani, L., Donnini, S., Giachetti, A., Christofori, G., & Ziche, M. (2018). PGE2 mediates EGFR internalization and nuclear translocation via caveolin endocytosis promoting its transcriptional activity and proliferation in human NSCLC cells. *Oncotarget*, *9*(19), 14939–14958. https://doi.org/10.18632/oncotarget.24499.

Beguelin, W., Diaz Flaque, M. C., Proietti, C. J., Cayrol, F., Rivas, M. A., Tkach, M., et al. (2010). Progesterone receptor induces ErbB-2 nuclear translocation to promote breast cancer growth via a novel transcriptional effect: ErbB-2 function as a coactivator of Stat3. *Molecular and Cellular Biology*, *30*(23), 5456–5472. https://doi.org/10.1128/MCB.00012-10.

Bencheikh, L., Diop, M. K., Riviere, J., Imanci, A., Pierron, G., Souquere, S., et al. (2019). Dynamic gene regulation by nuclear colony-stimulating factor 1 receptor in human monocytes and macrophages. *Nature Communications*, *10*(1), 1935. https://doi.org/10.1038/s41467-019-09970-9.

Bertelsen, V., & Stang, E. (2014). The mysterious ways of ErbB2/HER2 trafficking. *Membranes (Basel, Switzerland)*, *4*(3), 424–446. https://doi.org/10.3390/membranes4030424.

Bhattacharya, P., Shetake, N. G., Pandey, B. N., & Kumar, A. (2018). Receptor tyrosine kinase signaling in cancer radiotherapy and its targeting for tumor radiosensitization. *International Journal of Radiation Biology*, *94*(7), 628–644. https://doi.org/10.1080/09553002.2018.1478160.

Björner, S., Rosendahl, A. H., Simonsson, M., Markkula, A., Jirström, K., Borgquist, S., et al. (2017). Nuclear expression of insulin receptor (InsR) adds prognostic information in primary breast cancer patients with high BMI [abstract]. In *Paper presented at the American Association for Cancer Research Annual Meeting 2017, Washington, DC. Philadelphia (PA)*.

Blazquez, C., Cook, N., Micklem, K., Harris, A. L., Gatter, K. C., & Pezzella, F. (2006). Phosphorylated KDR can be located in the nucleus of neoplastic cells. *Cell Research*, *16*(1), 93–98. https://doi.org/10.1038/sj.cr.7310012.

Bonacchi, A., Taddei, M. L., Petrai, I., Efsen, E., Defranco, R., Nosi, D., et al. (2008). Nuclear localization of TRK-A in liver cells. *Histology and Histopathology*, *23*(3), 327–340. https://doi.org/10.14670/HH-23.327.

Brand, T. M., Iida, M., Li, C., & Wheeler, D. L. (2011). The nuclear epidermal growth factor receptor signaling network and its role in cancer. *Discovery Medicine*, *12*(66), 419–432.

Brand, T. M., Iida, M., Luthar, N., Starr, M. M., Huppert, E. J., & Wheeler, D. L. (2013). Nuclear EGFR as a molecular target in cancer. *Radiotherapy and Oncology*, *108*(3), 370–377. https://doi.org/10.1016/j.radonc.2013.06.010.

Brand, T. M., Iida, M., Luthar, N., Wleklinski, M. J., Starr, M. M., & Wheeler, D. L. (2013). Mapping C-terminal transactivation domains of the nuclear HER family receptor tyrosine kinase HER3. *PLoS One*, *8*(8), e71518. https://doi.org/10.1371/journal.pone.0071518.

Bryant, D. M., & Stow, J. L. (2005). Nuclear translocation of cell-surface receptors: Lessons from fibroblast growth factor. *Traffic*, *6*(10), 947–954. https://doi.org/10.1111/j.1600-0854.2005.00332.x.

Bryant, D. M., Wylie, F. G., & Stow, J. L. (2005). Regulation of endocytosis, nuclear translocation, and signaling of fibroblast growth factor receptor 1 by E-cadherin. *Molecular Biology of the Cell*, *16*(1), 14–23. https://doi.org/10.1091/mbc.e04-09-0845.

Bueter, W., Dammann, O., Zscheppang, K., Korenbaum, E., & Dammann, C. E. (2006). ErbB receptors in fetal endothelium—A potential linkage point for inflammation-associated neonatal disorders. *Cytokine*, *36*(5–6), 267–275. https://doi.org/10.1016/j.cyto.2007.02.002.

Cai, J., Chen, Z., Ruan, Q., Han, S., Liu, L., Qi, X., et al. (2011). gamma-Secretase and presenilin mediate cleavage and phosphorylation of vascular endothelial growth factor receptor-1. *The Journal of Biological Chemistry*, *286*(49), 42514–42523. https://doi.org/10.1074/jbc.M111.296590.

Campion, C. G., Zaoui, K., Verissimo, T., Cossette, S., Matsuda, H., Solban, N., et al. (2018). COMMD5/HCaRG hooks endosomes on cytoskeleton and coordinates EGFR trafficking. *Cell Reports*, *24*(3), 670–684.e677. https://doi.org/10.1016/j.celrep.2018.06.056.

Carpenter, G., & Pozzi, A. (2012). Cell responses to growth factors: The role of receptor tyrosine kinase intracellular domain fragments. *Science Signaling*, *5*(243), pe42. https://doi.org/10.1126/scisignal.2003526.

Casaletto, J. B., & McClatchey, A. I. (2012). Spatial regulation of receptor tyrosine kinases in development and cancer. *Nature Reviews. Cancer*, *12*(6), 387–400. https://doi.org/10.1038/nrc3277.

Cerliani, J. P., Guillardoy, T., Giulianelli, S., Vaque, J. P., Gutkind, J. S., Vanzulli, S. I., et al. (2011). Interaction between FGFR-2, STAT5, and progesterone receptors in breast cancer. *Cancer Research*, *71*(10), 3720–3731. https://doi.org/10.1158/0008-5472.CAN-10-3074.

Chandarlapaty, S., Scaltriti, M., Angelini, P., Ye, Q., Guzman, M., Hudis, C. A., et al. (2010). Inhibitors of HSP90 block p95-HER2 signaling in Trastuzumab-resistant tumors and suppress their growth. *Oncogene*, *29*(3), 325–334. https://doi.org/10.1038/onc.2009.337.

Chang, H. Y., Chang, T. C., Huang, W. Y., Lee, C. T., Yen, C. J., Tsai, Y. S., et al. (2016). RON nuclear translocation under hypoxia potentiates chemoresistance to DNA double-strand break-inducing anticancer drugs. *Molecular Cancer Therapeutics*, *15*(2), 276–286. https://doi.org/10.1158/1535-7163.MCT-15-0311.

Chang, W. H., Choi, S. H., Moon, B. S., Cai, M., Lyu, J., Bai, J., et al. (2017). Smek1/2 is a nuclear chaperone and cofactor for cleaved Wnt receptor Ryk, regulating cortical neurogenesis. *Proceedings of the National Academy of Sciences of the United States of America*, *114*(50), E10717–E10725. https://doi.org/10.1073/pnas.1715772114.

Chang, H. Y., Liu, H. S., Lai, M. D., Tsai, Y. S., Tzai, T. S., Cheng, H. L., et al. (2014). Hypoxia promotes nuclear translocation and transcriptional function in the oncogenic tyrosine kinase RON. *Cancer Research*, *74*(16), 4549–4562. https://doi.org/10.1158/0008-5472.CAN-13-3730.

Chaudhary, S. C., Cho, M. G., Nguyen, T. T., Park, K. S., Kwon, M. H., & Lee, J. H. (2014). A putative pH-dependent nuclear localization signal in the juxtamembrane region of c-Met. *Experimental & Molecular Medicine*, *46*, e119. https://doi.org/10.1038/emm.2014.67.

Chen, L., Acciani, T., Le Cras, T., Lutzko, C., & Perl, A. K. (2012). Dynamic regulation of platelet-derived growth factor receptor alpha expression in alveolar fibroblasts during realveolarization. *American Journal of Respiratory Cell and Molecular Biology*, *47*(4), 517–527. https://doi.org/10.1165/rcmb.2012-0030OC.

Chen, M. K., Du, Y., Sun, L., Hsu, J. L., Wang, Y. H., Gao, Y., et al. (2019). H_2O_2 induces nuclear transport of the receptor tyrosine kinase c-MET in breast cancer cells via a membrane-bound retrograde trafficking mechanism. *The Journal of Biological Chemistry*, *294*(21), 8516–8528. https://doi.org/10.1074/jbc.RA118.005953.

Chen, M. K., & Hung, M. C. (2015). Proteolytic cleavage, trafficking, and functions of nuclear receptor tyrosine kinases. *The FEBS Journal*, *282*(19), 3693–3721. https://doi.org/10.1111/febs.13342.

Chen, M. K., & Hung, M. C. (2016). Regulation of therapeutic resistance in cancers by receptor tyrosine kinases. *American Journal of Cancer Research*, *6*(4), 827–842.

Cheng, Y., Sun, Y., Wang, L. Z., Yu, Y. C., & Ding, X. (2015). Cytoplasmic c-ros oncogene 1 receptor tyrosine kinase expression may be associated with the development of human oral squamous cell carcinoma. *Oncology Letters*, *10*(2), 934–940. https://doi.org/10.3892/ol.2015.3340.

Chioni, A. M., & Grose, R. (2012). FGFR1 cleavage and nuclear translocation regulates breast cancer cell behavior. *The Journal of Cell Biology*, *197*(6), 801–817. https://doi.org/10.1083/jcb.201108077.

Chiusa, M., Hu, W., Liao, H. J., Su, Y., Borza, C. M., de Caestecker, M. P., et al. (2019). The extracellular matrix receptor discoidin domain receptor 1 regulates collagen transcription by translocating to the nucleus. *Journal of the American Society of Nephrology, 30*(9), 1605–1624. https://doi.org/10.1681/ASN.2018111160.

Chou, R. H., Wang, Y. N., Hsieh, Y. H., Li, L. Y., Xia, W., Chang, W. C., et al. (2014). EGFR modulates DNA synthesis and repair through Tyr phosphorylation of histone H4. *Developmental Cell, 30*(2), 224–237. https://doi.org/10.1016/j.devcel.2014.06.008.

Codony-Servat, J., Cuatrecasas, M., Asensio, E., Montironi, C., Martinez-Cardus, A., Marin-Aguilera, M., et al. (2017). Nuclear IGF-1R predicts chemotherapy and targeted therapy resistance in metastatic colorectal cancer. *British Journal of Cancer, 117*(12), 1777–1786. https://doi.org/10.1038/bjc.2017.279.

Coleman, S. J., Chioni, A. M., Ghallab, M., Anderson, R. K., Lemoine, N. R., Kocher, H. M., et al. (2014). Nuclear translocation of FGFR1 and FGF2 in pancreatic stellate cells facilitates pancreatic cancer cell invasion. *EMBO Molecular Medicine, 6*(4), 467–481. https://doi.org/10.1002/emmm.201302698.

Cortes-Reynosa, P., Robledo, T., & Salazar, E. P. (2009). Epidermal growth factor promotes epidermal growth factor receptor nuclear accumulation by a pathway dependent on cytoskeleton integrity in human breast cancer cells. *Archives of Medical Research, 40*(5), 331–338. https://doi.org/10.1016/j.arcmed.2009.06.007.

Dantas-Barbosa, C., Lesluyes, T., Loarer, F. L., Chibon, F., Treilleux, I., Coindre, J. M., et al. (2017). Expression and role of TYRO3 and AXL as potential therapeutical targets in leiomyosarcoma. *British Journal of Cancer, 117*(12), 1787–1797. https://doi.org/10.1038/bjc.2017.354.

Davidson, B., Reich, R., Lazarovici, P., Nesland, J. M., Skrede, M., Risberg, B., et al. (2003). Expression and activation of the nerve growth factor receptor TrkA in serous ovarian carcinoma. *Clinical Cancer Research, 9*(6), 2248–2259.

De Angelis Campos, A. C., Rodrigues, M. A., de Andrade, C., de Goes, A. M., Nathanson, M. H., & Gomes, D. A. (2011). Epidermal growth factor receptors destined for the nucleus are internalized via a clathrin-dependent pathway. *Biochemical and Biophysical Research Communications, 412*(2), 341–346. https://doi.org/10.1016/j.bbrc.2011.07.100.

Diaz Flaque, M. C., Vicario, R., Proietti, C. J., Izzo, F., Schillaci, R., & Elizalde, P. V. (2013). Progestin drives breast cancer growth by inducing p21(CIP1) expression through the assembly of a transcriptional complex among Stat3, progesterone receptor and ErbB-2. *Steroids, 78*(6), 559–567. https://doi.org/10.1016/j.steroids.2012.11.003.

Dittmann, K., Mayer, C., Fehrenbacher, B., Schaller, M., Kehlbach, R., & Rodemann, H. P. (2011). Nuclear epidermal growth factor receptor modulates cellular radio-sensitivity by regulation of chromatin access. *Radiotherapy and Oncology, 99*(3), 317–322. https://doi.org/10.1016/j.radonc.2011.06.001.

Dittmann, K., Mayer, C., Fehrenbacher, B., Schaller, M., Raju, U., Milas, L., et al. (2005). Radiation-induced epidermal growth factor receptor nuclear import is linked to activation of DNA-dependent protein kinase. *The Journal of Biological Chemistry, 280*(35), 31182–31189. https://doi.org/10.1074/jbc.M506591200.

Domingues, I., Rino, J., Demmers, J. A., de Lanerolle, P., & Santos, S. C. (2011). VEGFR2 translocates to the nucleus to regulate its own transcription. *PLoS One, 6*(9). e25668https://doi.org/10.1371/journal.pone.0025668.

Dong, Q., Du, Y., Li, H., Liu, C., Wei, Y., Chen, M. K., et al. (2019). EGFR and c-MET cooperate to enhance resistance to PARP inhibitors in hepatocellular carcinoma. *Cancer Research, 79*(4), 819–829. https://doi.org/10.1158/0008-5472.CAN-18-1273.

Du, Z., & Lovly, C. M. (2018). Mechanisms of receptor tyrosine kinase activation in cancer. *Molecular Cancer, 17*(1), 58. https://doi.org/10.1186/s12943-018-0782-4.

Du, Y., Shen, J., Hsu, J. L., Han, Z., Hsu, M. C., Yang, C. C., et al. (2014). Syntaxin 6-mediated Golgi translocation plays an important role in nuclear functions of EGFR through microtubule-dependent trafficking. *Oncogene*, *33*(6), 756–770. https://doi.org/10.1038/onc.2013.1.

Du, Y., Yamaguchi, H., Wei, Y., Hsu, J. L., Wang, H. L., Hsu, Y. H., et al. (2016). Blocking c-Met-mediated PARP1 phosphorylation enhances anti-tumor effects of PARP inhibitors. *Nature Medicine*, *22*(2), 194–201. https://doi.org/10.1038/nm.4032.

Dunham-Ems, S. M., Lee, Y. W., Stachowiak, E. K., Pudavar, H., Claus, P., Prasad, P. N., et al. (2009). Fibroblast growth factor receptor-1 (FGFR1) nuclear dynamics reveal a novel mechanism in transcription control. *Molecular Biology of the Cell*, *20*(9), 2401–2412. https://doi.org/10.1091/mbc.E08-06-0600.

Dunham-Ems, S. M., Pudavar, H. E., Myers, J. M., Maher, P. A., Prasad, P. N., & Stachowiak, M. K. (2006). Factors controlling fibroblast growth factor receptor-1's cytoplasmic trafficking and its regulation as revealed by FRAP analysis. *Molecular Biology of the Cell*, *17*(5), 2223–2235. https://doi.org/10.1091/mbc.e05-08-0749.

El-Deeb, I. M., Yoo, K. H., & Lee, S. H. (2011). ROS receptor tyrosine kinase: A new potential target for anticancer drugs. *Medicinal Research Reviews*, *31*(5), 794–818. https://doi.org/10.1002/med.20206.

Elizalde, P. V., Cordo Russo, R. I., Chervo, M. F., & Schillaci, R. (2016). ErbB-2 nuclear function in breast cancer growth, metastasis and resistance to therapy. *Endocrine-Related Cancer*, *23*(12), T243–T257. https://doi.org/10.1530/ERC-16-0360.

Fan, Q. W., Cheng, C. K., Gustafson, W. C., Charron, E., Zipper, P., Wong, R. A., et al. (2013). EGFR phosphorylates tumor-derived EGFRvIII driving STAT3/5 and progression in glioblastoma. *Cancer Cell*, *24*(4), 438–449. https://doi.org/10.1016/j.ccr.2013.09.004.

Fang, X., Stachowiak, E. K., Dunham-Ems, S. M., Klejbor, I., & Stachowiak, M. K. (2005). Control of CREB-binding protein signaling by nuclear fibroblast growth factor receptor-1: A novel mechanism of gene regulation. *The Journal of Biological Chemistry*, *280*(31), 28451–28462. https://doi.org/10.1074/jbc.M504400200.

Faria, J., de Andrade, C., Goes, A. M., Rodrigues, M. A., & Gomes, D. A. (2016). Effects of different ligands on epidermal growth factor receptor (EGFR) nuclear translocation. *Biochemical and Biophysical Research Communications*, *478*(1), 39–45. https://doi.org/10.1016/j.bbrc.2016.07.097.

Farkkila, A., Pihlajoki, M., Tauriala, H., Butzow, R., Leminen, A., Unkila-Kallio, L., et al. (2011). Serum vascular endothelial growth factor A (VEGF) is elevated in patients with ovarian granulosa cell tumor (GCT), and VEGF inhibition by bevacizumab induces apoptosis in GCT in vitro. *The Journal of Clinical Endocrinology and Metabolism*, *96*(12), E1973–E1981. https://doi.org/10.1210/jc.2011-1812.

Feng, L., Liao, W. X., Luo, Q., Zhang, H. H., Wang, W., Zheng, J., et al. (2012). Caveolin-1 orchestrates fibroblast growth factor 2 signaling control of angiogenesis in placental artery endothelial cell caveolae. *Journal of Cellular Physiology*, *227*(6), 2480–2491. https://doi.org/10.1002/jcp.22984.

Feng, Y., Venema, V. J., Venema, R. C., Tsai, N., & Caldwell, R. B. (1999). VEGF induces nuclear translocation of Flk-1/KDR, endothelial nitric oxide synthase, and caveolin-1 in vascular endothelial cells. *Biochemical and Biophysical Research Communications*, *256*(1), 192–197. https://doi.org/10.1006/bbrc.1998.9790.

Florenes, V. A., Maelandsmo, G. M., Holm, R., Reich, R., Lazarovici, P., & Davidson, B. (2004). Expression of activated TrkA protein in melanocytic tumors: Relationship to cell proliferation and clinical outcome. *American Journal of Clinical Pathology*, *122*(3), 412–420. https://doi.org/10.1309/CHFH-EYAT-44WW-P7J3.

Fonseca-Gomes, J., Jeronimo-Santos, A., Lesnikova, A., Casarotto, P., Castren, E., Sebastiao, A. M., et al. (2019). TrkB-ICD fragment, originating from BDNF receptor cleavage, is translocated to cell nucleus and phosphorylates nuclear and axonal proteins. *Frontiers in Molecular Neuroscience*, *12*, 4. https://doi.org/10.3389/fnmol.2019.00004.

Formisano, L., Stauffer, K. M., Young, C. D., Bhola, N. E., Guerrero-Zotano, A. L., Jansen, V. M., et al. (2017). Association of FGFR1 with ERalpha maintains ligand-independent ER transcription and mediates resistance to estrogen deprivation in ER(+) breast cancer. *Clinical Cancer Research*, *23*(20), 6138–6150. https://doi.org/10.1158/1078-0432.CCR-17-1232.

Fox, S. B., Turley, H., Cheale, M., Blazquez, C., Roberts, H., James, N., et al. (2004). Phosphorylated KDR is expressed in the neoplastic and stromal elements of human renal tumours and shuttles from cell membrane to nucleus. *The Journal of Pathology*, *202*(3), 313–320. https://doi.org/10.1002/path.1520.

Friedmann, B. J., Caplin, M., Savic, B., Shah, T., Lord, C. J., Ashworth, A., et al. (2006). Interaction of the epidermal growth factor receptor and the DNA-dependent protein kinase pathway following gefitinib treatment. *Molecular Cancer Therapeutics*, *5*(2), 209–218. https://doi.org/10.1158/1535-7163.MCT-05-0239.

Goh, L. K., & Sorkin, A. (2013). Endocytosis of receptor tyrosine kinases. *Cold Spring Harbor Perspectives in Biology*, *5*(5), a017459. https://doi.org/10.1101/cshperspect.a017459.

Gomes, D. A., Rodrigues, M. A., Leite, M. F., Gomez, M. V., Varnai, P., Balla, T., et al. (2008). c-Met must translocate to the nucleus to initiate calcium signals. *The Journal of Biological Chemistry*, *283*(7), 4344–4351. https://doi.org/10.1074/jbc.M706550200.

Gong, A., Zhang, Z., Xiao, D., Yang, Y., Wang, Y., & Chen, Y. (2007). Localization of phosphorylated TrkA in carrier vesicles involved in its nuclear translocation in U251 cell line. *Science in China. Series C, Life Sciences*, *50*(2), 141–146. https://doi.org/10.1007/s11427-007-0022-4.

Gonzalez, M. E., Martin, E. E., Anwar, T., Arellano-Garcia, C., Medhora, N., Lama, A., et al. (2017). Mesenchymal stem cell-induced DDR2 mediates stromal-breast cancer interactions and metastasis growth. *Cell Reports*, *18*(5), 1215–1228. https://doi.org/10.1016/j.celrep.2016.12.079.

Hadzisejdic, I., Mustac, E., Jonjic, N., Petkovic, M., & Grahovac, B. (2010). Nuclear EGFR in ductal invasive breast cancer: Correlation with cyclin-D1 and prognosis. *Modern Pathology*, *23*(3), 392–403. https://doi.org/10.1038/modpathol.2009.166.

Han, W., Carpenter, R. L., Cao, X., & Lo, H. W. (2013). STAT1 gene expression is enhanced by nuclear EGFR and HER2 via cooperation with STAT3. *Molecular Carcinogenesis*, *52*(12), 959–969. https://doi.org/10.1002/mc.21936.

Hanada, N., Lo, H. W., Day, C. P., Pan, Y., Nakajima, Y., & Hung, M. C. (2006). Co-regulation of B-Myb expression by E2F1 and EGF receptor. *Molecular Carcinogenesis*, *45*(1), 10–17. https://doi.org/10.1002/mc.20147.

Hancock, M. L., Meyer, R. C., Mistry, M., Khetani, R. S., Wagschal, A., Shin, T., et al. (2019). Insulin receptor associates with promoters genome-wide and regulates gene expression. *Cell*, *177*(3), 722–736. e722. https://doi.org/10.1016/j.cell.2019.02.030.

Haskins, J. W., Nguyen, D. X., & Stern, D. F. (2014). Neuregulin 1-activated ERBB4 interacts with YAP to induce Hippo pathway target genes and promote cell migration. *Science Signaling*, 7(355), ra116. https://doi.org/10.1126/scisignal.2005770.

Haugsten, E. M., Sorensen, V., Brech, A., Olsnes, S., & Wesche, J. (2005). Different intracellular trafficking of FGF1 endocytosed by the four homologous FGF receptors. *Journal of Cell Science*, *118*(Pt. 17), 3869–3881. https://doi.org/10.1242/jcs.02509.

Haugsten, E. M., Sorensen, V., Kunova Bosakova, M., de Souza, G. A., Krejci, P., Wiedlocha, A., et al. (2016). Proximity labeling reveals molecular determinants of FGFR4 endosomal transport. *Journal of Proteome Research*, *15*(10), 3841–3855. https://doi.org/10.1021/acs.jproteome.6b00652.

Haugsten, E. M., Zakrzewska, M., Brech, A., Pust, S., Olsnes, S., Sandvig, K., et al. (2011). Clathrin- and dynamin-independent endocytosis of FGFR3–implications for signalling. *PLoS One*, *6*(7), e21708. https://doi.org/10.1371/journal.pone.0021708.

Hoeing, K., Zscheppang, K., Mujahid, S., Murray, S., Volpe, M. V., Dammann, C. E., et al. (2011). Presenilin-1 processing of ErbB4 in fetal type II cells is necessary for control of fetal lung maturation. *Biochimica et Biophysica Acta*, *1813*(3), 480–491. https://doi.org/10.1016/j.bbamcr.2010.12.017.

Hoshino, M., Fukui, H., Ono, Y., Sekikawa, A., Ichikawa, K., Tomita, S., et al. (2007). Nuclear expression of phosphorylated EGFR is associated with poor prognosis of patients with esophageal squamous cell carcinoma. *Pathobiology*, *74*(1), 15–21. https://doi.org/10.1159/000101047.

Hossain, M. B., Shifat, R., Johnson, D. G., Bedford, M. T., Gabrusiewicz, K. R., Cortes-Santiago, N., et al. (2016). TIE2-mediated tyrosine phosphorylation of H4 regulates DNA damage response by recruiting ABL1. *Science Advances*, *2*(4). e1501290. https://doi.org/10.1126/sciadv.1501290.

Hsu, S. C., & Hung, M. C. (2007). Characterization of a novel tripartite nuclear localization sequence in the EGFR family. *The Journal of Biological Chemistry*, *282*(14), 10432–10440. https://doi.org/10.1074/jbc.M610014200.

Hsueh, Y. S., Chang, H. H., Shan, Y. S., Sun, H. S., Fletcher, J. A., Li, C. F., et al. (2019). Nuclear KIT induces a NFKBIB-RELA-KIT autoregulatory loop in imatinib-resistant gastrointestinal stromal tumors. *Oncogene*, *38*(38), 6550–6565. https://doi.org/10.1038/s41388-019-0900-9.

Hu, Y., Fang, X., Dunham, S. M., Prada, C., Stachowiak, E. K., & Stachowiak, M. K. (2004). 90-kDa ribosomal S6 kinase is a direct target for the nuclear fibroblast growth factor receptor 1 (FGFR1): Role in FGFR1 signaling. *The Journal of Biological Chemistry*, *279*(28), 29325–29335. https://doi.org/10.1074/jbc.M311144200.

Huang, W. C., Chen, Y. J., Li, L. Y., Wei, Y. L., Hsu, S. C., Tsai, S. L., et al. (2011). Nuclear translocation of epidermal growth factor receptor by Akt-dependent phosphorylation enhances breast cancer-resistant protein expression in gefitinib-resistant cells. *The Journal of Biological Chemistry*, *286*(23), 20558–20568. https://doi.org/10.1074/jbc.M111.240796.

Hung, L. Y., Tseng, J. T., Lee, Y. C., Xia, W., Wang, Y. N., Wu, M. L., et al. (2008). Nuclear epidermal growth factor receptor (EGFR) interacts with signal transducer and activator of transcription 5 (STAT5) in activating Aurora-A gene expression. *Nucleic Acids Research*, *36*(13), 4337–4351. https://doi.org/10.1093/nar/gkn417.

Huo, L., Wang, Y. N., Xia, W., Hsu, S. C., Lai, C. C., Li, L. Y., et al. (2010). RNA helicase A is a DNA-binding partner for EGFR-mediated transcriptional activation in the nucleus. *Proceedings of the National Academy of Sciences of the United States of America*, *107*(37), 16125–16130. https://doi.org/10.1073/pnas.1000743107.

Hwang, G., Kim, H., Yoon, H., Song, C., Lim, D. K., Sim, T., et al. (2017). In situ imaging of quantum dot-AZD4547 conjugates for tracking the dynamic behavior of fibroblast growth factor receptor 3. *International Journal of Nanomedicine*, *12*, 5345–5357. https://doi.org/10.2147/IJN.S141595.

Jean, S., Mikryukov, A., Tremblay, M. G., Baril, J., Guillou, F., Bellenfant, S., et al. (2010). Extended-synaptotagmin-2 mediates FGF receptor endocytosis and ERK activation in vivo. *Developmental Cell*, *19*(3), 426–439. https://doi.org/10.1016/j.devcel.2010.08.007.

Jeronimo-Santos, A., Vaz, S. H., Parreira, S., Rapaz-Lerias, S., Caetano, A. P., Buee-Scherrer, V., et al. (2015). Dysregulation of TrkB receptors and BDNF function by amyloid-beta peptide is mediated by Calpain. *Cerebral Cortex*, *25*(9), 3107–3121. https://doi.org/10.1093/cercor/bhu105.

Johnson, C., Segovia, B., & Kandpal, R. P. (2016). EPHA7 and EPHA10 physically interact and differentially co-localize in normal breast and breast carcinoma cell lines, and the co-localization pattern is altered in EPHB6-expressing MDA-MB-231 cells. *Cancer Genomics Proteomics*, *13*(5), 359–368.

Johnston, C. L., Cox, H. C., Gomm, J. J., & Coombes, R. C. (1995). Fibroblast growth factor receptors (FGFRs) localize in different cellular compartments. A splice variant of FGFR-3 localizes to the nucleus. *The Journal of Biological Chemistry*, *270*(51), 30643–30650. https://doi.org/10.1074/jbc.270.51.30643.

Kamio, T., Shigematsu, K., Sou, H., Kawai, K., & Tsuchiyama, H. (1990). Immunohistochemical expression of epidermal growth factor receptors in human adrenocortical carcinoma. *Human Pathology*, *21*(3), 277–282. https://doi.org/10.1016/0046-8177(90)90227-v.

Karoulia, Z., Gavathiotis, E., & Poulikakos, P. I. (2017). New perspectives for targeting RAF kinase in human cancer. *Nature Reviews. Cancer*, *17*(11), 676–691. https://doi.org/10.1038/nrc.2017.79.

Kesten, D., Horovitz-Fried, M., Brutman-Barazani, T., & Sampson, S. R. (2018). Insulin-induced translocation of IR to the nucleus in insulin responsive cells requires a nuclear translocation sequence. *Biochimica et Biophysica Acta, Molecular Cell Research*, *1865*(4), 551–559. https://doi.org/10.1016/j.bbamcr.2018.01.004.

Kim, H. P., Yoon, Y. K., Kim, J. W., Han, S. W., Hur, H. S., Park, J., et al. (2009). Lapatinib, a dual EGFR and HER2 tyrosine kinase inhibitor, downregulates thymidylate synthase by inhibiting the nuclear translocation of EGFR and HER2. *PLoS One*, *4*(6). e5933https://doi.org/10.1371/journal.pone.0005933.

Knittle, A. M., Helkkula, M., Johnson, M. S., Sundvall, M., & Elenius, K. (2017). SUMOylation regulates nuclear accumulation and signaling activity of the soluble intracellular domain of the ErbB4 receptor tyrosine kinase. *The Journal of Biological Chemistry*, *292*(48), 19890–19904. https://doi.org/10.1074/jbc.M117.794271.

Komuro, A., Nagai, M., Navin, N. E., & Sudol, M. (2003). WW domain-containing protein YAP associates with ErbB-4 and acts as a co-transcriptional activator for the carboxyl-terminal fragment of ErbB-4 that translocates to the nucleus. *The Journal of Biological Chemistry*, *278*(35), 33334–33341. https://doi.org/10.1074/jbc.M305597200.

Koumakpayi, I. H., Diallo, J. S., Le Page, C., Lessard, L., Gleave, M., Begin, L. R., et al. (2006). Expression and nuclear localization of ErbB3 in prostate cancer. *Clinical Cancer Research*, *12*(9), 2730–2737. https://doi.org/10.1158/1078-0432.CCR-05-2242.

Kuo, H. Y., Huang, Y. S., Tseng, C. H., Chen, Y. C., Chang, Y. W., Shih, H. M., et al. (2014). PML represses lung cancer metastasis by suppressing the nuclear EGFR-mediated transcriptional activation of MMP2. *Cell Cycle*, *13*(19), 3132–3142. https://doi.org/10.4161/15384101.2014.949212.

Kuroda, C., Kubota, S., Kawata, K., Aoyama, E., Sumiyoshi, K., Oka, M., et al. (2008). Distribution, gene expression, and functional role of EphA4 during ossification. *Biochemical and Biophysical Research Communications*, *374*(1), 22–27. https://doi.org/10.1016/j.bbrc.2008.06.089.

LaCombe, R. M. (2019). *Atypical localization of Eph-RTKs in childhood cancer rhabdomyosarcoma*. [Doctor of Philosophy Dissertation] University of Missouri. 2019 MU dissertations. Retrieved from https://hdl.handle.net/10355/69925.

Latha, K., Li, M., Chumbalkar, V., Gururaj, A., Hwang, Y., Dakeng, S., et al. (2013). Nuclear EGFRvIII-STAT5b complex contributes to glioblastoma cell survival by direct activation of the Bcl-XL promoter. *International Journal of Cancer*, *132*(3), 509–520. https://doi.org/10.1002/ijc.27690.

Laufman, O., Hong, W., & Lev, S. (2011). The COG complex interacts directly with syntaxin 6 and positively regulates endosome-to-TGN retrograde transport. *The Journal of Cell Biology*, *194*(3), 459–472. https://doi.org/10.1083/jcb.201102045.

Lee, H.-J., Lan, L., Peng, G., Chang, W.-C., Hsu, M.-C., Wang, Y.-N., et al. (2015). Tyrosine 370 phosphorylation of ATM positively regulates DNA damage response. *Cell Research*, *25*, 225. https://doi.org/10.1038/cr.2015.8.

Lee, T. H., Seng, S., Sekine, M., Hinton, C., Fu, Y., Avraham, H. K., et al. (2007). Vascular endothelial growth factor mediates intracrine survival in human breast carcinoma cells through internally expressed VEGFR1/FLT1. *PLoS Medicine*, *4*(6), e186. https://doi.org/10.1371/journal.pmed.0040186.

Lee, J. E., Shin, S. H., Shin, H. W., Chun, Y. S., & Park, J. W. (2019). Nuclear FGFR2 negatively regulates hypoxia-induced cell invasion in prostate cancer by interacting with HIF-1 and HIF-2. *Scientific Reports*, *9*(1), 3480. https://doi.org/10.1038/s41598-019-39843-6.

Lee, O. H., Xu, J., Fueyo, J., Fuller, G. N., Aldape, K. D., Alonso, M. M., et al. (2006). Expression of the receptor tyrosine kinase Tie2 in neoplastic glial cells is associated with integrin beta1-dependent adhesion to the extracellular matrix. *Molecular Cancer Research*, *4*(12), 915–926. https://doi.org/10.1158/1541-7786.MCR-06-0184.

Lemmon, M. A., & Schlessinger, J. (2010). Cell signaling by receptor tyrosine kinases. *Cell*, *141*(7), 1117–1134. https://doi.org/10.1016/j.cell.2010.06.011.

Li, L. Y., Chen, H., Hsieh, Y. H., Wang, Y. N., Chu, H. J., Chen, Y. H., et al. (2011). Nuclear ErbB2 enhances translation and cell growth by activating transcription of ribosomal RNA genes. *Cancer Research*, *71*(12), 4269–4279. https://doi.org/10.1158/0008-5472.CAN-10-3504.

Li, X., Kuang, J., Shen, Y., Majer, M. M., Nelson, C. C., Parsawar, K., et al. (2012). The atypical histone macroH2A1.2 interacts with HER-2 protein in cancer cells. *The Journal of Biological Chemistry*, *287*(27), 23171–23183. https://doi.org/10.1074/jbc.M112.379412.

Li, M., Seki, Y., Freitas, P. H., Nagata, M., Kojima, T., Sultana, S., et al. (2010). FGFR3 down-regulates PTH/PTHrP receptor gene expression by mediating JAK/STAT signaling in chondrocytic cell line. *Journal of Electron Microscopy*, *59*(3), 227–236. https://doi.org/10.1093/jmicro/dfq002.

Liao, H. J., & Carpenter, G. (2007). Role of the Sec61 translocon in EGF receptor trafficking to the nucleus and gene expression. *Molecular Biology of the Cell*, *18*(3), 1064–1072. https://doi.org/10.1091/mbc.e06-09-0802.

Liccardi, G., Hartley, J. A., & Hochhauser, D. (2011). EGFR nuclear translocation modulates DNA repair following cisplatin and ionizing radiation treatment. *Cancer Research*, *71*(3), 1103–1114. https://doi.org/10.1158/0008-5472.CAN-10-2384.

Lin, S. Y., Makino, K., Xia, W., Matin, A., Wen, Y., Kwong, K. Y., et al. (2001). Nuclear localization of EGF receptor and its potential new role as a transcription factor. *Nature Cell Biology*, *3*(9), 802–808. https://doi.org/10.1038/ncb0901-802.

Linggi, B., & Carpenter, G. (2006). ErbB-4 s80 intracellular domain abrogates ETO2-dependent transcriptional repression. *The Journal of Biological Chemistry*, *281*(35), 25373–25380. https://doi.org/10.1074/jbc.M603998200.

Liu, H. S., Hsu, P. Y., Lai, M. D., Chang, H. Y., Ho, C. L., Cheng, H. L., et al. (2010). An unusual function of RON receptor tyrosine kinase as a transcriptional regulator in cooperation with EGFR in human cancer cells. *Carcinogenesis*, *31*(8), 1456–1464. https://doi.org/10.1093/carcin/bgq100.

Lo, H. W., Ali-Seyed, M., Wu, Y., Bartholomeusz, G., Hsu, S. C., & Hung, M. C. (2006). Nuclear-cytoplasmic transport of EGFR involves receptor endocytosis, importin beta1 and CRM1. *Journal of Cellular Biochemistry*, *98*(6), 1570–1583. https://doi.org/10.1002/jcb.20876.

Lo, H. W., Cao, X., Zhu, H., & Ali-Osman, F. (2010). Cyclooxygenase-2 is a novel transcriptional target of the nuclear EGFR-STAT3 and EGFRvIII-STAT3 signaling axes. *Molecular Cancer Research*, *8*(2), 232–245. https://doi.org/10.1158/1541-7786.MCR-09-0391.

Lo, H. W., Hsu, S. C., Ali-Seyed, M., Gunduz, M., Xia, W., Wei, Y., et al. (2005). Nuclear interaction of EGFR and STAT3 in the activation of the iNOS/NO pathway. *Cancer Cell*, 7(6), 575–589. https://doi.org/10.1016/j.ccr.2005.05.007.

Lo, H. W., Xia, W., Wei, Y., Ali-Seyed, M., Huang, S. F., & Hung, M. C. (2005). Novel prognostic value of nuclear epidermal growth factor receptor in breast cancer. *Cancer Research*, *65*(1), 338–348.

Long, W., Wagner, K. U., Lloyd, K. C., Binart, N., Shillingford, J. M., Hennighausen, L., et al. (2003). Impaired differentiation and lactational failure of Erbb4-deficient mammary glands identify ERBB4 as an obligate mediator of STAT5. *Development*, *130*(21), 5257–5268. https://doi.org/10.1242/dev.00715.

Lowe, A. R., Tang, J. H., Yassif, J., Graf, M., Huang, W. Y., Groves, J. T., et al. (2015). Importin-beta modulates the permeability of the nuclear pore complex in a Ran-dependent manner. *eLife 4*, e04052. https://doi.org/10.7554/eLife.04052.

Lu, Y., Wan, J., Yang, Z., Lei, X., Niu, Q., Jiang, L., et al. (2017). Regulated intramembrane proteolysis of the AXL receptor kinase generates an intracellular domain that localizes in the nucleus of cancer cells. *The FASEB Journal*, *31*(4), 1382–1397. https://doi.org/10.1096/fj.201600702R.

Luberg, K., Park, R., Aleksejeva, E., & Timmusk, T. (2015). Novel transcripts reveal a complex structure of the human TRKA gene and imply the presence of multiple protein isoforms. *BMC Neuroscience*, *16*(1), 78. https://doi.org/10.1186/s12868-015-0215-x.

Lyu, J., Wesselschmidt, R. L., & Lu, W. (2009). Cdc37 regulates Ryk signaling by stabilizing the cleaved Ryk intracellular domain. *The Journal of Biological Chemistry*, *284*(19), 12940–12948. https://doi.org/10.1074/jbc.M900207200.

Lyu, J., Yamamoto, V., & Lu, W. (2008). Cleavage of the Wnt receptor Ryk regulates neuronal differentiation during cortical neurogenesis. *Developmental Cell*, *15*(5), 773–780. https://doi.org/10.1016/j.devcel.2008.10.004.

Martin, V., Xu, J., Pabbisetty, S. K., Alonso, M. M., Liu, D., Lee, O. H., et al. (2009). Tie2-mediated multidrug resistance in malignant gliomas is associated with upregulation of ABC transporters. *Oncogene*, *28*(24), 2358–2363. https://doi.org/10.1038/onc.2009.103.

Martins, A. S., Ordonez, J. L., Amaral, A. T., Prins, F., Floris, G., Debiec-Rychter, M., et al. (2011). IGF1R signaling in Ewing sarcoma is shaped by clathrin −/caveolin-dependent endocytosis. *PLoS One*, *6*(5), e19846. https://doi.org/10.1371/journal.pone.0019846.

Matsuda, R., Takahashi, T., Nakamura, S., Sekido, Y., Nishida, K., Seto, M., et al. (1993). Expression of the c-kit protein in human solid tumors and in corresponding fetal and adult normal tissues. *The American Journal of Pathology*, *142*(1), 339–346.

Matsumoto, K., Umitsu, M., De Silva, D. M., Roy, A., & Bottaro, D. P. (2017). Hepatocyte growth factor/MET in cancer progression and biomarker discovery. *Cancer Science*, *108*(3), 296–307. https://doi.org/10.1111/cas.13156.

Matteucci, E., Bendinelli, P., & Desiderio, M. A. (2009). Nuclear localization of active HGF receptor Met in aggressive MDA-MB231 breast carcinoma cells. *Carcinogenesis*, *30*(6), 937–945. https://doi.org/10.1093/carcin/bgp080.

Merilahti, J. A. M., & Elenius, K. (2019). Gamma-secretase-dependent signaling of receptor tyrosine kinases. *Oncogene*, *38*(2), 151–163. https://doi.org/10.1038/s41388-018-0465-z.

Mertens-Walker, I., Lisle, J. E., Nyberg, W. A., Stephens, C. R., Burke, L., Rutkowski, R., et al. (2015). EphB4 localises to the nucleus of prostate cancer cells. *Experimental Cell Research*, *333*(1), 105–115. https://doi.org/10.1016/j.yexcr.2015.02.015.

Miaczynska, M. (2013). Effects of membrane trafficking on signaling by receptor tyrosine kinases. *Cold Spring Harbor Perspectives in Biology*, *5*(11), a009035. https://doi.org/10.1101/cshperspect.a009035.

Migdall-Wilson, J., Bates, C., Schlegel, J., Brandao, L., Linger, R. M., DeRyckere, D., et al. (2012). Prolonged exposure to a Mer ligand in leukemia: Gas6 favors expression of a partial Mer glycoform and reveals a novel role for Mer in the nucleus. *PLoS One*, 7(2), e31635. https://doi.org/10.1371/journal.pone.0031635.

Mori, K., & Kato, H. (2002). A putative nuclear receptor coactivator (TMF/ARA160) associates with hbrm/hSNF2 alpha and BRG-1/hSNF2 beta and localizes in the Golgi apparatus. *FEBS Letters*, *520*(1–3), 127–132. https://doi.org/10.1016/s0014-5793(02)02803-x.

Moughal, N. A., Waters, C., Sambi, B., Pyne, S., & Pyne, N. J. (2004). Nerve growth factor signaling involves interaction between the Trk A receptor and lysophosphatidate receptor 1 systems: Nuclear translocation of the lysophosphatidate receptor 1 and Trk A receptors in pheochromocytoma 12 cells. *Cellular Signalling*, *16*(1), 127–136. https://doi.org/10.1016/j.cellsig.2003.08.004.

Muraoka-Cook, R. S., Sandahl, M., Husted, C., Hunter, D., Miraglia, L., Feng, S. M., et al. (2006). The intracellular domain of ErbB4 induces differentiation of mammary epithelial cells. *Molecular Biology of the Cell*, *17*(9), 4118–4129. https://doi.org/10.1091/mbc.e06-02-0101.

Myers, J. M., Martins, G. G., Ostrowski, J., & Stachowiak, M. K. (2003). Nuclear trafficking of FGFR1: A role for the transmembrane domain. *Journal of Cellular Biochemistry*, *88*(6), 1273–1291. https://doi.org/10.1002/jcb.10476.

Na, H. W., Shin, W. S., Ludwig, A., & Lee, S. T. (2012). The cytosolic domain of protein-tyrosine kinase 7 (PTK7), generated from sequential cleavage by a disintegrin and metalloprotease 17 (ADAM17) and gamma-secretase, enhances cell proliferation and migration in colon cancer cells. *The Journal of Biological Chemistry*, *287*(30), 25001–25009. https://doi.org/10.1074/jbc.M112.348904.

Ness, J. K., Skiles, A. A., Yap, E.-H., Fajardo, E. J., Fiser, A., & Tapinos, N. (2016). Nuc-ErbB3 regulates H3K27me3 levels and HMT activity to establish epigenetic repression during peripheral myelination. *Glia*, *64*(6), 977–992. https://doi.org/10.1002/glia.22977.

Ni, C. Y., Murphy, M. P., Golde, T. E., & Carpenter, G. (2001). gamma-Secretase cleavage and nuclear localization of ErbB-4 receptor tyrosine kinase. *Science*, *294*(5549), 2179–2181. https://doi.org/10.1126/science.1065412.

Nitulescu, G. M., Margina, D., Juzenas, P., Peng, Q., Olaru, O. T., Saloustros, E., et al. (2016). Akt inhibitors in cancer treatment: The long journey from drug discovery to clinical use (Review). *International Journal of Oncology*, *48*(3), 869–885. https://doi.org/10.3892/ijo.2015.3306.

Offterdinger, M., Schofer, C., Weipoltshammer, K., & Grunt, T. W. (2002). c-erbB-3: A nuclear protein in mammary epithelial cells. *The Journal of Cell Biology*, *157*(6), 929–939. https://doi.org/10.1083/jcb.200109033.

Oron, G., Ao, A., Friedman, O., Fisch, B., Zhang, X. Y., Ben-Haroush, A., et al. (2011). Expression of neurotrophin 3 and its tropomyosin-related kinase receptor C in human preantral follicles. *Fertility and Sterility*, *95*(6), 2056–2062. https://doi.org/10.1016/j.fertnstert.2011.02.012.

Ortega, J., Li, J. Y., Lee, S., Tong, D., Gu, L., & Li, G. M. (2015). Phosphorylation of PCNA by EGFR inhibits mismatch repair and promotes misincorporation during DNA synthesis. *Proceedings of the National Academy of Sciences of the United States of America*, *112*(18), 5667–5672. https://doi.org/10.1073/pnas.1417711112.

Orth, J. D., Krueger, E. W., Weller, S. G., & McNiven, M. A. (2006). A novel endocytic mechanism of epidermal growth factor receptor sequestration and internalization. *Cancer Research*, *66*(7), 3603–3610. https://doi.org/10.1158/0008-5472.CAN-05-2916.

Paatero, I., Jokilammi, A., Heikkinen, P. T., Iljin, K., Kallioniemi, O. P., Jones, F. E., et al. (2012). Interaction with ErbB4 promotes hypoxia-inducible factor-1alpha signaling. *The Journal of Biological Chemistry*, *287*(13), 9659–9671. https://doi.org/10.1074/jbc.M111.299537.

Papadopoulos, N., Lennartsson, J., & Heldin, C. H. (2018). PDGFRbeta translocates to the nucleus and regulates chromatin remodeling via TATA element-modifying factor 1. *The Journal of Cell Biology*, *217*(5), 1701–1717. https://doi.org/10.1083/jcb.201706118.

Pareja, F., Pines, G., & Yarden, Y. (2015). The EGFR/ERBB receptor family. In D. L. Wheeler & Y. Yarden (Eds.), *Receptor tyrosine kinases: Family and subfamilies* (pp. 107–164). Cham: Springer International Publishing.

Peng, H., Myers, J., Fang, X., Stachowiak, E. K., Maher, P. A., Martins, G. G., et al. (2002). Integrative nuclear FGFR1 signaling (INFS) pathway mediates activation of the tyrosine hydroxylase gene by angiotensin II, depolarization and protein kinase C. *Journal of Neurochemistry*, *81*(3), 506–524. https://doi.org/10.1046/j.1471-4159.2002.00833.x.

Perillo, E. P., Liu, Y. L., Huynh, K., Liu, C., Chou, C. K., Hung, M. C., et al. (2015). Deep and high-resolution three-dimensional tracking of single particles using nonlinear and multiplexed illumination. *Nature Communications*, *6*, 7874. https://doi.org/10.1038/ncomms8874.

Piccione, E. C., Lieu, T. J., Gentile, C. F., Williams, T. R., Connolly, A. J., Godwin, A. K., et al. (2012). A novel epidermal growth factor receptor variant lacking multiple domains directly activates transcription and is overexpressed in tumors. *Oncogene*, *31*(24), 2953–2967. https://doi.org/10.1038/onc.2011.465.

Platta, H. W., & Stenmark, H. (2011). Endocytosis and signaling. *Current Opinion in Cell Biology*, *23*(4), 393–403. https://doi.org/10.1016/j.ceb.2011.03.008.

Podlecki, D. A., Smith, R. M., Kao, M., Tsai, P., Huecksteadt, T., Brandenburg, D., et al. (1987). Nuclear translocation of the insulin receptor. A possible mediator of insulin's long term effects. *The Journal of Biological Chemistry*, *262*(7), 3362–3368.

Porebska, N., Latko, M., Kucinska, M., Zakrzewska, M., Otlewski, J., & Opalinski, L. (2018). Targeting cellular trafficking of fibroblast growth factor receptors as a strategy for selective cancer treatment. *Journal of Clinical Medicine*, *8*(1), 7. https://doi.org/10.3390/jcm8010007.

Pozner-Moulis, S., Pappas, D. J., & Rimm, D. L. (2006). Met, the hepatocyte growth factor receptor, localizes to the nucleus in cells at low density. *Cancer Research*, *66*(16), 7976–7982. https://doi.org/10.1158/0008-5472.CAN-05-4335.

Psyrri, A., Yu, Z., Weinberger, P. M., Sasaki, C., Haffty, B., Camp, R., et al. (2005). Quantitative determination of nuclear and cytoplasmic epidermal growth factor receptor expression in oropharyngeal squamous cell cancer by using automated quantitative analysis. *Clinical Cancer Research*, *11*(16), 5856–5862. https://doi.org/10.1158/1078-0432.CCR-05-0420.

Qi, Y., Lang, J., Li, L., Feng, M., & Huang, Y. (2017). Inhibition of EGFR nuclear translocation attenuate radioresistance through decreased p-DNA-PK expression in cervical cancer cells. *Journal of Clinical Oncology*, 35(15_Suppl), e17011. https://doi.org/10.1200/JCO.2017.35.15_suppl.e17011.

Reif, R., Adawy, A., Vartak, N., Schroder, J., Gunther, G., Ghallab, A., et al. (2016). Activated ErbB3 translocates to the nucleus via clathrin-independent endocytosis, which is associated with proliferating cells. *The Journal of Biological Chemistry*, *291*(8), 3837–3847. https://doi.org/10.1074/jbc.M115.686782.

Reilly, J. F., & Maher, P. A. (2001). Importin beta-mediated nuclear import of fibroblast growth factor receptor: Role in cell proliferation. *The Journal of Cell Biology*, *152*(6), 1307–1312. https://doi.org/10.1083/jcb.152.6.1307.

Reilly, J. F., Mizukoshi, E., & Maher, P. A. (2004). Ligand dependent and independent internalization and nuclear translocation of fibroblast growth factor (FGF) receptor 1. *DNA and Cell Biology*, *23*(9), 538–548. https://doi.org/10.1089/dna.2004.23.538.

Rokicki, J., Das, P. M., Giltnane, J. M., Wansbury, O., Rimm, D. L., Howard, B. A., et al. (2010). The ERalpha coactivator, HER4/4ICD, regulates progesterone receptor expression in normal and malignant breast epithelium. *Molecular Cancer*, *9*, 150. https://doi.org/10.1186/1476-4598-9-150.

Sanchez, J. N., Wang, T., & Cohen, M. S. (2018). BRAF and MEK inhibitors: Use and resistance in BRAF-mutated cancers. *Drugs*, *78*(5), 549–566. https://doi.org/10.1007/s40265-018-0884-8.

Santoro, M., Melillo, R. M., Carlomagno, F., Vecchio, G., & Fusco, A. (2004). Minireview: RET: Normal and abnormal functions. *Endocrinology*, *145*(12), 5448–5451. https://doi.org/10.1210/en.2004-0922.

Santos, S. C., & Dias, S. (2004). Internal and external autocrine VEGF/KDR loops regulate survival of subsets of acute leukemia through distinct signaling pathways. *Blood*, *103*(10), 3883–3889. https://doi.org/10.1182/blood-2003-05-1634.

Santos, S. C., Miguel, C., Domingues, I., Calado, A., Zhu, Z., Wu, Y., et al. (2007). VEGF and VEGFR-2 (KDR) internalization is required for endothelial recovery during wound healing. *Experimental Cell Research*, *313*(8), 1561–1574. https://doi.org/10.1016/j.yexcr.2007.02.020.

Sardi, S. P., Murtie, J., Koirala, S., Patten, B. A., & Corfas, G. (2006). Presenilin-dependent ErbB4 nuclear signaling regulates the timing of astrogenesis in the developing brain. *Cell*, *127*(1), 185–197. https://doi.org/10.1016/j.cell.2006.07.037.

Sarfstein, R., Pasmanik-Chor, M., Yeheskel, A., Edry, L., Shomron, N., Warman, N., et al. (2012). Insulin-like growth factor-I receptor (IGF-IR) translocates to nucleus and autoregulates IGF-IR gene expression in breast cancer cells. *The Journal of Biological Chemistry*, *287*(4), 2766–2776. https://doi.org/10.1074/jbc.M111.281782.

Schillaci, R., Guzman, P., Cayrol, F., Beguelin, W., Diaz Flaque, M. C., Proietti, C. J., et al. (2012). Clinical relevance of ErbB-2/HER2 nuclear expression in breast cancer. *BMC Cancer*, *12*, 74. https://doi.org/10.1186/1471-2407-12-74.

Schlessinger, J. (2014). Receptor tyrosine kinases: Legacy of the first two decades. *Cold Spring Harbor Perspectives in Biology*, *6*(3), a008912. https://doi.org/10.1101/cshperspect.a008912.

Schmahl, J., Kim, Y., Colvin, J. S., Ornitz, D. M., & Capel, B. (2004). Fgf9 induces proliferation and nuclear localization of FGFR2 in Sertoli precursors during male sex determination. *Development*, *131*(15), 3627–3636. https://doi.org/10.1242/dev.01239.

Schmidt, J. F., Agapova, O. A., Yang, P., Kaufman, P. L., & Hernandez, M. R. (2007). Expression of ephrinB1 and its receptor in glaucomatous optic neuropathy. *The British Journal of Ophthalmology*, *91*(9), 1219–1224. https://doi.org/10.1136/bjo.2006.112185.

Sehat, B., Tofigh, A., Lin, Y., Trocme, E., Liljedahl, U., Lagergren, J., et al. (2010). SUMOylation mediates the nuclear translocation and signaling of the IGF-1 receptor. *Science Signaling*, *3*(108), ra10. https://doi.org/10.1126/scisignal.2000628.

Seymour, P. A., Shih, H. P., Patel, N. A., Freude, K. K., Xie, R., Lim, C. J., et al. (2012). A Sox9/Fgf feed-forward loop maintains pancreatic organ identity. *Development*, *139*(18), 3363–3372. https://doi.org/10.1242/dev.078733.

Shamloo, B., & Usluer, S. (2019). p21 in Cancer Research. *Cancers*, *11*(8), 1178. https://doi.org/10.3390/cancers11081178.

Shay-Salit, A., Shushy, M., Wolfovitz, E., Yahav, H., Breviario, F., Dejana, E., et al. (2002). VEGF receptor 2 and the adherens junction as a mechanical transducer in vascular endothelial cells. *Proceedings of the National Academy of Sciences of the United States of America*, *99*(14), 9462–9467. https://doi.org/10.1073/pnas.142224299.

Shi, Y., Liu, N., Lai, W., Yan, B., Chen, L., Liu, S., et al. (2018). Nuclear EGFR-PKM2 axis induces cancer stem cell-like characteristics in irradiation-resistant cells. *Cancer Letters*, *422*, 81–93. https://doi.org/10.1016/j.canlet.2018.02.028.

Solomon-Zemler, R., Sarfstein, R., & Werner, H. (2017). Nuclear insulin-like growth factor-1 receptor (IGF1R) displays proliferative and regulatory activities in non-malignant cells. *PLoS One*, *12*(9), e0185164. https://doi.org/10.1371/journal.pone.0185164.

Sorokin, A., Mohammadi, M., Huang, J., & Schlessinger, J. (1994). Internalization of fibroblast growth factor receptor is inhibited by a point mutation at tyrosine 766. *The Journal of Biological Chemistry*, *269*(25), 17056–17061.

Stachowiak, M. K., Maher, P. A., & Stachowiak, E. K. (2007). Integrative nuclear signaling in cell development—a role for FGF receptor-1. *DNA and Cell Biology*, *26*(12), 811–826. https://doi.org/10.1089/dna.2007.0664.

Staquicini, F. I., Qian, M. D., Salameh, A., Dobroff, A. S., Edwards, J. K., Cimino, D. F., et al. (2015). Receptor tyrosine kinase EphA5 is a functional molecular target in human lung cancer. *The Journal of Biological Chemistry*, *290*(12), 7345–7359. https://doi.org/10.1074/jbc.M114.630525.

Stewart, M., Turley, H., Cook, N., Pezzella, F., Pillai, G., Ogilvie, D., et al. (2003). The angiogenic receptor KDR is widely distributed in human tissues and tumours and relocates intracellularly on phosphorylation. An immunohistochemical study. *Histopathology*, *43*(1), 33–39. https://doi.org/10.1046/j.1365-2559.2003.01644.x.

Tan, X., Lambert, P. F., Rapraeger, A. C., & Anderson, R. A. (2016). Stress-induced EGFR trafficking: Mechanisms, functions, and therapeutic implications. *Trends in Cell Biology*, *26*(5), 352–366. https://doi.org/10.1016/j.tcb.2015.12.006.

Tey, S. K., Tse, E. Y. T., Mao, X., Ko, F. C. F., Wong, A. S. T., Lo, R. C., et al. (2017). Nuclear Met promotes hepatocellular carcinoma tumorigenesis and metastasis by upregulation of TAK1 and activation of NF-kappaB pathway. *Cancer Letters*, *411*, 150–161. https://doi.org/10.1016/j.canlet.2017.09.047.

Thompson, M., Lauderdale, S., Webster, M. J., Chong, V. Z., McClintock, B., Saunders, R., et al. (2007). Widespread expression of ErbB2, ErbB3 and ErbB4 in non-human primate brain. *Brain Research*, *1139*, 95–109. https://doi.org/10.1016/j.brainres.2006.11.047.

Tian, T., Li, X., & Zhang, J. (2019). mTOR Signaling in Cancer and mTOR Inhibitors in Solid Tumor Targeting Therapy. *International Journal of Molecular Sciences*, *20*(3), 755. https://doi.org/10.3390/ijms20030755.

Tourette, C., Farina, F., Vazquez-Manrique, R. P., Orfila, A. M., Voisin, J., Hernandez, S., et al. (2014). The Wnt receptor Ryk reduces neuronal and cell survival capacity by repressing FOXO activity during the early phases of mutant huntingtin pathogenicity. *PLoS Biology*, *12*(6), e1001895. https://doi.org/10.1371/journal.pbio.1001895.

Traynor, A. M., Weigel, T. L., Oettel, K. R., Yang, D. T., Zhang, C., Kim, K., et al. (2013). Nuclear EGFR protein expression predicts poor survival in early stage non-small cell lung cancer. *Lung Cancer*, *81*(1), 138–141. https://doi.org/10.1016/j.lungcan.2013.03.020.

Tremblay, M. G., Herdman, C., Guillou, F., Mishra, P. K., Baril, J., Bellenfant, S., et al. (2015). Extended synaptotagmin interaction with the fibroblast growth factor receptor depends on receptor conformation, not catalytic activity. *The Journal of Biological Chemistry*, *290*(26), 16142–16156. https://doi.org/10.1074/jbc.M115.656918.

Tsai, S.-J., Chien, C.-W., Chen, S.-Y., Lin, S.-C., Lin, B.-W., & Lee, J.-C. (2018). Abstract 2488: Functional characterization of nuclear TYRO3 in colorectal cancer. In *Paper presented at the American Association for Cancer Research Annual Meeting 2018, Chichago, IL.*

Tseng, H. C., Kao, H. W., Ho, M. R., Chen, Y. R., Lin, T. W., Lyu, P. C., et al. (2011). Cytoskeleton network and cellular migration modulated by nuclear-localized receptor tyrosine kinase ROR1. *Anticancer Research*, *31*(12), 4239–4249.

Tseng, H. C., Lyu, P. C., & Lin, W. C. (2010). Nuclear localization of orphan receptor protein kinase (Ror1) is mediated through the juxtamembrane domain. *BMC Cell Biology*, *11*, 48. https://doi.org/10.1186/1471-2121-11-48.

Turkington, R. C., Longley, D. B., Allen, W. L., Stevenson, L., McLaughlin, K., Dunne, P. D., et al. (2014). Fibroblast growth factor receptor 4 (FGFR4): A targetable regulator of drug resistance in colorectal cancer. *Cell Death & Disease, 5*, e1046. https://doi.org/10.1038/cddis.2014.10.

Tyrsina, E., Nikulitskiy, S., Inshakov, A., Ryabaya, O., & Tyrsin, D. (2019). Detection and quantification of VEGFR-1 in the nuclei of tumor cells by a new flow cytometry-based method. *Journal of Immunotoxicology, 16*(1), 74–81. https://doi.org/10.1080/1547691X.2019.1598522.

Vardi, A., Bosviel, R., Rabiau, N., Adjakly, M., Satih, S., Dechelotte, P., et al. (2010). Soy phytoestrogens modify DNA methylation of GSTP1, RASSF1A, EPH2 and BRCA1 promoter in prostate cancer cells. In *Vivo, 24*(4), 393–400.

Venturutti, L., Romero, L. V., Urtreger, A. J., Chervo, M. F., Cordo Russo, R. I., Mercogliano, M. F., et al. (2016). Stat3 regulates ErbB-2 expression and co-opts ErbB-2 nuclear function to induce miR-21 expression, PDCD4 downregulation and breast cancer metastasis. *Oncogene, 35*(17), 2208–2222. https://doi.org/10.1038/onc.2015.281.

Vincent, L., Jin, D. K., Karajannis, M. A., Shido, K., Hooper, A. T., Rashbaum, W. K., et al. (2005). Fetal stromal-dependent paracrine and intracrine vascular endothelial growth factor-a/vascular endothelial growth factor receptor-1 signaling promotes proliferation and motility of human primary myeloma cells. *Cancer Research, 65*(8), 3185–3192. https://doi.org/10.1158/0008-5472.CAN-04-3598.

Volinsky, N., & Kholodenko, B. N. (2013). Complexity of receptor tyrosine kinase signal processing. *Cold Spring Harbor Perspectives in Biology, 5*(8), a009043. https://doi.org/10.1101/cshperspect.a009043.

Wang, S. C., & Hung, M. C. (2009). Nuclear translocation of the epidermal growth factor receptor family membrane tyrosine kinase receptors. *Clinical Cancer Research, 15*(21), 6484–6489. https://doi.org/10.1158/1078-0432.CCR-08-2813.

Wang, Y. N., & Hung, M. C. (2012). Nuclear functions and subcellular trafficking mechanisms of the epidermal growth factor receptor family. *Cell & Bioscience, 2*(1), 13. https://doi.org/10.1186/2045-3701-2-13.

Wang, Y. N., Lee, H. H., Lee, H. J., Du, Y., Yamaguchi, H., & Hung, M. C. (2012). Membrane-bound trafficking regulates nuclear transport of integral epidermal growth factor receptor (EGFR) and ErbB-2. *The Journal of Biological Chemistry, 287*(20), 16869–16879. https://doi.org/10.1074/jbc.M111.314799.

Wang, S. C., Lien, H. C., Xia, W., Chen, I. F., Lo, H. W., Wang, Z., et al. (2004). Binding at and transactivation of the COX-2 promoter by nuclear tyrosine kinase receptor ErbB-2. *Cancer Cell, 6*(3), 251–261. https://doi.org/10.1016/j.ccr.2004.07.012.

Wang, S.-C., Nakajima, Y., Yu, Y.-L., Xia, W., Chen, C.-T., Yang, C.-C., et al. (2006). Tyrosine phosphorylation controls PCNA function through protein stability. *Nature Cell Biology, 8*(12), 1359–1368. https://doi.org/10.1038/ncb1501.

Wang, Y. N., Wang, H., Yamaguchi, H., Lee, H. J., Lee, H. H., & Hung, M. C. (2010). COPI-mediated retrograde trafficking from the Golgi to the ER regulates EGFR nuclear transport. *Biochemical and Biophysical Research Communications, 399*(4), 498–504. https://doi.org/10.1016/j.bbrc.2010.07.096.

Wang, Y. N., Yamaguchi, H., Huo, L., Du, Y., Lee, H. J., Lee, H. H., et al. (2010). The translocon Sec61beta localized in the inner nuclear membrane transports membrane-embedded EGF receptor to the nucleus. *The Journal of Biological Chemistry, 285*(49), 38720–38729. https://doi.org/10.1074/jbc.M110.158659.

Waraky, A., Lin, Y., Warsito, D., Haglund, F., Aleem, E., & Larsson, O. (2017). Nuclear insulin-like growth factor 1 receptor phosphorylates proliferating cell nuclear antigen and rescues stalled replication forks after DNA damage. *The Journal of Biological Chemistry, 292*(44), 18227–18239. https://doi.org/10.1074/jbc.M117.781492.

Warsito, D., Sjostrom, S., Andersson, S., Larsson, O., & Sehat, B. (2012). Nuclear IGF1R is a transcriptional co-activator of LEF1/TCF. *EMBO Reports*, *13*(3), 244–250. https://doi.org/10.1038/embor.2011.251.

Williams, C. C., Allison, J. G., Vidal, G. A., Burow, M. E., Beckman, B. S., Marrero, L., et al. (2004). The ERBB4/HER4 receptor tyrosine kinase regulates gene expression by functioning as a STAT5A nuclear chaperone. *The Journal of Cell Biology*, *167*(3), 469–478. https://doi.org/10.1083/jcb.200403155.

Wong, V. C. (2018). Nuclear EGFR and integrator/super elongation complex concurrently binds to immediate early genes for gene transactivation. *Journal of Cancer*, *9*(1), 108–116. https://doi.org/10.7150/jca.21925.

Wu, G. H., Muthaiyan Shanmugam, M., Bhan, P., Huang, Y. H., & Wagner, O. I. (2016). Identification and characterization of LIN-2(CASK) as a regulator of kinesin-3 UNC-104(KIF1A) motility and clustering in neurons. *Traffic*, *17*(8), 891–907. https://doi.org/10.1111/tra.12413.

Xie, Y., & Hung, M. C. (1994). Nuclear localization of p185neu tyrosine kinase and its association with transcriptional transactivation. *Biochemical and Biophysical Research Communications*, *203*(3), 1589–1598. https://doi.org/10.1006/bbrc.1994.2368.

Xie, Y., Lu, W., Liu, S., Yang, Q., Carver, B. S., Li, E., et al. (2014). Crosstalk between nuclear MET and SOX9/beta-catenin correlates with castration-resistant prostate cancer. *Molecular Endocrinology*, *28*(10), 1629–1639. https://doi.org/10.1210/me.2014-1078.

Xie, Y., Nurkesh, A. A., Ibragimova, N., Zhanzak, Z., Meyerbekova, A., Alexeyeva, Z., et al. (2019). Systematic analysis of NLMP suggests nuclear localization of RTK/MET kinases resemble cancer cell clearance. *Journal of Experimental & Clinical Cancer Research*, *38*(1), 43. https://doi.org/10.1186/s13046-018-1004-z.

Yang, C. C., Lin, L. C., Lin, Y. W., Tian, Y. F., Lin, C. Y., Sheu, M. J., et al. (2019). Higher nuclear EGFR expression is a better predictor of survival in rectal cancer patients following neoadjuvant chemoradiotherapy than cytoplasmic EGFR expression. *Oncology Letters*, *17*(2), 1551–1558. https://doi.org/10.3892/ol.2018.9756.

Yang, K., Ren, X., Tao, L., Wang, P., Jiang, H., Shen, L., et al. (2019). Prognostic implications of epidermal growth factor receptor variant III expression and nuclear translocation in Chinese human gliomas. *Chinese Journal of Cancer Research*, *31*(1), 188–202. https://doi.org/10.21147/j.issn.1000-9604.2019.01.14.

Yu, Y. L., Chou, R. H., Wu, C. H., Wang, Y. N., Chang, W. J., Tseng, Y. J., et al. (2012). Nuclear EGFR suppresses ribonuclease activity of polynucleotide phosphorylase through DNAPK-mediated phosphorylation at serine 776. *The Journal of Biological Chemistry*, *287*(37), 31015–31026. https://doi.org/10.1074/jbc.M112.358077.

Zammit, C., Barnard, R., Gomm, J., Coope, R., Shousha, S., Coombes, C., et al. (2001). Altered intracellular localization of fibroblast growth factor receptor 3 in human breast cancer. *The Journal of Pathology*, *194*(1), 27–34. https://doi.org/10.1002/path.846.

Zhang, M., Sun, H., Deng, Y., Su, M., Wei, S., Wang, P., et al. (2019). COPI-mediated nuclear translocation of EGFRvIII promotes STAT3 phosphorylation and PKM2 nuclear localization. *International Journal of Biological Sciences*, *15*(1), 114–126. https://doi.org/10.7150/ijbs.28679.

Zhong, J., Kim, H. T., Lyu, J., Yoshikawa, K., Nakafuku, M., & Lu, W. (2011). The Wnt receptor Ryk controls specification of GABAergic neurons versus oligodendrocytes during telencephalon development. *Development*, *138*(3), 409–419. https://doi.org/10.1242/dev.061051.

Zhou, L., Yao, L. T., Liang, Z. Y., Zhou, W. X., You, L., Shao, Q. Q., et al. (2015). Nuclear translocation of fibroblast growth factor receptor 3 and its significance in pancreatic cancer. *International Journal of Clinical and Experimental Pathology*, *8*(11), 14640–14648.

Zhu, Y., Sullivan, L. L., Nair, S. S., Williams, C. C., Pandey, A. K., Marrero, L., et al. (2006). Coregulation of estrogen receptor by ERBB4/HER4 establishes a growth-promoting autocrine signal in breast tumor cells. *Cancer Research*, *66*(16), 7991–7998. https://doi.org/10.1158/0008-5472.CAN-05-4397.

Zscheppang, K., Dork, T., Schmiedl, A., Jones, F. E., & Dammann, C. E. (2011). Neuregulin receptor ErbB4 functions as a transcriptional cofactor for the expression of surfactant protein B in the fetal lung. *American Journal of Respiratory Cell and Molecular Biology*, *45*(4), 761–767. https://doi.org/10.1165/rcmb.2010-0179OC.

CHAPTER THREE

HER family in cancer progression: From discovery to 2020 and beyond

Rakesh Kumar[a,b,c,*], Bijesh George[a,†], Marcia R. Campbell[d,†], Nandini Verma[e], Aswathy Mary Paul[a], Cecília Melo-Alvim[f], Leonor Ribeiro[f], M. Radhakrishna Pillai[a], Luis Marques da Costa[f,g], Mark M. Moasser[d,*]

[a]Cancer Research Program, Rajiv Gandhi Centre for Biotechnology, Trivandrum, Kerala, India
[b]Department of Medicine, Division of Hematology & Oncology, Rutgers New Jersey Medical School, Newark, NJ, United States
[c]Department of Human and Molecular Genetics, Virginia Commonwealth University, School of Medicine, Richmond, VA, United States
[d]Department of Medicine, Helen Diller Family Comprehensive Cancer Center, University of California, San Francisco, CA, United States
[e]Advanced Centre for Treatment, Research and Education in Cancer, Mumbai, India
[f]Medical Oncology Department, Hospital de Santa Maria, Centro Hospitalar Universitário Lisboa Norte, Lisbon, Portugal
[g]Instituto de Medicina Molecular, Faculdade de Medicina, Universidade de Lisboa, Lisbon, Portugal
*Corresponding authors: e-mail address: rakeshkumar@rgcb.res.in; mark.moasser@ucsf.edu

Contents

† Equal contribution.

Advances in Cancer Research, Volume 147
ISSN 0065-230X
https://doi.org/10.1016/bs.acr.2020.04.001
© 2020 Elsevier Inc.
All rights reserved.

Abstract

The human epidermal growth factor receptor (HER) family of receptor tyrosine kinases (RTKs) are among the first layer of molecules that receive, interpret, and transduce signals leading to distinct cancer cell phenotypes. Since the discovery of the tooth-lid factor—later characterized as the epidermal growth factor (EGF)—and its high-affinity binding EGF receptor, HER kinases have emerged as one of the commonly upregulated or hyperactivated or mutated kinases in epithelial tumors, thus allowing HER1–3 family members to regulate several hallmarks of cancer development and progression. Each member of the HER family exhibits shared and unique structural features to engage multiple receptor activation modes, leading to a range of overlapping and distinct phenotypes. EGFR, the founding HER family member, provided the roadmap for the development of the cell surface RTK-directed targeted cancer therapy by serving as a prototype/precursor for the currently used HER-directed cancer drugs. We herein provide a brief account of the discoveries, defining moments, and historical context of the HER family and guidepost advances in basic, translational, and clinical research that solidified a prominent position of the HER family in cancer research and treatment. We also discuss the significance of HER3 pseudokinase in cancer biology; its unique structural features that drive transregulation among HER1–3, leading to a superior proximal signaling response; and potential role of HER3 as a shared effector of acquired therapeutic resistance against diverse oncology drugs. Finally, we also narrate some of the current drawbacks of HER-directed therapies and provide insights into postulated advances in HER biology with extensive implications of these therapies in cancer research and treatment.

1. Introduction

Plasma membrane components serve as the first layer of molecules that receive, interpret, and transduce signals in response to dynamic changes in the extracellular milieu, including polypeptide factors, in healthy and cancer cells. One such family of the cell surface proteins are receptor tyrosine kinases (RTKs) with extracellular, transmembrane, and intracellular domains, each having highly specialized subdomains and amino acid residues (Yarden & Ullrich, 1988). Among RTKs, the human epidermal growth factor receptor (HER) family plays a major role in the development and progression of human cancer (Carpenter, 1987). HER family members are widely upregulated, mutated, and hyperactivated in human cancer; they regulate several hallmarks of human cancer and serve as a prototype for targeting the cell surface RTKs in cancer therapeutics. The founding HER family members—EGFR, HER2, HER3, and HER4—exhibit distinct as well as shared structural features that allow these receptors to recognize growth factors in a differential manner, utilize multiple mechanisms of receptor activation, and engage the components of proximal signaling in regulating cancerous phenotypes (Kumar & Mendelsohn, 1991). The present volume comprises chapters from many of the original contributors who were involved in the discovery, biochemistry, and biology of EGFR and HER2 research (i.e., Yossi Yarden and Mien-Chie Hung); HER mouse models (i.e., William Muller); and translational significance of HER-transregulation by other cancer-relevant RTKs such as IGF1R and MET (i.e., Doug Yee and Ravi Salgia). Herein we briefly discuss the broader aspects of HER functionality, followed by a special focus on the role of HER3 in cancer biology, current limitations in HER-directed therapies, and postulations for anticipated advances in the forthcoming decade.

2. HERs: The big picture

HER family members are activated in an autocrine, paracrine, or juxtacrine manner by its ligands (Fig. 1); HER receptor activation involves ligand-induced intramolecular conformational changes, which leads to homo- or heterodimerization of HERs in a ligand-dependent manner, stimulation of tyrosine residues presents on intracellular domains, and activation of distinct signaling and adaptor proteins such as p85/PI3K, GRB2, Shc, and NCK (Yarden & Pines, 2012; Yarden & Sliwkowski, 2001). Among HERs, HER2 does not possess a ligand, whereas HER3 presents no or extremely less autophosphorylation abilities. These unique features allow both these HERs to heterodimerize with the other HERs and regulate cellular signaling through a co-operative mechanism despite their elemental dissimilarities (Citri, Skaria, & Yarden, 2003; Moasser, 2007a, 2007b). Despite of huge structural similarities among HERs, each HER member exhibits unique features, which might confer nonredundant mechanisms of action. HER dimerization is profoundly influenced by the binding of its ligands with leucine-rich residues in extracellular subdomains I and III as well as with

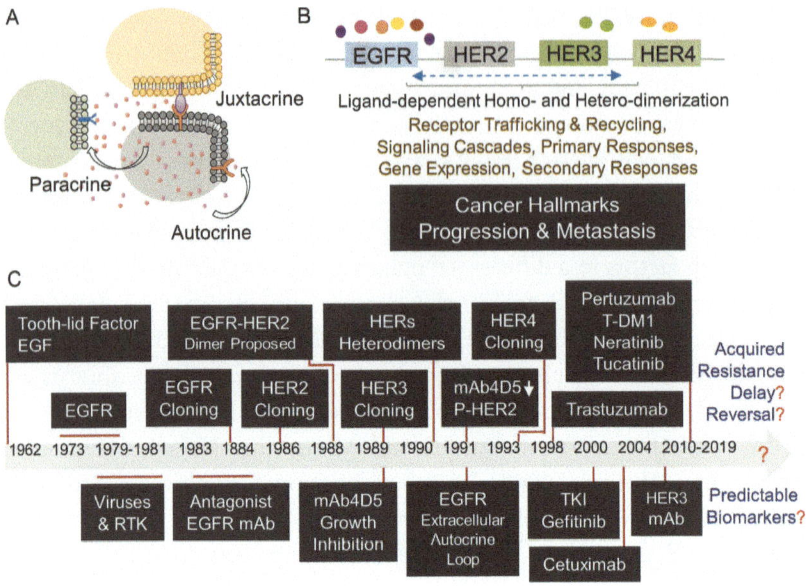

Fig. 1 EGFR family of growth factor receptor in oncobiology. (A) Illustration showing three modes of growth factor receptor activation by extracellular ligands. (B) Stimulation of HER family members in cancer cells by homo- and heterodimerization of HER family members. (C) Examples of representative turning points and anticipated advances in HER receptor biology. The timeline of events is not drawn to scale.

cysteine residues in extracellular subdomains II and IV, facilitating disulfide bond formation during receptor dimerization (Carpenter, 1987). For example, HER2 lacks a recognized ligand and is primarily activated through transregulation by other ligand-stimulated HERs or through extracellular domain (ECD) shedding. The ECDs of unliganded EGFR, HER3, and HER4 remain in a "tethered" intramolecular conformation, whereas the "dimerization motifs" of domain II are obscured by their interaction with domain IV of the same receptor molecule (Warren & Landgraf, 2006). Binding of the ligand to domains I and II triggers molecular conformational changes, permitting the release of the tethered domain II for binding to another ECD of an adjacent receptor monomer in the similar opened state. One of the unique features of the HER2 monomer is that it remains in an untethered open conformation constitutively in an unliganded form, leading to its activation in a ligand-independent manner. In addition to their role in the plasma membrane, each HER family member is proteolytically cleaved and accumulates in the nucleus, implying the role of HER or its specific fragments in noncanonical HER signaling.

3. The HER family: Discoveries, defining moments and historical context

Seeds of the HER family of RTKs and ligands in cancer biology and medical sciences were planted with the discovery of the tooth-lid factor approximately half a century ago. In 1962, Cohen observed that injecting extracts from the submaxillary gland into newborn mice accelerates opening of the eyelids and eruption of teeth; he named the putative factor responsible for this function as the tooth-lid factor (Cohen, 1962) (Fig. 1). This observation served as a precursor to the discovery of EGFR. The tooth-lid factor was later characterized as epidermal growth factor (EGF), and the quest for its target binding protein present in the plasma membrane led to the discovery of the EGF receptor (EGFR)—the first transmembrane RTK (Carpenter, King, & Cohen, 1978, 1979; Cohen, Carpenter, & King, 1980). The second defining moment for the HER family and HER-directed therapeutics was the observation that EGF phosphorylates EGFR on tyrosine residues (in addition to threonine and serine residues), similar to the cytoplasmic Src TK and that EGFR kinase could be recognized by antibodies to the first transforming TK—the Src (Chinkers & Cohen, 1981; Cohen, Carpenter, & King, 1981; Hunter, 2015)—providing evidence for the future significance of tyrosine residues in HER intracellular domains in signaling pathways of cancer cells (Fig. 1C). However, the concept of cellular communication and signaling cascades in signal transduction in response to dynamic changes in the extracellular milieu was just emerging then as the research field was engulfed with scientific joy, in leaps-and-bound, with cyclic AMP and cyclic GMP seconder messenger signaling (Krebs, 1989; Kuo & Greengard, 1970; Murad, 2006; SUTHERLAND & RALL, 1958). Researchers from Stanly Cohan's lab demonstrated the cyclic AMP- or cyclic GMP-independent nature of EGF-mediated EGFR phosphorylation. This study opened new avenues of intracellular signaling in normal and cancer cells. Following these remarkable discoveries in the field, on one hand, cancer researchers attempted to characterize the nature of intracellular events activated by the EGF-EGFR interaction and associate them with cellular phenotypes. Simultaneously, scientists were searching for additional secreted polypeptide factors for EGFR as well as transmembrane homologs of EGFR, which might be involved in the development and progression of human cancer. Because of its mitogenic and transforming nature, EGFR overexpression was believed to be involved in the growth regulation of various cancer cell lines, and this belief was confirmed by the findings of the first report on EGFR protein overexpression in lung squamous cell carcinomas (Hendler & Ozanne, 1984). These efforts eventually culminated in the discovery of three additional HER family members and numerous conceptual advances in our understanding of the biochemistry, biology, and therapeutic utility of the HER family in cancer therapeutics.

Research on EGF started in 1962, and the EGF-EGFR association was discovered in 1978. During those years, the cancer biology community was interested in the emerging recognition of transforming viruses, oncogenic and TK nature of virus-encoded proteins, and significance of cellular homologs of viral oncogenes having a TK domain in oncogenesis. Collectively, these studies have strengthened the notion of striking similarities between the action of the EGFR pathway and transforming gene products with TK activity (Bishop, 1983; Oppermann, Levinson, Varmus, Levintow, & Bishop, 1979; Schechter et al., 1984). These observations propelled the characterization of additional ErbB homologous clones. This led to the discovery of the *Neu* oncogene/HER2/ErbB-2 and its distinctiveness from EGFR/c-ErbB1 by several groups (Schechter et al., 1985; Semba, Kamata, Toyoshima, & Yamamoto, 1985), *HER2* amplification in mammary tumors (King, Kraus, & Aaronson, 1985), molecular cloning of *EGFR* and *HER2* complementary DNA (Coussens et al., 1985; Hung, Schechter, Chevray, Stern, & Weinberg, 1986; Lin et al., 1984), and demonstration of the transforming nature of HER2 overexpression in murine fibroblasts (Di Fiore et al., 1987). Studies on HER2 protein overexpression in breast cancer and its association with the clinical features of the disease in breast cancer patients in the late 1980s and subsequently, numerous other epithelial tumors have been considered one of the few defining events in breast cancer treatment (Slamon et al., 1989; van de Vijver et al., 1988). These studies have provided the second proof-of-principle to the idea of targeting the cell surface RTK for cancer therapeutics by highly specific monoclonal antibodies (mAbs), after the pioneering work of John Mendelson to target EGFR by highly specific anti-EGFR mAbs (Kawamoto et al., 1983; Sato et al., 1983). While attempting to search for ligands for HER2, several laboratories characterized the secreted or serum factors that could induce HER2 tyrosine phosphorylation independent of EGFR and in some cases, it could be blocked by an anti-HER3 mAb (Kumar, Shepard, & Mendelsohn, 1991; Lupu et al., 1990), and HER2 remained without a recognized ligand. These examples of early turning points for the HER family led to the notion that EGFR or HER2 overexpression in tumor cells causes hyperactivation of cellular pathways that control the growth stimulation, cell survival, invasiveness, metastasis, and relapse of numerous cancer types in patients.

The discovery of HER3 came on the coat-tails of efforts made for identifying additional TK oncogenes following the discovery of the two oncogenes EGFR and HER2 in the early 1980s. Low stringency hybridization of *v-erbB* or *EGFR* probes led to the identification of a third homologous member in the human genome—*erbB3* (Kraus, Issing, Miki, Popescu, & Aaronson, 1989; Plowman et al., 1990). ERBB3 RNA expression was found to be elevated in human breast cancers further confirming its crucial role in the ErbB family of receptors in human cancer pathogenesis (Kraus et al., 1989; Lemoine et al., 1992). An apparent oddity that the purified ErbB3 kinase domain lacks *in vitro* catalytic kinase activity became evident in the early years following ErbB3 receptor discovery. This peculiar finding violated the paradigm that tyrosine phosphorylation was the predominant biochemical activity and tumorigenic function of this oncogene family (Guy, Platko, Cantley, Cerione, & Carraway, 1994; Sierke, Cheng, Kim, & Koland, 1997). The subsequent cloning of ErbB4 and confirmation of its TK activity completed the characterization of this receptor family and established the uniqueness of ErbB3 as the catalytically inactive outlier in this family (Plowman et al., 1993). The nomenclature for the ErbB4 receptor family has evolved along duplicitous tracks, and although ErbB3 remains the official gene and protein name on public databases, the use of the HER3 nomenclature has become more common over the decades and is used almost exclusively in clinical and translational contexts.

As opposed to HER1–3, HER4 cDNA isolation was driven by the quest for additional homologs of the EGFR family by screening a cDNA library prepared from breast cancer cells with degenerated oligonucleotide primers against the exons of the conserved catalytic domains. Shortly after HER4 isolation, its distinct property, among the HER family members, of binding to both NRG and EGF family of ligands became evident (Plowman et al., 1993). Interestingly, HER4 signaling contributes to both differentiation and proliferative responses in mammary epithelial cells and tumors (Earp, Calvo, & Sartor, 2003). In contrast to HER1–3, HER4 is not overexpressed in breast tumors and its proteolytic fragment localizes in the nucleus in HER4-positive and estrogen receptor-positive breast tumors, which tend to be more differentiated (Jones, Welte, Fu, & Stern, 1999; Srinivasan, Gillett, Barnes, & Gullick, 2000). However, overexpression of HER4 and its coexpression with HER2 and NRG are common events in medulloblastoma and ependymoma (Gilbertson et al., 2002; Gilbertson, Perry, Kelly, Pearson, & Lunec, 1997).

In addition to the above examples, the story of HERs was immensely nurtured by the conceptual advances underlying the idea of HER-heterodimerization—the basis of our current understanding of HER signaling and therapeutics. While studying the details of EGF signaling, Stern and Kamps as well as Schlessinger's laboratory independently recognized the ability of EGF to stimulate HER2 tyrosine phosphorylation, leading to the proposal of receptor interactions (King, Borrello, Bellot, Comoglio, & Schlessinger, 1988; Stern & Kamps, 1988). This was followed the discovery of the HER2-EGFR heterodimerization by Greene's and Yarden's laboratories (Goldman, Ben-Levy, Peles, & Yarden, 1990; Wada, Qian, & Greene, 1990), and start of the notion of HER-transregulation among its family members and/or other cell surface RTKs.

4. Dysregulated expression and activity of HERs in cancer

Overexpression of the HER family is one of the most recognized cellular dysregulation in the pathobiology of solid tumors (Kumar & Mendelsohn, 1991; Yarden & Pines, 2012). Most of our current understanding of the role of the HER family in cancer biology is derived from the founding HER family member—EGFR—which is widely upregulated and/or hyperactivated in epithelial tumors due to gene amplification, transcription activation, protein overexpression, and mutations (Fig. 1C). Generally, human tumors with hyperactivated EGFR, irrespective of the underlying basis of receptor overexpression, correlate well with a poor clinical outcome and eventually with an acquired therapeutic resistance. EGFR exhibits pathological "hotspot" mutations related to cancer in the extracellular kinase domain or C-terminal cytoplasmic domains (Pines, Köstler, & Yarden, 2010). One of the most common activating point mutations in EGFR is L858R, accounting up to 45% mutations in the TK domain in some cancer types (Paez et al., 2004). EGFR with the L858R mutation in the activation loop acquires kinase activity approximately 50-fold higher than the wild-type EGFR due to destabilization of its auto-inhibitory conformation in the unbound receptor (Carey et al., 2006; Zhang, Gureasko, Shen, Cole, & Kuriyan, 2006).

Similar to EGFR, HER2 is widely overexpressed and amplified in human cancer, including in breast, gastric, esophagus, lung, endometrial, colon, pancreatic, and bladder cancers. HER2 overexpression in breast cancer correlates well with the aggressiveness and drug resistance of tumors and is an independent prognostic factor for a shorter period of disease-free survival (DFS) (Slamon et al., 1989; van de Vijver et al., 1988). In addition to gene amplification, HER2-driven oncogenesis is driven by activating somatic mutations in certain cancer types. A recent pan-cancer study found the highest frequency of HER2 mutations in the HER2 TK domain, followed by mutations in the ECD, and that many of these hotspot mutations are shared among many cancer types (Robichaux et al., 2019). As with other HER-directed therapies, despite substantial improvements in survival in patients with both early and advanced stages of breast cancer, some HER2-positive breast cancers on therapy exhibit disease progression due to innate or acquired resistance (Blackwell et al., 2010).

Despite its compromised kinase activity, HER3 plays a significant role in cancer progression, increased invasion, acquired therapeutic resistance, and drug resistance in epithelial tumors, and this is because of its strong capacity to form highly active heterodimers with HER2 and EGFR (see Sections 8–15). Among HERs, HER4 is the least overexpressed, amplified, and studied receptor in human cancer. Similar to other HERs, HER4 is activated upon binding to various polypeptides. However, unlike HER1–3 signaling, HER4 signaling promotes mammary epithelial cell differentiation, triggers breast cancer cell apoptosis, and stimulates cell growth under certain experimental conditions (Jones, 2008).

5. HER signaling: Off-mechanisms

The status of constitutive and ligand-induced HER signaling in cancer cells is a dynamic, tightly regulated coordinated process. The process and the extent of HER signaling are profoundly influenced by the extracellular milieu, spatial and temporal distribution of HERs, cytoskeleton remodeling, endocytosis and trafficking, dephosphorylation of crucial HER complexes, recycling to the surface, or degradation by the lysosomal machinery (Wang, 2017). Herein we briefly discuss few examples of off-mechanisms underlying dysregulated HER signaling in cancer cells.

Among RTKs, EGFR serves as a prototype cell surface receptor for understanding off-signaling mechanisms in growth factor-stimulated cancer cells. Upon binding of a ligand at the plasma membrane, EGFR is internalized using distinct routes of endocytotic pathways, which ultimately control the activity and fate of EGFR (Barbieri, Di Fiore, & Sigismund, 2016;

Bergeron, Di Guglielmo, Dahan, Dominguez, & Posner, 2016). In general, EGFR is internalized through clathrin-dependent endocytosis (CDE) and/or clathrin-independent endocytosis (CIE), and the nature of an engaged mode is determined by the type and concentration of the ligand. In addition to ligand concentration, the EGFR internalization mode is influenced by ligand nature. For example, binding of EGF or TGF-α predominantly utilizes the CDE mode, whereas HB-EGF or BTC engage both the CDE and CIE modes (Henriksen, Grandal, Knudsen, van Deurs, & Grøvdal, 2013). Once the ligand-induced receptor is internalized, ligand dissociation and phosphatase-mediated tyrosine dephosphorylation play vital roles in the deactivation of EGFR signaling. These observations suggest that dysregulated receptor endocytosis and intracellular trafficking actively contribute to enhanced and/or prolonged HER signaling (Mellman & Yarden, 2013). There are also situations where mutated HER could take over the endocytic machinery to support oncogenic processes (Sigismund et al., 2012).

Mechanistic details of HER2 trafficking are relatively less well understood compared to those of EGFR trafficking, partly, due to the lack of its ligand. The generally accepted limited endocytosis of HER2 is attributed to several of its unique features: the absence of an extracellular ligand-binding domain and intracellular internalization motifs (Baulida, Kraus, Alimandi, Di Fiore, & Carpenter, 1996; Shen, Lin, Childress, & Yang, 2008), inclusion of HER2 in membrane protrusions, reduced ability of HER2 to assemble clathrin-coated pits (Haslekås et al., 2005; Sorkin & von Zastrow, 2009), and binding of HER2 to heat shock protein 90 that stabilizes HER3 on the plasma membrane (Pedersen, Madshus, Haslekås, & Stang, 2008). In addition, HER2 was detected in EEA1-positive endosomal pools that are rapidly transported to the plasma membrane in breast cancer cells through retromer-dependent vesicle trafficking to reload the cell surface HER2 localization (Austin et al., 2004; Harari & Yarden, 2000; Hendriks, Opresko, Wiley, & Lauffenburger, 2003; Seaman, 2012). Interestingly, HER2 trafficking can augment the oncogenic potential of EGFR, as HER2/EGFR heterodimers are typically resistant to endocytosis-mediated degradation. This was thought to be due to the inability of HER2/EGFR heterodimers to interact with Cbl E3-ligage, which, in turn, promotes HER2 recycling to the cell membrane (Peschard & Park, 2003). In addition to influencing cellular responses, HER2 trafficking and its membrane localization majorly determine the therapeutic sensitivity of cancer cells to HER-directed therapies. For example, breast cancer cells refractory to lapatinib and trastuzumab exhibit increased intracellular HER2 levels compared with sensitive cells that predominantly express HER2 on the cell membrane (Pietilä et al., 2019).

The steady-state levels of HER3 on the cell surface are regulated by its recycling, N-linked glycosylation, or degradation in a ligand-independent or -dependent manner (Wiley, 2003). HER3 undergoes endocytosis at a slower rate than EGFR because of its decreased ubiquitination and inefficient targeting to the lysosome (Waterman, Alroy, Strano, Seger, & Yarden, 1999) (see Sections 8–15). As opposed to HER1–3, extremely few studies have examined the internalization, trafficking, and recycling of HER4 in cancer cell lines. This might be, partly, due to the proteolytic cleavage of HER4 ECDs and intracellular domains and generation of the 4ICD fragment, which tend to accumulate in the nucleus and mitochondrial compartments, in addition to the cytoplasm. Studies regarding the fate of the cleaved cytoplasmic HER4 fragment have stated that the nuclear 4ICD comprises D-box sequences that allow the degradation of certain proteins through the cyclosome ubiquitin ligase pathway (Strunk et al., 2007).

6. Cleaved HERs and therapeutic response

In addition to the autocrine, paracrine, or juxtracrine mode of activation, HER signaling could be effectively modulated by the proteolytic cleavage of its specific domains in physiological settings (Brinckerhoff & Matrisian, 2002; Maramotti et al., 2016; Mullooly et al., 2015). Because soluble ECDs of HERs retain a high affinity for the ligand, several soluble ECDs are speculated to compete with the interaction of the cell surface HERs with agonists or antagonists, modulating the kinase activity of remaining full-length membranous and/or truncated HERs on the cell membrane. For example, a soluble EGFR (sEGFR) could be generated through the alternative splicing of receptor mRNA or through the site-specific proteolytic cleavage of the receptor that releases the ECD (Perez-Torres et al., 2008; Reiter et al., 2001). As sEGFR retains the high-affinity binding sites for the ligand, it can inhibit the growth of certain cancer cells expressing the full-length EGFR, presumably, due to competition for ligand binding with the wild-type EGFR by suppressing transmembrane EGFR signaling in addition to other mechanisms (Zhou et al., 1993). In general, circulating levels of sEGFR negatively correlate with the progression of breast, ovarian, and lung cancers (Asgeirsson et al., 2007; Sandri et al., 2007). For example, metastatic progression and relapse of breast cancer is accomplished by a reduction in the level of sEGFR compared with the levels at the time of first diagnosis, suggesting the protective role of sEGFR in breast cancer. In addition to these reports, some studies have suggested a positive correlation of serum sEGFR levels with primary and metastatic cervical and gastric cancers (Choi et al., 1997; Oh et al., 2000).

Similar to EGFR, cleavage of the HER2 ECD by MMP and ADAM generates a truncated form of HER2, that is, p95HER2—a transforming protein superior to full-length HER2 (Duffy, McKiernan, O'Donovan, & McGowan, 2009; Sanderson, Dempsey, & Dunbar, 2006); HER2 (Christianson et al., 1998). Biochemically, cleavage-mediated shedding of the ECD releases inhibitory constrain on the catalytic domain, leading to hyperstimulation of the p95HER2 kinase activity. In general, soluble HER2 is suggested as a surrogate biomarker for monitoring the relapse rate, progression, or metastasis and worst prognosis of cancer or assessing the aggressiveness of cancer at the time of presentation (Hayes et al., 2001; Lam et al., 2012; Reix et al., 2016; Tsé, Gauchez, Jacot, & Lamy, 2012; Zhang et al., 2015). Because the circulating HER2 ECD comprises trastuzumab-binding epitopes, it could neutralize the therapeutic activity of trastuzumab and contribute to acquired resistance to trastuzumab (Scaltriti et al., 2007). ECD shedding may also prevent HER2 down-regulation from the cell surface, allowing hyperactivated truncated p95HER2 to remain active at the cell membrane and possibly contribute to mitogenic signaling. Moreover, breast cancer cells and tumors expressing p95HER2 are refractory to trastuzumab treatment; however, such cells were found to be sensitive to kinase inhibitor lapatinib for unknown reasons (Scaltriti et al., 2007). These studies have suggested that the suppression of constitutive kinase activity of the cell membrane p95/HER2 might be critical for restoring the sensitivity of trastuzumab-resistant cancer cells to other treatment modalities such as lapatinib (Vazquez-Martin et al., 2011).

Similar to HER2, sHER3 could be generated from five alternative splice variants encoding five secreted sHER3 isoforms or through proteolytic cleavage of HER3 at the cell surface (Katoh, Yazaki, Sugimura, & Terada, 1993; Lee & Maihle, 1998). Because p85-sHER3 retains NRG-binding sites, it acts as an antagonist of NRG-induced HER stimulation as well as of resulting downstream signaling kinases in cancer cells (Lee, Akita, Sliwkowski, & Maihle, 2001). Consistent with these studies, elevated levels of soluble HER3 in patients with prostate and bladder cancers were found to be associated with prolonged survival of the patients (Lin et al., 2008)—see Sections 8–15. Among HERs, HER4 is considered distinct as its proteolysis generates an ECD fragment as well as intracellular fragment which, in turn, could act in a nuclear or mitochondrial signaling module in a context-dependent manner (Jones, 2008; Sartor et al., 2001; Wali et al., 2014). Particularly, activated HER4 could undergo a regulated cleavage of its ECD by the tumor necrosis factor-α-converting enzyme (TACE), which in turn permits intramembrane proteolysis of HER4, releasing its intracellular domain 4ICD in a γ-secretase-dependent manner (Jones, 2008). The released 4ICD translocates to the nucleus and acts as a gene expression coactivator. These functions involve 4ICD recruitment onto target gene promoters due to its interaction with DNA-binding factors such as ER-α, STAT5a, or Yes-associated protein (Han, Sfondouris, & Jones, 2016; Haskins, Nguyen, & Stern, 2014; Naresh et al., 2008; Williams et al., 2004). In addition to nuclear signaling, 4ICD contributes to apoptosis of breast cancer cells in a BAK-dependent manner due to its translocation to the mitochondria and promotes cytochrome-*c* efflux (Naresh et al., 2008). However, the manner in which activated HER4 organizes its own proteolysis and the role that other HERs play in this process remain less well understood.

7. Ligand-induced HER3 dimerization and signaling

Ligands initiate signaling in the HER family by inducing receptor dimerization, typical of most receptor tyrosine kinase families. The ligands that bind HER3 include various isoform products of the NRG1 and NRG2 genes, collectively called neuregulins (NRG) (Falls, 2003; Hayes & Gullick, 2008). The structural basis for ligand-induced activation of HER family signaling has largely been discovered through the study of EGFR, but the concepts are similar across the entire HER family. These bivalent ligands bind domains I and III of ECD inducing a conformational re-arrangement that exposes domain II of the ECD containing a dimerization arm, mediating interactions with an analogous dimerization arm in a partner receptor (Ferguson, 2008; Linggi & Carpenter, 2006). The structure of the HER3 ECD has only been resolved in the unliganded mode, confirming the tethered conformation with the dimerization arm buried within (Cho & Leahy, 2002), but is thought to adopt open conformation when ligand-bound, similar to EGFR (Fig. 2). While this structural model only describes a general mode of receptor dimerization, there are preferences in dimerization partnerships that are apparent in cell-based studies. In particular, HER3 is a preferred dimerization partner for HER2 with a complex stability and

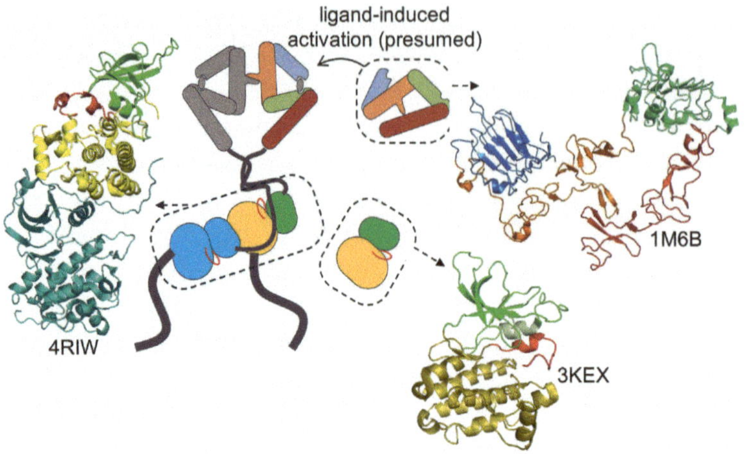

Fig. 2 Proposed mode of HER3 activation: The presumed mode of ligand-induced activation and dimerization of HER3 is shown in the central cartoon. The components that have been directly confirmed by existing crystal structure evidence are highlighted by the dashed outlines. These are the isolated kinase domain of HER3 (3KEX) (Jura et al., 2009), the unliganded extracellular domain of HER3 (1M6B) (Cho & Leahy, 2002), and the HER3-EGFR kinase domain dimer (4RIW) (Littlefield et al., 2014). The presumed model of ligand-induced extracellular domain dimerization is inferred from the known structures of the ligand-induced EGFR extracelluar domain dimerization. Structures are from the Protein Data Bank (PDB).

signaling intensity that is among the highest in the plethora of homo- and hetero-dimeric complexes possible in this four-member family (Duneau, Vegh, & Sturgis, 2007; Graus-Porta, Beerli, Daly, & Hynes, 1997; Tzahar et al., 1996; Wallasch et al., 1995).

The HER family of RTKs are highly homologous across most of the structured domains spanning the ECD, transmembrane, and intracellular kinase domains. However differential ligand-binding activities and entirely divergent C-terminal signaling tails account for enormous diversification in biologic signaling functions that remain difficult to sort out even as of today. HER3 shares ligand-binding specificities with HER4 but not with EGFR (Roskoski, 2014); however, all three of these ligand-bound receptors can dimerize with HER2 which lacks any ligand-binding activity but is a favored dimerization partner. How specific ligands induce preferred dimerization patterns and downstream signaling attributes remains poorly understood, but it is clear that different ligands for the same receptor induce different signaling dynamics with different biologic effects. This is best described for the many ligands that activate EGFR (Wilson et al., 2012), but there is also evidence that HER3 ligands similarly have ligand-specific biologic activities (Crovello, Lai, Cantley, & Carraway, 1998; Sweeney et al., 2001). The C-terminal tail of HER3 encodes at least 10 tyrosines that when phosphorylated are thought to participate in recruiting various SH2 or PTB containing adapter proteins (Jones, Gordus, Krall, & MacBeath, 2006; Schulze, Deng, & Mann, 2005). Some of these such as Shc or Grb2 seem redundant with the other HER family members. But what is unique in HER3 is an abundance of phosphotyrosine motifs for binding the p85 regulatory subunit of PI3K. There are a total of six tyrosines predicted to bind PI3K and to this day it is felt that this potent ability to activate PI3K imparts a unique identity and signaling activity to HER3 (Prigent & Gullick, 1994; Soltoff, Carraway, Prigent, Gullick, & Cantley, 1994).

8. Function of the inactive HER3 kinase domain

Although the lack of catalytic activity in the kinase domain of HER3 was unusual when initially revealed in the mid-1990s, the subsequent decades have brought about deeper insights into functional diversity within the human kinome and it is now recognized that approximately 10% of the proteins in the human kinome are similarly catalytically inactive (Boudeau, Miranda-Saavedra, Barton, & Alessi, 2006). These pseudokinases perform functions including scaffolding functions, allosteric regulation of interacting kinases and other proteins, and transcriptional functions (Kung & Jura, 2016; Rauch, Volinsky, Romano, & Kolch, 2011; Shaw, Kornev, Hu, Ahuja, & Taylor, 2014). The study of HER3 kinase domain has been at the forefront of the attempts to decipher the functions of pseudokinases. The breakthrough came when the mechanism of dimerization-induced kinase activation within the HER family was determined in 2006. While in most RTK families, the activation of the kinase domain is induced by phosphorylation of the activation segment (Nolen, Taylor, & Ghosh, 2004), this had been known to be unnecessary in the HER family (Gotoh, Tojo, Hino, Yazaki, & Shibuya, 1992). In this family of RTKs, the kinase domains dimerize in an asymmetric configuration such that the C-lobe activating interface of one kinase engages the N-lobe receiving interface of the other, thereby inducing allosteric changes that activate the catalytic function of the receiver kinase (Zhang et al., 2006). This resembles the mechanism by which cyclins activate CDKs such that there is no requirement for catalytic activity in the activator kinase which plays a role analogous to the cyclin protein. The C-lobe activating interface is conserved across the four members of the HER family including HER3, while the N-lobe receiver interface is conserved in EGFR, HER2, and HER4, but not HER3, confirming the role of the HER3 KD as an exclusive activator kinase in this family (Zhang et al., 2006). The expected similar asymmetric dimerizations have since been confirmed in HER4 homodimers (Qiu et al., 2008) and EGFR-HER3 heterodimers (Littlefield et al., 2014) (Fig. 1). Therefore, the function of the HER3 KD as an exclusive allosteric activator is now widely recognized.

9. HER3 localization and trafficking

The negative regulation of HER3 signaling appears to be distinct from that of EGFR. EGFR undergoes endocytic internalization and lysosomal degradation following activation by some ligands whereas other ligands promote recycling to the plasma membrane (Tomas, Futter, & Eden, 2014; Waterman, Sabanai, Geiger, & Yarden, 1998). The endocytic routing of HER3 predominantly proceeds along a recycling route (Baulida et al., 1996; Tomas et al., 2014; Waterman et al., 1998) and lysosomal degradation does not appear to be a major negative regulator of HER3 signaling. Rather, HER3 signaling is negatively regulated by ubiquitin-mediated proteosomal degradation, mediated through the E3 ligase Nrdp1 (Cao, Wu, Yen, Sweeney, & Carraway, 2007; Qiu & Goldberg, 2002). A body of literature has been suggesting that members of the HER family including HER3 undergo translocation to the nucleus for regulating transcription. This remains a contentious area of science and readers are referred to other reviews of this topic for coverage of this field (Carpenter & Liao, 2013; Song, Rosen, & Corfas, 2013; Wang & Hung, 2012).

10. HER3 in HER2-amplified cancers

There is significant evidence implicating HER3 in the pathogenesis of some human cancers and in conferring resistance to therapy in some human cancers. The evidence is solid in some cancer subtypes and more speculative in others. The most established is the role of HER3 in HER2-driven breast cancers. Approximately 20% of breast cancers are driven by amplification and overexpression of HER2 and a large body of experimental evidence from mouse transgenic models and from gene-targeting experiments in HER2-amplified breast cancer cell lines confirms that these tumors are driven by HER2 and addicted to continued HER2 signaling (Moasser, 2007a, 2007b; Ursini-Siegel, Schade, Cardiff, & Muller, 2007). A number of experimental model systems collectively confirm that HER3 is a requisite partner for HER2 in initiating and driving this type of cancer. The development of tumors in transgenic mice due to overexpression of a HER2 transgene in mammary tissue is significantly averted by Cre-mediated deletion of the endogenous *ERBB3* gene in mammary tissue (Vaught et al., 2012). In HER2-amplified breast cancer cell lines, the experimental transcriptional repression of HER3 expression stops tumor cell growth and proliferation (Holbro et al., 2003). The *in vivo* tumorigenic growth of HER2-amplified breast cancers can be stopped using shRNA knockdown of HER3 in the tumors (Campbell et al., 2016; Lee-Hoeflich et al., 2008) or Crispr-mediated knockout of HER3 in the tumors (Ruiz-Saenz et al., 2018). EGFR or HER4 do not seem to supplant the requirement for HER3 as the requisite partner for HER2 in these tumors.

The exclusive requirement for HER3 in this tumor type may be due to its competency in activating PI3K/Akt signaling downstream of HER2 but the experimental models are not conclusive in this regard. Targeting the PI3K-binding activity of HER3 appears to have little effect on tumor growth in one model (Vaught et al., 2012) but substantially retards tumor growth in another model (Lahlou, Müller, Sanguin-Gendreau, Birchmeier, & Muller, 2012). The requirement for HER3 in HER2-amplified tumors is not absolute and tumorigenic subclones of HER2-amplified breast cancers do emerge despite Crispr-mediated knockout of HER3 (Ruiz-Saenz et al., 2018). Mechanistic exploration of these tumors suggests both PI3K-dependent and independent modes of overcoming HER3 dependency.

HER3 is not only important in the natural biology of HER2-amplified breast cancers, but also in mitigating response to HER2-targeting therapies in this disease. Unlike HER2 which is constitutively expressed at high levels in these cancers, the expression of HER3 is quite variable, highly dynamic, and subject to feedback regulation. Upon exposure to HER2 inhibitors, the expression of HER3 is induced in these cancers and its overall phosphorylation and signaling activity can be restored despite a significant suppression of HER2 catalytic activity (Sergina et al., 2007). The compensatory increase in HER3 signaling can occur through a multitude of mechanisms including transcriptional, post-transcriptional, translational, post-translational and localization controls (Amin, Sergina, Lim, Goga, & Moasser, 2012; Garrett et al., 2011). The negative feedback regulation of HER3 is linked with downstream Akt signaling and can also be triggered through the inhibition of downstream PI3K, Akt, or mTor (Amin, Ahuja, Yaswen, & Moasser, 2015; Amin, Ruiz-Saenz, Gulizia, & Moasser, 2015; Amin et al., 2012; Chakrabarty, Sánchez, Kuba, Rinehart, & Arteaga, 2012). The requisite role of HER3 and the compensatory dynamic nature of HER3 signaling in HER2-amplified tumors has redefined HER2-HER3 signaling as the functionally relevant oncogenic driver of this disease with the ability to increase the output of its signaling about two logs (Amin et al., 2010) placing it outside the therapeutic index of all current HER2-targeting therapies and underlying the limited clinical efficacies of current HER2 kinase inhibitors (Blackwell et al., 2010). The highly effective treatment of this disease requires effective suppression of both HER2 and HER3 and awaits the development of effective inhibitors of HER3 function.

11. HER3 in other cancers

There is a fair amount of descriptive and experimental evidence suggesting a role for HER3 in the biology of other types of cancers not driven by amplification of HER2. In a mouse genetic model of colon cancer, the development of APC-linked tumorigenesis is averted by intestinal ERBB3 gene deletion (Lee et al., 2009). The correlation between HER3 and HER4 expression in this model suggests that HER4 may be the catalytic dimerization partner for HER3. On the other hand, a similar suppression of tumorigenesis is seen in this mouse model with a hypomorphic EGFR allele implicating a functional role for EGFR in this disease (Roberts et al., 2002). A number of experimental and association studies suggest that HER3 mediates resistance to anti-EGFR therapies in RAS wild-type colon cancers (Bosch-Vilaró et al., 2017; Seligmann et al., 2018) and to MEK inhibitors in Ras mutant colon cancers (Sun et al., 2014). A number of studies suggest that HER3 mediates resistance to anti-EGFR therapies in head and neck cancers, particularly in the EBV positive subtype (Erjala et al., 2006). A similar body of work implicates HER3 in the proliferative and metastatic attributes of melanomas (Buac et al., 2009; Tiwary et al., 2014) and in melanoma resistance to Raf/MEK inhibitors (Abel et al., 2013; Fattore et al., 2013). Although the physiologic function of HER3 is predominantly as a partner for its HER family siblings, there is evidence that it is also phosphorylated by c-MET when overexpressed in some tumors (Engelman et al., 2007; Frazier, Brand, Gordan, Grandis, & Jura, 2019). There are numerous reports implicating a role for HER3 in the biology of other types of cancer including prostate cancer (Chen et al., 2010; Gregory et al., 2005; Soler et al., 2009), ovarian cancer

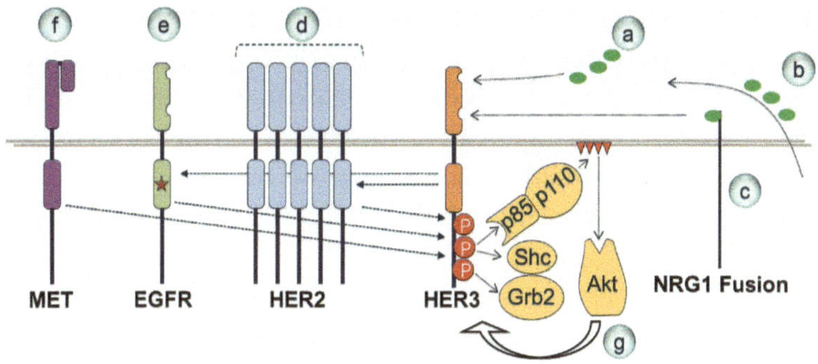

Fig. 3 Proposed modes of HER3 role in human cancer: Increased HER3 phosphorylation and signaling in cancers can be mediated through ligand activation in tumors in a paracrine manner (A) or autocrine manner (B), or through expression of mutant NRG1-fusion proteins (C). HER3 can be constitutively phosphorylated in cancers with HER2 overexpression (D) or in cancers with mutationally activated EGFR (E) and can also undergo promiscuous phosphorylation in cancers with overexpression of c-MET (F). HER3 is a strong activator of PI3K and its expression is also tightly linked with downstream PI3K/Akt signaling creating a negative feedback loop. In many types of cancers, inhibition of oncogenic PI3K/Akt signaling by various targeted therapies can lead to the compensatory upregulation of HER3 (G).

(Chung et al., 2019) and other cancers. Whether HER3 plays a driving role in these cancer subtypes remains to be determined.

Of particular relevance to HER3 is the discovery of NRG1 fusion events in a small percentage of lung cancers, pancreatic cancers, and other cancers (Dhanasekaran et al., 2014; Fernandez-Cuesta et al., 2014; Jones et al., 2017; Jonna et al., 2019). These fusion proteins preserve the EGF-like domains of NRG1, typically expressed at the membrane surface, and lead to HER3 activation and phosphorylation (Fernandez-Cuesta et al., 2014; Heining et al., 2018). The driving role of this NRG1 genetic event is supported by anectodal reports of response to HER family kinase inhibitors or anti-HER3 antibodies in these patients (Drilon et al., 2018; Gay et al., 2017). The various mechanisms suggested by existing literature that invoke HER3 in the pathogenesis of human cancers is schematically summarized in Fig. 3.

12. HER3 mutations in cancers

Some have suggested that as much as 11% of gastric and colon cancers harbor activating mutations of HER3 and some *in vitro* studies have been provided to support these claims (Choi et al., 2014; Jaiswal et al., 2013;

Mishra, Patel, Alanazi, Yuan, & Garrett, 2018). But these *in vitro* studies require HER2 overexpression to produce biologic transformation and tumorigenicity, mitigating any conclusions that the alteration in HER3 observed in tumors lacking HER2 overexpression is an oncogenic event. It is likely that these reported mutations are passenger mutations misclassified as driver mutations. Consistent with this, other large-scale tumor sequencing efforts have not reported such prevalent HER3 driver mutations in gastric or colon cancers (Lieu et al., 2019; Ye, Yu, Aggarwal, & Reinhard, 2016) except for the subset of microsatellite unstable gastric cancers with exceedingly high passenger mutation burdens (2014). A genomic-driven clinical study of the HER family inhibitor neratinib showed good clinical efficacy in cancers with activating HER2 mutations, but there was not a single response in any cancers with one of the reported HER3 mutations (Hyman et al., 2018) consistent with the non-driver nature of such mutations. The only HER3 alteration that is observed in some reports at a frequency consistent with a hotspot event is at Val104 in the extracellular domain. Although this HER3 mutant is not transforming on its own (Jaiswal et al., 2013), it is possible that such alterations in HER3 can have a selective advantage and be enriched as a companion to another primary oncogenic driver event in HER2 or other HER family members. Alternatively, this too may be a passenger mutation. It is now well recognized that many cancer mutations previously thought to be driver events on the basis of recurrence at hotspots are actually passenger mutations that concentrate at hotspots due to the non-random mechanistic nature of the cancer mutational process (Buisson et al., 2019; Hess et al., 2019; Seton-Rogers, 2019).

13. HER as a prototype for RTK-directed targeted cancer therapy

Over the years, a tremendous progress has been made in cancer research and treatment, leading to the prolonged survival of cancer patients (and infrequently, to cancer cure) after initial diagnosis. In addition to superior modalities of radio and chemotherapeutics, a large part of this progress is attributed to targeted cancer therapy wherein a designed drug is intended for a particular target, activity, or pathway in cancer cells (Kumar et al., 2019; Mendelsohn, 2000; Sawyers, 2004). The founding HER family member—EGFR—plays a crucial role in targeted therapy, as it provided the prototype for developing the concept of HER-directed therapies (Kumar, 2001).

In this context, the hypothesis of targeting the EGFR/HER1 pathway by highly specific anti-EGFR blocking antibodies—which function by impairing an extracellular autocrine loop (Van de Vijver, Kumar, & Mendelsohn, 1991)—by John Mendelsohn led to the development of cetuximab for epithelial tumors with increased EGFR levels (Mendelsohn, 2000). The belief that the functionality of HER ECDs could be blocked with mAbs was realized for HER2 because anti-HER2 mAb 4D5 could block HER2 tyrosine phosphorylation in serum-stimulated breast cancer cells with high HER2 levels (Kumar et al., 1991). In addition, mAb 4D5 also prevented the levels of HRG-induced HER2 tyrosine phosphorylation as well as HRG-stimulated HER3-HER2 interaction and resulting activation of PI-3 kinase pathway in breast cancer cells without HER2 overexpression (Adam et al., 1998). The anti-HER2 mAb4D5 was humanized and developed as trastuzumab or herceptin for breast cancer patients with HER2 (Baselga et al., 1999; Carter et al., 1992). These early developments in targeting cell surface RTKs provided a roadmap to the scientific community and pharmaceutical industry for developing additional highly selective Abs, TK inhibitors (TKIs), or conjugates to target distinct epitopes on the ECDs of HER1, HER2, or HER3; transmembrane regions involved in HER-heterodimerization to prevent resulting signaling; or specific target sites in cytoplasmic kinase domains. As the HER signaling outcome is also influenced by other over-activated RTKs, such as IGF1R, MET, or VEGFR, the concept of co-targeting such RTKs, has offered another avenue to potentiate the benefits of HER-directed therapies in certain cancer types (Fig. 4).

14. Development of HER3 inhibitors

The small molecule ATP analog class of tyrosine kinase inhibitors (TKIs) are not effective against HER3 since its KD is catalytically inactive. Although inhibiting the signaling functions of the HER3 KD may not be possible with standard ATP-competitive small molecules, it remains possible that allosteric class small molecules that interfere with the activating functions of its KD or ATP-competitive small molecules that somehow disturb the structural integrity of its KD may be able to disrupt the signaling functions of the HER3 KD. Another approach that can be used to target pseudokinases such as HER3 is through the design of proteolysis targeting chimeric (PROTACT) drugs that induce its degradation and such attempts have been reported (Lim et al., 2015; Xie et al., 2014). The membrane localization of HER3 requires processing through the Sec61 translocon and

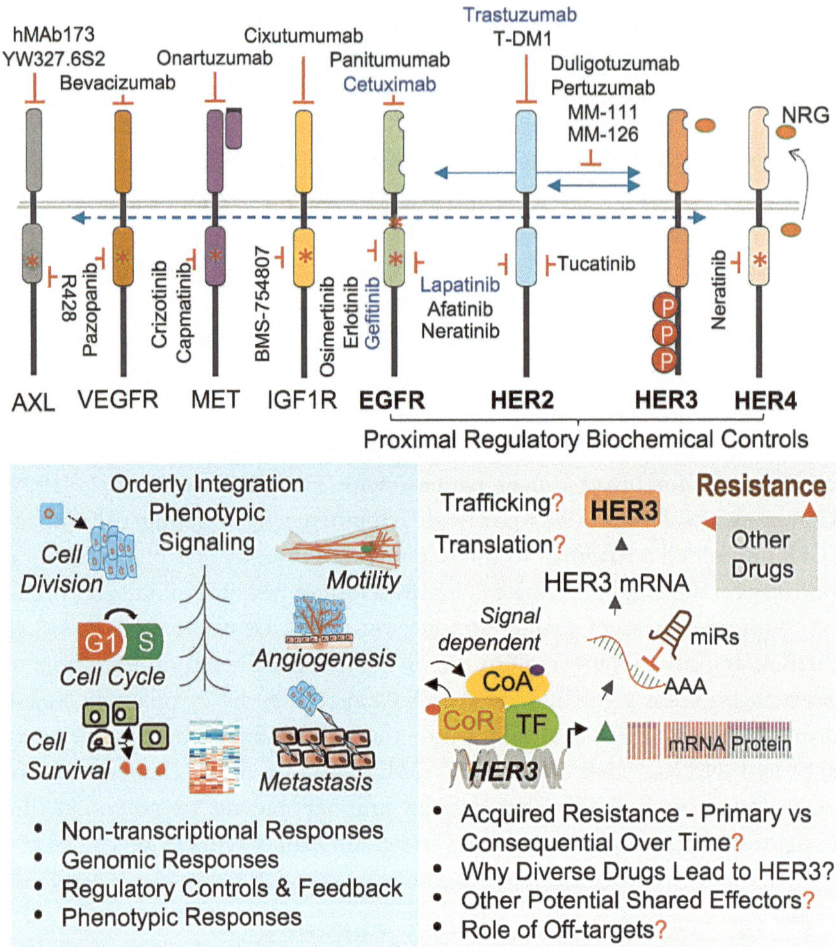

Fig. 4 HERs in phenotypic signaling and therapeutic resistance. Homo- and heterodimerization of HER among the HER family of other cell surface RTK family members trigger signaling cascades in a coordinated manner, leading to examples of phenotypic changes. Examples of HER-directed targeted therapeutics and some of phenotypic effects are shown here. HER-directed therapies eventually lead to acquired therapeutic development. An example of HER3 upregulation by diverse therapies is depicted as a potential effector of therapeutic resistance. Therapeutic names in blue shown to upregulate HER3 an/or NRG.

drugs targeting this channel can interfere with HER3 processing and induce its degradation with a remarkable degree of selectivity (Ruiz-Saenz et al., 2015). Some have suggested that inhibitors of Hsp90 can induce HER3 degradation (Gerbin & Landgraf, 2010) although the broad client profile of Hsp90 has made it a difficult target for clinical translation. These and other strategies to disrupt HER3 function by small molecules are underway, but no agent has yet emerged suitable for clinical testing.

Monoclonal antibodies (mAbs) can be generated to target virtually any surface protein with remarkable selectivity and many such agents have been developed against the HER3 ECD. HER3-targeting mAbs have been developed with different design principles. Most are designed to inhibit the ligand-induced activation of HER3 either by interfering with ligand binding or by interfering with the conformation changes induced by ligand binding while others have been reported to induce HER3 internalization and downregulation, and yet others optimized for maximal immunologic effects (Gaborit, Lindzen, & Yarden, 2016; Mishra et al., 2018). At least one antibody-drug conjugate (ADC) has also been developed targeting HER3 (Koganemaru et al., 2019). Many of these HER3-targeting mAbs have advanced to clinical testing. The clinical activity of HER3-targeting mAb drugs has been modest at best and none have yet been approved for marketing. The challenge is to identify cancers suitable for HER3-targeting and to develop drugs that can disrupt the mode of function of HER3 in these cancers. The cancer type with the most solid evidence implicating HER3 in its pathogenesis is the HER2-amplified subtype of cancers. However in these cancers massive overexpression of HER2 drives HER2-HER3 signaling in a ligand-independent manner and HER3 mAbs designed to block ligand-driven signaling are unlikely to inactivate constitutive HER2-HER3 signaling. Antibodies that can induce internalization and degradation of HER3 can, in principle, be effective against these cancers, but such mechanistic claims are often not well substantiated or not sufficiently potent. Clinical studies of anti-HER3 mAbs in HER2-amplified cancers are underway and no data have yet been reported.

Outside of the context of HER2-amplified cancers, it remains a difficult task to identify cancers that are known to be driven by HER3 signaling. An assumption that has presided over the clinical development of most HER3-targeting mAbs has been that many human cancers are driven by neuregulin-HER3 signaling, providing a mechanistic rationale for testing mAbs that disrupt ligand-driven HER3 signaling with high expectations. This assumption is fueled by many studies of tumor tissues which report widespread expression of neuregulins in human cancers (Esteva et al., 2001; Lyne et al., 1997; Moustafa, Alaoui-Jamali, Paterson, & O'Connor-McCourt, 1999). But the clinical activities of HER3 mAbs have not lived up to these expectations with minimal to modest single agent activity seen across this class of drugs and minimal to modest incremental activity when added to standard chemotherapeutics, and several agents have already been pulled from the developmental pipelines.

The reasons why HER3 mAbs have underperformed expectations is a matter of ongoing debate, but in my opinion the assumption that most or many human cancers are driven by neuregulin signaling was weak to begin with and this assumption is now in serious jeopardy. Tissue studies reporting widespread neuregulin expression in tumors are over-interpreted. Soluble ligands act at very low concentrations and are difficult to detect in tissue stroma by immunostaining, and detected ligands are not necessarily bioavailable due to stromal sinks that regulate their bioavailability. Cellular markers such as RNA expression or intracellular staining are not reliable surrogates since there are multiple trafficking and proteolytic events that regulate the bioavailability of ligands that these analyses do not account for. The exception to this are the rare cancers with NRG fusion events. The hypothesis that these tumors are driven by the fusion oncoprotein is supported by the genetic history of the tumor and there are anectodal reports of response to anti-HER3 mAbs in some of such cases. The quest continues for biomarkers that can identify HER3-driven tumors other than HER2-amplified cancers and NRG1-fusion cancers.

15. HER signaling and therapeutic sensitivity

The HER family members regulate cellular processes in both kinase-dependent and -independent manner. For example, noncanonical HER signaling components promote gene expression through the nuclear HER2, HER2, or HER4 fragment and autophagy in the case of EGFR or promote apoptosis during translocation of the HER4 fragment 4IC to the mitochondria. As growth factor signaling stimulates cancer cell progression through the S-phase of the cell cycle, when cells are actively involved in *de novo* DNA synthesis and DNA damage repair, and as radiation therapy and chemotherapy largely act by targeting the DNA repair machinery in actively dividing cells, therapeutic approaches aimed to co-interrupt HER signaling and DNA repair pathways emerged as effective treatment modalities in numerous human cancer types. Mechanistically, interrupting EGFR signaling sensitizes cancer cells to radiation therapy and chemotherapy due to the impairment of repair activity in the nucleus at multiple levels, including maintenance of the DNA-dependent protein kinase (DNA-PK)—an enzyme speculated to be involved in double-strand break repair and V(D) J recombination—in the nucleus (Bandyopadhyay, Mandal, Adam,

Mendelsohn, & Kumar, 1998; Harari & Huang, 2001). Similarly, targeting HER2 inhibits the repair of DNA interstand crosslinks produced by chemotherapeutic agents in breast cancer cells; inhibiting HER3 in cancer cells, or targeting HER3 in pancreatic cancer cells through an anti-HER3 antibody conjugate sensitizes these cells to growth inhibition or apoptosis by DNA-damaging agents (Bourillon et al., 2019).

16. NRG-HER3 axis: A shared potential effector of acquired drug resistance

HER3 transduces the strongest cell survival and invasive signaling among HERs due to the presence of six p85/PI3K-binding motifs in its cytoplasmic domain (Hellyer, Cheng, & Koland, 1998) and due to its ability to preferably form HER3/HER2 heterodimers in HRG/NRG-stimulated cancer cells (Yarden & Pines, 2012) and induce profound cytoskeleton remodeling favoring motility and invasion through specific signaling nodes (Adam et al., 1998). Among HERs, HER3 is overexpressed in luminal A/B breast cancers. A large body of work using different model systems and sometimes, tumor specimens, suggests that transcriptional and post-transcriptional upregulation of HER3 expression is central to the development of acquired drug resistance against a diverse family of anticancer agents in multiple cancer types (Fig. 4). In addition to HER2 levels, sensitivity of breast cancer to endocrine therapy has been shown to be also dependent on HER3 levels (Tovey et al., 2005). For example, breast cancer treatment with fulvestrant, an antiestrogenic agent, upregulates HER3 expression as well as PI3K signaling, whereas targeted inactivation/depletion of HER3 in fulvestrant-treated breast cancer cells inhibits PI3K signaling and potentiates growth inhibition (Huang et al., 2010; Li et al., 2018; Liu et al., 2007). HER3 overexpression in ER-positive breast cancer minimizes the beneficial effects of antiestrogenic therapy due to the development of hormone independence and could contribute to cancer relapse (Frogne et al., 2009). Interestingly, fulvestrant treatment also increases HER phosphorylation levels, presumably, due to NRG-induced transregulation among HERs (Morrison et al., 2013). However, the mechanism of fulvestrant-mediated increased HER3 phosphorylation remains to be delineated completely.

In addition to antiestrogenic resistance, HER3 upregulation is a common mechanism of acquired drug resistance conferred by nearly all types of HER-directed therapies (Kumar, 2007). Despite the remarkable clinical success of gefitinib and trastuzumab in the treatment of EGFR-mutant non-small-cell lung cancer (NSCLC) and HER2-positive breast and gastric cancers, respectively, the benefits of response do not sustain for long due to relapses. Verma et al. and others have revealed that EGFR TKIs such as gefitinib and erlotinib upregulate HER3 expression in triple-negative breast cancer and lung carcinomas (Verma et al., 2017). There are also examples wherein resistance to *cetuximab therapy could be associated with increased levels of circulating NRG, the ligand for triggering HER3-HER signaling* (Yonesaka et al., 2011). Similarly, lapatinib-mediated inhibition of HER2 activity is accompanied by increased HER3 expression in a PI3K-FOXO3a-dependent manner as well as increased HER3 phosphorylation, presumably, due to the *involvement of ligand-induced HER3 activation* (Morrison et al., 2013). The *efficacy of approaches aimed at inhibiting the PI3K-Akt pathway is compromised due to HER3 and IGF1R upregulation in a FOXO-dependent manner* (Chandarlapaty et al., 2011). *Because HER3 upregulation and phosphorylation in a 3D culture is induced in a PI3K-independent manner (*Pickl & Ries, 2009), *FOXO3a may also be activated by kinases other than Akt.*

In addition to breast cancer, HER3 and NRG1 overexpression are common events in ovarian cancer (Davies et al., 2014; Mills & Yarden, 2010), and NRG1/HER3-signaling has emerged as a major determinant of therapeutic responsiveness (Gollamudi, Nethery, Liu, & Kern, 2004; Lorusso et al., 2019; Nagumo et al., 2009; Sheng et al., 2010; Tanner et al., 2006). Increased HER3 expression and resulting HER3/HER2 signaling are vital in the development of hormone-independent prostate cancer (Agus et al., 2002; Chen et al., 2010; Jathal, Chen, Mudryj, & Ghosh, 2011). HER upregulation signaling also contributes through cross-talks with other RTKs, such as IGF1R, MET, or AXL, and feeds into therapeutic resistance in cancer cells (Camblin et al., 2019; Engelman et al., 2007; Jia et al., 2013; Lu, Zi, Zhao, Mascarenhas, & Pollak, 2001; Torka et al., 2014). Moreover, HER3 levels are considerably upregulated in radioresistant pancreatic cancer cell lines act as determinants of HSP inhibitor-induced radio-sensitization (Dote, Cerna, Burgan, Camphausen, & Tofilon, 2005), and are associated with therapeutic resistance against paclitaxel in cancer cells (Teixidó et al., 2012), and doxorubicin-resistant ovarian cancer cells (Bezler, Hengstler, & Ullrich, 2012).

A broader significance of HER3 upregulation in therapeutic resistance is evident through its increased expression in tumors with mutant BRAF, such as melanoma, thyroid, and colon subtypes, which develop resistance to BRAF inhibitors (Herr et al., 2018; Montero-Conde et al., 2013). In such cases, combination of RAF inhibitors with anti-HER therapies confer a superior growth inhibitory response in various model systems (Kugel, Hartsough, Davies, Setiady, & Aplin, 2014). Mechanistically, the BRAF inhibitor vemurafenib stimulates *HER3* transcription through FOXD3 and ZNF217 DNA-binding factors in tumor cells (Abel et al., 2013; Krig et al., 2010). In thyroid cancer cells, BRAF inhibitors induce *HER3* transcription due to the loss of its transcriptional repression by CtBP1 and CtBP2 corepressors and increased NRG1-HER3 signaling axis (Montero-Conde et al., 2013). The acquired resistance reported in these studies could be counteracted by trastuzumab, which implies the involvement of NRG-mediated HER3/HER2 dimer formation. Additionally, *HER3* transcription could be positively regulated by additional DNA-binding factors, such as SOX10, FOXO1 and FOXO3a, and negatively modulated by the MITF corepressor (Alver et al., 2016; Prasad et al., 2011).

In addition to *HER3* transcription, HER3 levels are upregulated in human cancers, including in breast cancer, in the absence of detectable HER3 mRNA upregulation. Thus, nontranscriptional mechanisms of HER3 upregulation may exist, such as microRNA control of HER3 mRNA stability and/or regulation of the steady-state level of HER3 protein by components of trafficking or degradation. It is worth noting that HER3 mRNA comprises seed motifs for miR125a, miR125b, and miR205 and could be downregulated by entinostat (a pan HDAC inhibitor) in an miR-dependent manner (Wang et al., 2013). As transcription in growth factor-stimulated cancer cells is tightly regulated through an orderly formation of coactivator/transcription factor complexes on the target promoter chromatin (Kumar, Wang, & Barnes, 2004; O'Malley & Kumar, 2009) as well as orchestrated by proximal signaling (Kumar & Hung, 2005), *HER3* promoter chromatin might be regulated through a coordinated differential recruitment or derecruitment of coactivator and corepressor complexes through chromatin remodeling complexes. However, the role of pathways in controlling HER3 expression during acquired drug resistance and the manner in which signals from diverse anticancer agents converge into molecules/pathways responsible for modulating the *HER3 chromatin,* HER3 mRNA, or HER3 protein stability remain unknown; studies for understanding these functions are currently pursued in Kumar's laboratory. We envision that underlying shared mechanisms are likely to involve the engagement of shared proximal signaling kinases, physical translocation of activated kinases to the nucleus, inactivation of transcriptional corepressors, activation of coactivators and DNA-binding factors, and recruitment of chromatin remodeling complexes onto HER3 and

NRG1/2 promoters. In brief, these examples suggest that compensatory increased HER3 expression and resulting positive feedback regulation of the NRG-HER3 axis might be one of the shared mechanisms of therapeutic resistance to diverse anticancer drugs and highlight the potential benefits of co-targeting HER3, along with other target therapy modalities in several cancer types (Kennedy et al., 2019; Liu et al., 2019).

17. Acquired therapeutic resistance: A model for consideration

In general, epithelial tumors treated with HER-directed targeted therapies develop therapeutic resistance. This is partly due to cellular effects of therapeutic agents, such as accumulation of somatic mutations and other genomic alterations, as well as rewiring of critical cellular pathways supporting growth, survival, metabolic requirements, and oncogenic processes. Therefore, the effectiveness of HER-directed therapies in most epithelial cancers may depend on the timepoint of treatment at which a tumor starts developing resistance and starts counteracting the beneficial effects of the drug. This suggests that the development of cancer therapeutic resistance is inevitable, and an effective targeting strategy must consider the role of molecules beyond the obvious intended target(s) or targets. Because resistance cannot be avoided, one potential approach to prolong the beneficial effects of RTK-directed therapy is to develop strategies to delay the development of resistance and use this window of time to completely impair the functionality of the intended target(s) (Fig. 5). In addition, currently, there is a lack of predictable biomarkers that could be used to accurately determine the time to resistance before its development. Because the underlying

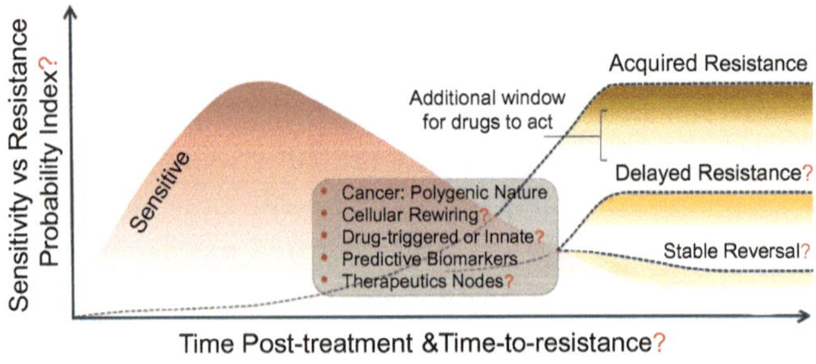

Fig. 5 A proposed model to overcome acquired therapeutic resistance against RTK-directed therapies.

resistance mechanisms are believed to be shared among multiple targeted therapeutics, as is the case of HER3-NRG upregulation, developing a comprehensive molecular portrait of a tumor as it starts drifting from a drug-sensitive phenotype may be prove beneficial. This portrait can provide an improved understanding of the nature of cellular changes in a physiologically relevant model system as a function of time-to-treatment before the development of full-blown resistance. Such a postulated working model is expected to offer a battery of dynamic molecular nodes of predictable biomarkers as well as allow switching or co-targeting of such molecules during the course of treatment to ward-off or delay the development of therapeutic resistance. At present, we do not know if such an approach would be superior to the current model of co-targeting the molecules once the resistance has already developed. In brief, achieving stable reversal of full-blown acquired resistance is likely to be more challenging than attempting to further delay or reverse the process before it is completely developed. This is clearly a less-appreciated area of HER research at present and is expected to be at the forefront in the coming decade.

18. HER-directed therapies: Potential drawbacks and future scope

The development of HER-directed mAbs and TKI provided several mechanistic insights into the biology of HER family and extensive benefits for the well-being of cancer patients globally. This development has allowed us to achieve a high cure rate in the early stages of cancer and an impressive extension of progression-free survival (PFS) and overall survival in a metastatic setting, particularly in breast cancer (Parakh et al., 2017; Swain et al., 2015). Despite the successful therapeutic application of existing HER-directed therapies, benefits of personalized treatment with new biomarkers in offering a superior tailoring of anti-HER therapies and approaches to overcome acquired resistance to such agents remains a great challenge (Parakh et al., 2017). We here discuss HER2-directed therapies as an example to highlight some of potential drawbacks of these therapies as future objectives are being planned (Table 1).

HER2 is overexpressed, amplified, or mutated in several solid tumors, including those in breast, NSCL, and gastric cancers. However, these alterations are not clearly differentiated in terms of shared, specific, or guiding principles of patient care. For example, HER2 amplification and mutations do not overlap and thus may represent distinct clinical entities and

Table 1 Representative clinical trials of HER-directed therapies in human cancer patients.

	Drug/ Tumor Type/ Pivotal Study	Tratment arms	Endpoint	Overall response rate (%)	Progression-free survival (months)	Overall survival
Metastatic	Trastuzumab/ Gastric/ ToGA Bang et al (2010)	Trastuzumab + Chemotherapy vs. Chemotherapy alone	OS 13.8 vs. 11.1 months HR 0·74 (p=0·0046)	47.3 vs. 34.5	6.7 vs. 5.5 HR 0.71 (p=0.002)	
	Pertuzumab/ Breast/ CLEOPATRA Tabernero et al (2018)	Trastuzumab + Docetaxel vs. Trastuzumab + Docetaxel + Pertuzumab	PFS 18.7 vs. 12.4 months HR 0.68 (p<0.0001)			56.5 vs. 40.8 months HR=0.68 (p=0.0002)
	T-DM1/ Breast/ EMILIA Diérias et al (2017)	T-DM1 + Capecitabine vs. Lapatinib + Capecitabine	PFS 9.6 vs 6.4 months HR 0.65 (p<0.001)	43.6 vs 30.8 (p<0.001)		30.9 vs 25.1 months HR 0.68 (p<0.001)
	T-DM1/ Breast/ TH3RESA Perez et al (2017)	T-DM1 vs. Physician choice			6.2 vs. 3.3 HR 0.528 (p<0·0001)	68.6% vs. 56.9%
	T-DM1/ Breast/ MARIANNE Geyer et al (2006)	T-DM1 plus Pertuzumab T-DM1 plus placebo Trastuzumab plus taxane	PFS (months) 15.2 14.1 13.7	64.2 59.7 67.9		
	Tucatinib/ Breast/ Rashmi K. Murthy et al (2020)	Trastuzumab + Capecitabine + Tucatinib Trastuzumab + Capecitabine + Placebo	PFS (months) 7.8 vs. 5.6.7 HR 0.54 (p<0.001)			21.9 vs. 17.4 months HR=0.66 (p=0.005)
	Lapatinib/ Breast/ Gianni et al (2012)	Lapatinib + Capecitabine vs. Capecitabine	TTP 8.4 vs 4.4 months HR 0.47 (p<0.001)	22% vs 14% (p=0.09)		35 deaths on Capecitabine vs. 36 Lapatinib + Capecitabine (p=0.72)

Neoadjuvant	Pertuzumab /Breast / Gianni et al (2012)	Trastuzumab + Docetaxel Pertuzumab + Trastuzumab + Docetaxel Pertuzumab + Trastuzumab Pertuzumab + Docetaxel	DFS 81% 86% 73% 73%	pCR 29 45.8 16.8 24	
Adjuvant	Trastuzumab /Breast / HERA Piccart-Gebhart et al (2005)	Trastuzumab 1 year vs. Trastuzumab 2 years	DFS 75.8 vs. 76% HR 0.99 (p=0.86)		86.9% vs. 88.7% HR 1.05 (p=0.63)
	T-DM1/ Breas/ KATHERINE Minckwitz et al (2019)	T-DM1 vs. Trastuzumab	3-year DFS 88.3% vs. 77% (p<0.001)		36 deaths with T-DM1 vs. 42 with Trastuzumab (p=0.08)
	Pertuzumab /Breast / APHINITY Minchwitz et al (2017)	Pertuzumab + Trastuzumab + Chemotherapy vs. Placebo + Trastuzumab + Chemotherapy	3-year DFS (N+disease) 94.2% vs. 93.2% (p=0.02)		

therapeutic targets (Li et al., 2016). In addition, the criteria of HER2 status in decision-making slightly differ between breast and gastric cancers due to inherent biological differences between the two cancer types, for example, basolateral membrane staining is used in gastric cancer and apical staining in breast cancer (Yan, Parker, Schwab, & Kurzrock, 2014). These examples emphasize the need to formulate better approaches for validating the status and spatial temporal distribution of HER2, as this, in turn, might explain the divergent results of several completed trials describing the ineffectiveness of combining pertuzumab with trastuzumab and chemotherapy for the first-line treatment of HER2-positive metastatic gastric cancer, as opposed to breast cancer (Li et al., 2016; Swain et al., 2015; Tabernero et al., 2018; Yan et al., 2014).

Although primary resistance to anti-HER2 agents is unavoidable, most therapeutic failures are driven by acquired resistance due to progressive selection of clones during the treatment. Several potential mechanisms of primary or secondary resistance to anti-HER2 agents have been identified, mostly involving genetic or epigenetic alterations that could result in the overexpression or constitutive activation of HERs or other cell surface RTKs (i.e., MET and FGFR1) or their downstream effectors, as well as an altered immune response affecting the antibody-dependent cellular cytotoxicity response. Irrespective of the specific mechanism in-effect, the reactivation of the cell survival signaling axis is crucial for inducing and maintaining the status of anti-HER2 therapy resistance. Similarly, during T-DM1 (ado-trastuzumab emtansine) resistance, mechanisms involving lysosomal function and/or drug internalization may play an important role in the development of T-DM1 resistance (Parakh et al., 2017; Vernieri et al., 2019).

Currently, no validated biomarkers can completely predict the responsiveness to anti-HER2 therapies. Analyses from the CLEOPATRA study led to the identification of high levels of HER2 protein, HER2 mRNA, HER3 protein, and wild-type PIK3CA as molecules associated with a superior prognosis. Of these, the PIK3CA mutation status revealed the highest prognostic effect, with the median PFS longer in patients with the wild-type PIK3CA than in those with mutant PIK3CA (Parakh et al., 2017; Swain et al., 2015). A similar predictive effect of the PIK3CA mutational status was reported in the CHER-LOB study on gastric cancer (Díaz-Serrano et al., 2016; Guarneri et al., 2015). However, the association of mutated PIK3CA with favorable prognosis in hormone receptor-positive breast tumors was also reported; T-DM1 was found to be effective in tumors with both the wild-type and mutated PI3KCA (Baselga et al., 2016; Kim et al., 2016; Parakh et al., 2017). A recent meta-analysis designed to understand the effect of PIK3CA-activating mutations on neoadjuvant therapy in HER2-positive breast tumors revealed the association of PIK3CA mutations with lower pathologic complete response rates that are significantly higher in hormone receptor-positive cancers (Brandão et al., 2018; Loibl et al., 2016; Parakh et al., 2017).

Other promising biomarkers in advanced HER2-positive cancers are the molecules capable of providing a global view of tumor heterogeneity by quantifying HER2 expression in all metastatic sites (Brandão et al., 2018). One way to achieve this objective is by using 89Zr-trastuzumab-PET/CT to quantify the HER2-receptor expression status at baseline. In the ZEPHIR trial, such an approach, in conjunction with FDG-PET/CT, could clearly discriminate between trastuzumab-resistant patients with >50% tumor burden expressing the HER2-receptor and those with a >50% "negative" tumor load (Gebhart et al., 2016; Parakh et al., 2017). In the similar context, the IMPACT-MBC trial is also exploring the aforementioned imaging modality by assessing its predictive value and comparing 18F-FES-PET/CT with circulating tumor DNA and overall cost-effectiveness. Another approach is to use circulating tumor DNA to serially assess genomic alterations in all metastatic sites and eventually retrieve an image of an HER2-positive "tumor load" in a patient. However, this method yet-to-be prospectively validated in HER2-positive cancers (Ma et al., 2016).

Similar to the triple-negative breast cancer, HER2-positive tumors are immunogenic; several studies have revealed the expression of tumor infiltrating lymphocytes (TILs) in metastatic tumors as well as increased mutational burden in these tumors compared with luminal tumors (Brandão et al., 2018). For example, HER2-positive patients with increased baseline TILs experience greater benefit with trastuzumab plus chemotherapy in the adjuvant setting (Loi et al., 2014). On the other hand, results of a meta-analysis of trials in 1256 patients treated with neoadjuvant chemotherapy plus trastuzumab and lapatinib, either alone or in combination, confirmed a strong association between the high TILs at the base line and increased pathologic complete response rates (OR: 2.46, 95%; CI: 1.36–4.43) (Brandão et al., 2018; Solinas et al., 2017). In the advanced disease setting, the PANACEA trial revealed that patients with high-TIL tumors have a greater overall response rate with immunotherapy/trastuzumab (Loi et al., 2019). However, in the CLEOPATRA study, TILs were found to be prognostic but not a better predictive of a response to dual HER2-blockade (Swain et al., 2015). Overall, these observations suggest that HER2-positive tumors and their interaction with the host immune system may not be always reflected only by TILs, and hence, more functional assays should be performed to precisely determine the true prognostic and predictive role of infiltrating lymphocytes in HER2-positive cancers (Khoury et al., 2018).

The efficacy of anti-HER2 therapies in brain metastases is poorly understood. The administration of anti-HER2 mAbs considerably delays the onset of symptomatic brain cancer; however, the limited penetration of mAbs into the central nervous system hinders their efficacy. As TKIs can cross the blood-brain barrier, such small molecules are expected to be more effective for treating brain metastases in HER2-positive breast cancer. However, currently available TKIs exhibit several limitations, including toxicity, resistance, and a poor pharmacokinetic profile (Singla, Munshi, Banipal, & Kumar, 2018). Novel HER2 inhibitors with an improved pharmacological profile need to be developed. Tucatinib—an oral selective HER2 inhibitor—possesses a superior intracranial activity than lapatinib and neratinib in a HER2-positive breast cancer xenograft model (Brandão et al., 2018; Parakh et al., 2017). In brief, with improved knowledge of the biology of and acquired resistance to anti-HER2 therapies, combinations with drugs such as PI3K/Akt/mTOR inhibitors, CDK4/6 inhibitors, and anti-PD(L)1-antibodies and endocrine therapies and new anti-HER2 agents (including in low HER2 level cancers) are extensively examined in clinical trials (Brandão et al., 2018).

19. Future anticipated advances

Targeted cancer therapy is yet to realize its complete potential in suppressing the target tumor in majority of patients, if such individuals, presumed to be benefitted by a given drug, are selected on the basis of an active target for a specific cancer drug. A new approach may provide additional significant gains and benefits for cancer patients. However, considering the progress made since the discovery of the HER family, the authors anticipate that the HER family will continue to serve as a prototypic RTK and influence cancer research and treatment in the forthcoming decade. The following is a partial list of postulated advances in HER-centered cancer research in the forthcoming decade.

- Improved understanding of HER overexpression, amplification, and mutation differences in difficult-to-treat cancers

- Development of predictable biomarkers of therapeutic responsiveness beyond the mere presence or upregulation of the intended target
- Improved understanding of the determinants of off-target toxicity as well as beneficial effects of a given HER-directed therapy
- Development of treatment strategies allowing a superior tailoring of anti-HER-directed therapies while considering the complete portrait of the HER family
- Improved understanding of dormant pathways serving as precursors for yet-to-be manifested therapeutic resistance and development of predictable biomarkers of acquired resistance
- Increased consideration of noncanonical, kinase-independent HER pathways while designing translational research questions
- Improved understanding of trafficking pathways responsible for the accumulation of HERs or their cleaved fragments in distinct intracellular compartments and the nature of off-mechanisms involved in such HER fragments
- Optimization of intracranial activity of anti-HER2 therapies
- Improved understanding of bidirectional regulatory cross-talk between the epithelial tumor cells and the components of the host immune system
- Start considering the mechanistic significance of circulating HERs in cellular functions and HER-directed clinical trials

20. Conclusions

Since the discovery of the founding HER family member, the EGFR, approximately four decades ago, the HER field has witnessed a tremendous progress in all three areas of HER research: basic, translational, and clinical. The central role of HERs in cancer biology has become more apparent, and this understanding has fueled some of the biggest advances in cancer therapeutics. Understanding the HER3 biology lagged initially because some of the paradigms applied to EGFR, HER2, and RTK do not apply to HER3, which is a pseudokinase receptor. Examples of such HER-centered advances include, wide-ranging acceptance of the notion of signal-dependent transregulation among HERs in phenotypic signaling; role of HER signaling in cytoskeleton remodeling and invasion; unprecedented insights into the details of the HER structure-function relationship, leading to the development of superior anti-HER-directed therapeutics; emergence of the kinase domain-independent noncanonical HER signaling; mechanistic role of compensatory stimulation of the HER3 pathway in the development of resistance to HER2-directed therapies; and significance of circulating cleaved HER fragments in therapeutic sensitivity. These advances led to the acceptance of the HER family as the most preferred therapeutic target in a large number of epithelial tumor types.

Acknowledgments

Due to paucity of space, the authors apologize to many of their colleagues for not including many other deserving observations in this article. Bijesh George and Aswathy Mary Paul are graduate students registered with the Manipal University Ph.D. Program.

References

Abel, E. V., Basile, K. J., Kugel, C. H., 3rd, Witkiewicz, A. K., Le, K., Amaravadi, R. K., et al. (2013). Melanoma adapts to RAF/MEK inhibitors through FOXD3-mediated upregulation of ERBB3. *The Journal of Clinical Investigation*, *123*, 2155–2168.

Adam, L., Vadlamudi, R., Kondapaka, S. B., Chernoff, J., Mendelsohn, J., & Kumar, R. (1998). Heregulin regulates cytoskeletal reorganization and cell migration through the p21-activated kinase-1 via phosphatidylinositol-3 kinase. *The Journal of Biological Chemistry*, *273*, 28238–28246.

Agus, D. B., Akita, R. W., Fox, W. D., Lewis, G. D., Higgins, B., Pisacane, P. I., et al. (2002). Targeting ligand-activated ErbB2 signaling inhibits breast and prostate tumor growth. *Cancer Cell*, *2*, 127–137.

Alver, T. N., Lavelle, T. J., Longva, A. S., Øy, G. F., Hovig, E., & Bøe, S. L. (2016). MITF depletion elevates expression levels of ERBB3 receptor and its cognate ligand NRG1-beta in melanoma. *Oncotarget*, *7*, 55128–55140.

Amin, D. N., Ahuja, D., Yaswen, P., & Moasser, M. M. (2015). A TORC2-Akt feed-forward topology underlies HER3 resiliency in HER2-amplified cancers. *Molecular Cancer Therapeutics*, *14*, 2805–2817.

Amin, D. N., Ruiz-Saenz, A., Gulizia, N., & Moasser, M. M. (2015). Chemical probing of HER2-amplified cancer cells identifies TORC2 as a particularly effective secondary target for combination with lapatinib. *Oncotarget*, *6*, 41123–41133.

Amin, D. N., Sergina, N., Ahuja, D., McMahon, M., Blair, J. A., Wang, D., et al. (2010). Resiliency and vulnerability in the HER2-HER3 tumorigenic driver. *Science Translational Medicine*, *2*, 16ra7.

Amin, D. N., Sergina, N., Lim, L., Goga, A., & Moasser, M. M. (2012). HER3 signalling is regulated through a multitude of redundant mechanisms in HER2-driven tumour cells. *The Biochemical Journal*, *447*, 417–425.

Asgeirsson, K. S., Agrawal, A., Allen, C., Hitch, A., Ellis, I. O., Chapman, C., et al. (2007). Serum epidermal growth factor receptor and HER2 expression in primary and metastatic breast cancer patients. *Breast Cancer Research*, *9*, R75.

Austin, C. D., De Mazière, A. M., Pisacane, P. I., van Dijk, S. M., Eigenbrot, C., Sliwkowski, M. X., et al. (2004). Endocytosis and sorting of ErbB2 and the site of action of cancer therapeutics trastuzumab and geldanamycin. *Molecular Biology of the Cell*, *15*, 5268–5282.

Bandyopadhyay, D., Mandal, M., Adam, L., Mendelsohn, J., & Kumar, R. (1998). Physical interaction between epidermal growth factor receptor and DNA-dependent protein kinase in mammalian cells. *The Journal of Biological Chemistry*, *273*, 1568–1573.

Bang, Y.-J., Van Cutsem, E., Feyereislova, A., Chung, H. C., Shen, L., Sawaki, A., et al. (2010). Trastuzumab in combination with chemotherapy versus chemotherapy alone for treatment of HER2-positive advanced gastric or gastro-oesophageal junction cancer

(ToGA): A phase 3, open-label, randomised controlled trial. *Lancet (London, England), 376*, 687–697.

Barbieri, E., Di Fiore, P. P., & Sigismund, S. (2016). Endocytic control of signaling at the plasma membrane. *Current Opinion in Cell Biology, 39*, 21–27.

Baselga, J., Lewis Phillips, G. D., Verma, S., Ro, J., Huober, J., Guardino, A. E., et al. (2016). Relationship between tumor biomarkers and efficacy in EMILIA, a phase III study of trastuzumab emtansine in HER2-positive metastatic breast cancer. *Clinical Cancer Research: An Official Journal of the American Association for Cancer Research, 22*, 3755–3763.

Baselga, J., Tripathy, D., Mendelsohn, J., Baughman, S., Benz, C. C., Dantis, L., et al. (1999). Phase II study of weekly intravenous trastuzumab (Herceptin) in patients with HER2/neu-overexpressing metastatic breast cancer. *Seminars in Oncology, 26*, 78–83.

Baulida, J., Kraus, M. H., Alimandi, M., Di Fiore, P. P., & Carpenter, G. (1996). All ErbB receptors other than the epidermal growth factor receptor are endocytosis impaired. *The Journal of Biological Chemistry, 271*, 5251–5257.

Bergeron, J. J. M., Di Guglielmo, G. M., Dahan, S., Dominguez, M., & Posner, B. I. (2016). Spatial and temporal regulation of receptor tyrosine kinase activation and intracellular signal transduction. *Annual Review of Biochemistry, 85*, 573–597.

Bezler, M., Hengstler, J. G., & Ullrich, A. (2012). Inhibition of doxorubicin-induced HER3-PI3K-AKT signalling enhances apoptosis of ovarian cancer cells. *Molecular Oncology, 6*, 516–529.

Bishop, J. M. (1983). Cellular oncogenes and retroviruses. *Annual Review of Biochemistry, 52*, 301–354.

Blackwell, K. L., Burstein, H. J., Storniolo, A. M., Rugo, H., Sledge, G., Koehler, M., et al. (2010). Randomized study of lapatinib alone or in combination with trastuzumab in women with ErbB2-positive, trastuzumab-refractory metastatic breast cancer. *Journal of Clinical Oncology: Official Journal of the American Society of Clinical Oncology, 28*, 1124–1130.

Bosch-Vilaró, A., Jacobs, B., Pomella, V., Abbasi Asbagh, L., Kirkland, R., Michel, J., et al. (2017). Feedback activation of HER3 attenuates response to EGFR inhibitors in colon cancer cells. *Oncotarget, 8*, 4277–4288.

Boudeau, J., Miranda-Saavedra, D., Barton, G. J., & Alessi, D. R. (2006). Emerging roles of pseudokinases. *Trends in Cell Biology, 16*, 443–452.

Bourillon, L., Bourgier, C., Gaborit, N., Garambois, V., Llès, E., Zampieri, A., et al. (2019). An auristatin-based antibody-drug conjugate targeting HER3 enhances the radiation response in pancreatic cancer. *International Journal of Cancer, 145*, 1838–1851.

Brandão, M., Pondé, N. F., Poggio, F., Kotecki, N., Salis, M., Lambertini, M., et al. (2018). Combination therapies for the treatment of HER2-positive breast cancer: Current and future prospects. *Expert Review of Anticancer Therapy, 18*, 629–649.

Brinckerhoff, C. E., & Matrisian, L. M. (2002). Matrix metalloproteinases: A tail of a frog that became a prince. *Nature Reviews. Molecular Cellular Biology, 3*, 207–214.

Buac, K., Xu, M., Cronin, J., Weeraratna, A. T., Hewitt, S. M., & Pavan, W. J. (2009). NRG1/ERBB3 signaling in melanocyte development and melanoma: inhibition of differentiation and promotion of proliferation. *Pigment Cell & Melanoma Research, 22*, 773–784.

Buisson, R., Langenbucher, A., Bowen, D., Kwan, E. E., Benes, C. H., Zou, L., et al. (2019). Passenger hotspot mutations in cancer driven by APOBEC3A and mesoscale genomic features. *Science (New York, N.Y.), 364*, eaaw2872.

Camblin, A. J., Tan, G., Curley, M. D., Yannatos, I., Iadevaia, S., Rimkunas, V., et al. (2019). Dual targeting of IGF-1R and ErbB3 as a potential therapeutic regimen for ovarian cancer. *Scientific Reports, 9*, 16832.

Campbell, M. R., Zhang, H., Ziaee, S., Ruiz-Saenz, A., Gulizia, N., Oeffinger, J., et al. (2016). Effective treatment of HER2-amplified breast cancer by targeting HER3 and β1 integrin. *Breast Cancer Research and Treatment, 155*, 431–440.

Cao, Z., Wu, X., Yen, L., Sweeney, C., & Carraway, K. L., 3rd. (2007). Neuregulin-induced ErbB3 downregulation is mediated by a protein stability cascade involving the E3 ubiquitin ligase Nrdp1. *Molecular and Cellular Biology, 27*, 2180–2188.

Carey, K. D., Garton, A. J., Romero, M. S., Kahler, J., Thomson, S., Ross, S., et al. (2006). Kinetic analysis of epidermal growth factor receptor somatic mutant proteins shows increased sensitivity to the epidermal growth factor receptor tyrosine kinase inhibitor, erlotinib. *Cancer Research, 66*, 8163–8171.

Carpenter, G. (1987). Receptors for epidermal growth factor and other polypeptide mitogens. *Annual Review of Biochemistry, 56*, 881–914.

Carpenter, G., King, L., Jr., & Cohen, S. (1978). Epidermal growth factor stimulates phosphorylation in membrane preparations in vitro. *Nature, 276*, 409–410.

Carpenter, G., King, L., Jr., & Cohen, S. (1979). Rapid enhancement of protein phosphorylation in A-431 cell membrane preparations by epidermal growth factor. *The Journal of Biological Chemistry, 254*, 4884–4891. https://pubmed.ncbi.nlm.nih.gov/312292.

Carpenter, G., & Liao, H.-J. (2013). Receptor tyrosine kinases in the nucleus. *Cold Spring Harbor Perspectives in Biology, 5*, a008979.

Carter, P., Presta, L., Gorman, C. M., Ridgway, J. B., Henner, D., Wong, W. L., et al. (1992). Humanization of an anti-p185HER2 antibody for human cancer therapy. *Proceedings of the National Academy of Sciences of the United States of America, 89*, 4285–4289.

Chakrabarty, A., Sánchez, V., Kuba, M. G., Rinehart, C., & Arteaga, C. L. (2012). Feedback upregulation of HER3 (ErbB3) expression and activity attenuates antitumor effect of PI3K inhibitors. *Proceedings of the National Academy of Sciences of the United States of America, 109*, 2718–2723.

Chandarlapaty, S., Sawai, A., Scaltriti, M., Rodrik-Outmezguine, V., Grbovic-Huezo, O., Serra, V., et al. (2011). AKT inhibition relieves feedback suppression of receptor tyrosine kinase expression and activity. *Cancer Cell, 19*, 58–71.

Chen, L., Siddiqui, S., Bose, S., Mooso, B., Asuncion, A., Bedolla, R. G., et al. (2010). Nrdp1-mediated regulation of ErbB3 expression by the androgen receptor in androgen-dependent but not castrate-resistant prostate cancer cells. *Cancer Research, 70*, 5994–6003.

Chinkers, M., & Cohen, S. (1981). Purified EGF receptor-kinase interacts specifically with antibodies to Rous sarcoma virus transforming protein. *Nature, 290*, 516–519.

Cho, H.-S., & Leahy, D. J. (2002). Structure of the extracellular region of HER3 reveals an interdomain tether. *Science (New York, N.Y.), 297*, 1330–1333.

Choi, M. R., An, C. H., Chung, Y. J., Choi, Y. J., Yoo, N. J., & Lee, S. H. (2014). Mutational and expressional analysis of ERBB3 gene in common solid cancers. *APMIS (Acta Pathologica, Microbiologica et Immunologica Scandinavica), 122*, 1207–1212.

Choi, J. H., Oh, J. Y., Ryu, S. K., Kim, S. J., Lee, N. Y., Kim, Y. S., et al. (1997). Detection of epidermal growth factor receptor in the serum of gastric carcinoma patients. *Cancer, 79*, 1879–1883.

Christianson, T. A., Doherty, J. K., Lin, Y. J., Ramsey, E. E., Holmes, R., Keenan, E. J., et al. (1998). NH2-terminally truncated HER-2/neu protein: Relationship with shedding of the extracellular domain and with prognostic factors in breast cancer. *Cancer Research, 58*, 5123–5129. https://pubmed.ncbi.nlm.nih.gov/9823322.

Chung, Y. W., Kim, S., Hong, J. H., Lee, J. K., Lee, N. W., Lee, Y. S., et al. (2019). Overexpression of HER2/HER3 and clinical feature of ovarian cancer. *Journal of Gynecologic Oncology, 30*, e75.

Citri, A., Skaria, K. B., & Yarden, Y. (2003). The deaf and the dumb: The biology of ErbB-2 and ErbB-3. *Experimental Cell Research, 284*, 54–65.

Cohen, S. (1962). Isolation of a mouse submaxillary gland protein accelerating incisor eruption and eyelid opening in the new-born animal. *The Journal of Biological Chemistry, 237*, 1555–1562. https://pubmed.ncbi.nlm.nih.gov/13880319.

Cohen, S., Carpenter, G., & King, L., Jr. (1980). Epidermal growth factor-receptor-protein kinase interactions. Co-purification of receptor and epidermal growth factor-enhanced phosphorylation activity. *The Journal of Biological Chemistry, 255*, 4834–4842. https://pubmed.ncbi.nlm.nih.gov/6246084.

Cohen, S., Carpenter, G., & King, L., Jr. (1981). Epidermal growth factor-receptor-protein kinase interactions. *Progress in Clinical and Biological Research, 66*(Pt. A), 557–567. https://pubmed.ncbi.nlm.nih.gov/6273929.

Coussens, L., Yang-Feng, T. L., Liao, Y. C., Chen, E., Gray, A., McGrath, J., et al. (1985). Tyrosine kinase receptor with extensive homology to EGF receptor shares chromosomal location with neu oncogene. *Science (New York, N.Y.), 230*, 1132–1139.

Crovello, C. S., Lai, C., Cantley, L. C., & Carraway, K. L., 3rd. (1998). Differential signaling by the epidermal growth factor-like growth factors neuregulin-1 and neuregulin-2. *The Journal of Biological Chemistry, 273*, 26954–26961.

Davies, S., Holmes, A., Lomo, L., Steinkamp, M. P., Kang, H., Muller, C. Y., et al. (2014). High incidence of ErbB3, ErbB4, and MET expression in ovarian cancer. *International Journal of Gynecological Pathology: Official Journal of the International Society of Gynecological Pathologists, 33*, 402–410.

Dhanasekaran, S. M., Balbin, O. A., Chen, G., Nadal, E., Kalyana-Sundaram, S., Pan, J., et al. (2014). Transcriptome meta-analysis of lung cancer reveals recurrent aberrations in NRG1 and Hippo pathway genes. *Nature Communications, 5*, 5893.

Di Fiore, P. P., Pierce, J. H., Kraus, M. H., Segatto, O., King, C. R., & Aaronson, S. A. (1987). erbB-2 is a potent oncogene when overexpressed in NIH/3T3 cells. *Science (New York, N.Y.), 237*, 178–182.

Díaz-Serrano, A., Lopez-Rios, F., Dominguez, C., Fonseca, P. J., Angulo, B., Pazo Cid, R. A., et al. (2016). PI3K upregulation as a negative predictive factor of survival in HER2 amplified gastric cancer treated with Trastuzumab. *Journal of Clinical Oncology, 34*, e15567.

Diéras, V., Miles, D., Verma, S., Pegram, M., Welslau, M., Baselga, J., et al. (2017). Trastuzumab emtansine versus capecitabine plus lapatinib in patients with previously treated HER2-positive advanced breast cancer (EMILIA): A descriptive analysis of final overall survival results from a randomised, open-label, phase 3 trial. *The Lancet Oncology, 18*, 732–742.

Dote, H., Cerna, D., Burgan, W. E., Camphausen, K., & Tofilon, P. J. (2005). ErbB3 expression predicts tumor cell radiosensitization induced by Hsp90 inhibition. *Cancer Research, 65*, 6967–6975.

Drilon, A., Somwar, R., Mangatt, B. P., Edgren, H., Desmeules, P., Ruusulehto, A., et al. (2018). Response to ERBB3-directed targeted therapy in NRG1-rearranged cancers. *Cancer Discovery, 8*, 686–695.

Duffy, M. J., McKiernan, E., O'Donovan, N., & McGowan, P. M. (2009). Role of ADAMs in cancer formation and progression. *Clinical Cancer Research: An Official Journal of the American Association for Cancer Research, 15*, 1140–1144.

Duneau, J.-P., Vegh, A. P., & Sturgis, J. N. (2007). A dimerization hierarchy in the transmembrane domains of the HER receptor family. *Biochemistry, 46*, 2010–2019.

Earp, H. S., 3rd, Calvo, B. F., & Sartor, C. I. (2003). The EGF receptor family—multiple roles in proliferation, differentiation, and neoplasia with an emphasis on HER4. *Transactions of the American Clinical and Climatological Association, 114*, 315–334. https://pubmed.ncbi.nlm.nih.gov/12813928.

Engelman, J. A., Zejnullahu, K., Mitsudomi, T., Song, Y., Hyland, C., Park, J. O., et al. (2007). MET amplification leads to gefitinib resistance in lung cancer by activating ERBB3 signaling. *Science (New York, N.Y.), 316*, 1039–1043.

Erjala, K., Sundvall, M., Junttila, T. T., Zhang, N., Savisalo, M., Mali, P., et al. (2006). Signaling via ErbB2 and ErbB3 associates with resistance and epidermal growth factor receptor (EGFR) amplification with sensitivity to EGFR inhibitor gefitinib in head and neck squamous cell carcinoma cells. *Clinical Cancer Research: An Official Journal of the American Association for Cancer Research*, *12*, 4103–4111.

Esteva, F. J., Hortobagyi, G. N., Sahin, A. A., Smith, T. L., Chin, D. M., Liang, S. Y., et al. (2001). Expression of erbB/HER receptors, heregulin and P38 in primary breast cancer using quantitative immunohistochemistry. *Pathology Oncology Research*, *7*, 171–177.

Falls, D. L. (2003). Neuregulins: Functions, forms, and signaling strategies. *Experimental Cell Research*, *284*, 14–30.

Fattore, L., Marra, E., Pisanu, M. E., Noto, A., de Vitis, C., Belleudi, F., et al. (2013). Activation of an early feedback survival loop involving phospho-ErbB3 is a general response of melanoma cells to RAF/MEK inhibition and is abrogated by anti-ErbB3 antibodies. *Journal of Translational Medicine*, *11*, 180.

Ferguson, K. M. (2008). Structure-based view of epidermal growth factor receptor regulation. *Annual Review of Biophysics*, *37*, 353–373.

Fernandez-Cuesta, L., Plenker, D., Osada, H., Sun, R., Menon, R., Leenders, F., et al. (2014). CD74-NRG1 fusions in lung adenocarcinoma. *Cancer Discovery*, *4*, 415–422.

Frazier, N. M., Brand, T., Gordan, J. D., Grandis, J., & Jura, N. (2019). Overexpression-mediated activation of MET in the Golgi promotes HER3/ERBB3 phosphorylation. *Oncogene*, *38*, 1936–1950.

Frogne, T., Benjaminsen, R. V., Sonne-Hansen, K., Sorensen, B. S., Nexo, E., Laenkholm, A.-V., et al. (2009). Activation of ErbB3, EGFR and Erk is essential for growth of human breast cancer cell lines with acquired resistance to fulvestrant. *Breast Cancer Research and Treatment*, *114*, 263–275.

Gaborit, N., Lindzen, M., & Yarden, Y. (2016). Emerging anti-cancer antibodies and combination therapies targeting HER3/ERBB3. *Human Vaccines & Immunotherapeutics*, *12*, 576–592.

Garrett, J. T., Olivares, M. G., Rinehart, C., Granja-Ingram, N. D., Sánchez, V., Chakrabarty, A., et al. (2011). Transcriptional and posttranslational up-regulation of HER3 (ErbB3) compensates for inhibition of the HER2 tyrosine kinase. *Proceedings of the National Academy of Sciences of the United States of America*, *108*, 5021–5026.

Gay, N. D., Wang, Y., Beadling, C., Warrick, A., Neff, T., Corless, C. L., et al. (2017). Durable response to Afatinib in lung adenocarcinoma harboring NRG1 gene fusions. *Journal of Thoracic Oncology: Official Publication of the International Association for the Study of Lung Cancer*, *12*, e107–e110.

Gebhart, G., Lamberts, L. E., Wimana, Z., Garcia, C., Emonts, P., Ameye, L., et al. (2016). Molecular imaging as a tool to investigate heterogeneity of advanced HER2-positive breast cancer and to predict patient outcome under trastuzumab emtansine (T-DM1): The ZEPHIR trial. *Annals of Oncology: Official Journal of the European Society for Medical Oncology*, *27*, 619–624.

Gerbin, C. S., & Landgraf, R. (2010). Geldanamycin selectively targets the nascent form of ERBB3 for degradation. *Cell Stress & Chaperones*, *15*, 529–544.

Geyer, C. E., Forster, J., Lindquist, D., Chan, S., Romieu, C. G., Pienkowski, T., et al. (2006). Lapatinib plus capecitabine for HER2-positive advanced breast cancer. *The New England Journal of Medicine*, *355*, 2733–2743.

Gianni, L., Pienkowski, T., Im, Y.-H., Roman, L., Tseng, L.-M., Liu, M.-C., et al. (2012). Efficacy and safety of neoadjuvant pertuzumab and trastuzumab in women with locally advanced, inflammatory, or early HER2-positive breast cancer (NeoSphere): A randomised multicentre, open-label, phase 2 trial. *The Lancet Oncology*, *13*(1), 25–32.

Gilbertson, R. J., Bentley, L., Hernan, R., Junttila, T. T., Frank, A. J., Haapasalo, H., et al. (2002). ERBB receptor signaling promotes ependymoma cell proliferation and represents a potential novel therapeutic target for this disease. *Clinical Cancer Research*, *8*, 3054–3064.

Gilbertson, R. J., Perry, R. H., Kelly, P. J., Pearson, A. D., & Lunec, J. (1997). Prognostic significance of HER2 and HER4 coexpression in childhood medulloblastoma. *Cancer Research, 57*, 3272–3280.

Goldman, R., Ben-Levy, R., Peles, E., & Yarden, Y. (1990). Heterodimerization of the erbB-1 and erbB-2 receptors in human breast carcinoma cells: A mechanism for receptor transregulation. *Biochemistry, 29*, 11024–11028.

Gollamudi, M., Nethery, D., Liu, J., & Kern, J. A. (2004). Autocrine activation of ErbB2/ ErbB3 receptor complex by NRG-1 in non-small cell lung cancer cell lines. *Lung Cancer (Amsterdam, Netherlands), 43*, 135–143.

Gotoh, N., Tojo, A., Hino, M., Yazaki, Y., & Shibuya, M. (1992). A highly conserved tyrosine residue at codon 845 within the kinase domain is not required for the transforming activity of human epidermal growth factor receptor. *Biochemical and Biophysical Research Communications, 186*, 768–774.

Graus-Porta, D., Beerli, R. R., Daly, J. M., & Hynes, N. E. (1997). ErbB-2, the preferred heterodimerization partner of all ErbB receptors, is a mediator of lateral signaling. *The EMBO Journal, 16*, 1647–1655.

Gregory, C. W., Whang, Y. E., McCall, W., Fei, X., Liu, Y., Ponguta, L. A., et al. (2005). Heregulin-induced activation of HER2 and HER3 increases androgen receptor transactivation and CWR-R1 human recurrent prostate cancer cell growth. *Clinical Cancer Research: An Official Journal of the American Association for Cancer Research, 11*, 1704–1712.

Guarneri, V., Dieci, M. V., Frassoldati, A., Maiorana, A., Ficarra, G., Bettelli, S., et al. (2015). Prospective biomarker analysis of the randomized CHER-LOB study evaluating the dual anti-HER2 treatment with trastuzumab and lapatinib plus chemotherapy as neoadjuvant therapy for HER2-positive breast cancer. *The Oncologist, 20*, 1001–1010.

Guy, P. M., Platko, J. V., Cantley, L. C., Cerione, R. A., & Carraway, K. L., 3rd. (1994). Insect cell-expressed p180erbB3 possesses an impaired tyrosine kinase activity. *Proceedings of the National Academy of Sciences of the United States of America, 91*, 8132–8136.

Han, W., Sfondouris, M. E., & Jones, F. E. (2016). Direct coupling of the HER4 intracellular domain (4ICD) and STAT5A signaling is required to induce mammary epithelial cell differentiation. *Biochemistry and Biophysics Reports, 7*, 323–327.

Harari, P. M., & Huang, S. M. (2001). Radiation response modification following molecular inhibition of epidermal growth factor receptor signaling. *Seminars in Radiation Oncology, 11*, 281–289.

Harari, D., & Yarden, Y. (2000). Molecular mechanisms underlying ErbB2/HER2 action in breast cancer. *Oncogene, 19*, 6102–6114.

Haskins, J. W., Nguyen, D. X., & Stern, D. F. (2014). Neuregulin 1-activated ERBB4 interacts with YAP to induce Hippo pathway target genes and promote cell migration. *Science Signaling, 7*, ra116.

Haslekås, C., Breen, K., Pedersen, K. W., Johannessen, L. E., Stang, E., & Madshus, I. H. (2005). The inhibitory effect of ErbB2 on epidermal growth factor-induced formation of clathrin-coated pits correlates with retention of epidermal growth factor receptor-ErbB2 oligomeric complexes at the plasma membrane. *Molecular Biology of the Cell, 16*, 5832–5842.

Hayes, N. V. L., & Gullick, W. J. (2008). The neuregulin family of genes and their multiple splice variants in breast cancer. *Journal of Mammary Gland Biology and Neoplasia, 13*, 205–214.

Hayes, D. F., Yamauchi, H., Broadwater, G., Cirrincione, C. T., Rodrigue, S. P., Berry, D. A., et al. (2001). Circulating HER-2/erbB-2/c-neu (HER-2) extracellular domain as a prognostic factor in patients with metastatic breast cancer: Cancer and Leukemia Group B Study 8662. *Clinical Cancer Research: An Official Journal of the American Association for Cancer Research, 7*, 2703–2711. https://pubmed.ncbi.nlm.nih.gov/11555582.

Heining, C., Horak, P., Uhrig, S., Codo, P. L., Klink, B., Hutter, B., et al. (2018). NRG1 fusions in KRAS wild-type pancreatic cancer. *Cancer Discovery*, *8*, 1087–1095.

Hellyer, N. J., Cheng, K., & Koland, J. G. (1998). ErbB3 (HER3) interaction with the p85 regulatory subunit of phosphoinositide 3-kinase. *The Biochemical Journal*, *333*(Pt. 3), 757–763.

Hendler, F. J., & Ozanne, B. W. (1984). Human squamous cell lung cancers express increased epidermal growth factor receptors. *The Journal of Clinical Investigation*, *74*, 647–651.

Hendriks, B. S., Opresko, L. K., Wiley, H. S., & Lauffenburger, D. (2003). Coregulation of epidermal growth factor receptor/human epidermal growth factor receptor 2 (HER2) levels and locations: Quantitative analysis of HER2 overexpression effects. *Cancer Research*, *63*, 1130–1137. https://pubmed.ncbi.nlm.nih.gov/12615732.

Henriksen, L., Grandal, M. V., Knudsen, S. L. J., van Deurs, B., & Grøvdal, L. M. (2013). Internalization mechanisms of the epidermal growth factor receptor after activation with different ligands. *PLoS One*, *8*, e58148.

Herr, R., Halbach, S., Heizmann, M., Busch, H., Boerries, M., & Brummer, T. (2018). BRAF inhibition upregulates a variety of receptor tyrosine kinases and their downstream effector Gab2 in colorectal cancer cell lines. *Oncogene*, *37*, 1576–1593.

Hess, J. M., Bernards, A., Kim, J., Miller, M., Taylor-Weiner, A., Haradhvala, N. J., et al. (2019). Passenger hotspot mutations in cancer. *Cancer Cell*, *36*, 288–301.e14.

Holbro, T., Beerli, R. R., Maurer, F., Koziczak, M., Barbas, C. F., 3rd, & Hynes, N. E. (2003). The ErbB2/ErbB3 heterodimer functions as an oncogenic unit: ErbB2 requires ErbB3 to drive breast tumor cell proliferation. *Proceedings of the National Academy of Sciences of the United States of America*, *100*, 8933–8938.

Huang, X., Gao, L., Wang, S., McManaman, J. L., Thor, A. D., Yang, X., et al. (2010). Heterotrimerization of the growth factor receptors erbB2, erbB3, and insulin-like growth factor-i receptor in breast cancer cells resistant to herceptin. *Cancer Research*, *70*, 1204–1214.

Hung, M. C., Schechter, A. L., Chevray, P. Y., Stern, D. F., & Weinberg, R. A. (1986). Molecular cloning of the neu gene: Absence of gross structural alteration in oncogenic alleles. *Proceedings of the National Academy of Sciences of the United States of America*, *83*, 261–264.

Hunter, T. (2015). Discovering the first tyrosine kinase. *Proceedings of the National Academy of Sciences of the United States of America*, *112*, 7877–7882.

Hyman, D. M., Piha-Paul, S. A., Won, H., Rodon, J., Saura, C., Shapiro, G. I., et al. (2018). HER kinase inhibition in patients with HER2- and HER3-mutant cancers. *Nature*, *554*, 189–194.

Jaiswal, B. S., Kljavin, N. M., Stawiski, E. W., Chan, E., Parikh, C., Durinck, S., et al. (2013). Oncogenic ERBB3 mutations in human cancers. *Cancer Cell*, *23*, 603–617.

Jathal, M. K., Chen, L., Mudryj, M., & Ghosh, P. M. (2011). Targeting ErbB3: The new RTK(id) on the prostate cancer block. *Immunology, Endocrine & Metabolic Agents in Medicinal Chemistry*, *11*, 131–149.

Jia, Y., Zhang, Y., Qiao, C., Liu, G., Zhao, Q., Zhou, T., et al. (2013). IGF-1R and ErbB3/HER3 contribute to enhanced proliferation and carcinogenesis in trastuzumab-resistant ovarian cancer model. *Biochemical and Biophysical Research Communications*, *436*, 740–745.

Jones, F. E. (2008). HER4 intracellular domain (4ICD) activity in the developing mammary gland and breast cancer. *Journal of Mammary Gland Biology and Neoplasia*, *13*, 247–258.

Jones, R. B., Gordus, A., Krall, J. A., & MacBeath, G. (2006). A quantitative protein interaction network for the ErbB receptors using protein microarrays. *Nature*, *439*, 168–174.

Jones, M. R., Lim, H., Shen, Y., Pleasance, E., Ch'ng, C., Reisle, C., et al. (2017). Successful targeting of the NRG1 pathway indicates novel treatment strategy for metastatic cancer. *Annals of Oncology: Official Journal of the European Society for Medical Oncology*, *28*, 3092–3097.

Jones, F. E., Welte, T., Fu, X. Y., & Stern, D. F. (1999). ErbB4 signaling in the mammary gland is required for lobuloalveolar development and Stat5 activation during lactation. *The Journal of Cell Biology*, *147*, 77–88.

Jonna, S., Feldman, R. A., Swensen, J., Gatalica, Z., Korn, W. M., Borghaei, H., et al. (2019). Detection of NRG1 gene fusions in solid tumors. *Clinical Cancer Research: An Official Journal of the American Association for Cancer Research*, *25*, 4966–4972.

Katoh, M., Yazaki, Y., Sugimura, T., & Terada, M. (1993). c-erbB3 gene encodes secreted as well as transmembrane receptor tyrosine kinase. *Biochemical and Biophysical Research Communications*, *192*, 1189–1197.

Kawamoto, T., Sato, J. D., Le, A., Polikoff, J., Sato, G. H., & Mendelsohn, J. (1983). Growth stimulation of A431 cells by epidermal growth factor: Identification of high-affinity receptors for epidermal growth factor by an anti-receptor monoclonal antibody. *Proceedings of the National Academy of Sciences of the United States of America*, *80*, 1337–1341.

Kennedy, S. P., Han, J. Z. R., Portman, N., Nobis, M., Hastings, J. F., Murphy, K. J., et al. (2019). Targeting promiscuous heterodimerization overcomes innate resistance to ERBB2 dimerization inhibitors in breast cancer. *Breast Cancer Research*, *21*, 43.

Khoury, T., Nagrale, V., Opyrchal, M., Peng, X., Wang, D., & Yao, S. (2018). Prognostic significance of stromal versus intratumoral infiltrating lymphocytes in different subtypes of breast cancer treated with cytotoxic neoadjuvant chemotherapy. *Applied Immunohistochemistry & Molecular Morphology: AIMM*, *26*, 523–532.

Kim, S.-B., Wildiers, H., Krop, I. E., Smitt, M., Yu, R., Lysbet de Haas, S., et al. (2016). Relationship between tumor biomarkers and efficacy in TH3RESA, a phase III study of trastuzumab emtansine (T-DM1) vs. treatment of physician's choice in previously treated HER2-positive advanced breast cancer. *International Journal of Cancer*, *139*, 2336–2342.

King, C. R., Borrello, I., Bellot, F., Comoglio, P., & Schlessinger, J. (1988). Egf binding to its receptor triggers a rapid tyrosine phosphorylation of the erbB-2 protein in the mammary tumor cell line SK-BR-3. *The EMBO Journal*, 7, 1647–1651.

King, C. R., Kraus, M. H., & Aaronson, S. A. (1985). Amplification of a novel v-erbB-related gene in a human mammary carcinoma. *Science (New York, N.Y.)*, *229*, 974–976.

Koganemaru, S., Kuboki, Y., Koga, Y., Kojima, T., Yamauchi, M., Maeda, N., et al. (2019). U3-1402, a novel HER3-targeting antibody-drug conjugate, for the treatment of colorectal cancer. *Molecular Cancer Therapeutics*, *18*, 2043–2050.

Kraus, M. H., Issing, W., Miki, T., Popescu, N. C., & Aaronson, S. A. (1989). Isolation and characterization of ERBB3, a third member of the ERBB/epidermal growth factor receptor family: Evidence for overexpression in a subset of human mammary tumors. *Proceedings of the National Academy of Sciences of the United States of America*, *86*, 9193–9197.

Krebs, E. G. (1989). The Albert Lasker Medical Awards. Role of the cyclic AMP-dependent protein kinase in signal transduction. *JAMA*, *262*, 1815–1818.

Krig, S. R., Miller, J. K., Frietze, S., Beckett, L. A., Neve, R. M., Farnham, P. J., et al. (2010). ZNF217, a candidate breast cancer oncogene amplified at 20q13, regulates expression of the ErbB3 receptor tyrosine kinase in breast cancer cells. *Oncogene*, *29*, 5500–5510.

Kugel, C. H., 3rd, Hartsough, E. J., Davies, M. A., Setiady, Y. Y., & Aplin, A. E. (2014). Function-blocking ERBB3 antibody inhibits the adaptive response to RAF inhibitor. *Cancer Research*, *74*(15), 4122–4132.

Kumar, R. (2001). Targeting epidermal growth factor receptor family members for treatment of breast cancer. *Biological Therapy of Breast Cancer*, *3*, 3–6.

Kumar, R. (2007). ErbB-dependent signaling as a determinant of trastuzumab resistance. *Clinical Cancer Research: An Official Journal of the American Association for Cancer Research*, *13*, 4657–4659.

Kumar, R., & Hung, M.-C. (2005). Signaling intricacies take center stage in cancer cells. *Cancer Research, 65*, 2511–2515.

Kumar, R., & Mendelsohn, J. (1991). Polypeptide growth factors in the regulation of human tumor cell proliferation. *Current Opinion in Oncology, 3*, 70–74.

Kumar, R., Shepard, H. M., & Mendelsohn, J. (1991). Regulation of phosphorylation of the c-erbB-2/HER2 gene product by a monoclonal antibody and serum growth factor(s) in human mammary carcinoma cells. *Molecular and Cellular Biology, 11*, 979–986.

Kumar, R., Van de Vijver, M., Tortora, G., Ciardiello, F., Goldkorn, T., Miller, W. H., et al. (2019). A tribute to John Mendelsohn: A pioneer in targeted cancer therapy. *Cancer Research, 79*, 4315–4323.

Kumar, R., Wang, R.-A., & Barnes, C. J. (2004). Coregulators and chromatin remodeling in transcriptional control. *Molecular Carcinogenesis, 41*, 221–230.

Kung, J. E., & Jura, N. (2016). Structural basis for the non-catalytic functions of protein kinases. *Structure (London, England: 1993), 24*, 7–24.

Kuo, J. F., & Greengard, P. (1970). Cyclic nucleotide-dependent protein kinases. VI. Isolation and partial purification of a protein kinase activated by guanosine 3',5'-monophosphate. *The Journal of Biological Chemistry, 245*, 2493–2498.

Lahlou, H., Müller, T., Sanguin-Gendreau, V., Birchmeier, C., & Muller, W. J. (2012). Uncoupling of PI3K from ErbB3 impairs mammary gland development but does not impact on ErbB2-induced mammary tumorigenesis. *Cancer Research, 72*, 3080–3090.

Lam, L., McAndrew, N., Yee, M., Fu, T., Tchou, J. C., & Zhang, H. (2012). Challenges in the clinical utility of the serum test for HER2 ECD. *Biochimica et Biophysica Acta, 1826*, 199–208.

Lee, H., Akita, R. W., Sliwkowski, M. X., & Maihle, N. J. (2001). A naturally occurring secreted human ErbB3 receptor isoform inhibits heregulin-stimulated activation of ErbB2, ErbB3, and ErbB4. *Cancer Research, 61*, 4467–4473. https://pubmed.ncbi.nlm.nih.gov/11389077.

Lee, H., & Maihle, N. J. (1998). Isolation and characterization of four alternate c-erbB3 transcripts expressed in ovarian carcinoma-derived cell lines and normal human tissues. *Oncogene, 16*, 3243–3252.

Lee, D., Yu, M., Lee, E., Kim, H., Yang, Y., Kim, K., et al. (2009). Tumor-specific apoptosis caused by deletion of the ERBB3 pseudo-kinase in mouse intestinal epithelium. *The Journal of Clinical Investigation, 119*, 2702–2713.

Lee-Hoeflich, S. T., Crocker, L., Yao, E., Pham, T., Munroe, X., Hoeflich, K. P., et al. (2008). A central role for HER3 in HER2-amplified breast cancer: Implications for targeted therapy. *Cancer Research, 68*, 5878–5887.

Lemoine, N. R., Barnes, D. M., Hollywood, D. P., Hughes, C. M., Smith, P., Dublin, E., et al. (1992). Expression of the ERBB3 gene product in breast cancer. *British Journal of Cancer, 66*, 1116–1121.

Li, B. T., Ross, D. S., Aisner, D. L., Chaft, J. E., Hsu, M., Kako, S. L., et al. (2016). HER2 amplification and HER2 mutation are distinct molecular targets in lung cancers. *Journal of Thoracic Oncology, 11*, 414–419.

Li, X., Xu, Y., Ding, Y., Li, C., Zhao, H., Wang, J., et al. (2018). Posttranscriptional upregulation of HER3 by HER2 mRNA induces trastuzumab resistance in breast cancer. *Molecular Cancer, 17*, 113.

Lieu, C. H., Golemis, E. A., Serebriiskii, I. G., Newberg, J., Hemmerich, A., Connelly, C., et al. (2019). Comprehensive genomic landscapes in early and later onset colorectal cancer. *Clinical Cancer Research, 25*, 5852–5858.

Lim, S. M., Xie, T., Westover, K. D., Ficarro, S. B., Tae, H. S., Gurbani, D., et al. (2015). Development of small molecules targeting the pseudokinase Her3. *Bioorganic & Medicinal Chemistry Letters, 25*, 3382–3389.

Lin, C. R., Chen, W. S., Kruiger, W., Stolarsky, L. S., Weber, W., Evans, R. M., et al. (1984). Expression cloning of human EGF receptor complementary DNA: Gene amplification and three related messenger RNA products in A431 cells. *Science (New York, N.Y.), 224*, 843–848.

Lin, S.-H., Lee, Y.-C., Choueiri, M. B., Wen, S., Mathew, P., Ye, X., et al. (2008). Soluble ErbB3 levels in bone marrow and plasma of men with prostate cancer. *Clinical Cancer Research, 14*, 3729–3736.

Linggi, B., & Carpenter, G. (2006). ErbB receptors: New insights on mechanisms and biology. *Trends in Cell Biology, 16*, 649–656.

Littlefield, P., Liu, L., Mysore, V., Shan, Y., Shaw, D. E., & Jura, N. (2014). Structural analysis of the EGFR/HER3 heterodimer reveals the molecular basis for activating HER3 mutations. *Science Signaling*, 7, ra114.

Liu, X., Liu, S., Lyu, H., Riker, A. I., Zhang, Y., & Liu, B. (2019). Development of Effective Therapeutics Targeting HER3 for Cancer Treatment. *Biological Procedures Online, 21*, 5.

Liu, B., Ordonez-Ercan, D., Fan, Z., Edgerton, S. M., Yang, X., & Thor, A. D. (2007). Downregulation of erbB3 abrogates erbB2-mediated tamoxifen resistance in breast cancer cells. *International Journal of Cancer, 120*, 1874–1882.

Loi, S., Giobbie-Hurder, A., Gombos, A., Bachelot, T., Hui, R., Curigliano, G., et al. (2019). Pembrolizumab plus trastuzumab in trastuzumab-resistant, advanced, HER2-positive breast cancer (PANACEA): A single-arm, multicentre, phase 1b-2 trial. *The Lancet Oncology, 20*, 371–382.

Loi, S., Michiels, S., Salgado, R., Sirtaine, N., Jose, V., Fumagalli, D., et al. (2014). Tumor infiltrating lymphocytes are prognostic in triple negative breast cancer and predictive for trastuzumab benefit in early breast cancer: Results from the FinHER trial. *Annals of Oncology: Official Journal of the European Society for Medical Oncology, 25*, 1544–1550.

Loibl, S., Majewski, I., Guarneri, V., Nekljudova, V., Holmes, E., Bria, E., et al. (2016). PIK3CA mutations are associated with reduced pathological complete response rates in primary HER2-positive breast cancer: Pooled analysis of 967 patients from five prospective trials investigating lapatinib and trastuzumab. *Annals of Oncology: Official Journal of the European Society for Medical Oncology, 27*, 1519–1525.

Lorusso, D., Hilpert, F., González Martin, A., Rau, J., Ottevanger, P., Greimel, E., et al. (2019). Patient-reported outcomes and final overall survival results from the randomized phase 3 PENELOPE trial evaluating pertuzumab in low tumor human epidermal growth factor receptor 3 (HER3) mRNA-expressing platinum-resistant ovarian cancer. *International Journal of Gynecological Cancer, 29*, 1141–1147.

Lu, Y., Zi, X., Zhao, Y., Mascarenhas, D., & Pollak, M. (2001). Insulin-like growth factor-I receptor signaling and resistance to trastuzumab (herceptin). *Journal of the National Cancer Institute, 93*, 1852–1857.

Lupu, R., Colomer, R., Zugmaier, G., Sarup, J., Shepard, M., Slamon, D., et al. (1990). Direct interaction of a ligand for the erbB2 oncogene product with the EGF receptor and p185erbB2. *Science (New York, N.Y.), 249*, 1552–1555.

Lyne, J. C., Melhem, M. F., Finley, G. G., Wen, D., Liu, N., Deng, D. H., et al. (1997). Tissue expression of neu differentiation factor/heregulin and its receptor complex in prostate cancer and its biologic effects on prostate cancer cells in vitro. *The Cancer Journal from Scientific American, 3*, 21–30.

Ma, F., Zhu, W., Guan, Y., Yang, L., Xia, X., Chen, S., et al. (2016). ctDNA dynamics: A novel indicator to track resistance in metastatic breast cancer treated with anti-HER2 therapy. *Oncotarget*, 7, 66020–66031.

Maramotti, S., Paci, M., Manzotti, G., Rapicetta, C., Gugnoni, M., Galeone, C., et al. (2016). Soluble epidermal growth factor receptors (sEGFRs) in cancer: Biological aspects and clinical relevance. *International Journal of Molecular Sciences, 17*, 593.

Mellman, I., & Yarden, Y. (2013). Endocytosis and cancer. *Cold Spring Harbor Perspectives in Biology*, *5*, a016949.

Mendelsohn, J. (2000). Jeremiah Metzger lecture. Targeted cancer therapy. *Transactions of the American Clinical and Climatological Association*, *111*, 95–111.

Mills, G. B., & Yarden, Y. (2010). The rebirth of a phoenix: Ovarian cancers are addicted to ErbB-3. *Cancer Cell*, *17*, 217–218.

Mishra, R., Patel, H., Alanazi, S., Yuan, L., & Garrett, J. T. (2018). HER3 signaling and targeyted therapy in cancer. *Oncology Reviews*, *12*, 45–61.

Moasser, M. M. (2007a). The oncogene HER2: Its signaling and transforming functions and its role in human cancer pathogenesis. *Oncogene*, *26*, 6469–6487.

Moasser, M. M. (2007b). Targeting the function of the HER2 oncogene in human cancer therapeutics. *Oncogene*, *26*, 6577–6592.

Montero-Conde, C., Ruiz-Llorente, S., Dominguez, J. M., Knauf, J. A., Viale, A., Sherman, E. J., et al. (2013). Relief of feedback inhibition of HER3 transcription by RAF and MEK inhibitors attenuates their antitumor effects in BRAF-mutant thyroid carcinomas. *Cancer Discovery*, *3*, 520–533.

Morrison, M. M., Hutchinson, K., Williams, M. M., Stanford, J. C., Balko, J. M., Young, C., et al. (2013). ErbB3 downregulation enhances luminal breast tumor response to antiestrogens. *The Journal of Clinical Investigation*, *123*, 4329–4343.

Moustafa, A. E., Alaoui-Jamali, M., Paterson, J., & O'Connor-McCourt, M. (1999). Expression of P185erbB-2, P160erbB-3, P180erbB-4, and heregulin alpha in human normal bronchial epithelial and lung cancer cell lines. *Anticancer Research*, *19*, 481–486.

Mullooly, M., McGowan, P. M., Kennedy, S. A., Madden, S. F., Crown, J., & O' Donovan, N., et al. (2015). ADAM10: A new player in breast cancer progression? *British Journal of Cancer*, *113*, 945–951.

Murad, F. (2006). Shattuck lecture. Nitric oxide and cyclic GMP in cell signaling and drug development. *The New England Journal of Medicine*, *355*, 2003–2011.

Murthy, R. K., Loi, S., Okines, A., Paplomata, E., Hamilton, E., Hurvitz, S. A., et al. (2020). Tucatinib, trastuzumab, and capecitabine for HER2-positive metastatic breast cancer. *The New England Journal of Medicine*, *382*(7), 597–609.

Nagumo, Y., Faratian, D., Mullen, P., Harrison, D. J., Hasmann, M., & Langdon, S. P. (2009). Modulation of HER3 is a marker of dynamic cell signaling in ovarian cancer: Implications for pertuzumab sensitivity. *Molecular Cancer Research*, 7, 1563–1571.

Naresh, A., Thor, A. D., Edgerton, S. M., Torkko, K. C., Kumar, R., & Jones, F. E. (2008). The HER4/4ICD estrogen receptor coactivator and BH3-only protein is an effector of tamoxifen-induced apoptosis. *Cancer Research*, *68*, 6387–6395.

Nolen, B., Taylor, S., & Ghosh, G. (2004). Regulation of protein kinases; controlling activity through activation segment conformation. *Molecular Cell*, *15*, 661–675.

O'Malley, B. W., & Kumar, R. (2009). Nuclear receptor coregulators in cancer biology. *Cancer Research*, *69*, 8217–8222.

Oh, M. J., Choi, J. H., Kim, I. H., Lee, Y. H., Huh, J. Y., Park, Y. K., et al. (2000). Detection of epidermal growth factor receptor in the serum of patients with cervical carcinoma. *Clinical Cancer Research*, *6*, 4760–4763.

Oppermann, H., Levinson, A. D., Varmus, H. E., Levintow, L., & Bishop, J. M. (1979). Uninfected vertebrate cells contain a protein that is closely related to the product of the avian sarcoma virus transforming gene (src). *Proceedings of the National Academy of Sciences of the United States of America*, *76*, 1804–1808.

Paez, J. G., Jänne, P. A., Lee, J. C., Tracy, S., Greulich, H., Gabriel, S., et al. (2004). EGFR mutations in lung cancer: Correlation with clinical response to gefitinib therapy. *Science (New York, N.Y.)*, *304*, 1497–1500.

Parakh, S., Gan, H. K., Parslow, A. C., Burvenich, I. J. G., Burgess, A. W., & Scott, A. M. (2017). Evolution of anti-HER2 therapies for cancer treatment. *Cancer Treatment Reviews*, *59*, 1–21.

Pedersen, N. M., Madshus, I. H., Haslekås, C., & Stang, E. (2008). Geldanamycin-induced down-regulation of ErbB2 from the plasma membrane is clathrin dependent but proteasomal activity independent. *Molecular Cancer Research*, *6*, 491–500.

Perez, E. A., Barrios, C., Eiermann, W., Toi, M., Im, Y.-H., Conte, P., et al. (2017). Trastuzumab emtansine with or without pertuzumab versus trastuzumab plus taxane for human epidermal growth factor receptor 2-positive, Advanced Breast Cancer: Primary results from the phase III MARIANNE study. *Journal of Clinical Oncology : Official Journal of the American Society of Clinical Oncology*, *35*, 141–148.

Perez-Torres, M., Valle, B. L., Maihle, N. J., Negron-Vega, L., Nieves-Alicea, R., & Cora, E. M. (2008). Shedding of epidermal growth factor receptor is a regulated process that occurs with overexpression in malignant cells. *Experimental Cell Research*, *314*, 2907–2918.

Peschard, P., & Park, M. (2003). Escape from Cbl-mediated downregulation: A recurrent theme for oncogenic deregulation of receptor tyrosine kinases. *Cancer Cell*, *3*, 519–523.

Piccart-Gebhart, M. J., Procter, M., Leyland-Jones, B., Goldhirsch, A., Untch, M., Smith, I., et al. (2005). Trastuzumab after adjuvant chemotherapy in HER2-positive breast cancer. *The New England Journal of Medicine*, *353*, 1659–1672.

Pickl, M., & Ries, C. H. (2009). Comparison of 3D and 2D tumor models reveals enhanced HER2 activation in 3D associated with an increased response to trastuzumab. *Oncogene*, *28*, 461–468.

Pietilä, M., Sahgal, P., Peuhu, E., Jäntti, N. Z., Paatero, I., Närvä, E., et al. (2019). SORLA regulates endosomal trafficking and oncogenic fitness of HER2. *Nature Communications*, *10*, 2340.

Pines, G., Köstler, W. J., & Yarden, Y. (2010). Oncogenic mutant forms of EGFR: Lessons in signal transduction and targets for cancer therapy. *FEBS Letters*, *584*, 2699–2706.

Plowman, G. D., Culouscou, J. M., Whitney, G. S., Green, J. M., Carlton, G. W., Foy, L., et al. (1993). Ligand-specific activation of HER4/p180erbB4, a fourth member of the epidermal growth factor receptor family. *Proceedings of the National Academy of Sciences of the United States of America*, *90*, 1746–1750.

Plowman, G. D., Whitney, G. S., Neubauer, M. G., Green, J. M., McDonald, V. L., Todaro, G. J., et al. (1990). Molecular cloning and expression of an additional epidermal growth factor receptor-related gene. *Proceedings of the National Academy of Sciences of the United States of America*, *87*, 4905–4909.

Prasad, M. K., Reed, X., Gorkin, D. U., Cronin, J. C., McAdow, A. R., Chain, K., et al. (2011). SOX10 directly modulates ERBB3 transcription via an intronic neural crest enhancer. *BMC Developmental Biology*, *11*, 40.

Prigent, S. A., & Gullick, W. J. (1994). Identification of c-erbB-3 binding sites for phosphatidylinositol 3'-kinase and SHC using an EGF receptor/c-erbB-3 chimera. *The EMBO Journal*, *13*, 2831–2841.

Qiu, X.-B., & Goldberg, A. L. (2002). Nrdp1/FLRF is a ubiquitin ligase promoting ubiquitination and degradation of the epidermal growth factor receptor family member, ErbB3. *Proceedings of the National Academy of Sciences of the United States of America*, *99*, 14843–14848.

Qiu, C., Tarrant, M. K., Choi, S. H., Sathyamurthy, A., Bose, R., Banjade, S., et al. (2008). Mechanism of activation and inhibition of the HER4/ErbB4 kinase. *Structure (London, England: 1993)*, *16*, 460–467.

Rauch, J., Volinsky, N., Romano, D., & Kolch, W. (2011). The secret life of kinases: Functions beyond catalysis. *Cell Communication and Signaling: CCS*, *9*, 23.

Reiter, J. L., Threadgill, D. W., Eley, G. D., Strunk, K. E., Danielsen, A. J., Schehl Sinclair, C., et al. (2001). Comparative genomic sequence analysis and isolation of human and mouse alternative EGFR transcripts encoding truncated receptor isoforms. *Genomics, 71*, 1–20.

Reix, N., Malina, C., Chenard, M.-P., Bellocq, J.-P., Delpous, S., Molière, S., et al. (2016). A prospective study to assess the clinical utility of serum HER2 extracellular domain in breast cancer with HER2 overexpression. *Breast Cancer Research and Treatment, 160*, 249–259.

Roberts, R. B., Min, L., Washington, M. K., Olsen, S. J., Settle, S. H., Coffey, R. J., et al. (2002). Importance of epidermal growth factor receptor signaling in establishment of adenomas and maintenance of carcinomas during intestinal tumorigenesis. *Proceedings of the National Academy of Sciences of the United States of America, 99*, 1521–1526.

Robichaux, J. P., Elamin, Y. Y., Vijayan, R. S. K., Nilsson, M. B., Hu, L., He, J., et al. (2019). Pan-cancer landscape and analysis of ERBB2 mutations identifies poziotinib as a clinically active inhibitor and enhancer of T-DM1 activity. *Cancer Cell, 36*, 444-457.e7.

Roskoski, R., Jr. (2014). The ErbB/HER family of protein-tyrosine kinases and cancer. *Pharmacological Research, 79*, 34–74.

Ruiz-Saenz, A., Dreyer, C., Campbell, M. R., Steri, V., Gulizia, N., & Moasser, M. M. (2018). HER2 amplification in tumors activates PI3K/Akt signaling independent of HER3. *Cancer Research, 78*, 3645–3658.

Ruiz-Saenz, A., Sandhu, M., Carrasco, Y., Maglathlin, R. L., Taunton, J., & Moasser, M. M. (2015). Targeting HER3 by interfering with its Sec61-mediated cotranslational insertion into the endoplasmic reticulum. *Oncogene, 34*, 5288–5294.

Sanderson, M. P., Dempsey, P. J., & Dunbar, A. J. (2006). Control of ErbB signaling through metalloprotease mediated ectodomain shedding of EGF-like factors. *Growth Factors (Chur, Switzerland), 24*, 121–136.

Sandri, M. T., Johansson, H. A., Zorzino, L., Salvatici, M., Passerini, R., Maisonneuve, P., et al. (2007). Serum EGFR and serum HER-2/neu are useful predictive and prognostic markers in metastatic breast cancer patients treated with metronomic chemotherapy. *Cancer, 110*, 509–517.

Sartor, C. I., Zhou, H., Kozlowska, E., Guttridge, K., Kawata, E., Caskey, L., et al. (2001). Her4 mediates ligand-dependent antiproliferative and differentiation responses in human breast cancer cells. *Molecular and Cellular Biology, 21*, 4265–4275.

Sato, J. D., Kawamoto, T., Le, A. D., Mendelsohn, J., Polikoff, J., & Sato, G. H. (1983). Biological effects in vitro of monoclonal antibodies to human epidermal growth factor receptors. *Molecular Biology & Medicine, 1*, 511–529.

Sawyers, C. (2004). Targeted cancer therapy. *Nature, 432*, 294–297.

Scaltriti, M., Rojo, F., Ocaña, A., Anido, J., Guzman, M., Cortes, J., et al. (2007). Expression of p95HER2, a truncated form of the HER2 receptor, and response to anti-HER2 therapies in breast cancer. *Journal of the National Cancer Institute, 99*, 628–638.

Schechter, A. L., Hung, M. C., Vaidyanathan, L., Weinberg, R. A., Yang-Feng, T. L., Francke, U., et al. (1985). The neu gene: An erbB-homologous gene distinct from and unlinked to the gene encoding the EGF receptor. *Science (New York, N.Y.), 229*, 976–978.

Schechter, A. L., Stern, D. F., Vaidyanathan, L., Decker, S. J., Drebin, J. A., Greene, M. I., et al. (1984). The neu oncogene: An erb-B-related gene encoding a 185,000-Mr tumour antigen. *Nature, 312*, 513–516.

Schulze, W. X., Deng, L., & Mann, M. (2005). Phosphotyrosine interactome of the ErbB-receptor kinase family. *Molecular Systems Biology, 1*, 2005.0008.

Seaman, M. N. J. (2012). The retromer complex—endosomal protein recycling and beyond. *Journal of Cell Science, 125*, 4693–4702.

Seligmann, J. F., Hatch, A. J., Richman, S. D., Elliott, F., Jacobs, B., Brown, S., et al. (2018). Association of tumor HER3 messenger RNA expression with panitumumab efficacy in advanced colorectal cancer. *JAMA Oncology*, *4*, 564–568.

Semba, K., Kamata, N., Toyoshima, K., & Yamamoto, T. (1985). A v-erbB-related protooncogene, c-erbB-2, is distinct from the c-erbB-1/epidermal growth factor-receptor gene and is amplified in a human salivary gland adenocarcinoma. *Proceedings of the National Academy of Sciences of the United States of America*, *82*, 6497–6501.

Sergina, N. V., Rausch, M., Wang, D., Blair, J., Hann, B., Shokat, K. M., et al. (2007). Escape from HER-family tyrosine kinase inhibitor therapy by the kinase-inactive HER3. *Nature*, *445*, 437–441.

Seton-Rogers, S. (2019). Passengers masquerading as cancer drivers. *Nature Reviews. Cancer*, *19*, 485.

Shaw, A. S., Kornev, A. P., Hu, J., Ahuja, L. G., & Taylor, S. S. (2014). Kinases and pseudokinases: Lessons from RAF. *Molecular and Cellular Biology*, *34*, 1538–1546.

Shen, F., Lin, Q., Childress, C., & Yang, W. (2008). Identification of the domain in ErbB2 that restricts ligand-induced degradation. *Cellular Signalling*, *20*, 779–786.

Sheng, Q., Liu, X., Fleming, E., Yuan, K., Piao, H., Chen, J., et al. (2010). An activated ErbB3/NRG1 autocrine loop supports in vivo proliferation in ovarian cancer cells. *Cancer Cell*, *17*, 298–310.

Sierke, S. L., Cheng, K., Kim, H. H., & Koland, J. G. (1997). Biochemical characterization of the protein tyrosine kinase homology domain of the ErbB3 (HER3) receptor protein. *The Biochemical Journal*, *322*(Pt. 3), 757–763.

Sigismund, S., Confalonieri, S., Ciliberto, A., Polo, S., Scita, G., & Di Fiore, P. P. (2012). Endocytosis and signaling: Cell logistics shape the eukaryotic cell plan. *Physiological Reviews*, *92*, 273–366.

Singla, H., Munshi, A., Banipal, R. P. S., & Kumar, V. (2018). Recent updates on the therapeutic potential of HER2 tyrosine kinase inhibitors for the treatment of breast cancer. *Current Cancer Drug Targets*, *18*, 306–327.

Slamon, D. J., Godolphin, W., Jones, L. A., Holt, J. A., Wong, S. G., Keith, D. E., et al. (1989). Studies of the HER-2/neu proto-oncogene in human breast and ovarian cancer. *Science (New York, N.Y.)*, *244*, 707–712.

Soler, M., Mancini, F., Meca-Cortés, O., Sánchez-Cid, L., Rubio, N., López-Fernández, S., et al. (2009). HER3 is required for the maintenance of neuregulin-dependent and -independent attributes of malignant progression in prostate cancer cells. *International Journal of Cancer*, *125*, 2565–2575.

Solinas, C., Ceppi, M., Lambertini, M., Scartozzi, M., Buisseret, L., Garaud, S., et al. (2017). Tumor-infiltrating lymphocytes in patients with HER2-positive breast cancer treated with neoadjuvant chemotherapy plus trastuzumab, lapatinib or their combination: A meta-analysis of randomized controlled trials. *Cancer Treatment Reviews*, *57*, 8–15.

Soltoff, S. P., Carraway, K. L., 3rd, Prigent, S. A., Gullick, W. G., & Cantley, L. C. (1994). ErbB3 is involved in activation of phosphatidylinositol 3-kinase by epidermal growth factor. *Molecular and Cellular Biology*, *14*, 3550–3558.

Song, S., Rosen, K. M., & Corfas, G. (2013). Biological function of nuclear receptor tyrosine kinase action. *Cold Spring Harbor Perspectives in Biology*, *5*, a009001.

Sorkin, A., & von Zastrow, M. (2009). Endocytosis and signalling: Intertwining molecular networks. *Nature Reviews. Molecular Cellular Biology*, *10*, 609–622.

Srinivasan, R., Gillett, C. E., Barnes, D. M., & Gullick, W. J. (2000). Nuclear expression of the c-erbB-4/HER-4 growth factor receptor in invasive breast cancers. *Cancer Research*, *60*, 1483–1487.

Stern, D. F., & Kamps, M. P. (1988). EGF-stimulated tyrosine phosphorylation of p185neu: A potential model for receptor interactions. *The EMBO Journal*, 7, 995–1001.

Strunk, K. E., Husted, C., Miraglia, L. C., Sandahl, M., Rearick, W. A., Hunter, D. M., et al. (2007). HER4 D-box sequences regulate mitotic progression and degradation of the nuclear HER4 cleavage product s80HER4. *Cancer Research*, *67*, 6582–6590.

Sun, C., Hobor, S., Bertotti, A., Zecchin, D., Huang, S., Galimi, F., et al. (2014). Intrinsic resistance to MEK inhibition in KRAS mutant lung and colon cancer through transcriptional induction of ERBB3. *Cell Reports*, 7, 86–93.

SUTHERLAND, E. W., & RALL, T. W. (1958). Fractionation and characterization of a cyclic adenine ribonucleotide formed by tissue particles. *The Journal of Biological Chemistry*, *232*, 1077–1091.

Swain, S. M., Baselga, J., Kim, S.-B., Ro, J., Semiglazov, V., Campone, M., et al. (2015). Pertuzumab, trastuzumab, and docetaxel in HER2-positive metastatic breast cancer. *The New England Journal of Medicine*, *372*, 724–734.

Sweeney, C., Fambrough, D., Huard, C., Diamonti, A. J., Lander, E. S., Cantley, L. C., et al. (2001). Growth factor-specific signaling pathway stimulation and gene expression mediated by ErbB receptors. *The Journal of Biological Chemistry*, *276*, 22685–22698.

Tabernero, J., Hoff, P. M., Shen, L., Ohtsu, A., Shah, M. A., Cheng, K., et al. (2018). Pertuzumab plus trastuzumab and chemotherapy for HER2-positive metastatic gastric or gastro-oesophageal junction cancer (JACOB): Final analysis of a double-blind, randomised, placebo-controlled phase 3 study. *The Lancet Oncology*, *19*, 1372–1384.

Tanner, B., Hasenclever, D., Stern, K., Schormann, W., Bezler, M., Hermes, M., et al. (2006). ErbB-3 predicts survival in ovarian cancer. *Journal of Clinical Oncology: Official Journal of the American Society of Clinical Oncology*, *24*, 4317–4323.

Teixidó, C., Arguelaguet, E., Pons, B., Aracil, M., Jimeno, J., Somoza, R., et al. (2012). ErbB3 expression predicts sensitivity to elisidepsin treatment: In vitro synergism with cisplatin, paclitaxel and gemcitabine in lung, breast and colon cancer cell lines. *International Journal of Oncology*, *41*, 317–324.

Tiwary, S., Preziosi, M., Rothberg, P. G., Zeitouni, N., Corson, N., & Xu, L. (2014). ERBB3 is required for metastasis formation of melanoma cells. *Oncogenesis*, *3*, e110.

Tomas, A., Futter, C. E., & Eden, E. R. (2014). EGF receptor trafficking: Consequences for signaling and cancer. *Trends in Cell Biology*, *24*, 26–34.

Torka, R., Pénzes, K., Gusenbauer, S., Baumann, C., Szabadkai, I., Őrfi, L., et al. (2014). Activation of HER3 interferes with antitumor effects of Axl receptor tyrosine kinase inhibitors: Suggestion of combination therapy. *Neoplasia (New York, N.Y.)*, *16*, 301–318.

Tovey, S., Dunne, B., Witton, C. J., Forsyth, A., Cooke, T. G., & Bartlett, J. M. S. (2005). Can molecular markers predict when to implement treatment with aromatase inhibitors in invasive breast cancer? *Clinical Cancer Research*, *11*, 4835–4842.

Tsé, C., Gauchez, A.-S., Jacot, W., & Lamy, P.-J. (2012). HER2 shedding and serum HER2 extracellular domain: Biology and clinical utility in breast cancer. *Cancer Treatment Reviews*, *38*, 133–142.

Tzahar, E., Waterman, H., Chen, X., Levkowitz, G., Karunagaran, D., Lavi, S., et al. (1996). A hierarchical network of interreceptor interactions determines signal transduction by neu differentiation factor/neuregulin and epidermal growth factor. *Molecular and Cellular Biology*, *16*, 5276–5287.

Ursini-Siegel, J., Schade, B., Cardiff, R. D., & Muller, W. J. (2007). Insights from transgenic mouse models of ERBB2-induced breast cancer. *Nature Reviews. Cancer*, 7, 389–397.

Van de Vijver, M. J., Kumar, R., & Mendelsohn, J. (1991). Ligand-induced activation of A431 cell epidermal growth factor receptors occurs primarily by an autocrine pathway that acts upon receptors on the surface rather than intracellularly. *The Journal of Biological Chemistry*, *266*, 7503–7508.

van de Vijver, M. J., Peterse, J. L., Mooi, W. J., Wisman, P., Lomans, J., Dalesio, O., et al. (1988). Neu-protein overexpression in breast cancer. Association with comedo-type ductal carcinoma in situ and limited prognostic value in stage II breast cancer. *The New England Journal of Medicine, 319*, 1239–1245.

Vaught, D. B., Stanford, J. C., Young, C., Hicks, D. J., Wheeler, F., Rinehart, C., et al. (2012). HER3 is required for HER2-induced preneoplastic changes to the breast epithelium and tumor formation. *Cancer Research, 72*, 2672–2682.

Vazquez-Martin, A., Oliveras-Ferraros, C., Cufí, S., Del Barco, S., Martin-Castillo, B., & Menendez, J. A. (2011). Lapatinib, a dual HER1/HER2 tyrosine kinase inhibitor, augments basal cleavage of HER2 extracellular domain (ECD) to inhibit HER2-driven cancer cell growth. *Journal of Cellular Physiology, 226*, 52–57.

Verma, N., Müller, A.-K., Kothari, C., Panayotopoulou, E., Kedan, A., Selitrennik, M., et al. (2017). Targeting of PYK2 synergizes with EGFR antagonists in basal-like TNBC and circumvents HER3-associated resistance via the NEDD4-NDRG1 axis. *Cancer Research, 77*, 86–99.

Vernieri, C., Milano, M., Brambilla, M., Mennitto, A., Maggi, C., Cona, M. S., et al. (2019). Resistance mechanisms to anti-HER2 therapies in HER2-positive breast cancer: Current knowledge, new research directions and therapeutic perspectives. *Critical Reviews in Oncology/Hematology, 139*, 53–66.

von Minckwitz, G., Huang, C.-S., Mano, M. S., Loibl, S., Mamounas, E. P., Untch, M., et al.Investigators, K., (2019). Trastuzumab emtansine for residual invasive HER2-positive breast cancer. *The New England Journal of Medicine, 380*, 617–628.

von Minckwitz, G., Procter, M., de Azambuja, E., Zardavas, D., Benyunes, M., Viale, G., et al. (2017). Adjuvant pertuzumab and trastuzumab in early HER2-positive breast cancer. *The New England Journal of Medicine, 377*, 122–131.

Wada, T., Qian, X., & Greene, M. I. (1990). Intermolecular association of the p185 protein and EGF receptor modulates EGF receptor function. *Cell, 61*, 1339–1347.

Wali, V. B., Haskins, J. W., Gilmore-Hebert, M., Platt, J. T., Liu, Z., & Stern, D. F. (2014). Convergent and divergent cellular responses by ErbB4 isoforms in mammary epithelial cells. *Molecular Cancer Research, 12*, 1140–1155.

Wallasch, C., Weiss, F. U., Niederfellner, G., Jallal, B., Issing, W., & Ullrich, A. (1995). Heregulin-dependent regulation of HER2/neu oncogenic signaling by heterodimerization with HER3. *The EMBO Journal, 14*, 4267–4275.

Wang, Z. (2017). ErbB receptors and cancer. *Methods in Molecular Biology (Clifton, N.J.), 1652*, 3–35.

Wang, S., Huang, J., Lyu, H., Lee, C.-K., Tan, J., Wang, J., et al. (2013). Functional cooperation of miR-125a, miR-125b, and miR-205 in entinostat-induced downregulation of erbB2/erbB3 and apoptosis in breast cancer cells. *Cell Death & Disease, 4*, e556.

Wang, Y.-N., & Hung, M.-C. (2012). Nuclear functions and subcellular trafficking mechanisms of the epidermal growth factor receptor family. *Cell & Bioscience, 2*, 13.

Warren, C. M., & Landgraf, R. (2006). Signaling through ERBB receptors: Multiple layers of diversity and control. *Cellular Signalling, 18*, 923–933.

Waterman, H., Alroy, I., Strano, S., Seger, R., & Yarden, Y. (1999). The C-terminus of the kinase-defective neuregulin receptor ErbB-3 confers mitogenic superiority and dictates endocytic routing. *The EMBO Journal, 18*, 3348–3358.

Waterman, H., Sabanai, I., Geiger, B., & Yarden, Y. (1998). Alternative intracellular routing of ErbB receptors may determine signaling potency. *The Journal of Biological Chemistry, 273*, 13819–13827.

Wiley, H. S. (2003). Trafficking of the ErbB receptors and its influence on signaling. *Experimental Cell Research, 284*, 78–88.

Williams, C. C., Allison, J. G., Vidal, G. A., Burow, M. E., Beckman, B. S., Marrero, L., et al. (2004). The ERBB4/HER4 receptor tyrosine kinase regulates gene expression by functioning as a STAT5A nuclear chaperone. *The Journal of Cell Biology, 167*, 469–478.

Wilson, K. J., Mill, C., Lambert, S., Buchman, J., Wilson, T. R., Hernandez-Gordillo, V., et al. (2012). EGFR ligands exhibit functional differences in models of paracrine and autocrine signaling. *Growth Factors (Chur, Switzerland)*, *30*, 107–116.

Xie, T., Lim, S. M., Westover, K. D., Dodge, M. E., Ercan, D., Ficarro, S. B., et al. (2014). Pharmacological targeting of the pseudokinase Her3. *Nature Chemical Biology*, *10*, 1006–1012.

Yan, M., Parker, B. A., Schwab, R., & Kurzrock, R. (2014). HER2 aberrations in cancer: Implications for therapy. *Cancer Treatment Reviews*, *40*, 770–780.

Yarden, Y., & Pines, G. (2012). The ERBB network: At last, cancer therapy meets systems biology. *Nature Reviews. Cancer*, *12*, 553–563.

Yarden, Y., & Sliwkowski, M. X. (2001). Untangling the ErbB signalling network. *Nature Reviews. Molecular and Cellular Biology*, *2*, 127–137.

Yarden, Y., & Ullrich, A. (1988). Growth factor receptor tyrosine kinases. *Annual Review of Biochemistry*, *57*, 443–478.

Ye, X. S., Yu, C., Aggarwal, A., & Reinhard, C. (2016). Genomic alterations and molecular subtypes of gastric cancers in Asians. *Chinese Journal of Cancer*, *35*, 42.

Yonesaka, K., Zejnullahu, K., Okamoto, I., Satoh, T., Cappuzzo, T., Souglakos, J., et al. (2011). Activation of ERBB2 signaling causes resistance to the EGFR-directed therapeutic antibody cetuximab. *Science Translational Medicine*, *3*, 99ra86.

Zhang, K., Cui, J., Xi, H., Bian, S., Ma, L., Shen, W., et al. (2015). Serum HER2 is a potential surrogate for tissue HER2 status in gastric cancer: A systematic review and meta-analysis. *PLoS One*, *10*, e0136322.

Zhang, X., Gureasko, J., Shen, K., Cole, P. A., & Kuriyan, J. (2006). An allosteric mechanism for activation of the kinase domain of epidermal growth factor receptor. *Cell*, *125*, 1137–1149.

Zhou, M., Rubinstein, M., Lax, I., Schlessinger, J., Felder, S., Hurwitz, D. R., et al. (1993). Real-time measurements of kinetics of EGF binding to soluble EGF receptor monomers and dimers support the dimerization model for receptor activation. *Biochemistry*, *32*, 8193–8198.

CHAPTER FOUR

EGFR: An essential receptor tyrosine kinase-regulator of cancer stem cells

Sarmistha Talukdar[a,b], Luni Emdad[a,b,c], Swadesh K. Das[a,b,c], Paul B. Fisher[a,b,c,*]

[a]Department of Human and Molecular Genetics, Virginia Commonwealth University, School of Medicine, Richmond, VA, United States
[b]VCU Institute of Molecular Medicine, Virginia Commonwealth University, School of Medicine, Richmond, VA, United States
[c]VCU Massey Cancer Center, Virginia Commonwealth University, School of Medicine, Richmond, VA, United States
[*]Corresponding author: e-mail address: paul.fisher@vcuhealth.org

Contents

Abstract

The Epidermal Growth Factor Receptor (EGFR) is frequently expressed at elevated levels in different forms of cancer and expression often correlates positively with cancer progression and poor prognosis. Different mutant forms of this protein also contribute to cancer heterogeneity. A constitutively active form of EGFR, EGFRvIII is one of the most important variants. EGFR is responsible for the maintenance and functions of cancer stem cells (CSCs), including stemness, metabolism, immunomodulatory-activity, dormancy and therapy-resistance. EGFR regulates these pathways through several signaling cascades, and often cooperates with other RTKs to exert further control. Inhibitors of EGFR have been extensively studied and display some anticancer efficacy. However, CSCs can also acquire resistance to EGFR inhibitors making effective therapy even more difficult. To ameliorate this limitation of EGFR inhibitors when used as single agents, it may be of value to simultaneously combine multiple EGFR inhibitors or use EGFR inhibitors with regulators of other important cancer phenotype regulating molecules,

Advances in Cancer Research, Volume 147
ISSN 0065-230X
https://doi.org/10.1016/bs.acr.2020.04.003
© 2020 Elsevier Inc.
All rights reserved.

such as STAT3, or involved in important processes such as DNA repair. These combinatorial approaches require further experimental confirmation, but if successful would expand and improve therapeutic outcomes employing EGFR inhibitors as one arm of the therapy.

1. Introduction

Receptor tyrosine kinases (RTKs) are cell surface receptors with high affinity for many polypeptide growth factors, cytokines, and hormones. Receptor tyrosine kinases have emerged as seminal regulators of major cellular processes, including cell proliferation, differentiation, survival, metabolism, migration and cell cycle control (Blume-Jensen & Hunter, 2001; Ullrich & Schlessinger, 1990). All RTKs include a ligand-binding region in the extracellular domain, a single transmembrane helix, and a cytoplasmic region that contains the protein tyrosine kinase (TK) domain plus additional carboxy (C-) terminal and juxtamembrane regulatory regions (Lemmon & Schlessinger, 2010). This is true for all the 20 subfamilies of RTKs, which are EGFR, InsR, PDGFR, VEGFR, FGFR, PTK7/CCK4, Trk, Ror, MuSK, MET, Axl, Tie, EphA/B, Ret, Ryk, DDR1/2, Ros, LMR, ALK and SuRTK106/STYK1. The intracellular domain of RTKs is the region that possesses tyrosine kinase activity, thus referred to as the tyrosine kinase domain or TKD. This tyrosine kinase domain can phosphorylate tyrosine residues on proteins either in the *cis* (within the same molecule) or in the *trans* (residing on a different molecule) conformation. This consensus structure of RTKs is conserved across evolution and mutations in RTKs lead to development of structural abnormalities in various disorders (Butti et al., 2018). In general, growth factor binding activates RTKs by inducing receptor dimerization (Ullrich & Schlessinger, 1990). Irrespective of the "inactive" state configuration, whether monomeric or oligomeric, activation of the receptor requires the bound ligand to stabilize a specific relationship between individual receptor molecules in an "active" dimer or oligomer (Lemmon & Schlessinger, 2010) (Fig. 1). The method of receptor activation further divides RTKs into three subclasses. Receptor activation may occur either by (1) binding of monomeric ligands resulting in a conformational change of the extracellular domain and dimer formation (subclass I), (2) interaction of the ligand with a disulfide stabilized receptor dimer and ensuing intracomplex conformational change (subclass II), or (3) regulation of dimer formation through a dimeric ligand (subclass III) (Ullrich & Schlessinger, 1990) (Fig. 2).

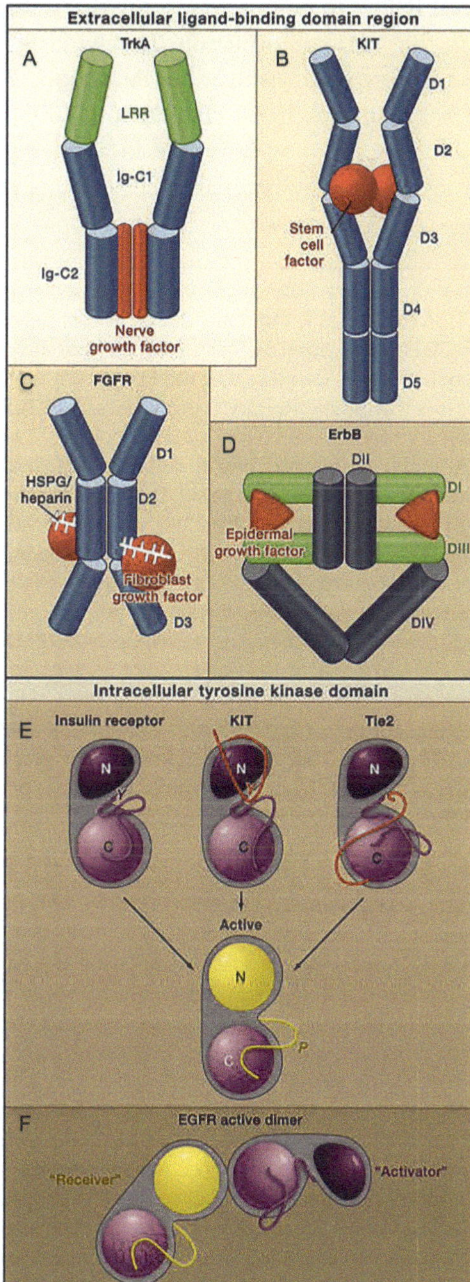

Fig. 1 Receptor tyrosine kinase dimerization and kinase activation. Top: In general, receptor tyrosine kinases (RTKs) associate into dimers when ligand (red) binds to their extracellular regions. The bound ligand, which can form all, a portion, or none of the dimer interface, activates the receptors by stabilizing a specific relationship between *(Continued)*

Studies of several RTKs in vitro demonstrate that autophosphorylation of RTKs occurs in the *trans* conformation (Favelyukis, Till, Hubbard, & Miller, 2001; Furdui, Lew, Schlessinger, & Anderson, 2006; Honegger, Kris, Ullrich, & Schlessinger, 1989; Till et al., 2002) and that these autophosphorylation sites are phosphorylated in a precise temporal order. Each consecutive event has a substantial effect on catalytic properties of

Fig. 1—Cont'd two individual receptor molecules. (A) A nerve growth factor dimer (red) cross-links two TrkA molecules without any direct contact between the two receptors (Wehrman et al., 2007). (B) A stem cell factor dimer (red) also cross links two KIT molecules. In addition, two Ig-like domains (D4 and D5), which reorient upon receptor activation, interact across the dimer interface (Yuzawa et al., 2007). Thus, KIT combines ligand-mediated and receptor mediated dimerization modes. (C) Two fibroblast growth factor receptor (FGFR) molecules contact one another through the Ig-like domain D2, and the accessory molecule heparin or heparin sulfate proteoglycans (white sticks) also contacts this domain. In addition, each fibroblast growth factor molecule (red) contacts Ig-like domains D2 and D3 of both FGFR molecules. (D) Dimerization of ErbB receptors is mediated entirely by the receptor. Binding simultaneously to two sites (DI and DIII) within the same receptor molecule, the ligand drives conformational changes in epidermal growth factor receptor (EGFR) that expose a previously-occluded dimerization site in Domain II. Bottom: Dimerization of the extracellular regions of RTKs activates the intracellular tyrosine kinase domains (TKDs), which contain a C-lobe (light purple or yellow), N-lobe (dark purple or yellow in the inactive and active states), and an activation loop (dark purple or yellow in the inactive and active states, respectively). Although the crystal structures of the activated TKDs are very similar, structures of inactive TKDs differ substantially among the receptors (top row), reflecting the diversity in their regulatory mechanisms. However, many receptors are inhibited by a set of intramolecular (or *cis*) interactions: (E) Insulin receptor-like (activation loop inhibition): In FGFR, insulin receptor, and IGF1-receptor the activation loop interacts directly with the active site of the kinase and blocks access to protein substrates (in FGFR) or to both ATP and protein substrates (in insulin and IGF1 receptors). Phosphorylation of key tyrosines (Y) disrupts these autoinhibitory interactions and allows the kinase to "relax" to the Active state. KIT-like (juxtamembrane inhibition): In KIT, PDGFR, and Eph receptors the juxtamembrane region (red) interacts with elements within the active site of the kinase (including the αC helix and the activation loop) to stabilize an inactive conformation. Phosphorylation of key tyrosines in the juxtamembrane region destabilizes these autoinhibitory interactions and allows the TKD to assume an active conformation. Tie2-like (C-terminal tail inhibition): In Tie1 and Tie2 (and possibly Met and Ron), the C-terminal tail (red) interacts with the active site of the TKD to stabilize an inactive conformation. (F) The EGFR TKD is allosterically activated by direct contacts between the C-lobe of one TKD, the "Activator," and the N-lobe of another TKD, "Receiver." The Activator TKD destabilizes autoinhibitory interactions that involve the activation loop of the Receiver TKD. No activation loop phosphorylation is required in this mechanism. *From Lemmon, M. A., & Schlessinger, J. (2010). Cell signaling by receptor tyrosine kinases.* Cell, 141, *1117–1134.*

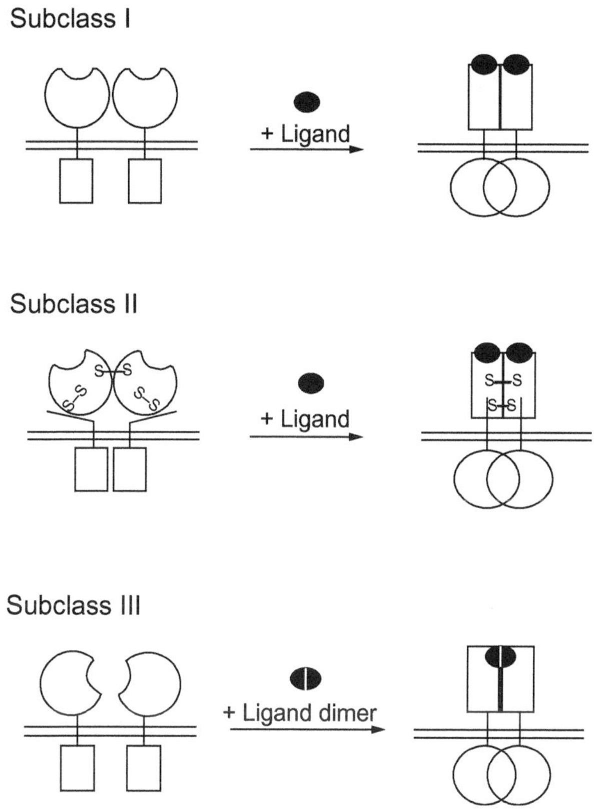

Fig. 2 Models of receptor subclass-specific variations of the mechanism of activation by dimerization. Receptor activation may occur by binding of monomeric ligands resulting in a conformational change of the extracellular domain and dimer formation (subclass I), by interaction of the ligand with a disulfide-stabilized receptor dimer and subsequent intracomplex conformational change (subclass II), or by mediation of dimer formation through a dimeric ligand (subclass III). *From Ullrich, A., & Schlessinger, J. (1990). Signal transduction by receptors with tyrosine kinase activity.* Cell, 61, *203–212.*

the reaction by destabilizing *cis*-autoinhibitory interactions (Lemmon & Schlessinger, 2010). Activation of RTKs, such as the fibroblast growth factor receptors, epidermal growth factor receptors, platelet-derived growth factor receptors and insulin-like growth factor receptors (IGF1R) regulate the developmental responses that are elicited in stem and progenitor cells (Annenkov, 2014). Intracellular inhibitors of RTK signaling are also capable of regulating the various developmental pathways in stem and progenitor cells and these often govern the developmental outcomes of RTK activation (Annenkov, 2014; Chen, Song, & Amato, 2015). Moreover, cellular responses of stem cells to the activation of a particular RTK are also significantly regulated by the extent of signal strength (Annenkov, 2014; Chen, Song, & Amato, 2015). Cellular responsiveness also plays an important role in determining the developmental outcomes of a given RTK signaling event. In combination, these mechanisms are responsible for maintaining the balance between stem cell renewal, maintenance and differentiation, during the multiple stages of development and adulthood (Annenkov, 2014; Chen, Song, & Amato, 2015). How particular developmental responses of stem/progenitor cells conform to specific pathways of RTK signaling is also becoming increasingly illuminated (Annenkov, 2014; Chen, Song, & Amato, 2015). Co-activation of several RTKs in developing stem cells is a very common occurrence, and analysis of cooperation between their different signaling pathways can significantly shift the perception of RTK-based stem-mediated regulation in development (Annenkov, 2014; Chen, Song, & Amato, 2015). Similar to the role of regulation of RTKs in stem cells, RTKs are also capable of regulating cancer stem cells (CSCs) (Chen, Song, & Amato, 2015). Elevated levels of RTKs have been found to correlate significantly with increased cancer aggressiveness, progression and decreased overall prognosis and disease-free survival (Butti et al., 2018; Templeton et al., 2014). In addition, RTKs also regulate transdifferentiation of cancer cells to endothelial cells, to give rise to new blood vessels through a process known as vasculogenic mimicry (Butti et al., 2018; D'Ippolito et al., 2016; Plantamura et al., 2014). RTKs are also known to regulate several signaling pathways such as MAPK, PI3K/AKT and JAK/STAT which play a crucial role in the regulation of cancer stemness, angiogenesis and metastasis (Butti et al., 2018; Ibrahim et al., 2017; Wise & Zolkiewska, 2017). Accordingly, co-targeting different RTKs in CSCs results in a significant decrease in the growth of recurrent tumors (Qazi et al., 2018). In tumor and stromal cells, RTK signaling pathways are also involved in the regulation of different CSC phenotypes (Bussard, Mutkus, Stumpf, Gomez-Manzano, & Marini, 2016; Butti et al., 2018; Liu et al., 2014; Shibuya, 2011) such as self-renewal capacity, viability, invasiveness, and tumorigenicity (Chen, Song, & Amato, 2015).

2. EGFR and cancer

The epidermal growth factor receptor (EGFR) belongs to a family of receptor tyrosine kinases that comprises three other members (ErbB2/HER-2, ErbB3/HER-3, and ErbB4/HER-4) (Hynes & Lane, 2005;

Scaltriti & Baselga, 2006 ; Yarden & Sliwkowski, 2001). EGFR is activated either by receptor overexpression (a condition which is frequently observed in cancer) or through ligand-dependent and ligand-independent mechanisms. EGFRs possess seven different cognate ligands that are capable of activating it, such as EGF, TGFα, betacellulin (BTC), heparin-binding EGF, amphiregulin (AREG), epiregulin, and epigen (Butti et al., 2018; Macdonald-Obermann & Pike, 2014; Wilson, Gilmore, Foley, Lemmon, & Riese, 2009). The EGFR family is made up of EGFR1 (EGFR, HER1, c-ErbB1), EGFR2 (HER2, c-ErbB2), EGFR3 (c-ErbB3, HER3) and EGFR4 (c-ErbB4, HER4) proteins (Butti et al., 2018; Macdonald-Obermann & Pike, 2014; Wilson et al., 2009). Binding of the ligand induces a conformational change of the receptor ectodomain which assists receptor dimerization as well as autophosphorylation of several tyrosine residues within the COOH-terminal tail of the receptors (Burgess et al., 2003; Hubbard, 2005; Scaltriti & Baselga, 2006). Activation of the receptor results in phosphorylation of crucial tyrosine residues within the COOH-terminal portion of EGFR thereby allowing formation of specific docking sites for cytoplasmic proteins containing Src homology 2 and phosphotyrosine-binding domains (Scaltriti & Baselga, 2006; Yarden & Sliwkowski, 2001). These proteins bind to specific phosphotyrosine residues in EGFR and thus initiate intracellular signaling via the following main pathways (Scaltriti & Baselga, 2006; Sigismund, Avanzato, & Lanzetti, 2018):

a. RAS/RAF/mitogen-activated protein kinase pathway: After EGFR is phosphorylated, the GRB2/SOS/SHC complex attaches to specific docking sites on EGFR, leading to a conformational change in SOS that results in RAS and RAF-1 activation, which subsequently phosphorylates the mitogen-activated protein kinases (MAPK) and extracellular signal-regulated kinases 1 and 2 (Scaltriti & Baselga, 2006).

b. Phosphatidylinositol 3-kinase/Akt pathway: Dimerization of EGFR with HER-3 catalyzes the PI3K pathway. This pathway is very important for cell growth, apoptosis resistance, invasion, and migration (Scaltriti & Baselga, 2006).

c. Phospholipase Cγ: Phospholipase Cγ binds with activated EGFR leading to the hydrolysis of phosphatidylinositol 4,5-diphosphate yielding inositol 1,3,5-triphosphate, which aids in protein kinase C activation and in turn MAPK and c-Jun activation (Scaltriti & Baselga, 2006).

d. Signal transducers and activators of transcription pathway: EGFR, regulates and enhances constitutive STAT3 activation, which subsequently regulates tumor formation and progression (Scaltriti & Baselga, 2006).

e. Src kinase pathway: The relationship between Src and EGFR is complicated and intricate. Src assists EGFR activation, but it also supports resistance to EGFR therapies through other receptor-mediated signaling pathways (Scaltriti & Baselga, 2006).

f. Endocytotic pathway: EGFR is internalized through multiple endocytic pathways and once internalized, EGFRs reach the early endosomes. This allows regulation of EGFR signaling in a different manner, i.e., the number of EGFR clusters per endosome can change endosome fission/fusion rate thereby regulating EGFR signaling output, such as deciding between proliferation and differentiation (Sigismund et al., 2018).

g. Stress pathway: When cells are stressed (through hypoxia, UV radiation) EGFR endocytosis and trafficking in a ligand-independent fashion is promoted (Sigismund et al., 2018).

h. Autophagy pathway: EGFR is a very important regulator of autophagy. In nutrient-rich conditions, EGFR signaling promotes cell proliferation and inhibits autophagy. However under nutrient-starved conditions, inactive EGFR helps in the promotion of autophagy (Sigismund et al., 2018).

Overexpression of EGFR is a common phenomenon in several cancer tissues and has been reported to be associated with higher aggressiveness and poor clinical outcomes in breast, lung, ovarian, cervical, bladder, esophageal, brain as well as head and neck cancers (Butti et al., 2018; Evangelopoulos, Weis, & Kruttgen, 2005; Lee et al., 2015; Liao et al., 2015; Nicholson, Gee, & Harper, 2001; Park et al., 2014; Zhu et al., 2009). EGFR is overexpressed in these human tumors, often as a consequence of gene amplification (Gan, Cvrljevic, & Johns, 2013). Tumors with EGFR gene amplification often contain various EGFR gene rearrangements, the most common extracellular domain mutation being the expression of EGFRvIII variant protein (Gan et al., 2013). This mutation results in a deletion of exons 2–7 of the EGFR gene and leaves the mutant receptor unable to bind appropriate ligands required for activation (Gan et al., 2013). Regardless of this change, EGFRvIII is capable of low-level constitutive sustained signaling that is further amplified by reduced internalization and downregulation (Gan et al., 2013; Schmidt, Furnari, Cavenee, & Bogler, 2003). Aberrant EGFRvIII expression is important in regulating cancer progression and its expression often correlates with poor prognosis (Gan et al., 2013). The outcomes of EGFRvIII signaling is additionally augmented through interaction with several other receptor tyrosine kinases, such as c-MET, platelet-derived growth factor receptor β (PDGFRβ) and VEGF receptor 2 (VEGFR2)

(Gan et al., 2013). EGFRvIII can form a heterodimer with c-MET, resulting in the direct activation and increased cellular response to the c-MET ligand hepatocyte growth factor (Greenall & Johns, 2016). Additionally, EGFRvIII can regulate STAT3 signaling (Greenall & Johns, 2016; Jahani-Asl et al., 2016), an important signaling pathway that controls stemness (Galoczova, Coates, & Vojtesek, 2018; Raz, Lee, Cannizzaro, d'Eustachio, & Levy, 1999). Anticancer agents have been designed by different groups that target the extracellular ligand-binding region of the receptor. These include monoclonal antibodies, immunotoxins and ligand-binding cytotoxic agents as well as agents that target the intracellular tyrosine kinase region such as small molecule tyrosine kinase inhibitors (TKIs), which act by interfering with ATP binding to the receptor, and multiple other compounds that either perform at substrate-binding regions or downstream components of the signaling pathway (Raymond, Faivre, & Armand, 2000; Shinozaki et al., 2017; Yang et al., 2013).

3. Role of EGFR in regulation of cancer stem cells (CSCs)

3.1 Stemness regulation

The EGFR signaling pathway acts as a crucial regulator for terminal differentiation of various lineage-committed progenitor and stem cells (Barberan & Cebria, 2019). EGFR is required as a cofactor for Wnt9a–Fzd9b signaling, a key regulator for hematopoietic stem cells (Grainger et al., 2019). EGFR assists the phosphorylation of a tyrosine residue on the Fzd9b intracellular region in response to Wnt9a, which promotes internalization of the Wnt9a–Fzd9b–LRP signalosome and initiation of the succeeding signal transduction pathway in stem cells (Grainger et al., 2019). EGFR is also reported to play an important role in differentiation of embryonic stem cells (Wu & Adamson, 1996). Inhibition of EGFR results in a significant shift from asymmetric to symmetric division modes, while EGF treatment intensifies asymmetric divisions, and therefore self-renewal in stem cells. EGFR activation results in the orientation of mitotic centrosomes through a protein called Aurka, and inhibiting Aurka decreases EGF induced asymmetric division. Treatment of mdx mice (a mouse model used to study Duchenne muscular dystrophy) with EGF also increases asymmetric divisions of dystrophin-deficient satellite cells, thereby resulting in increased progenitor cells, regeneration, and restoration of muscle strength. Thus, EGFR regulates a crucial polarity pathway that is capable of functionally rescuing dystrophin-deficient satellite cells and muscle force generation (Wang et al., 2019). Wnt and EGFR pathways co-operate to promote follicle stem cell self-renewal (Kim-Yip & Nystul, 2018). Wnt works upstream of EGFR signaling, hence activation by EGFR ligand Spitz, as well as constitutive activation of EGFR, partially rescues the loss of self-renewal resulting from the loss of Wnt signaling. EGFR and MET also collaborate to regulate the self-renewal properties of hepatic progenitor cells through activation of the extracellular signal-regulated kinase (ERK) pathway (Kitade et al., 2013). EGFR also regulates NOTCH1 to promote neural stem cell self-renewal (Aguirre, Rubio, & Gallo, 2010) as well as cholangiocyte specification and branching morphogenesis while concomitantly suppressing hepatocyte commitment (Kitade et al., 2013). It has been reported that liver-specific conditional loss of *Egfr,* promotes liver regeneration, by switching progenitor cell differentiation toward the hepatocyte lineage (Kitade et al., 2013). Similarly, EGFR also regulates stemness in several types of cancer. Treatment with EGFR tyrosine kinase inhibitors regulates the acquired resistance and the appearance of stemness in lung cancer cells (Shien et al., 2013). SALL4, an important transcriptional factor and stem cell regulator expressed in hematopoietic stem cells and various malignancies, is induced by the activation of EGFR through the extracellular signal-regulated kinase 1/2 (ERK1/2) signaling pathway in lung cancer (Du et al., 2018). EGFR also induces ILF3 and G9A to maintain OCT4-mediated stemness phenotypes in EGFR-positive cancers (Cheng et al., 2017). Activation of EGFR signaling, following the addition of epidermal growth factor (EGF) or expression of the constitutively-active EGFR mutant variant EGFRvIII, is reported to increase SOX2 expression and thereby self-renewal of prostate cancer stem cells (Rybak & Tang, 2013). EGFRvIII is reported to regulate cancer stemness in glioma stem cells through MEK/ERK pathway (Kwon et al., 2017). EGFRvIII can also regulate STAT3 signaling (Greenall & Johns, 2016; Jahani-Asl et al., 2016), a pathway that controls stemness in both normal and malignant tissues (Galoczova et al., 2018; Raz et al., 1999). EGFR activation status in glioma stem cells varies between patients and is reported to be an important determinant of responsiveness to anti-EGFR therapy. EGFR+/PTEN+ glioma stem cells are reported to be susceptible to anti-EGFR therapy in vitro (Howard et al., 2010). EGFR pathway deregulation assists the acquisition of stem-like properties, and in EGFR-mutation-positive patients, the cancer stem cells were observed to be resistant to EGFR inhibition (Codony-Servat et al., 2019). EGFR also regulates the inflammatory breast cancer cell population that expresses cancer stem-like cell markers through COX-2, a key mediator of inflammation whose expression correlates with worse outcome in inflammatory breast cancer (Wang et al., 2017). CSCs are also reported to rise from a common progenitor with signature genetic mutations in CDKN2A/B, EGFR and TP53 (Sottoriva et al., 2013). Tumor associated macrophages can also promote cancer stem cell-like phenotypes in murine breast cancer cells by upregulating their expression of SOX2 through the EGFR/STAT3/SOX2 signaling pathway (Yang et al., 2013) suggesting the EGFR activity even in the stroma could regulate stemness in cancers (Figs. 3–4).

3.2 Resistance to therapy

EGFR is known to contribute to resistance to various forms of cancer therapy. Constitutive activation through expression of EGFRvIII increases clonogenic survival by almost two- and fourfold after 4-Gy of radiation and 4-days of anoxia, respectively (Weppler et al., 2007). Radiation-induced EGFR-signaling can also regulate DNA-double strand break repair (Rodemann, Dittmann, & Toulany, 2007) contributing to its resistance. EGFRvIII was also

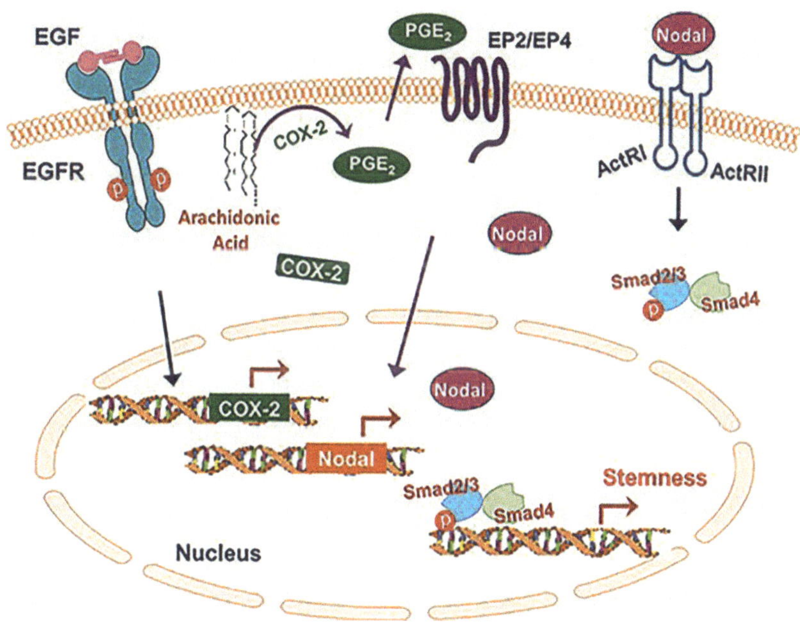

Fig. 3 Schematic summarizing the role of the EGFR/COX-2/Nodal signaling axis in regulating IBC stemness. EGFR signaling upregulates the expression of COX-2, which subsequently increases the production of prostaglandins such as PGE2 and PGF2α in IBC cells; the activation of the COX-2 pathway upregulates the transcription of Nodal and activates Nodal signaling, which subsequently regulates IBC stemness. *From Wang, X., Reyes, M. E., Zhang, D., Funakoshi, Y., Trape, A. P., Gong, Y., et al. (2017). Egfr signaling promotes inflammation and cancer stem-like activity in inflammatory breast cancer.* Oncotarget, 8, *67904–67917.*

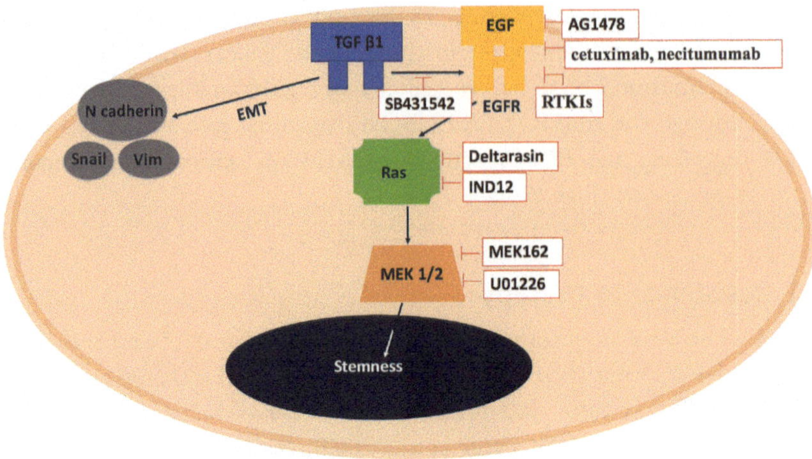

Fig. 4 Schema showing how EGFR/RAS/MEK signaling pathway regulates TGF-β1-induced stemness and tumorigenicity. The activation of TGF-β pathway triggers cells into EMT, concurrently inducing a stem cell-like and tumorigenic state. TGF-β1 induces an EGFR/RAS gene expression signature marked by increased EGF expression. The activation of EGFR/RAS pathway promotes stemness and tumorigenicity in a MEK1/2-dependent manner that did not involve increased EMT. *Modified from Voon, D. C., Wang, H., Koo, J. K., Chai, J. H., Hor, Y. T., Tan, T. Z., et al. (2013). EMT-induced stemness and tumorigenicity are fueled by the EGFR/Ras pathway,* PLoS One, *8, e70427.*

associated with a substantially bigger colony size after anoxic treatment (Weppler et al., 2007). When translocated to the nucleus, EGFR functions as a co-transcription factor for several genes involved in cell proliferation and angiogenesis, as well as a tyrosine kinase to activate and stabilize proliferating cell nuclear antigen and DNA dependent protein kinase (Brand et al., 2013). Nuclear localization of EGFR is highly associated with cancer progression, poor prognosis and significantly increased resistance to radiation, chemotherapy, and anti-EGFR therapies such as gefitinib and cetuximab, in several types of cancers (Brand et al., 2013).

EGFR tyrosine kinase inhibitors (EGFR-TKIs) regulate wt EGFR positive cancer stem- cells (CSCs) by inhibiting autophosphorylation of EGFR and downstream signaling proteins, such as β-catenin, a stemness regulator (Yang, Li, Liu, You, & Zhou, 2018), and are often used as chemotherapeutic strategies and for these reasons in clinical settings. Wt EGFR + glioma stem cells are reported to be susceptible to anti-EGFR therapy in vitro (Howard et al., 2010). Inhibition of EGFR increases the chemo- and radio-sensitivity of both canine and human glioma CSCs, thereby helping the targeting of this unique, rare, resistant, and tumor initiating population of cells

(Pang, Saunders, & Argyle, 2017). However, EGFR-TKI exposure can lead to acquired resistance and enrichment of cancer stem cells (Shien et al., 2013). The effect of EGFR inhibition on EGFRvIII cells is controversial. In some studies EGFR-vIII-positive patients' CSCs are reported to be resistant to EGFR inhibition (Codony-Servat et al., 2019), while some report that the EGFR mutant cells are susceptible (An, Aksoy, Zheng, Fan, & Weiss, 2018).

The genetic and molecular heterogeneity of lung cancer has varied consequences in terms of possible pathways of regulating either the intrinsic or acquired resistance to EGFR-targeted inhibitors. Several mechanisms of resistance to EGFR-TKIs in preclinical models and in non-small cell lung cancer patients have been reported, such as the occurrence of secondary mutations (T790M, C797S), the activation of alternative signaling pathways (MET, HGF, AXL, Hh, IGF1R), the deregulation of the downstream pathways (AKT mutations, loss of PTEN), the inhibition of EGFR-TKIs-facilitated apoptosis pathway (BCL2-like 11/BIM deletion polymorphism) and histological transformations (Morgillo, Della Corte, Fasano, & Ciardiello, 2016) (Fig. 5 and Table 1). The inherent heterogeneity of EGFR mutation-positive non-small-cell lung cancer enhances clonal expansion of resistant tumors via multiple molecular mechanisms (Kohsaka, Petronczki, Solca, & Maemondo, 2019). First-generation TKIs, such as gefitinib and erlotinib, can non-covalently and reversibly bind to EGFR (both wild-type and mutated forms). They inhibit EGFR signaling by competing with ATP at the intracellular kinase domain of the receptor, thus reducing dimer autophosphorylation and diminishing intracellular signaling (Kohsaka et al., 2019; Sharma, Bell, Settleman, & Haber, 2007; Spicer & Rudman, 2010). The second-generation EGFR-TKIs, such as afatinib, works differently to erlotinib and gefitinib. It covalently and irreversibly binds not only to EGFR but also to other members of the ErbB family of receptors, such as HER2 and HER4, therefore blocking transphosphorylation of tyrosine residues at the C-terminus of all the three receptors (Modjtahedi, Cho, Michel, & Solca, 2014; Solca et al., 2012). It is also capable of blocking the transphosphorylation of the kinase inactive HER3 protein via its partner in HER3-containing heterodimers (Kohsaka et al., 2019; Modjtahedi et al., 2014). Third-generation EGFR-TKIs such as osimertinib are used for the first-line treatment of EGFR mutation-positive NSCLC. In the patients treated with osimertinib that showed cancer progression, T790M was observed to be present in 50% patients and lost in 50% of the patients. EGFR C797 and L792 (26%) mutations were the most

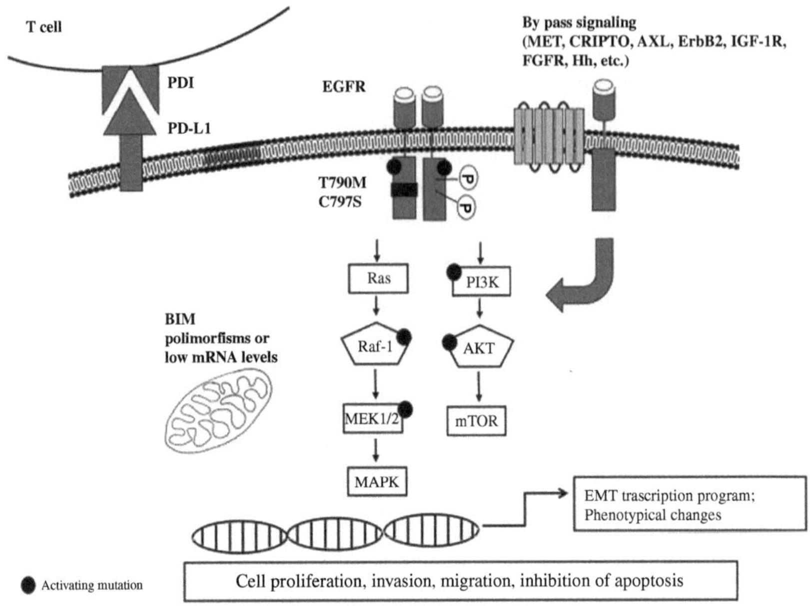

Fig. 5 Schematic representation of main EGFR-TKIs resistance mechanisms. Resistance to EGFR-TKIs can occur through different mechanisms either intrinsic or acquired. Known mechanisms are secondary resistance mutations occurring in the ATP-binding domain (such as T790M and C797S), mutation or amplification of bypass signaling (such as AXL, Hh, ErbB2, CRIPTO, etc.), activating mutations in the downstream pathways (PI3K, AKT, MEK, RAF), low levels of mRNA or polymorphisms of the pro-apoptotic protein BIM, induction of a transcription program for EMT and phenotypical changes, or induction of elevated tumor PD-L1 levels. EGFR, epidermal growth factor receptor; EMT, epithelial-to-mesenchymal transition; mRNA, messenger RNA; PD-1, programmed death receptor-1; PD-L1, programmed death ligand-1; TKI, tyrosine kinase inhibitor. *From Morgillo, F., Della Corte, C. M., Fasano, M., & Ciardiello, F. (2016). Mechanisms of resistance to egfr-targeted drugs: Lung cancer.* ESMO Open, 1, *e000060.*

common resistance mechanism and were observed exclusively in T790M-preserved cases (Le et al., 2018). Osimertinib resistance is associated with varied, and mostly EGFR-independent genomic alterations (Le et al., 2018). Kegelman et al. (2017) reported that the small molecule *PDZ1i* (113B7), identified through NMR-guided fragment-based drug design, inhibited binding of MDA-9/Syntenin (SDCBP), an adaptor protein, to EGFRvIII. They reported that both genetic (sh*mda-9*) and pharmacological (*PDZ1i*) targeting of MDA-9/Syntenin reduced invasion gains in GBM cells following radiation. This treatment did not affect normal astrocyte survival when combined with radiation, however, *PDZ1i* radio-sensitized GBM cells in spite of constitutive EGFR activity through EGFRvIII (Kegelman et al., 2017).

Table 1 Phase III studies of EGFR-TKI as first-line treatment of patients with EGFR mutated NSCLC.

Study	EGFR-TKI	Chemotherapy	Mutation	Median PFS (months)	RR (%)	Median OS (months)
IPASS[a]	Gefitinib	Carboplatin–paclitaxel	All	9.5 vs 6.3	71.2 vs 47.3	21.6 vs 21.9
WJTOG3405[b]	Gefitinib	Cisplatin–docetaxel	L858R Ex19D	9.2 vs 6.3	62.1 vs 32.2	36 vs 39
NEJ002[c]	Gefitinib	Carboplatin–paclitaxel	L858R Ex19D	10.8 vs 5.4	73.7 vs 30.7	27.7 vs 26.6
OPTIMAL[d]	Erlotinib	Gemcitabine–carboplatin	L858R Ex19D	13.1 vs 4.6	83 vs 36	22.7 vs 28.9
First-Signal[e]	Gefitinib	Gemcitabine–cisplatin	All	8.0 vs 6.3	84.6 vs 37.5	27.2 vs 25.6
EURTAC[f]	Erlotinib	Cisplatin–docetaxel/gemcitabine	L858R Ex19D	9.7 vs 5.2	58 vs 15	19.3 vs 19.5
LUX-Lung 3[g,h]	Afatinib	Cisplatin–pemetrexed	All	11.1 vs 6.9	56 vs 23	31.6 vs 28.2
LUX-Lung 6[i]	Afatinib	Gemcitabine–cisplatin	All	11.0 vs 5.6	66.9 vs 23	23.6 vs 23.511

[a]https://www.ncbi.nlm.nih.gov/pmc/articles/PMC5070275/#R4.
[b]https://www.ncbi.nlm.nih.gov/pmc/articles/PMC5070275/#R5.
[c]https://www.ncbi.nlm.nih.gov/pmc/articles/PMC5070275/#R6.
[d]https://www.ncbi.nlm.nih.gov/pmc/articles/PMC5070275/#R7.
[e]https://www.ncbi.nlm.nih.gov/pmc/articles/PMC5070275/#R8.
[f]https://www.ncbi.nlm.nih.gov/pmc/articles/PMC5070275/#R9.
[g]https://www.ncbi.nlm.nih.gov/pmc/articles/PMC5070275/#R10.
[h]https://www.ncbi.nlm.nih.gov/pmc/articles/PMC5070275/#R12.
[i]https://www.ncbi.nlm.nih.gov/pmc/articles/PMC5070275/#R11.

EGFR-TKI, epidermal growth factor receptor tyrosine kinase inhibitor; NSCLC, non-small cell lung cancer; OS, overall survival; PFS, progression-free survival; RR, response rate.

From Morgillo, F., Della Corte, C. M., Fasano, M., & Ciardiello, F. (2016). Mechanisms of resistance to egfr-targeted drugs: Lung cancer. *ESMO Open*, *1*, e000060.

3.3 Metabolism

EGFR signaling has been reported to regulate numerous metabolic processes that are important for cancer progression such as the biosynthesis of fatty acids, generation of pyrimidines, lipid remodeling and even glucose catabolism (Bi et al., 2019; Guo et al., 2009; Makinoshima et al., 2014; Sigismund et al., 2018). EGFR positively regulates these metabolic pathways either through direct phosphorylation of the rate-limiting enzymes (Lim et al., 2016; Sigismund et al., 2018; Zhang, Song, et al., 2017), or indirectly through activation of the MYC and of the AKT signaling pathways (Babic et al., 2013; DeBerardinis & Chandel, 2016; Guo et al., 2009; Makinoshima et al., 2014; Masui, Cavenee, & Mischel, 2014; Sigismund et al., 2018). EGFR can promote aerobic glycolysis through both kinase-dependent and kinase-independent mechanisms. Physical association of EGFR with SGLT1 stabilizes the sodium-glucose cotransporter at the cell surface resulting in an increase of glucose influx (Sigismund et al., 2018; Weihua et al., 2008). This kinase-independent function provides survival advantages to cells, helping them escape autophagic cell death when grown in the presence of low glucose concentrations (Sigismund et al., 2018; Weihua et al., 2008). EGFR and glutaminase inhibition can result in metabolic crisis in EGFR mutant lung cancer (Momcilovic et al., 2017). NSCLC cells have been reported to have a greater dependency on the urea cycle to sustain their central carbon, pyrimidine, and arginine metabolism to meet cellular energetics requirements post-EGFR inhibition (Pham-Danis et al., 2019).

After binding to its ligand EGF, EGFR is capable of regulating the expression of hexokinase (HK1) and phosphorylation of the pyruvate kinase M2 (PKM2), two glycolytic enzymes that catalyze main steps in glucose metabolism, thereby increasing aerobic glycolysis in breast cancer cells (Lim et al., 2016; Sigismund et al., 2018). Increased aerobic glycolysis results in the production of high levels of lactate, which inhibits the cytotoxic activity of T cells, supporting the immune escape of these breast cancer cells (Lim et al., 2016; Sigismund et al., 2018). It is usually assumed that the EGFR only acts at the plasma membrane and the vesicles of the endosomal compartment. However, it can also localize to the nucleus and mitochondria (Sigismund et al., 2018). The role of EGFR in mitochondria is not completely known but some studies suggest connections with antiapoptotic and metabolic functions. Furthermore, the translocation of both wild-type EGFR and EGFRvIII mutant proteins to the mitochondria has also been reported in cells treated with kinase inhibitors, or other proapoptotic treatments (Cao, Zhu, Ali-Osman, & Lo, 2011; Sigismund et al., 2018).

This translocation is significantly associated with resistance to cell death and EGFR targeting treatments (Cao et al., 2011; Sigismund et al., 2018). This phenomenon is possibly due to the ability of both wild-type EGFR and EGFRvIII to constitutively bind to p53-upregulated modulator of apoptosis (PUMA), a proapoptotic member of the BCL-2 family of proteins mainly localized in the mitochondria (Sigismund et al., 2018; Zhu et al., 2009). Treatment of cancer cells with EGF also results in palmitoylation of mitochondrial EGFR, which, subsequently favors fusion of mitochondria (Bollu et al., 2014; Sigismund et al., 2018). Independent of its tyrosine kinase activity, EGFR, can interact with the fatty acid synthase enzyme, causing de novo synthesis of palmitate (Bollu et al., 2014; Sigismund et al., 2018). Because of these important regulatory reasons there are several ongoing clinical trials of EGFR targeted strategies to regulate metabolic pathways in cancer (Gao, Lei, & Ye, 2013). Since these metabolic pathways regulate the survival of CSCs it is highly likely that EGFR can regulate the metabolism of CSCs. CSCs are more dependent on glycolysis as well as autophagy for survival and EGFR is possibly involved in their regulation (Jutten & Rouschop, 2014; Mao et al., 2013; Talukdar et al., 2018; Vlashi et al., 2011; Vlashi & Pajonk, 2015). Protective autophagy is a survival mechanism in cancer stem cells during periods of metabolic and environmental stress, where the self-digestive process provides an alternative energy source and aids in the removal of unfolded proteins (Jutten & Rouschop, 2014). EGFR activity is regulated by MDA-9/Syntenin which controls protective autophagy in anoikis resistant glioma stem cells (Talukdar et al., 2018). Cancer stem cells with constitutive EGFR activity due to the expression of EGFRvIII show higher levels of basal autophagy (Talukdar et al., 2018) and this is also observed in EGFRvIII-expressing tumors which show higher activation of autophagy during metabolic stress (Guo et al., 2009; Jutten & Rouschop, 2014).

3.4 Dormancy

Dormancy is a very interesting phenomenon observed in CSCs, and is reported to aid in relapse in cancer (Manjili, 2017). It is a process when cells remain alive but stop proliferating. When a majority of a cancer population undergoes this unique phenomenon, the process is called tumor dormancy (Manjili, 2017), but when a single cancer cell undergoes this process, it is referred to as quiescence (Chen, Dong, Haiech, Kilhoffer, & Zeniou, 2016). CSCs share several overlapping characteristics and an ability to slip into quiescence/dormancy is one of these properties. Cancer cells can be broadly grouped into dormancy-competent CSCs (DCCs), cancer-repopulating cells (CRCs), dormancy-incompetent CSCs and disseminated tumor cells (DTCs) (Crea, Nur Saidy, Collins, & Wang, 2015). The conditions which stimulate cancer cells to progress into the dormancy phase to survive and adapt are: (i) primary cancer dormancy; (ii) metastatic dormancy; (iii) therapy-induced dormancy; and (iv) immunologic dormancy. Dormancy, therapy-resistance and plasticity of CSCs are essentially interlinked, organized processes which are regulated through mechanisms involving reversible genetic modifications (Chen et al., 2016; Enderling, 2013; Kleffel & Schatton, 2013; Talukdar et al., 2019).

EGFR is reported to regulate in vivo dormancy of D-HEp3 cells via uPAR (Liu, Aguirre Ghiso, Estrada, & Ossowski, 2002). A combination of EGFR inhibitors with TSP-1 could possibly provide a potential therapeutic strategy in glioblastoma treatment, since it is capable of converting fast growing glioblastoma cells into a dormant-like state (Tiram et al., 2018). The combined inhibition of EGFR and FGFR1 signaling decreased the collagen-enrichment of the tumor stroma, which significantly delayed tumor recurrence. This stromal remodeling is important not only for tumor recurrence, but also for tumor dormancy (Holdman et al., 2015). EGFR downregulation also induces tumor dormancy in ovarian cancer via MED12 signaling (Luo, Deng, et al., 2018). Dormant lung cancer cells were found to be resistant to EGFR tyrosine kinase inhibitor (TKI) treatment. These dormant cells possessed a high MIG6 expression level which could be one of the causes of EGFR-TKI resistance in EGFR mutant lung cancer cells (Endo et al., 2017). EGFR appears also to be involved in radiation-mediated quiescence (Ahsan, Hiniker, Davis, Lawrence, & Nyati, 2009). There are also reports that FBXW7 could possibly help in maintenance of quiescence in gefitinib-resistant lung CSCs in EGFR mutation-positive NSCLC (Hidayat et al., 2019). These studies indicate that EGFR could also play a crucial role in CSC quiescence.

3.5 Immune escape

Immunotherapy has become a crucial strategy for the treatment of non-small-cell lung cancer. PD-1 and PD-L1, have emerged as critical targets for immunotherapy. PD-L1/PD-1 interaction plays an important role in the decrease of specialized T cell-induced apoptosis, immune response to cancer cells, as well as immune evasion of cancer cells (Ji et al., 2015; Kashani-Sabet, 2010).

The inhibition of PD-1/PD-L1 signaling cascade can disturb the proliferation of activated effector T cells, resulting in tumor evasion from cytotoxic T-lymphocytes, causing a weakening of anti-tumor immune response in the host (Duraiswamy, Kaluza, Freeman, & Coukos, 2013; Ji et al., 2015). This antitumor phenomenon has been confirmed by both in vitro and in vivo studies (Ji et al., 2015). CSCs utilizing their immunomodulating abilities, are capable of evading immunosurveillance by maintaining their status as quiescent cells and this phenomenon is possibly responsible for resistance to immunotherapeutic treatments. Cancer cells commonly evade immune detection by downregulating several important components of the antigen processing and presentation pathways, thereby preventing the exposure of neo-antigens. This phenomenon is highly heightened in the CSCs of several cancer types (Sultan et al., 2016). These unique immunoediting mechanisms, in addition to recruitment of protumorigenic/anti-inflammatory immune cells are also well utilized by CSCs (Sultan et al., 2016).

Several studies report the contribution of EGFR in immune escape. Activation of the PD-1 pathway is observed to contribute to immune escape in EGFR-driven lung tumors (Akbay et al., 2013). Upregulation of PD-L1 by EGFR activation is reported to regulate immune escape in EGFR-driven NSCLC (Chen, Fang, et al., 2015). EGFR can also regulate immune evasion through the expression of B7-H5 in non-small cell lung carcinoma (Dong, Zhang, Xu, & Zhang, 2018). Modulating EGFR signaling can increase aerobic glycolysis in triple-negative breast cancer cells to enhance cancer growth and immune escape (Lim et al., 2016). In addition BIN1 can reverse PD-L1-mediated immune escape through c-MYC and EGFR/MAPK signaling pathway deactivation in NSCLC (Wang et al., 2017). PD-L1 positive status was significantly associated with presence of EGFR mutations ($P=0.001$) (D'Incecco et al., 2015). Furthermore, in addition, it was shown that patients with PD-L1 positive expression showed higher sensitivity to EGFR-TKIs, longer time to progression and OS ($P=0.09$) than PD-L1 negative patients. However some studies report that in stage I NSCLC patients there was no significant correlation between PD-L1 expression and EGFR expression (Mu, Huang, Chen, Chen, & Zhang, 2011). No significant relations have also been reported between PD-L1 expression and EGFR/KRAS expression in lung adenocarcinoma (Zhang, Li, et al., 2017). Thus, the correlation between PD-1/PD-L1 expression and EGFR expression is controversial and needs more evidences to elucidate better strategies to navigate interactions of PD-1/PD-L1 inhibitors and tyrosine kinase inhibitors (Ji et al., 2015).

4. Conclusion

EGFR is a very important receptor tyrosine kinase that regulates multiple aspects necessary for the initiation, maintenance and survival of cancer stem cells (Fig. 6). The EGFR family is made up of EGFR1 (EGFR, HER1, c-ErbB1), EGFR2 (HER2, c-ErbB2), EGFR3 (c-ErbB3, HER3) and EGFR4 (c-ErbB4, HER4) proteins and possess the ability to bind and be activated by seven different ligands. However, several mutant forms of EGFR also exist, most prominent being EGFRvIII, which can be constitutively activated without the need of ligand binding. Overexpression of EGFR is frequently observed and has been reported to be associated with higher aggressiveness and poorer clinical outcomes in several forms of cancer. EGFR is possibly responsible for the poor prognosis based on how it regulates stemness, metabolism, resistance to therapy, dormancy and immune escape of tumor initiating cells in various types of tumors (Fig. 3). Moreover, the heterogeneity of its expression, combined with its ability to co-operate with other receptors and signaling pathways, makes it an extremely worthwhile target. Different generations of EGFR-TKIs have been designed, and while they can inhibit EGFR in different ways, they can also contribute to acquired resistance and further complications when combined with immunotherapeutic agents. For these reasons, it is very important to further analyze the ways in which EGFR can be targeted to prevent recurrence and relapse in patients.

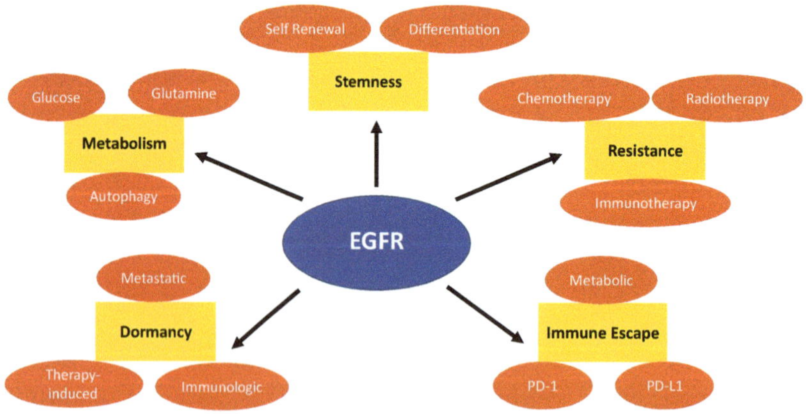

Fig. 6 Schema showing how EGFR regulates self-renewal, metabolism, dormancy, immune escape and therapy resistance in cancer stem cells.

5. Future prospective

The importance of representative model systems of EGFR-amplified tumors for research is highlighted by recent reports which describe the occurrence of circulating tumor cells (which often exhibit stem-like phenotypes) in the blood (Liffers, Lamszus, & Schulte, 2015; Macarthur et al., 2014; Müller et al., 2014). Since EGFR works in conjunction with other RTKs such as c-MET, a potentially viable strategy would be to use EGFR inhibitors with other RTK inhibitors to eliminate these CTCs as well as overcome EGFR resistance in various cancers. EGFR is also known to promote STAT3 signaling and thereby contribute to the maintenance and resistance of cancer stem cells, hence STAT3 inhibitors could be used in conjunction with EGFR inhibitors to avoid the enrichment of CSCs post-EGFR and post-RTKI treatment. As EGFR contributes to radiation resistance, and radiation also activates EGFR signaling and DSB repair, inhibitors of DSB repair with EGFR inhibitors could sensitize CSCs to irradiation. Use of small molecules like *PDZ1i* for the treatment of irradiated glioma cells seems promising. CSCs frequently pump out various therapeutic drugs, which could also apply to various RTKIs, and combining these with drug efflux inhibitors or regulators of MDR genes could provide a viable direction to explore for cancer therapy. As EGFR is reported to be involved in regulation of protective autophagy through MDA-9/Syntenin, it is important to assess its effects with *PDZ1i* (small molecule pharmacological inhibitor of MDA-9/Syntenin), and autophagy inhibitors, both in vivo and in vitro, in various cancer types. EGFR is implicated in the differentiation of stem cells and transdifferentiation of cancer cells, and this could represent another promising therapeutic strategy to target cancer stem cells.

References

Aguirre, A., Rubio, M. E., & Gallo, V. (2010). Notch and egfr pathway interaction regulates neural stem cell number and self-renewal. *Nature*, *467*, 323–327.

Ahsan, A., Hiniker, S. M., Davis, M. A., Lawrence, T. S., & Nyati, M. K. (2009). Role of cell cycle in epidermal growth factor receptor inhibitor-mediated radiosensitization. *Cancer Research*, *69*, 5108–5114.

Akbay, E. A., Koyama, S., Carretero, J., Altabef, A., Tchaicha, J. H., Christensen, C. L., et al. (2013). Activation of the pd-1 pathway contributes to immune escape in egfr-driven lung tumors. *Cancer Discovery*, *3*, 1355–1363.

An, Z., Aksoy, O., Zheng, T., Fan, Q. W., & Weiss, W. A. (2018). Epidermal growth factor receptor and egfrviii in glioblastoma: Signaling pathways and targeted therapies. *Oncogene*, *37*, 1561–1575.

Annenkov, A. (2014). Receptor tyrosine kinase (rtk) signalling in the control of neural stem and progenitor cell (nspc) development. *Molecular Neurobiology*, *49*, 440–471.
Babic, I., Anderson, E. S., Tanaka, K., Guo, D., Masui, K., Li, B., et al. (2013). Egfr mutation-induced alternative splicing of max contributes to growth of glycolytic tumors in brain cancer. *Cell Metabolism*, *17*, 1000–1008.
Barberan, S., & Cebria, F. (2019). The role of the egfr signaling pathway in stem cell differentiation during planarian regeneration and homeostasis. *Seminars in Cell & Developmental Biology*, *87*, 45–57.
Bi, J., Ichu, T.-A., Zanca, C., Yang, H., Zhang, W., Gu, Y., et al. (2019). Oncogene amplification in growth factor signaling pathways renders cancers dependent on membrane lipid remodeling. *Cell Metabolism*, *30*, 525–538.e528.
Blume-Jensen, P., & Hunter., T. (2001). Oncogenic kinase signalling. *Nature*, *411*, 355–365.
Bollu, L. R., Ren, J., Blessing, A. M., Katreddy, R. R., Gao, G., Xu, L., et al. (2014). Involvement of de novo synthesized palmitate and mitochondrial egfr in egf induced mitochondrial fusion of cancer cells. *Cell Cycle*, *13*, 2415–2430.
Brand, T. M., Iida, M., Luthar, N., Starr, M. M., Huppert, E. J., & Wheeler, D. L. (2013). Nuclear egfr as a molecular target in cancer. *Radiotherapy and Oncology*, *108*, 370–377.
Burgess, A. W., Cho, H. S., Eigenbrot, C., Ferguson, K. M., Garrett, T. P., Leahy, D. J., et al. (2003). An open-and-shut case? Recent insights into the activation of egf/erbb receptors. *Molecular Cell*, *12*, 541–552.
Bussard, K. M., Mutkus, L., Stumpf, K., Gomez-Manzano, C., & Marini, F. C. (2016). Tumor-associated stromal cells as key contributors to the tumor microenvironment. *Breast Cancer Research*, *18*, 84.
Butti, R., Das, S., Gunasekaran, V. P., Yadav, A. S., Kumar, D., & Kundu, G. C. (2018). Receptor tyrosine kinases (rtks) in breast cancer: Signaling, therapeutic implications and challenges. *Molecular Cancer*, *17*, 34.
Cao, X., Zhu, H., Ali-Osman, F., & Lo, H. W. (2011). Egfr and egfrviii undergo stress- and egfr kinase inhibitor-induced mitochondrial translocalization: A potential mechanism of egfr-driven antagonism of apoptosis. *Molecular Cancer*, *10*, 26.
Chen, W., Dong, J., Haiech, J., Kilhoffer, M. C., & Zeniou, M. (2016). Cancer stem cell quiescence and plasticity as major challenges in cancer therapy. *Stem Cells International*, *2016*, 1740936.
Chen, N., Fang, W., Zhan, J., Hong, S., Tang, Y., Kang, S., et al. (2015). Upregulation of pd-l1 by egfr activation mediates the immune escape in egfr-driven nsclc: Implication for optional immune targeted therapy for nsclc patients with egfr mutation. *Journal of Thoracic Oncology*, *10*, 910–923.
Chen, J., Song, W., & Amato, K. (2015). Eph receptor tyrosine kinases in cancer stem cells. *Cytokine & Growth Factor Reviews*, *26*, 1–6.
Cheng, C.-C., Chang, J., Lin, H.-C., Ho, A.-S., Lim, K.-H., Chang, C.-C., et al. (2017). Abstract 934: Egfr induces ilf3 and g9a expressions to maintain oct4-mediated stemness property in lung cancer. *Cancer Research*, *77*, 934.
Codony-Servat, J., Codony-Servat, C., Cardona, A. F., Gimenez-Capitan, A., Drozdowskyj, A., Berenguer, J., et al. (2019). Cancer stem cell biomarkers in egfr-mutation-positive non-small-cell lung cancer. *Clinical Lung Cancer*, *20*, 167–177.
Crea, F., Nur Saidy, N. R., Collins, C. C., & Wang, Y. (2015). The epigenetic/noncoding origin of tumor dormancy. *Trends in Molecular Medicine*, *21*, 206–211.
DeBerardinis, R. J., & Chandel, N. S. (2016). Fundamentals of cancer metabolism. *Science Advances*, *2*, e1600200.
D'Incecco, A., Andreozzi, M., Ludovini, V., Rossi, E., Capodanno, A., Landi, L., et al. (2015). Pd-1 and pd-l1 expression in molecularly selected non-small-cell lung cancer patients. *British Journal of Cancer*, *112*, 95–102.

D'Ippolito, E., Plantamura, I., Bongiovanni, L., Casalini, P., Baroni, S., Piovan, C., et al. (2016). Mir-9 and mir-200 regulate pdgfrbeta-mediated endothelial differentiation of tumor cells in triple-negative breast cancer. *Cancer Research*, *76*, 5562–5572.

Dong, Z., Zhang, L., Xu, W., & Zhang, G. (2018). Egfr may participate in immune evasion through regulation of b7h5 expression in nonsmall cell lung carcinoma. *Molecular Medicine Reports*, *18*, 3769–3779.

Du, W., Ni, L., Liu, B., Wei, Y., Lv, Y., Qiang, S., et al. (2018). Upregulation of sall4 by egfr activation regulates the stemness of cd44-positive lung cancer. *Oncogene*, *7*, 36.

Duraiswamy, J., Kaluza, K. M., Freeman, G. J., & Coukos, G. (2013). Dual blockade of pd-1 and ctla-4 combined with tumor vaccine effectively restores t-cell rejection function in tumors. *Cancer Research*, *73*, 3591–3603.

Enderling, H. (2013). In H. Enderling, N. Almog, & L. Hlatky (Eds.), *Systems biology of tumor dormancy* (pp. 55–71). New York, NY: Springer New York.

Endo, H., Okami, J., Okuyama, H., Nishizawa, Y., Imamura, F., & Inoue, M. (2017). The induction of mig6 under hypoxic conditions is critical for dormancy in primary cultured lung cancer cells with activating egfr mutations. *Oncogene*, *36*, 2824–2834.

Evangelopoulos, M. E., Weis, J., & Kruttgen, A. (2005). Signalling pathways leading to neuroblastoma differentiation after serum withdrawal: Hdl blocks neuroblastoma differentiation by inhibition of egfr. *Oncogene*, *24*, 3309–3318.

Favelyukis, S., Till, J. H., Hubbard, S. R., & Miller, W. T. (2001). Structure and autoregulation of the insulin-like growth factor 1 receptor kinase. *Nature Structural Biology*, *8*, 1058–1063.

Furdui, C. M., Lew, E. D., Schlessinger, J., & Anderson, K. S. (2006). Autophosphorylation of fgfr1 kinase is mediated by a sequential and precisely ordered reaction. *Molecular Cell*, *21*, 711–717.

Galoczova, M., Coates, P., & Vojtesek, B. (2018). Stat3, stem cells, cancer stem cells and p63. *Cellular and Molecular Biology Letters*, *23*, 12.

Gan, H. K., Cvrljevic, A. N., & Johns, T. G. (2013). The epidermal growth factor receptor variant iii (egfrviii): Where wild things are altered. *The FEBS Journal*, *280*, 5350–5370.

Gao, Q., Lei, T., & Ye, F. (2013). Therapeutic targeting of egfr-activated metabolic pathways in glioblastoma. *Expert Opinion on Investigational Drugs*, *22*, 1023–1040.

Grainger, S., Nguyen, N., Richter, J., Setayesh, J., Lonquich, B., Oon, C. H., et al. (2019). Egfr is required for wnt9a-fzd9b signalling specificity in haematopoietic stem cells. *Nature Cell Biology*, *21*, 721–730.

Greenall, S. A., & Johns, T. G. (2016). Egfrviii: The promiscuous mutation. *Cell Death Discovery*, *2*, 16049.

Guo, D., Prins, R. M., Dang, J., Kuga, D., Iwanami, A., Soto, H., et al. (2009). Egfr signaling through an akt-srebp-1-dependent, rapamycin-resistant pathway sensitizes glioblastomas to antilipogenic therapy. *Science Signaling*, *2*, ra82.

Hidayat, M., Mitsuishi, Y., Takahashi, F., Tajima, K., Yae, T., Miyahara, K., et al. (2019). Role of fbxw7 in the quiescence of gefitinib-resistant lung cancer stem cells in egfr-mutant non-small cell lung cancer. *Bosnian Journal of Basic Medical Sciences*, *19*, 355–367.

Holdman, X. B., Welte, T., Rajapakshe, K., Pond, A., Coarfa, C., Mo, Q., et al. (2015). Upregulation of egfr signaling is correlated with tumor stroma remodeling and tumor recurrence in fgfr1-driven breast cancer. *Breast Cancer Research*, *17*, 141.

Honegger, A. M., Kris, R. M., Ullrich, A., & Schlessinger, J. (1989). Evidence that autophosphorylation of solubilized receptors for epidermal growth factor is mediated by intermolecular cross-phosphorylation. *Proceedings of the National Academy of Sciences of the United States of America*, *86*, 925–929.

Howard, B. M., Gursel, D. B., Bleau, A. M., Beyene, R. T., Holland, E. C., & Boockvar, J. A. (2010). Egfr signaling is differentially activated in patient-derived glioblastoma stem cells. *Journal of Experimental Therapeutics and Oncology*, *8*, 247–260.

Hubbard, S. R. (2005). Egf receptor inhibition: Attacks on multiple fronts. *Cancer Cell*, *7*, 287–288.

Hynes, N. E., & Lane, H. A. (2005). Erbb receptors and cancer: The complexity of targeted inhibitors. *Nature Reviews Cancer*, *5*, 341–354.

Ibrahim, S. A., Gadalla, R., El-Ghonaimy, E. A., Samir, O., Mohamed, H. T., Hassan, H., et al. (2017). Syndecan-1 is a novel molecular marker for triple negative inflammatory breast cancer and modulates the cancer stem cell phenotype via the il-6/stat3, notch and egfr signaling pathways. *Molecular Cancer*, *16*, 57.

Jahani-Asl, A., Yin, H., Soleimani, V. D., Haque, T., Luchman, H. A., Chang, N. C., et al. (2016). Control of glioblastoma tumorigenesis by feed-forward cytokine signaling. *Nature Neuroscience*, *19*, 798–806.

Ji, M., Liu, Y., Li, Q., Li, X. D., Zhao, W. Q., Zhang, H., et al. (2015). Pd-1/pd-l1 pathway in non-small-cell lung cancer and its relation with egfr mutation. *Journal of Translational Medicine*, *13*, 5.

Jutten, B., & Rouschop, K. M. (2014). Egfr signaling and autophagy dependence for growth, survival, and therapy resistance. *Cell Cycle*, *13*, 42–51.

Kashani-Sabet, M. (2010). Tumor progression by immune evasion in melanoma. *Cancer*, *116*, 1623–1625.

Kegelman, T. P., Wu, B., Das, S. K., Talukdar, S., Beckta, J. M., Hu, B., et al. (2017). Inhibition of radiation-induced glioblastoma invasion by genetic and pharmacological targeting of mda-9/syntenin. *Proceedings of the National Academy of Sciences of the United States of America*, *114*, 370–375.

Kim-Yip, R. P., & Nystul, T. G. (2018). Wingless promotes egfr signaling in follicle stem cells to maintain self-renewal. *Development*, *145*, dev168716.

Kitade, M., Factor, V. M., Andersen, J. B., Tomokuni, A., Kaji, K., Akita, H., et al. (2013). Specific fate decisions in adult hepatic progenitor cells driven by met and egfr signaling. *Genes & Development*, *27*, 1706–1717.

Kleffel, S., & Schatton, T. (2013). In H. Enderling, N. Almog, & L. Hlatky (Eds.), *Systems biology of tumor dormancy* (pp. 145–179). New York, NY: Springer New York.

Kohsaka, S., Petronczki, M., Solca, F., & Maemondo, M. (2019). Tumor clonality and resistance mechanisms in egfr mutation-positive non-small-cell lung cancer: Implications for therapeutic sequencing. *Future Oncology*, *15*, 637–652.

Kwon, S. J., Kwon, O. S., Kim, K. T., Go, Y. H., Yu, S. I., Lee, B. H., et al. (2017). Role of mek partner-1 in cancer stemness through mek/erk pathway in cancerous neural stem cells, expressing egfrviii. *Molecular Cancer*, *16*, 140.

Le, X., Puri, S., Negrao, M. V., Nilsson, M. B., Robichaux, J., Boyle, T., et al. (2018). Landscape of egfr-dependent and -independent resistance mechanisms to osimertinib and continuation therapy beyond progression in egfr-mutant nsclc. *Clinical Cancer Research*, *24*, 6195–6203.

Lee, H. J., Seo, A. N., Kim, E. J., Jang, M. H., Kim, Y. J., Kim, J. H., et al. (2015). Prognostic and predictive values of egfr overexpression and egfr copy number alteration in her2-positive breast cancer. *British Journal of Cancer*, *112*, 103–111.

Lemmon, M. A., & Schlessinger, J. (2010). Cell signaling by receptor tyrosine kinases. *Cell*, *141*, 1117–1134.

Liao, B. C., Lee, J. H., Lin, C. C., Chen, Y. F., Chang, C. H., Ho, C. C., et al. (2015). Epidermal growth factor receptor tyrosine kinase inhibitors for non-small-cell lung cancer patients with leptomeningeal carcinomatosis. *Journal of Thoracic Oncology*, *10*, 1754–1761.

Liffers, K., Lamszus, K., & Schulte, A. (2015). Egfr amplification and glioblastoma stem-like cells. *Stem Cells International*, *2015*, 427518.

Lim, S. O., Li, C. W., Xia, W., Lee, H. H., Chang, S. S., Shen, J., et al. (2016). Egfr signaling enhances aerobic glycolysis in triple-negative breast cancer cells to promote tumor growth and immune escape. *Cancer Research*, *76*, 1284–1296.

Liu, D., Aguirre Ghiso, J., Estrada, Y., & Ossowski, L. (2002). Egfr is a transducer of the urokinase receptor initiated signal that is required for in vivo growth of a human carcinoma. *Cancer Cell*, *1*, 445–457. Jun.

Liu, S., Cong, Y., Wang, D., Sun, Y., Deng, L., Liu, Y., et al. (2014). Breast cancer stem cells transition between epithelial and mesenchymal states reflective of their normal counterparts. *Stem Cell Reports*, *2*, 78–91.

Luo, X.-L., Deng, C., Su, X.-D., Wang, F., Chen, Z., Wu, X.-P., et al. (2018). Loss of med12 induces tumor dormancy in human epithelial ovarian cancer via downregulation of egfr. *Cancer Research*, *78*, 3532–3543. 0134.2018.

Macarthur, K. M., Kao, G. D., Chandrasekaran, S., Alonso-Basanta, M., Chapman, C., Lustig, R. A., et al. (2014). Detection of brain tumor cells in the peripheral blood by a telomerase promoter-based assay. *Cancer Research*, *74*, 2152–2159.

Macdonald-Obermann, J. L., & Pike, L. J. (2014). Different epidermal growth factor (egf) receptor ligands show distinct kinetics and biased or partial agonism for homodimer and heterodimer formation. *Journal of Biological Chemistry*, *289*, 26178–26188.

Makinoshima, H., Takita, M., Matsumoto, S., Yagishita, A., Owada, S., Esumi, H., et al. (2014). Epidermal growth factor receptor (egfr) signaling regulates global metabolic pathways in egfr-mutated lung adenocarcinoma. *Journal of Biological Chemistry*, *289*, 20813–20823.

Manjili, M. H. (2017). Tumor dormancy and relapse: From a natural byproduct of evolution to a disease state. *Cancer Research*, *77*, 2564–2569.

Mao, P., Joshi, K., Li, J., Kim, S. H., Li, P., Santana-Santos, L., et al. (2013). Mesenchymal glioma stem cells are maintained by activated glycolytic metabolism involving aldehyde dehydrogenase 1a3. *Proceedings of the National Academy of Sciences of the United States of America*, *110*, 8644–8649.

Masui, K., Cavenee, W. K., & Mischel, P. S. (2014). Mtorc2 in the center of cancer metabolic reprogramming. *Trends in Endocrinology and Metabolism*, *25*, 364–373.

Modjtahedi, H., Cho, B. C., Michel, M. C., & Solca, F. (2014). A comprehensive review of the preclinical efficacy profile of the erbb family blocker afatinib in cancer. *Naunyn-Schmiedeberg's Archives of Pharmacology*, *387*, 505–521.

Momcilovic, M., Bailey, S. T., Lee, J. T., Fishbein, M. C., Magyar, C., Braas, D., et al. (2017). Targeted inhibition of egfr and glutaminase induces metabolic crisis in egfr mutant lung cancer. *Cell Reports*, *18*, 601–610.

Morgillo, F., Della Corte, C. M., Fasano, M., & Ciardiello, F. (2016). Mechanisms of resistance to egfr-targeted drugs: Lung cancer. *ESMO Open*, *1*, e000060.

Mu, C.-Y., Huang, J.-A., Chen, Y., Chen, C., & Zhang, X.-G. (2011). High expression of pd-l1 in lung cancer may contribute to poor prognosis and tumor cells immune escape through suppressing tumor infiltrating dendritic cells maturation. *Medical Oncology*, *28*, 682–688.

Müller, C., Holtschmidt, J., Auer, M., Heitzer, E., Lamszus, K., Schulte, A., et al. (2014). Hematogenous dissemination of glioblastoma multiforme. *Science Translational Medicine*, *6*, 247ra101.

Nicholson, R. I., Gee, J. M., & Harper, M. E. (2001). Egfr and cancer prognosis. *European Journal of Cancer*, *37*, S9–15.

Pang, L. Y., Saunders, L., & Argyle, D. J. (2017). Epidermal growth factor receptor activity is elevated in glioma cancer stem cells and is required to maintain chemotherapy and radiation resistance. *Oncotarget*, *8*, 72494–72512.

Park, H. S., Jang, M. H., Kim, E. J., Kim, H. J., Lee, H. J., Kim, Y. J., et al. (2014). High egfr gene copy number predicts poor outcome in triple-negative breast cancer. *Modern Pathology*, *27*, 1212–1222.

Pham-Danis, C., Gehrke, S., Danis, E., Rozhok, A. I., Daniels, M. W., Gao, D., et al. (2019). Urea cycle sustains cellular energetics upon egfr inhibition in egfr-mutant nsclc. *Molecular Cancer Research*, *17*, 1351–1364.

Plantamura, I., Casalini, P., Dugnani, E., Sasso, M., D'Ippolito, E., Tortoreto, M., et al. (2014). Pdgfrbeta and fgfr2 mediate endothelial cell differentiation capability of triple negative breast carcinoma cells. *Molecular Oncology*, *8*, 968–981.

Qazi, M. A., Vora, P., Venugopal, C., Adams, J., Singh, M., Hu, A., et al. (2018). Cotargeting ephrin receptor tyrosine kinases a2 and a3 in cancer stem cells reduces growth of recurrent glioblastoma. *Cancer Research*, *78*, 5023–5037.

Raymond, E., Faivre, S., & Armand, J. P. (2000). Epidermal growth factor receptor tyrosine kinase as a target for anticancer therapy. *Drugs*, *60*, 15–23.

Raz, R., Lee, C. K., Cannizzaro, L. A., d'Eustachio, P., & Levy, D. E. (1999). Essential role of stat3 for embryonic stem cell pluripotency. *Proceedings of the National Academy of Sciences of the United States of America*, *96*, 2846–2851.

Rodemann, H. P., Dittmann, K., & Toulany, M. (2007). Radiation-induced egfr-signaling and control of DNA-damage repair. *International Journal of Radiation Biology*, *83*, 781–791.

Rybak, A. P., & Tang, D. (2013). Sox2 plays a critical role in egfr-mediated self-renewal of human prostate cancer stem-like cells. *Cellular Signalling*, *25*, 2734–2742.

Scaltriti, M., & Baselga, J. (2006). The epidermal growth factor receptor pathway: A model for targeted therapy. *Clinical Cancer Research*, *12*, 5268–5272.

Schmidt, M. H. H., Furnari, F. B., Cavenee, W. K., & Bogler, O. (2003). Epidermal growth factor receptor signaling intensity determines intracellular protein interactions, ubiquitination, and internalization. *Proceedings of the National Academy of Sciences of the United States of America*, *100*, 6505–6510.

Sharma, S. V., Bell, D. W., Settleman, J., & Haber, D. A. (2007). Epidermal growth factor receptor mutations in lung cancer. *Nature Reviews Cancer*, *7*, 169–181.

Shibuya, M. (2011). Vascular endothelial growth factor (vegf) and its receptor (vegfr) signaling in angiogenesis: A crucial target for anti- and pro-angiogenic therapies. *Genes & Cancer*, *2*, 1097–1105.

Shien, K., Toyooka, S., Yamamoto, H., Soh, J., Jida, M., Thu, K. L., et al. (2013). Acquired resistance to egfr inhibitors is associated with a manifestation of stem cell-like properties in cancer cells. *Cancer Research*, *73*, 3051–3061.

Shinozaki, E., Yoshino, T., Yamazaki, K., Muro, K., Yamaguchi, K., Nishina, T., et al. (2017). Clinical significance of braf non-v600e mutations on the therapeutic effects of anti-egfr monoclonal antibody treatment in patients with pretreated metastatic colorectal cancer: The biomarker research for anti-egfr monoclonal antibodies by comprehensive cancer genomics (breac) study. *British Journal of Cancer*, *117*, 1450–1458.

Sigismund, S., Avanzato, D., & Lanzetti, L. (2018). Emerging functions of the egfr in cancer. *Molecular Oncology*, *12*, 3–20.

Solca, F., Dahl, G., Zoephel, A., Bader, G., Sanderson, M., Klein, C., et al. (2012). Target binding properties and cellular activity of afatinib (bibw 2992), an irreversible erbb family blocker. *Journal of Pharmacology and Experimental Therapeutics*, *343*, 342–350.

Sottoriva, A., Spiteri, I., Piccirillo, S. G., Touloumis, A., Collins, V. P., Marioni, J. C., et al. (2013). Intratumor heterogeneity in human glioblastoma reflects cancer evolutionary dynamics. *Proceedings of the National Academy of Sciences of the United States of America*, *110*, 4009–4014.

Spicer, J. F., & Rudman, S. M. (2010). Egfr inhibitors in non-small cell lung cancer (nsclc): The emerging role of the dual irreversible egfr/her2 inhibitor bibw 2992. *Targeted Oncology*, *5*, 245–255.

Sultan, M., Coyle, K. M., Vidovic, D., Thomas, M. L., Gujar, S., & Marcato, P. (2016). Hide-and-seek: The interplay between cancer stem cells and the immune system. *Carcinogenesis*, *38*, 107–118.

Talukdar, S., Bhoopathi, P., Emdad, L., Das, S., Sarkar, D., & Fisher, P. B. (2019). In C. I. Civin, T. J. Kingsbury, M. Kim, & P. B. Fisher (Eds.), *Vol. 141. Advances in cancer research* (pp. 43–84). Academic Press.

Talukdar, S., Pradhan, A. K., Bhoopathi, P., Shen, X.-N., August, L. A., Windle, J. J., et al. (2018). Mda-9/syntenin regulates protective autophagy in anoikis-resistant glioma stem cells. *Proceedings of the National Academy of Sciences*, *115*, 5768–5773.

Templeton, A. J., Diez-Gonzalez, L., Ace, O., Vera-Badillo, F., Seruga, B., Jordan, J., et al. (2014). Prognostic relevance of receptor tyrosine kinase expression in breast cancer: A meta-analysis. *Cancer Treatment Reviews*, *40*, 1048–1055.

Till, J. H., Becerra, M., Watty, A., Lu, Y., Ma, Y., Neubert, T. A., et al. (2002). Crystal structure of the musk tyrosine kinase: Insights into receptor autoregulation. *Structure*, *10*, 1187–1196.

Tiram, G., Ferber, S., Ofek, P., Eldar-Boock, A., Ben-Shushan, D., Yeini, E., et al. (2018). Reverting the molecular fingerprint of tumor dormancy as a therapeutic strategy for glioblastoma. *The FASEB Journal*, *32*, 5835–5850.

Ullrich, A., & Schlessinger, J. (1990). Signal transduction by receptors with tyrosine kinase activity. *Cell*, *61*, 203–212.

Vlashi, E., Lagadec, C., Vergnes, L., Matsutani, T., Masui, K., Poulou, M., et al. (2011). Metabolic state of glioma stem cells and nontumorigenic cells. *Proceedings of the National Academy of Sciences of the United States of America*, *108*, 16062–16067.

Vlashi, E., & Pajonk, F. (2015). The metabolic state of cancer stem cells-a valid target for cancer therapy? *Free Radical Biology & Medicine*, *79*, 264–268.

Wang, Y. X., Feige, P., Brun, C. E., Hekmatnejad, B., Dumont, N. A., Renaud, J.-M., et al. (2019). Egfr-aurka signaling rescues polarity and regeneration defects in dystrophin-deficient muscle stem cells by increasing asymmetric divisions. *Cell Stem Cell*, *24*, 419–432.e416.

Wang, X., Reyes, M. E., Zhang, D., Funakoshi, Y., Trape, A. P., Gong, Y., et al. (2017). Egfr signaling promotes inflammation and cancer stem-like activity in inflammatory breast cancer. *Oncotarget*, *8*, 67904–67917.

Wehrman, T., He, X., Raab, B., Dukipatti, A., Blau, H., & Garcia, K. C. (2007). Structural and mechanistic insights into nerve growth factor interactions with the TrkA and p75 receptors. *Neuron*, *53*, 5–38.

Weihua, Z., Tsan, R., Huang, W. C., Wu, Q., Chiu, C. H., Fidler, I. J., et al. (2008). Survival of cancer cells is maintained by egfr independent of its kinase activity. *Cancer Cell*, *13*, 385–393.

Weppler, S. A., Li, Y., Dubois, L., Lieuwes, N., Jutten, B., Lambin, P., et al. (2007). Expression of egfr variant viii promotes both radiation resistance and hypoxia tolerance. *Radiotherapy and Oncology*, *83*, 333–339.

Wilson, K. J., Gilmore, J. L., Foley, J., Lemmon, M. A., & Riese, D. J., 2nd. (2009). Functional selectivity of egf family peptide growth factors: Implications for cancer. *Pharmacology and Therapeutics*, *122*, 1–8.

Wise, R., & Zolkiewska, A. (2017). Metalloprotease-dependent activation of egfr modulates cd44(+)/cd24(−) populations in triple negative breast cancer cells through the mek/erk pathway. *Breast Cancer Research and Treatment*, *166*, 421–433.

Wu, J. X., & Adamson, E. D. (1996). Kinase-negative mutant epidermal growth factor receptor (egfr) expression during embryonal stem cell differentiation favours egfr-independent lineages. *Development*, *122*, 3331–3342.

Yang, F., Li, Y., Liu, B., You, J., & Zhou, Q. (2018). Cancer stem cell-like population is preferentially suppressed by egfr-tkis in egfr-mutated pc-9 tumor models. *Experimental Cell Research*, *362*, 195–202.

Yang, J., Liao, D., Chen, C., Liu, Y., Chuang, T. H., Xiang, R., et al. (2013). Tumor-associated macrophages regulate murine breast cancer stem cells through a novel paracrine egfr/stat3/sox-2 signaling pathway. *Stem Cells*, *31*, 248–258.

Yarden, Y., & Sliwkowski, M. X. (2001). Untangling the erbb signalling network. *Nature Reviews Molecular Cell Biology*, *2*, 127–137.

Yuzawa, S., Opatowsky, Y., Zhang, Z., Mandiyan, V., Lax, I., & Schlessinger, J. (2007). Structural basis for activation of the receptor tyrosine kinase KIT by stem cell factor. *Cell*, *130*, 323–334.

Zhang, M., Li, G., Wang, Y., Wang, Y., Zhao, S., Haihong, P., et al. (2017). Pd-l1 expression in lung cancer and its correlation with driver mutations: A meta-analysis. *Science Reports*, *7*, 10255.

Zhang, J., Song, F., Zhao, X., Jiang, H., Wu, X., Wang, B., et al. (2017). Egfr modulates monounsaturated fatty acid synthesis through phosphorylation of scd1 in lung cancer. *Molecular Cancer*, *16*, 127.

Zhu, H., Acquaviva, J., Ramachandran, P., Boskovitz, A., Woolfenden, S., Pfannl, R., et al. (2009). Oncogenic egfr signaling cooperates with loss of tumor suppressor gene functions in gliomagenesis. *Proceedings of the National Academy of Sciences of the United States of America*, *106*, 2712–2716.

CHAPTER FIVE

In vivo modeling of the EGFR family in breast cancer progression and therapeutic approaches

Alexandra M. Simond[a,b], **William J. Muller**[a,b,c,*]
[a]Rosalind and Morris Goodman Cancer Research Center, McGill University, Montreal, QC, Canada
[b]Department of Biochemistry, McGill University, Montreal, QC, Canada
[c]Faculty of Medicine, McGill University, Montreal, QC, Canada
*Corresponding author: e-mail address: william.muller@mcgill.ca

Contents

Abstract

Modeling breast cancer through the generation of genetically engineered mouse models (GEMMs) has become the gold standard in the study of human breast cancer. Notably, the *in vivo* modeling of the epidermal growth factor receptor (EGFR) family has been key to the development of therapeutics and has helped better understand the signaling pathways involved in cancer initiation, progression and metastasis.

Advances in Cancer Research, Volume 147
ISSN 0065-230X
https://doi.org/10.1016/bs.acr.2020.04.004
© 2020 Elsevier Inc.
All rights reserved.

The HER2/ErbB2 receptor is a member of the EGFR family and 20% of breast cancers are found to belong in the HER2-positive histological subtype. Historical and more recent advances in the field have shaped our understanding of HER2-positive breast cancer signaling and therapeutic approaches.

1. Background

1.1 Breast cancer history and incidence

Breast cancer is the most common cancer in women worldwide and is the cause of a quarter of cancer diagnoses. In 2018, breast cancer accounted for 24.3% of new cancer cases (approximately 2.1 million) and 15% of cancer related deaths (Bray et al., 2018). In their lifetime, approximately 1 in 8 women will be diagnosed with breast cancer and 1 in 31 will die from it. Although the 5-year survival rate is relatively high at 87% compared to lung (17%), colorectal (64%) and pancreatic (6%) cancer (Canadian Cancer Statistics Advisory Committee, 2018), there is still a need for the development of novel targeted therapies to help better understand and treat this leading cause of death in women in more than 100 countries (Bray et al., 2018). Hence, the need for appropriate pre-clinical models to further our knowledge of cancer subtypes and response to therapy.

1.2 Breast cancer classification

Breast cancer has always been classified histologically and, in the clinic, it is still classified as belonging to one of these three classes: (1) HER2/ErbB2 receptor positive, (2) Hormone receptor positive (estrogen receptor (ER) positive and/or progesterone receptor (PR) positive) and (3) Triple negative (lacking expression of ErbB2, ER and PR). The first histological subtype to have been identified was ER positive breast cancer which accounts for 70% of breast cancers diagnoses. To be considered as hormone receptor positive, 1% or more of the cells need to possess nuclear staining. In addition, the HER2-positive subtype accounts for 20–30% of cases and is diagnosed if the tissue stains positive for 10% or more cells. The remaining are negative for all three histological markers and fall in the triple-negative category which usually displays the worst prognosis (Hammond et al., 2010; Wolff et al., 2018).

Approximately 20 years ago, molecular classification of tumors based on gene expression rather than the presence or absence of certain proteins markers was developed. The Pam50 classification is based on expression of 50 genes that allows for the classification of breast cancers into molecular subtypes, which are: Luminal A, Luminal B, HER2-enriched and Basal. However, recently new molecular subtypes have been added, such as the Claudin-low and Normal-like subtype (Perou et al., 2000; Sørlie et al., 2001). Although histological and molecular classifications are very different, there does exist some overlap. For example, the majority of Triple-negative breast cancers are Basal-like. However, if a tumor is classified as HER2-positive histologically it may not always fall in the HER2-enriched molecular subtype.

1.3 Genetically engineered mouse models (GEMMs) of breast cancer

In the study of breast cancer, one key event that has led to the development of standardized therapies and targeted medicine is the use of pre-clinical models. Notably, genetically engineered mouse models (GEMMs) have been a powerful tool in the study of breast cancer cell signaling and understanding the mechanisms of therapeutic resistance, as well as the development of new drugs that are currently being used in the clinic. GEMMs have been instrumental in the study of breast cancer and breast cancer therapeutics. Mouse models can be inbred on different genetic backgrounds and tumors can be induced chemically, by random mutagenesis or by genetic engineering. GEMMs have allowed us to mimic the genetic changes that are known to occur in subtypes of human breast cancer and reproduce a living model of a particular subtype of interest. These models have then allowed us to study tumor onset, progression and metastasis on a molecular level and have been an important pre-clinical tool for the trial of novel drugs. They have also allowed us to understand in further depth the effect of therapy on cancer and have given us an ideal system to study therapy resistance. One advantage of using GEMMs over studying individual human tumors is that mice are inbred as to remove any genetic variability seen in human cancers. The importance of genetic background has been highlighted by observations in MMTV-PyVmT or MMTV-ErbB2 GEMMs where tumor penetrance onset and metastasis has been found to be influenced by genetic background (Davie et al., 2007; Gillgrass, Cardiff, Sharan, Kannan, & Muller, 2003).

1.4 EGFR family signaling

The epidermal growth factor receptor (EGFR) family are composed of four members: HER1/EGFR/ErbB1, HER2/ErbB2, HER3/ErbB3 and

HER4/ErbB4 that are activated by a wide variety of ligands. Upon ligand binding, the receptor hetero or homodimerizes leading to cross-phosphorylation of specific tyrosine residues and recruitment of adaptor proteins or kinases. All family members have binding sites for growth factor receptor bound protein 2 (Grb2) and Src-homology containing domain A (ShcA). Grb2 can bind to son of sevenless (SOS) which in turn activates the mitogen activated protein kinase (MAPK) pathway. In addition, Grb2 can also recruit Grb2 associated binding protein 1 (Gab1) and activate the phosphoinositide 3-kinase/v-akt thymoma viral oncogene homolog (PI3K/Akt) pathway through binding of the p85 regulatory subunit. Both ErbB3 and ErbB4 have binding sites for p85 (6,1 binding sites, respectively) directly activating the PI3K/Akt pathway (Schulze, Deng, & Mann, 2005).

Although, ErbB1 and ErbB4 can form active homodimers, heterodimer formation among the EGFR family is the most common way of activating multiple pathways at once (Harari & Yarden, 2000). The ErbB receptors are tyrosine kinases that all maintain a particular structure. They are considered type 1 transmembrane proteins composed of: an extracellular ligand binding domain, a transmembrane domain (single helix), an intracellular cytoplasmic tyrosine kinase domain and a C-terminal tail (Franklin et al., 2004). ErbB2 lacks a ligand binding domain and ErbB3 lacks an intracellular kinase domain and for that reason, they are one of the most commonly occurring and studied heterodimers. ErbB2 is also the preferred partner for heterodimerization of each family member and has been found to contribute to resistance to HER2-targeted therapies (Berghoff et al., 2014; Graus Porta, Beerli, Daly, & Hynes, 1997; Tzahar et al., 1996).

In this chapter, we will review the contributions that GEMMs have brought to our understanding of breast cancer progression and examine how these studies have aided the development of novel therapeutic targets.

2. GEMMs representing the Neu/HER2-positive human breast cancer subtype

In the early 1980s multiple groups identified human HER2 and its rat homolog, Neu, as potential oncogenes (Bargmann, Hung, & Weinberg, 1986a; Drebin, Stern, Link, Weinberg, & Greene, 1984; Padhy, Shih, Cowing, Finkelstein, & Weinberg, 1982; Schechter et al., 1984; Shih, Padhy, Murray, & Weinberg, 1981). In the rat Neu, activation was found to be associated with a point mutation in the transmembrane domain (valine to glutamic acid mutation at amino acid 664). NeuNT was generated where the entire normal *Neu* cDNA was cloned, only replacing 488 nucleotides by those from the transforming *Neu* cDNA (nucleotides 1899 and 2837). Mechanistically, this point mutation was found to function by inducing homodimerization of Neu resulting in constitutive activation (Fig. 1) (Bargmann, Hung, & Weinberg, 1986b; Weiner, Liu, Cohen, Williams, & Greene, 1989).

To directly demonstrate the oncogenic potential of Neu in the mammary epithelium, transgenic mice expressing NeuNT placed under the control of the mouse mammary tumor virus (MMTV) promoter were generated (Fig. 2A). MMTV is a retrovirus that is expressed in epithelial cells between 6 and 21 days post-partum, thus not interfering with normal development during gestation. Although, expression of this retrovirus is primarily limited to the mammary gland, lower levels can be found in other organs such as skin, salivary glands, ovaries and B and T lymphocytes, all dependent on where the transgene has been incorporated in the genome. In addition, the MMTV promotor has been found to be sensitive to testosterone and progesterone and has been a valuable tool in the study of

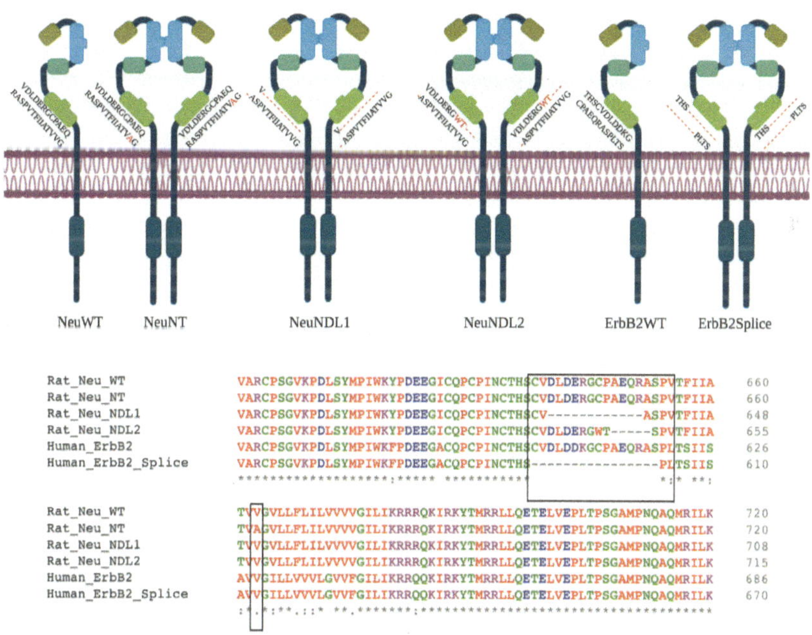

Fig. 1 ErbB2 models and mutations associated. (A) Schematic diagram of The Neu/ErbB2 receptors and where the mutations in each model are localized in the juxta-transmembrane domain. (B) Multiple sequence alignment showing wild-type and mutant Neu/ErbB2 protein sequences (created with BioRender).

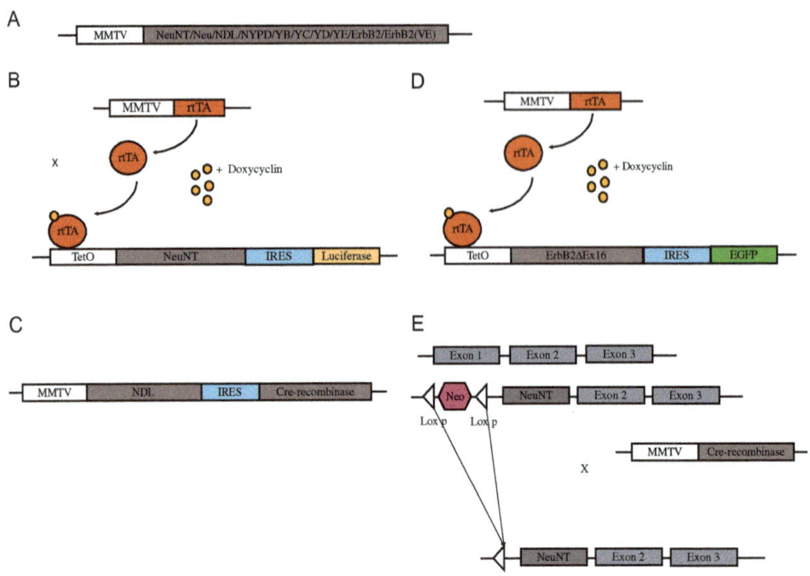

Fig. 2 Breeding strategy for ErbB2 GEMMs. (A) MMTV-driven Neu/ErbB2 (B) MMTV-driven rrTA which in the presence of doxycyclin can bind to the TET Operator and transcribe NeuNT and Luciferase that are also linked through an IRES. This allows for the inducible expression of NeuNT. (C) MMTV-driven NDL and Cre recombinase linked by IRES allowing for expression of both Neu and Cre in the same cell. (D) MMTV-driven rrTA which in the presence of doxycyclin can bind to the TET Operator and transcribe ErbB2 and EGFP that are also linked through an IRES. This allows for the inducible expression of ErbB2ΔEx16. (E) Endogenously expressed NeuNT replacing wild-type exon 1 of ErbB2 with a loxP flanked neomycin stop cassette found upstream. In the presence of MMTV-driven Cre recombinase, loxP sites recombine and NeuNT is expressed. This allowed for the expression of NeuNT in a physiologically relevant way.

specific expression of oncogenes to the mammary epithelium (Cardiff & Muller, 1993; Otten, Sanders, & McKnight, 1988; Ursini Siegel, Schade, Cardiff, & Muller, 2007; Wagner, Ward, Davis, Wiseman, & Hennighausen, 2001). Consistent with mammary epithelial expression of MMTV, both male and female mice display high levels of NeuNT in the mammary gland and rapid development of multifocal tumors with high penetrance that eventually metastasize to the lung (Muller, Sinn, Pattengale, Wallace, & Leder, 1988). Multiple groups went on to independently derive MMTV-driven NeuNT mice which were found to also develop mammary tumors rapidly. In one of these studies, mammary tumor onset was significantly delayed compared to the original study, displayed low penetrance and male mice remained tumor free (Table 1)

Table 1 Summary table of ErbB2 models and characteristics.

Publication	Mouse model	Penetrance	T50/average onset	Lung metastasis	Tumor pathology	Background
Muller et al. (1988)	MMTV-NeuNT	100%	89 days (T50)	(yes # not reported)	Adenocarcinoma	CDI, FVB
Bouchard, Lamarre, Tremblay, and Jolicoeur (1989)	MMTV-NeuNT	39% (when all founder lines averaged)	5–10 months (average)	3/5 60%	Adenocarcinoma	Balb-c
Guy et al. (1992)	MMTV-Neu	Approximately 70%	205 days (T50)	72%	Adenocarcinoma	FVB
Stöcklin, Botteri, and Groner (1993)	MMTV-ErbB2(VE)	N/A	N/A	N/A	N/A	C57BL/6, CB6F1
Guy, Cardiff, and Muller (1996)	MMTV-NeuNT	100%	90 days (average)	N/A	Adenocarcinoma	FVB
Guy et al. (1996)	Truncated MMTV-NeuNT	100%	95 days (average)	N/A	Adenocarcinoma	FVB
Siegel, Ryan, Cardiff, and Muller (1999)	MMTV-Neu-NDL	80%	161, 122 days (average)	67%, 60%	Adenocarcinoma	FVB
Andrechek et al. (2000)	NeuNT/ MMTV-Cre	45%	396 days (T50)	Approximately 20%	Comedeo-adenocarcinoma	CDI, Balb-c
Dankort et al. (2001)	MMTV-Neu-NDL NYPD	59%	252 nuliparious, 163 multiparious days (average)	38%	Neu-type	FVB

Continued

Table 1 Summary table of ErbB2 models and characteristics.—cont'd

Publication	Mouse model	Penetrance	T50/average onset	Lung metastasis	Tumor pathology	Background
Dankort et al. (2001)	MMTV-Neu-NDL YB	100%	152 days (average)	50%	Papillary	FVB
Dankort et al. (2001)	MMTV-Neu-NDL YD	100%	102 days (average)	10%	Solid comedeo	FVB
Moody et al. (2002)	Teto-NeuNT/MTB	100%	42 days (T50)	92% runted mice	Solid nodular carcinoma	FVB
Finkle et al. (2004)	MMTV-ErbB2	73%	252 days (average)	23%	Adenocarcinoma	FVB
Schade et al. (2007)	MMTV-Neu-NDL YC	100%	135 days (average)	34%	Solid nodular	FVB
Schade et al. (2007)	MMTV-Neu-NDL-YE	100%	124 days (average)	40%	Solid nodular	FVB
Ursini Siegel et al. (2008)	MMTV-NIC	100%	146 days (average)	Not reported	Solid nodular/ adenocarcinoma	FVB
Turpin et al. (2016)	Teto-ErbB2ΔEx16-GFP/MTB	>90%	10, 28 days (T50)	95%,74%	Glandular	FVB

Reported in chronological order. T50 corresponds to when 50% of mice have developed tumors and background references the reported genetic background of the mice.

(Bouchard et al., 1989; Guy et al., 1996). Discrepancies between these studies can most likely be explained by genetic background or integration site, level and temporal expression of the MMTV promoter. Taken together, these results argue that Neu/ErbB2 is an important oncogenic driver in the mammary epithelium.

Although these studies suggest that activated Neu/ErbB2 is necessary and sufficient for mammary tumor induction in these transgenic mice, the role of ErbB2 in the maintenance of the transformed phenotype was unknown. To address this issue, a doxycyclin inducible NeuNT GEMM was generated (Teo-NeuNT/MTB) (Fig. 2B). Like the original MMTV-NeuNT strain, inducible expression of NeuNT resulted in the rapid induction of multifocal mammary tumors. Once tumors reached tumor burden endpoint, doxycycline was removed and transgene expression was undetectable within 48 h. Following Neu de-induction, over 90% of tumors and their derived metastases regressed completely within a few weeks demonstrating that ErbB2 expression was required to maintain tumor development. However, all animals whose tumors had fully regressed eventually displayed Neu independent tumor growth (Table 1) (Moody et al., 2002). Indicating that even though tumors are initially dependent on ErbB2, they evolve to become completely independent of the initiating oncogene.

Despite the demonstrated role of Neu activating mutations in tumor induction in transgenic mice, human HER2 activating mutations have not been observed in breast cancer. To directly address this issue, Neu and human ErbB2 murine models were generated using the wild-type cDNA (MMTV-Neu, MMTV-ErbB2) and in contrast to the point activated form of Neu, wild-type Neu and ErbB2 driven tumors arose focally, after a long latency and with a decreased penetrance (Fig. 2A, Table 1) (Finkle et al., 2004; Guy et al., 1992; Stöcklin et al., 1993). There are multiple possible explanations for the delayed onset observed in the wild-type driven tumors. First, additional mutations may be required in cooperating genes, or second, mutational activation of ErbB2 itself may be the oncogenic requirement for tumor induction. In agreement with the later hypothesis, a large proportion of Neu and ErbB2 driven tumors displayed 7–12 amino acid deletions in the juxta-transmembrane domain which has been found to allow receptor homodimerization and constitutive activation through cysteine imbalance and disulfide bond formation (Fig. 1) (Guy et al., 1992; Siegel, Dankort, Hardy, & Muller, 1994; Weiner et al., 1989). Indeed, GEMMs expressing activated deletion mutants (MMTV-NDL strains) led to rapid development of multifocal tumors with incomplete penetrance that were metastatic (Fig. 2C, Table 1). Although these mutational events are rare in human tumors, an alternative splice isoform involving the removal of the exon immediately preceding the transmembrane domain could be detected. Significantly, exon 16 spans the same region where the spontaneous deletions were found to occur (Fig. 1) (Siegel et al., 1999). One possible explanation as to why a splice isoform of ErbB2 is common in human but absent in the transgenic mice stems from how the mice were generated. In the Neu/ErbB2 models, wild-type cDNA was used and since cDNA cannot undergo alternative splicing, activation of ErbB2 occurs through deletion mutations, giving rise to a similar phenotype to the alternatively spliced form of ErbB2. Both splicing and mutational activation of ErbB2 renders the receptor capable of homodimerization through a similar mechanism (Marchini et al., 2011). Indeed, driving inducible ErbB2ΔEx16 (Teto-ErbB2ΔEx16-GFP/MTB) specifically to the mammary epithelium resulted in the development of multifocal, metastatic tumors with a high penetrance and a rapid onset (Fig. 2D, Table 1). However, there were a few differences between the inducible model and previously generated Neu models in that tumors display a diverse histopathology co-expressing basal and luminal keratins, whereas, the conventional models mainly expressed luminal keratins. In addition, tumors exhibit an inflammatory immune microenvironment and possessed gene signatures involved in metastasis and stem cell renewal (Marchini et al., 2011; Turpin et al., 2016). Given that both wild-type ErbB2 and ErbB2ΔEx16 are found to be co-expressed in HER2-amplified breast cancer, distinguishing the specific roles of each isoform will provide us with valuable insights into ErbB2 tumor progression.

Although the MMTV promoter has allowed for the development of murine models with expression of particular genes specifically to the mammary epithelium, it is a strong viral promoter that is responsive to hormones and causes overexpression of the oncogene that it is driving, thus decreasing its physiological relevance. To bypass this and to allow for the study of endogenously expressed ErbB2, a MMTV-Cre inducible Neu (NeuNT) was introduced after the first coding exon of the ErbB2 endogenous gene flanked by loxP sites and a neomycin stop cassette ($ErbB2^{KI}$ strain). When Cre recombinase is present, the loxP sites recombine and this leads to the excision of the neomycin stop cassette which allows for expression of NeuNT on its endogenous promoter. In contrast to MMTV-NeuNT and MMTV-NDL models, the endogenously expressed $ErbB2^{KI}$ develop focal tumors with a long latency of approximately 14 months, low penetrance and metastasis as well as focal tumors that are highly heterogeneous (Fig. 2D, Table 1) (Andrechek et al., 2000). Unlike the MMTV-driven NeuNT and NDL models, endogenous expression of NeuNT does not seem to be sufficient for rapid mammary tumor induction and may require additional cooperating mutations to drive tumorigenesis as seen by the late onset. Indeed, genetic and molecular analyses of these tumors revealed selective amplification of *ErbB2*. Remarkably, the genes that are co-amplified and co-expressed (i.e., *Grb7*, *Cab1*) are identical to what is observed in *HER2*-amplified human breast cancer (Andrechek et al., 2003). However, the role of individual genes in the *ErbB2* amplicon in the induction of mammary tumorigenesis remains to be elucidated. Additional genetic analyses of this model identified a region that was frequently lost on chromosome 4 encoding *14-3-3σ*, that is a normal epithelial marker downregulated during tumor progression and in all cell lines over-expressing ErbB2, it has been found to be low or absent (Neve et al., 2006; Prasad, Valverius, McDuffie, & Cooper, 1992). Consistent with the tumor suppressor function of 14-3-3σ, mammary epithelial disruption of 14-3-3σ resulted in dramatic acceleration of tumor onset in the $ErbB2^{KI}$ model. Loss of 14.3.3σ also correlated with loss of epithelial polarity and the deletion of one allele of *14-3-3σ* led to loss of expression of the second allele by methylation. These observations further indicate a tumor suppressive role for 14.3.3σ and was one of the first studies demonstrating the potential application of the $ErbB2^{KI}$ model in identifying key genetic events in ErbB2 tumor progression (Ling, Su, Zuo, & Muller, 2012).

3. Specific recruitment of adaptor proteins to the ErbB2 receptor results in distinct *in vivo* phenotypes

Upon receptor hetero or homodimerization, cross-phosphorylation of specific tyrosine residues leads to the recruitment of kinases or adaptor proteins to Src-homology 2 (SH2) and protein tyrosine binding (PTB) domains. Some notable examples are ShcA, Grb2, Src family kinases (SFK) and the p85 subunit of PI3K (Dankort, Wang, Blackmore, Moran, & Muller, 1997; Schulze et al., 2005). To assess the importance of the recruitment of different adaptor proteins to phosphotyrosine sites on the ErbB2 receptor, multiple add back mutants were generated in an MMTV-driven activated ErbB2 model (MMTV-NDL). This model was modified to express phosphotyrosine sites for Grb2 (YB), cysteine-rich receptor-like protein kinase 11 (Crk11) (YC), ShcA (YD), ShcA (YE) and a phosphotyrosine-deficient (NYPD) mutant that lacked all tyrosine sites. Whereas all add back mutants developed mammary tumors rapidly that tended to be multifocal (except for YB which was more focal), NYPD expressing mice developed focal tumors after a long latency of over 9 months. All add backs had relatively low metastatic potential, apart from YB, that correlated with an enhanced metastatic phenotype. These results suggested that interaction with at least one adaptor protein is necessary for rapid ErbB2-mediated tumorigenesis and that activation of downstream signaling by different adaptors has a range of oncogenic potentials (Dankort et al., 2001; Schade et al., 2007). Consistent with the differences in oncogenic potential of these phosphotyrosine add back mutants, when expressing these constructs in mouse embryos, they led to distinct embryonic developmental phenotypes (Chan, Hardy, Dankort, Laing, & Muller, 2004).

Whereas these results provide clear evidence for the importance of downstream signaling in ErbB2-mediated tumorigenesis, the relative contribution of the individual pathways remained to be assessed. One drawback of Cre-mediated excision is that Cre may not be expressed in every cell leading to cells retaining conditional alleles, thus making the phenotype extremely variable. One way to circumvent this is to directly link the transgene with Cre recombinase through an internal ribosomal entry site (IRES) which in this case would be linked to an activated form of Neu (MMTV-NDL-IRES-Cre). Crossing the MMTV-NIC model to a conditional *ShcA* allele led to a complete block in mammary tumor progression (Fig. 2, Table 1) (Ursini Siegel et al., 2008). Studies using the GEMMs described above have been instrumental in understanding the complex ErbB2 signaling network and have allowed us to elucidate many proteins that influence ErbB2 directly and indirectly, some of which are outlined below (Fig. 3).

4. ErbB2 tumor development is dependent on activation of the PI3K signaling axis

A large proportion of HER2-positive patients develop ErbB2-targeted therapy resistance due to the formation of ErbB2-ErbB3 heterodimers, mutations in PI3K or loss of the PI3K regulator phosphatase and tensin homolog (PTEN), all of which potently activate the PI3K pathway. Studying downstream signaling would allow us to better understand resistance and potential therapeutic approaches (Berghoff et al., 2014; Hubalek, Brunner, Matthä, & Marth, 2010; Pohlmann, Mayer, & Mernaugh, 2009; Ross et al., 2009). PTEN is involved in hydrolyzing phosphatidylinositol (3,4,5)-trisphosphate (PIP3) to phosphatidylinositol 4,5-bisphosphate (PIP2). PIP3 is an essential second messenger that

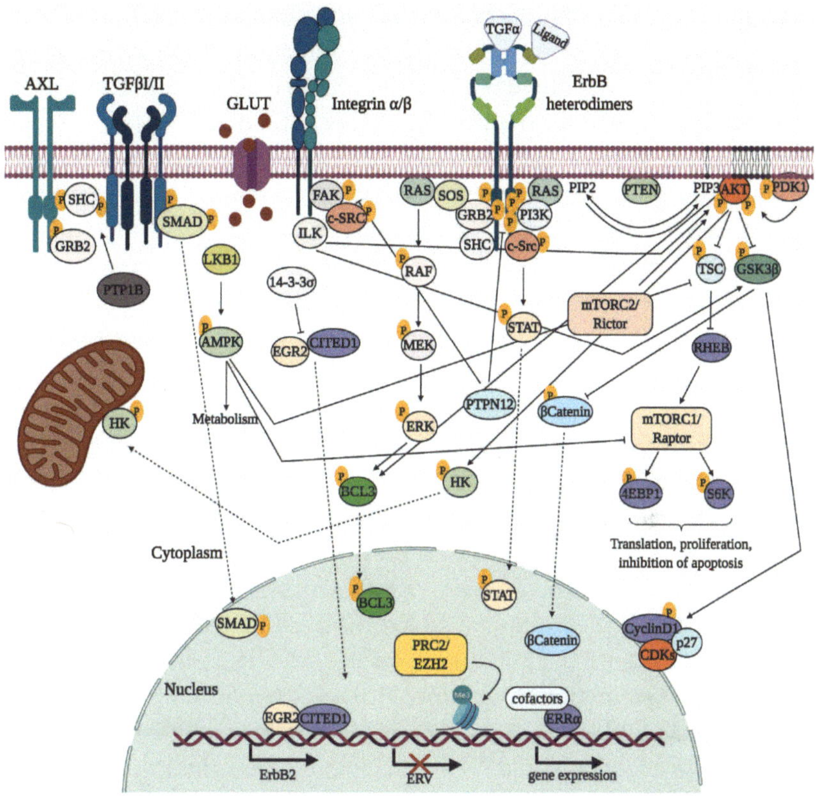

Fig. 3 ErbB2 signaling network. Downstream ErbB2 signaling and pathways affected in ErbB2-positive mouse models of breast cancer, as reported in this review. Solid lines with an arrow correspond to activation, solid lines with a bar-head indicate repression and dashed lines represent movement. Arrows located under the DNA strand correspond to activation or repression (red cross) of transcription. Yellow ovals marked with P indicate phosphorylation (created with BioRender).

recruits Akt and phosphoinositide-dependent kinase 1 (PDK1) to the membrane and allows for auto-phosphorylation of PDK1 and cross-phosphorylation and activation of downstream signaling (Wick et al., 2002). To address the importance of the PTEN lipid phosphatase, a conditional PTEN model was crossed to both the ErbB2^{KI} and the MMTV-NIC model. In both models loss of one or both of the alleles resulted in a dose dependent acceleration of mammary tumor onset. In the ErbB2^{KI} model, loss of one allele led to loss of heterozygosity of the second allele, which suggested an important dose dependent role for PTEN in mammary tumor onset in ErbB2-positive breast cancer (Schade et al., 2009).

In addition to PTEN loss of function mutations, hyperactivation of the PI3K pathway can directly occur through mutation of the PI3K catalytic subunit. PI3K is a class I lipid tyrosine kinase that is composed of a p110 catalytic subunit: α, β, γ or δ and a regulatory subunit p85, p101, p84/p87 (Kaplan et al., 1987). p110α and p110β are the only catalytic subunits to be ubiquitously expressed and p110α has been found to be mutated in approximately 40% of ErbB2-positive breast cancer and confers therapy resistance. These mutations usually occur in one of three hotspots two of which are located in the helical domain and one of which is found in the kinase domain. The most common of these mutations being the histidine to arginine point mutation at amino acid 1047 which confers constitutive activation (Rimawi et al., 2017). To investigate the role of point activated p110α in ErbB2-mediated tumorigenesis an inducible (teto-PI3KH1047R/MTB) model expressing the p110α^{HR} mutation in a doxycycline inducible manner was intercrossed with the MMTV-NIC model. Activation of p110α^{HR} dramatically accelerated tumor onset with a significant increase in tumor and metastatic burden. ErbB2 mammary tumors expressing mutated p110α^{HR} were resistant to ErbB2 small molecule inhibitors such as Lapatinib. Following de-induction of p110α^{HR} expression, tumor cells regained their sensitivity to Lapatinib. However tumors ultimately recurred through the MAPK pathway (Cheng et al., 2016).

Given the effect of p110α^{HR} on therapy response and ErbB2-mediated tumorigenesis, it was important to assess the effect of loss of p110α (mimicking inhibition) in ErbB2-positive breast cancer. Consistent with the role of p110α^{HR} in ErbB2-mediated tumorigenesis, inhibition of p110α could be predicted to severely impact ErbB2 mammary tumor onset. Thus, to test this hypothesis, a conditional p110α model was crossed to the MMTV-NIC GEMM. Mammary epithelial loss of p110α led to a significant delay in tumor onset as well as a decrease in tumor burden and incidence of metastasis (Simond, Rao, Zuo, Zhao, & Muller, 2017; Utermark et al., 2012). Interestingly, ablation of the ubiquitously expressed p110β subunit had little effect on mammary tumor onset consistent with the notion that p110α is the primary oncogenic driver of PI3K in ErbB2-positive tumors (Utermark et al., 2012). However, tumors eventually arose in the p110α-deficient model and were all sensitive to p110β inhibition, indicating that all tumors that have lost the p110α oncogenic driver are now dependent on p110β. Biochemical analysis revealed that 23% of these tumors displayed reduction in PTEN levels, indicating a role for PTEN in p110β specificity. Consistent with the role of PTEN downregulation in oncogenic PI3K isoform specificity, heterozygous ablation of one allele of *Pten* in a conditional model crossed to the p110α-deficient mice, completely rescued tumor onset and signaling was found to occur through the p110β subunit (Simond et al., 2017).

Another important PI3K dependent signaling pathway is activation of the Akt serine-threonine kinases (Akt1, 2 and 3) which leads to the induction of proliferation, growth and inhibition of apoptosis. Notably, Akt1 has been associated to poor prognosis in breast cancer (Pérez-Tenorio & Stål, 2002; Sun et al., 2001; Vivanco & Sawyers, 2002). To investigate the role of different family members in ErbB2 tumor progression, constitutively active Akt1 (AktDD) and Akt2 (Akt2DD) driven by the MMTV promoter were crossed to MMTV-NDL mice. Although, mammary epithelial expression of activated Akt1 and Akt2 did not suffice in inducing mammary tumors, co-expression of AktDD with activated ErbB2 resulted in dramatic acceleration of tumor onset, whereas co-expression with activated Akt2 had little impact on tumor induction. However, activation of Akt1 led to a severe impairment in lung metastasis as compared to the parental lines, whereas activation of Akt2 significantly increased metastatic potential. Consistent with the role of different Akt isoforms in metastatic potential of ErbB2-positive tumors, cells derived from the $ErbB2^{KI}$ identified as highly or lowly metastatic were found to correlate with phosphorylation levels of Akt1 in the lowly metastatic cells and Akt2 in the highly metastatic ones. Together, these data argue that activation of Akt isoforms have opposing effects on the metastatic potential of ErbB2-positve tumors (Dillon et al., 2009; Hutchinson, Jin, Cardiff, Woodgett, & Muller, 2004).

Akt activates mammalian target of rapamycin complex (mTORC1) and mTORC2 phosphorylates Akt at serine 473 (McDonald et al., 2008). mTORC1 and mTORC2 have a variety of different cofactors, such as regulatory associated protein of MTOR complex 1 (Raptor) and rapamycin-insensitive companion of mammalian target of rapamycin (Rictor), and are involved in different processes. mTORC1 is directly downstream of Akt, thus, a direct target of the PI3K pathway and is important in the process of cell growth and survival. However, the function of mTORC2 in breast tumorigenesis, more specifically ErbB2-mediated tumorigenesis, is less well understood. To investigate the role of mTORC2 in ErbB2-positive breast cancer, a conditional Rictor model was crossed to the MMTV-NIC mice. Conditionally ablating Rictor led to decreased cell survival and decreased serine phosphorylation of Akt as well as significantly delayed tumor onset with the majority of mice never developing mammary tumors. In addition, inhibition of ErbB2-positive human cancer cells with a dual mTORC1/2 inhibitor had a similar cellular phenotype but silencing Raptor did not recapitulate this phenotype (Joly et al., 2016). Another study also ablated Rictor in the MMTV-NIC model and found decrease in incidence of lung metastasis corroborated by loss of Rictor in human cell lines, which was found to be driven by Ras-related C3 botulinum toxin substrate 1 (Rac1) in parental cells (Joly et al., 2017). These data indicate the importance of the Rictor/mTORC2 complex in driving ErbB2 tumor progression and highlights the importance of developing mTORC2-specific inhibitors.

5. EGFR/integrin receptor crosstalk

ErbB2 has been the most extensively studied EGFR member in breast cancer; however, other ErbB receptors have been found to impact ErbB2 signaling and even how ErbB2-positive tumors respond to therapy. High levels of ErbB2 protein is found to correlate with high levels of ErbB3 protein (Siegel et al., 1999). When knocking-out ErbB3 in a system expressing MMTV-ErbB2 and MMTV-Cre, focal mammary tumors were found to arise after a long latency but all these tumors were positive for ErbB3, suggesting that these tumors had escaped ErbB3 ablation due to Cre and ErbB2 being driven on separate promoters. To circumvent this issue, the same mice were crossed to the MMTV-NIC model and loss of ErbB3 was found to completely abrogate tumor induction. These data argue for the critical importance for ErbB3 expression in mediating ErbB2 tumor induction. To then test if ErbB3 would be a functional therapeutic target, an inducible ErbB3 conditional model was used and when treating these mice with doxycyclin and with either a vehicle or Lapatinib, it was found that loss of ErbB3 sensitized ErbB2-positive tumors to Lapatinib which implies a potential therapeutic benefit of targeting these two receptors simultaneously (Vaught et al., 2012). Given this result, it was important to then test the reliance of ErbB3 on PI3K signaling in ErbB2-positive breast cancer. To accomplish this, an ErbB3 knock-in mouse model was generated lacking all the p85 binding sites, preventing the p85 regulatory subunit of PI3K from binding to ErbB3 (ErbB3$^{\Delta 85}$). The ErbB3$^{\Delta 85}$ conditional knock-in model was crossed to the MMTV-NDL mice. Mice expressing ErbB3$^{\Delta 85}$ had a modest delay of tumor onset and lung metastasis but 100% of female mice developed multifocal mammary tumors (Lahlou, Muller, Sanguin-Gendreau, Birchmeier, & Muller, 2012). These results argue that ErbB3 coupling to the PI3K pathway is dispensable for ErbB2-mediated tumorigenesis and metastasis and that ErbB3 has critical PI3K independent functions.

Another heterodimerization partner of ErbB2 is EGFR, a naturally occurring mutant that possesses a mutation in the kinase domain which decreases its ability to autophosphorylate and phosphorylate other proteins (Waved-2) was crossed to the MMTV-NDL model (Fowler et al., 1995; Luetteke et al., 1994). The Waved-2 expressing mice had impaired ErbB2-driven mammary tumor formation but no difference in metastatic potential (Gillgrass et al., 2003). This argues that the EGFR has an important role in ErbB2 tumor onset but is dispensable for tumor metastasis. Interestingly, when driving TGFα (ligand for EGFR) using the MMTV promoter and crossing it to MMTV-Neu model, the phenotype is modest but the bigenics never developed mutations in *Neu* as had been previously reported, arguing that activation of EGFR is sufficient to drive ErbB2-mediated tumorigenesis without the need for ErbB2 homodimerization (Muller et al., 1996). Loss of ErbB4 has no impact on ErbB2-driven tumorigenesis suggesting that ErbB4 is dispensable for tumor development (Jackson-Fisher et al., 2006). Collectively these data imply that each EGFR family member plays a distinct role in ErbB2 tumorigenesis.

The transforming growth factor β (TGFβ) receptors have been associated with epithelial to mesenchymal transition and expression often correlates with disease progression (Derynck, Akhurst, & Balmain, 2001; Letterio & Roberts, 1998). TGFβ comes in two forms, TGFβI and TGFβII. Ligand binds to TGFβII which then recruits TGFβI and activates downstream signaling (Wrana, Attisano, Wieser, Ventura, & Massagué, 1994). However, its role in breast cancer metastasis is less well understood. To address this issue, YB and YD were used and crossed to either an activated form of the TGFβI receptor rendering it constitutively active and a dominant negative form TGFβII making it incapable of phosphorylating the type I receptor (also driven by the MMTV promoter). Constitutive activation of TGFβI coupled to YB or YD delayed tumor onset compared to YB and YD alone and in YB, impairment of TGFβII accelerated onset compared to YB alone. Although the incidence of metastasis was unchanged, animals displayed more extravasation *versus* intravasation of pulmonary metastasis in the YD stain crossed with the constitutively active TGFβI than the YD strain alone. No difference was observed in the YB strain. Significant inhibition of extravascular lung metastasis was observed in the YB strain expressing the dominant negative form of TGFβII. These data argue that depending on which pathway is activated by phosphorylation of different tyrosine residues on the Neu receptor (either Grb2 or ShcA) varying effects will be observed during ErbB2 mammary tumor progression. In addition, these results imply that activation of TGFβI receptor in the mammary epithelium is tumor suppressive but once tumors have metastasized, it promotes extravasation of metastases (Siegel, Shu, Cardiff, Muller, & Massagué, 2003).

Integrin receptors are another important component of the extracellular matrix that have been found to regulate a variety of processes during mammary tumor formation, such as cell proliferation and survival. Integrins are cell surface receptors composed of an α and β subunit, that recognize different extracellular matrix components and in turn activate a number of downstream signaling pathways (Giancotti & Ruoslahti, 1999). It has been found that α6β4 can promote tumorigenesis by promoting receptor tyrosine kinase signaling (Mainiero et al., 1997; Shaw, Rabinovitz, Wang, Toker, & Mercurio, 1997). Phosphorylation of the cytoplasmic tail of β4-integrin is mediated by SFK's that then recruit ShcA and activates the MAPK and Akt pathways. ErbB2 has notably been found to associate to the α6β4 receptor and phosphorylate β4 (Falcioni et al., 1997; Mariotti et al., 2001; Trusolino, Bertotti, & Comoglio, 2001). To assess the importance of β4-integrin on ErbB2-mediated tumorigenesis, a β4 mutant model with targeted deletion of the signaling domain was crossed to the YD model and tumors were found to arise with a longer latency and significantly fewer metastatic lesions. The use of Gefitinib (inhibitor of EGFR and ErbB2) restored adherence and tight junctions in YD expressing cells. In the transgenic mice, use of Gefitinib induced regression of β4-deficient tumors but only slower growth in wild-type tumors, indicating that loss of β4 increases efficacy of ErbB2-targeted therapies and implies a potential therapeutic benefit of dual inhibition in the ErbB2-positive patients (Guo et al., 2006).

Integrin receptors are also important receptors involved in the activation of v-src sarcoma viral oncogene (c-Src), focal adhesion kinase (FAK), integrin-linked kinase (ILK) and β1 integrin expression has specifically been associated with poor prognosis in breast cancer patients (Gonzalez et al., 1999; Yao et al., 2007). To evaluate the importance of β1-integrin in ErbB2 mammary tumor progression, the NIC strain was intercrossed with a conditional β1-integrin model. The results showed that while mammary specific ablation of β1-integrin had little effect on mammary tumor onset or penetrance, metastatic progression was significantly decreased (Huck, Pontier, Zuo, & Muller, 2010). However, β1-integrin-deficient ErbB2 tumors upregulated the expression of other integrin members such as α5 and β3, suggesting that upregulation of other family members probably compensate for the loss of β1-integrin. To directly test whether upregulation of β3-integrin compensates for β1-loss, both β1 and β3 was conditionally ablated in the MMTV-NIC model (Huck et al., 2010). Consistent with microarray analysis of β1 and β3 single mutants, all cluster together but the double knockouts clustered independently, supporting the idea of functional redundancy between β1 and β3 (Bui et al., 2019).

Directly downstream of with β1 and β3 integrin, ILK interacts with their cytoplasmic tails. ILK then phosphorylates Akt1 and glycogen Synthase Kinase 3 Beta (GSK3β) and GSK3β phosphorylation inhibits its function and causes stabilization of β-catenin and cyclin D1(D'Amico et al., 2000; Delcommenne et al., 1998; Hannigan et al., 1996; Radeva et al., 1997). ILK has also been found to directly interact with Rictor in mTOR complex 2 suggesting a multifaceted role of this kinase in the process of tumor progression (McDonald et al., 2008; Troussard et al., 2006). To explore the role of ILK in ErbB2-positive breast cancer a conditional ILK mouse model was crossed to the MMTV-NIC model. NIC mice lacking ILK in the mammary epithelium had a drastically delayed tumor onset and penetrance. Tumors that arose were focal in nature and incidence of metastasis was decreased. These data argue that activation of ILK is critical in mediating ErbB2 tumorigenesis and cooperating mechanisms need to occur to allow for mammary tumor development (Pontier et al., 2019).

Another protein found to function downstream of the integrin receptors is focal adhesion kinase (FAK), that has notably been found to be implicated in the process of cell migration and metastasis, thus, unsurprisingly is associated with invasive cancers (Lightfoot et al., 2004; Oktay, Oktay, Hamele-Bena, Buyuk, & Koss, 2003; Sieg et al., 2000). Notably, levels of FAK have been found to correlate with levels of ErbB2 in human breast cancer (Lark et al., 2005; Nurcombe et al., 2000; Xu et al., 2000). To further investigate the role of FAK in ErbB2-mediated tumorigenesis, a conditional FAK murine model was crossed to the MMTV-Cre and MMTV-NDL model. Depletion of FAK led to a significant delay in tumor onset but penetrance remained complete. Some mice that had tumors arise in the conditional strain still retained FAK expression because of the escapee phenomenon due to Cre recombinase not being co-expressed in the same cells as the conditional FAK. In addition, the incidence of metastasis was unchanged, however, all metastatic lesions expressed FAK. To circumvent this issue, conditional FAK mice were crossed to the MMTV-NIC model and FAK depletion caused significant delay in tumor onset but tumors eventually developed and were found to occur due to functional redundancy between protein tyrosine kinase 2 beta (Pyk2) and FAK (Fan & Guan, 2011; Lahlou, Sanguin Gendreau, Frame, & Muller, 2012).

Protein tyrosine phosphatase non-receptor type 12 (PTPN12) is a protein tyrosine phosphatase that can dephosphorylate proteins involved with the cytoskeleton, such as Pyk2, FAK, Crk-associated substrate (Cas) as well as ErbB2 itself, and is thought to function as a tumor suppressor since loss has been reported in human breast cancers (Sun et al., 2011; Wu et al., 2013; Xunyi, Zhentao, Dandan, & Funian, 2012). To better understand the role of this potential tumor suppressor in ErbB2-positive breast cancer a conditional PTPN12 was crossed to the MMTV-NIC model. All mice developed multifocal tumors with a short latency and there was an increase in number of lung lesions observed. A Pyk2 and FAK inhibitor was used (PF431396) which caused a dose dependent inhibition of migration and proliferation. Collectively, these observations support the tumor suppressive role of PTPN12 in ErbB2-positive breast cancer and can influence the integrin dependent migration program (Li et al., 2015).

6. ErbB2 signaling in epigenetic regulation and metabolism

Downstream of the ErbB2 and integrin receptors, elevated levels and activity of c-Src has been found to play a critical role in human breast cancer progression and its levels has been found to be elevated in human breast cancers and plays a major role in the ErbB2 signaling pathway (Hennipman, Van Oirschot, Smits, Rijksen, & Staal, 1989; Muthuswamy, Siegel, Dankort, Webster, & Muller, 1994; Ottenhoff-Kalff et al., 1992). Consistent with the importance of c-Src in ErbB2-mediated tumorigenesis, SFKs have been found to be expressed in mammary tumors derived from MMTV-Neu/ErbB2 activated strains (Guy, Muthuswamy, Cardiff, Soriano, & Muller, 1994; Muthuswamy & Muller, 1994, 1995a, 1995b). In addition, c-Src can be directly activated by the ErbB2 catalytic domain by interacting with the c-Src kinase domain (Kim et al., 2005; Marcotte, Zhou, Kim, Roskelly, & Muller, 2009). Thus, to further explore the role of c-Src in ErbB2-driven mammary tumorigenesis, a conditional c-Src model was crossed to the MMTV-NIC mice. Mammary specific deletion of c-Src resulted in a modest delayed of tumor onset which was associated with a dramatic decrease in proliferation (Smith et al., 2019). Subsequent microarray analysis of c-Src-deficient tumors allowed for the identification of a downregulated polycomb repressor complex 2 (PRC2) signature. When performing CHIP-Seq analysis on c-Src-deficient and proficient MMTV-NIC tumors, it was found that there was global decrease in histone

3 lysine 27 (H3K27) trimethylation arguing loss in PRC2 function which correlated with lower protein levels of enhancer of zeste homolog 2 (Ezh2) and SUZ12 polycomb repressive complex 2 subunit (Suz12). PRC2 is one of the most important epigenetic regulators in ErbB2-positive breast cancer and is composed of a multiple subunits including Suz12 and Ezh2 which methylates H3 leading to transcriptional repression (Holm et al., 2012; Kleer et al., 2003). The low levels of Suz12 and Ezh2 also correlated with lower activation of mTORC1. mTORC1 inhibitors decreased Ezh2 and Suz12 protein levels in human ErbB2-positive cell lines but not in triple-negative human breast cancer lines. Consistent with the importance of the mTORC1/PRC2 signaling axis, mammary epithelial ablation of Ezh2 resulted in a complete block in ErbB2 mammary tumorigenesis. These data argue that the c-Src/Ezh2 signaling axis is an essential downstream component involved in ErbB2 mammary tumor progression. Further biochemical analyses of c-Src-deficient tumors revealed that loss of mTORC1 activity was associated with fragmentation of the mitochondrial network which resulted in loss of mitochondrial function. These observations indicate that epigenetic and metabolic regulation are co-regulated in ErbB2 signaling (Smith et al., 2019).

c-Src is not the only protein found to have an impact on metabolism in ErbB2-positive breast cancer. Liver kinase B1 (LKB1) phosphorylates adenosine monophosphate-activated kinase (AMPK) during energetic stress (Hawley et al., 2003; Shaw et al., 2004). This then leads to the inhibition of mTORC through TSC and Raptor. LKB1 has been found to regulate many processes including metabolism and cell polarity (Alessi, Sakamoto, & Bayascas, 2006). Using conditional LKB1 mice crossed to the MMTV-NIC, onset was found to be unchanged. When transplanting mice with LKB1 deficient cells, tumors grew slower and incidence of metastasis was similar but lung lesions were smaller than with the parental cells. LKB1-deficient cells displayed increased levels of intracellular glucose and lactate which are known to occur during the Warburg effect (Liberti & Locasale, 2016), as well as an increase in Creatine, ATP and ADP. When deregulating mTORC1, cells can gain a metabolic growth advantage but are also more sensitive to glucose deprivation which was the case in LKB1-deficient cells (Dupuy et al., 2013). Collectively, these results argue for an important role for LKB1 in metabolism, specifically in ErbB2-positive cells, and possibly a role in lung metastasis.

A proteomic study of ErbB2-positive human breast cancer identified proteins involved in lipid synthesis and glycolysis (Zhang et al., 2005).

When conducting gene expression profiling one important gene found to be upregulated in mouse ErbB2-positive transgenic tumors was 2, 4-dienoyl-CoA reductase (DecR1), involved in fatty acid β-oxidation and known to be downregulated during tumor progression (Landis, Seachrist, Abdul-Karim, & Keri, 2006; Landis, Seachrist, Montanez-Wiscovich, Danielpour, & Keri, 2005). DecR1 catalyzes the β-oxidation of polyunsaturated fatty acids and deficiency in DecR1 is lethal in humans (Roe et al., 1990). *DecR1* mRNA and protein levels were found to be downregulated in mammary tumors of MMTV-NDL and MMTV-YB and MMTV-YD as compared to normal mammary gland (Ursini Siegel, Rajput, et al., 2007). These data argue that cancer cells are indeed very dependent on the synthesis of lipids to support the high rate of cell division occurring during tumor growth.

Glucose is an important sugar involved in metabolism and enters the cell through the glucose transporter (GLUT) proteins transporters. GLUT1 has been found to be overexpressed in breast cancer and notably has been found to be associated with larger tumors and higher grade (Macheda, Rogers, & Best, 2005; Pinheiro et al., 2011). GLUT1 expression can notably be induced by a variety of events such as by expression of Akt1, hypoxia inducible factor 1 subunit alpha (HIF1α), hypoxia, c-Myc and aryl hydrocarbon receptor nuclear translocator (ARNT), as well as estrogen. To identify the role of GLUT1 in ErbB2-mediated tumorigenesis a GLUT1 conditional model was crossed to the MMTV-NIC model which abrogated mammary tumor development suggesting that loss of GLUT1 is sufficient to halt tumor progression (Wellberg et al., 2016).

Hexokinases (HK) catalyze the first step in glucose metabolism and four isoforms exist: HK1, HK2, HK3, HK4 (Robey & Hay, 2006). Notably, cancer cells have been found to express high levels of HK2 (Mathupala, Rempel, & Pedersen, 2001; Shinohara, Yamamoto, Kogure, Ichihara, & Terada, 1994). To investigate the role of HK2 in ErbB2-mediated tumorigenesis, conditional HK2 mice were crossed to the MMTV-NIC model. Consistent with the critical role of glucose metabolism in breast cancer, tumor onset was significantly delayed in the HK2-deficient ErbB2 tumor cells. However, the tumors that arose maintained expression of HK2, suggesting that cancer cells were arising from tumors cells where HK2 had not been efficiently excised (Patra et al., 2013). These data indicate that HK2 plays an essential role in ErbB2-mediated tumorigenesis and expression is necessary to drive ErbB2 tumors.

7. ErbB2 signaling impacts nuclear transcription factors

β-catenin is a critical multifunctional protein involved in tissue architecture and transcriptional regulation (Heuberger & Birchmeier, 2010). Expression of β-catenin is often associated with a more aggressive molecular subtype of breast cancer, such as HER2-enriched and Basal subtypes (Geyer et al., 2011; Schade et al., 2013). To explore the role of β-catenin in ErbB2 tumor progression, the conditional β-catenin mice were interbred with either the ErbB2^{KI} model or the NIC strain representing the Basal or Luminal ErbB2-positive human subtypes (Schade et al., 2013). Heterozygous loss of β-catenin in the ErbB2^{KI} mice caused a dramatic acceleration of tumor onset but homozygous deletion only very modestly accelerated onset compared to the wild-type mice due to the upregulation of a related Plakoglobin gene. By contrast to this basal ErbB2 model, abrogation of β-catenin function had little impact on the luminal ErbB2 model. These data argue that β-catenin has a critical role in both a Luminal and Basal model of breast cancer on tumor initiation and progression (Bui et al., 2017).

Estrogen receptor alpha (ERα) and paired box 2 (Pax2) have been found to regulate expression of ErbB2 and high levels of ErbB2 are observed in tumors resistant to Tamoxifen (Ali & Coombes, 2002; Berry et al., 2000; Kun et al., 2003; Shou et al., 2004). Estrogen related receptor alpha (ERRα) is a nuclear orphan receptor which has structural similarities to ERα. Expression of ERRα positively correlates with poor prognosis and expression of ErbB2 (Ariazi, Clark, & Mertz, 2002; Suzuki et al., 2004). To better understand the role of ERRα in the mammary gland as physiologically as possible, a conditional ERRα model was crossed to the ErbB2^{KI} model. Loss of ERRα modestly delayed tumor onset as well as slightly decrease tumor penetrance. Gene expression analysis identified that all the genes in the *ErbB2* amplicon were lower in the ERRα-deficient mice. Taken together, ERRα and ErbB2 were mostly found to be correlative and ERRα seems to control the level of expression of the *ErbB2* amplicon (Deblois et al., 2010).

Signal transducer and activator of transcription 3 (Stat3) has been found to be active downstream of many genes involved in mammary tumor formation, notably ErbB2 (Ren & Schaefer, 2002). To evaluate the contribution of Stat3 in ErbB2 tumor progression, the MMTV-NIC strain was crossed with a conditional Stat3 model. Although tumor onset was unaffected, there was a fourfold decrease in metastasis in the Stat3-deficient mice as well as a decrease in total tumor burden. Tail-vein injection of Stat3-proficient and deficient MMTV-NIC mice led to a 5-fold reduction in number and 12-fold reduction in size of the metastases. By gene expression profiling Stat3-deficient tumors lacked genes that activate acute phase response which most likely caused the decrease in metastasis, as tumor environment is less inflammatory (Ranger, Levy, Shahalizadeh, Hallett, & Muller, 2009). These data suggest a pivotal role for Stat3 in ErbB2 metastasis through modulating the immune microenvironment.

8. Cell cycle regulation in ErbB2-positive breast cancer

Around 90% of all breast cancers overexpress ErbB2 and protein tyrosine phosphatase 1B (PTP1B) (Wiener et al., 1994). PTP1B had been found to be a negative regulator in metabolic dysregulations (Bourdeau, Dubé, & Tremblay, 2005; Dubé & Tremblay, 2004; Östman & Böhmer, 2001). To address the unknown role of PTP1B in ErbB2-positive breast cancer, the MMTV-NDL model was crossed to PTP1B-deficient mice. Mammary tumor onset was significantly delayed and the incidence of lung metastasis was decreased.

In the absence of PTP1B the Ras/MAPK and PI3K pathways were attenuated. Since both pathways regulate expression of cyclin D1 and its regulator $p27^{KIP1}$. Levels of cyclinD1 and $p27^{KIP1}$ were assayed and it was found that cyclin D1 and $p27^{KIP1}$ levels were decreased in the PTP1B-deficient tumors. Since deletion of PTP1B seemed to delay ErbB2-mediated tumorigenesis, MMTV-NDL mice were treated with a PTP1B inhibitor and tumor onset was significantly delayed in the treated mice suggesting a potential therapeutic benefit of targeting PTP1B in ErbB2-positive breast cancer (Julien et al., 2007).

D-type cyclins all regulate cell cycle progression through CDK4 and CDK6 leading to phosphorylation of retinoblastoma (RB) which then drives cells through G1 (Sherr & Roberts, 1999). Cyclin D1 has also been shown to be overexpressed in MMTV-ErbB2 tumors (Lee et al., 2000). When crossing a cyclin D1-deficient strain to MMTV-Neu mice, no tumors arose which suggested a critical role for this cell cycle regulator in ErbB2-driven mammary tumorigenesis and potential therapeutic benefit of inhibiting cyclinD1 (Yu, Geng, & Sicinski, 2001). $p27^{KIP1}$ is a CDK inhibitor that can inhibit most cyclin/CDK complexes and has notably been associated with cell migration and poor prognosis (Carrano, Eytan, Hershko, & Pagano, 1999). In cell lines ErbB2 levels negatively correlates with $p27^{KIP1}$ expression (Yang, Zhou, Hung, & Lee, 2000). To further investigate this observation, MMTV-Neu mice were crossed to heterozygous $p27^{KIP1}$-deficient mice. Onset was significantly accelerated and resulted in 80% of $p27^{KIP1}$ expression. Therefore, it seems that the loss of $p27^{KIP1}$ displays a haploinsufficient phenotype and seems to affect latency of tumor onset in ErbB2-positive tumors (Hulit et al., 2006).

9. ErbB2 in immune regulation

There has been evidence of a link between ErbB2 and the interleukin 6 (IL6)/signal transducer and activator of transcription 3 (Stat3) signaling axis (Hartman et al., 2011). In a previous studies, B-cell lymphoma 3 (Bcl3) was found to be a Stat3 responsive gene and identified to be important in cell survival and proliferation (Clarkson et al., 2006; Massoumi, Chmielarska, Hennecke, Pfeifer, & Fässler, 2006). In addition, Bcl3 has also been shown to be a cofactor of nuclear factor kappa B (NF-κB) which has been associated with tumor progression in ErbB2-positive tumors, and Bcl3 has been hypothesized to potentially serve as a link between Stat3 and NF-κB (Aggarwal et al., 2005; Bours et al., 1993; Clarkson & Watson, 1999; Helbig et al., 2003; Park et al., 2007). To better understand the role of Bcl3 in ErbB2-mediated tumorigenesis, the MMTV-Neu model was crossed to a Bcl3 knock-out model and it was found that tumors that were deficient in Bcl3 had a decrease in incidence of lung metastasis. Mice given a Bcl3 inhibitor also displayed a reduced incidence of metastasis. These data implicate Bcl3 as a mediator in the process of metastasis and its potential use as a therapeutic target (Wakefield et al., 2013).

The TAM family of RTK's is composed of Tyro3, Axl, and Mer. And Axl is activated by its ligand growth arrest specific 6 (Gas6). Axl has been found to be expressed in many cancers including breast and is also associated with EMT and invasion, and is essential to inhibit innate immunity (Asiedu et al., 2014; Gjerdrum et al., 2010; Lemke & Rothlin, 2008; Rothlin, Ghosh, Zuniga, Oldstone, & Lemke, 2007; Vuoriluoto et al., 2011; Wilson et al., 2014). Axl levels were found to be present throughout all molecular subtypes when staining a human panel of breast tumor tissue. Therefore, as to study Axl in the context of a more aggressive breast cancer subtype, a conditional Axl was crossed to the MMTV-NIC model. All mice developed mammary tumors with a delayed tumor onset, as well as an increase in incidence and number of lung metastasis. Using proximity ligation assay it was found that ErbB2 directly interacts with Axl in cell lines. To identify the role of Axl in the process of lung metastasis all the steps in the metastatic cascade were analyzed and it was found that Axl contributes to intravasation, extravasation and survival at the metastatic site. Taken together, the role of Axl seems to be involved in every step of the metastatic cascade and directly interacts with the ErbB2 receptor (Goyette et al., 2018).

10. Use of mouse models in the development of HER2-targeted therapies

10.1 History of HER2-targeted therapies

In 1984 the first mAb against Neu was created using NIH3T3 cells transfected with DNA from transformed rat neuroblastoma cells. The antibodies recognized a phosphoprotein of approximately 185 kDa which turned out to be Neu (Schechter et al., 1984). When exposing transformed NIH3T3 cells with purified 7.16.4 monoclonal antibody, it was found that it binds to two molecules of Neu meaning that it targets the extracellular domain of Neu when the receptor is in a dimerized state and was found to potently inhibit this receptor. This was the first time that an antibody specific to Neu was found to have anti-oncogenic properties (Drebin, Link, Stern, Weinberg, & Greene, 1985).

In 1986, the Neu oncogene was injected in transformed NIH3T3 cells and then injected into athymic nude mice subcutaneously and it was found that tumors grew very rapidly. When treating these mice intraperitoneally with 7.16.4 antibody, growth was significantly inhibited. However, this monoclonal antibody does not have strong specificity for the human ErbB2 protein and thus, as good as it is to study downregulation of Neu, it could not be used as a potential drug applicable to human cancer (Drebin, Link, Weinberg, & Greene, 1986). Shortly thereafter, a new monoclonal antibody was developed that specifically recognized the extracellular domain of human ErbB2 (e.g., 4D5) generated in a similar way using NIH3T3 cells infected with human *ErbB2* cDNA. 4D5 was chosen for further characterization as it immunoprecipiated solely with ErbB2. In addition, in the presence of antibody, ErbB2 was degraded more rapidly than in control conditions (Hudziak et al., 1989).

To address the issue of giving a mouse raised antibody to human tumor cells, a humanized version of 4D5 was generated named (rhu)mAb 4D5, this antibody displayed three times the affinity to ErbB2 as the mouse version of 4D5: (mum)mAB 4D5 and this humanized version of 4D5 would go on to be termed Trastuzumab/Herceptin and is considered standard care for patients that have ErbB2-positive breast cancer (Carter et al., 1992; Lewis et al., 1993).

In the clinic, there are multiple types of tyrosine kinase inhibitors (TKI's), some reversible and some irreversible. The reversible ones compete with ATP in the ATP binding pocket of the kinase domain, thus upon removal of therapy, receptor function is restored. Whereas irreversible TKI's can alkalyate a particular cysteine residue and cause permanent remodeling of the receptor into an inactive state. Since TKI's are found to usually target more than one receptor, they are of critical importance in the clinic as ErbB family members have been found to heterodimerize and activate multiple pathways at once (Harari & Yarden, 2000).

Many reversible inhibitors have been developed such as Gefitinib and Erlotinib that are specific for EGFR and GW572016 (Lapatinib) that has equal sensitivity for ErbB2 and EGFR (Fry et al., 1994; Rusnak et al., 2001; Ward et al., 1994). Pan-TKI's have been found to be more efficient than single targeting antibodies or inhibitors. Lapatinib inhibits both ErbB2 and EGFR and leads to a significant increase in apoptosis. The advantage that Lapatinib has over Trastuzumab is that by targeting the kinase domain of both ErbB2 and EGFR, it renders its activity independent of PTEN as phosphorylation of the Akt and MAPK pathway depends on the kinase activity of the ErbB family members, whereas Trastuzumab depends on full activity of PTEN to prevent downstream signaling (Xia et al., 2007).

10.2 GEMMs used as pre-clinical models to test the efficacy of HER2-targeted therapy

Mouse models have been instrumental in understanding when and how to give ErbB2-targeted therapy to patients. When treating MMTV-NeuNT mice with 7.16.4 it was found that early treatment, before tumor onset (20–45 weeks), starting at 6 weeks of age by intraperitoneally injection led to 50% of mice not developing tumors even after 90 weeks (Katsumata et al., 1995). When administering mu4D5 to MMTV-ErbB2 mice, it was also found that early administration delayed tumor onset, and decreased incidence of lung metastasis. This demonstrated that early treatment of m4D5 significantly improved outcome and showed that there would be a therapeutic benefit of giving Trastuzumab to patients with early low grade disease rather than after displaying resistance to chemotherapy (Finkle et al., 2004).

As Ezh2 overexpression has been found to be associated with ErbB2-positive tumors and poor prognosis and resistance to ErbB2-targeted therapy has been largely focused on receptor interactions and genetic events causing constitutive downstream signaling, it was important to assess the potential role of Ezh2 in ErbB2 therapy resistance (Holm et al., 2012; Kleer et al., 2003; Loibl et al., 2014; Majewski et al., 2015; Wilken & Maihle, 2010). Examination of patient biopsies prior to treatment with Trastuzumab revealed that patients that developed resistance to therapy had higher levels of H3K27-trimethylation. Treatment of ErbB2-positive breast cancer cells with GSK126 (Ezh2 inhibitor) led to the identification of an IFNα/β signature by gene expression profiling. Many retrotransposon families were identified in treated cells, many of which were endogenous retroviruses (ERV's) which generate double stranded RNA that are important in activating the interferon signaling through the mitochondrial protein MAVS (mitochondrian anti-viral signaling protein). Consistent with the importance of Ezh2 in Trastuzumab resistance, combination of an Ezh2 inhibitor and an ErbB2-targeted antibody enhanced response in an immunocompetent model but only in tumors that were already sensitive to ErbB2 inhibition and did not re-sensitize tumors that were resistant. Future clinical trials will be important to assay the utility of this approach in treating Trastuzumab sensitive patients (Hirukawa et al., 2019).

These studies illustrate how GEMMs have been exploited to revolutionized pre-clinical testing of drugs that target ErbB2. In recent years, many more studies have assessed the effect of knocking out or knocking in genes downstream or upstream of ErbB2 in an attempt to better understand this subtype and suggest novel therapeutic approaches. All of which has been possible with the development of ErbB2-positive GEMMs as a means to model human disease.

References

Aggarwal, B. B., Shishodia, S., Takada, Y., Banerjee, S., Newman, R. A., Bueso-Ramos, C. E., et al. (2005). Curcumin suppresses the paclitaxel-induced nuclear factor-κB pathway in breast cancer cells and inhibits lung metastasis of human breast cancer in nude mice. *Clinical Cancer Research*, *11*, 7490.

Alessi, D. R., Sakamoto, K., & Bayascas, J. R. (2006). LKB1-dependent signaling pathways. *Annual Review of Biochemistry*, *75*, 137–163.

Ali, S., & Coombes, R. C. (2002). Endocrine-responsive breast cancer and strategies for combating resistance. *Nature Reviews. Cancer*, *2*, 101.

Andrechek, E. R., Hardy, W. R., Siegel, P. M., Rudnicki, M. A., Cardiff, R. D., & Muller, W. J. (2000). Amplification of the neu/erbB-2 oncogene in a mouse model of mammary tumorigenesis. *Proceedings of the National Academy of Sciences of the United States of America*, *97*, 3444–3449.

Andrechek, E. R., Laing, M. A., Girgis-Gabardo, A. A., Siegel, P. M., Cardiff, R. D., & Muller, W. J. (2003). Gene expression profiling of Neu-induced mammary tumors from transgenic mice reveals genetic and morphological similarities to ErbB2-expressing human breast cancers. *Cancer Research*, *63*, 4920.

Ariazi, E. A., Clark, G. M., & Mertz, J. E. (2002). Estrogen-related receptor α and estrogen-related receptor γ associate with unfavorable and favorable biomarkers, respectively, in human breast cancer. *Cancer Research*, *62*, 6510–6518.

Asiedu, M. K., Beauchamp-Perez, F. D., Ingle, J. N., Behrens, M. D., Radisky, D. C., & Knutson, K. L. (2014). AXL induces epithelial-to-mesenchymal transition and regulates the function of breast cancer stem cells. *Oncogene*, *33*, 1316.

Bargmann, C. I., Hung, M.-C., & Weinberg, R. A. (1986a). The neu oncogene encodes an epidermal growth factor receptor-related protein. *Nature*, *319*, 226.

Bargmann, C. I., Hung, M.-C., & Weinberg, R. A. (1986b). Multiple independent activations of the neu oncogene by a point mutation altering the transmembrane domain of p185. *Cell*, *45*, 649–657.

Berghoff, A. S., Bartsch, R., Preusser, M., Ricken, G., Steger, G. G., Bago-Horvath, Z., et al. (2014). Co-overexpression of HER2/HER3 is a predictor of impaired survival in breast cancer patients. *The Breast*, *23*, 637–643. https://doi.org/10.1016/j.breast.2014.06.011.

Berry, D. A., Muss, H. B., Thor, A. D., Dressler, L., Liu, E. T., Broadwater, G., et al. (2000). HER-2/neu and p53 expression versus tamoxifen resistance in estrogen receptor–positive, node-positive breast cancer. *Journal of Clinical Oncology*, *18*, 3471–3479.

Bouchard, L., Lamarre, L., Tremblay, P. J., & Jolicoeur, P. (1989). Stochastic appearance of mammary tumors in transgenic mice carrying the MMTV/c-neu oncogene. *Cell*, *57*, 931–936.

Bourdeau, A., Dubé, N., & Tremblay, M. L. (2005). Cytoplasmic protein tyrosine phosphatases, regulation and function: The roles of PTP1B and TC-PTP. *Current Opinion in Cell Biology*, *17*, 203–209.

Bours, V., Franzoso, G., Azarenko, V., Park, S., Kanno, T., Brown, K., et al. (1993). The oncoprotein Bcl-3 directly transactivates through κB motifs via association with DNA-binding p50B homodimers. *Cell*, *72*, 729–739.

Bray, F., Ferlay, J., Soerjomataram, I., Siegel, R. L., Torre, L. A., & Jemal, A. (2018). Global cancer statistics 2018: GLOBOCAN estimates of incidence and mortality worldwide for 36 cancers in 185 countries. *CA: A Cancer Journal for Clinicians*, *68*, 394–424. https://doi.org/10.3322/caac.21492.

Bui, T., Rennhack, J., Mok, S., Ling, C., Perez, M., Roccamo, J., et al. (2019). Functional redundancy between β1 and β3 integrin in activating the IR/Akt/mTORC1 signaling axis to promote ErbB2-driven breast cancer. *Cell Reports*, *29*, 589–602. https://doi.org/10.1016/j.celrep.2019.09.004.

Bui, T., Schade, B., Cardiff, R. D., Aina, O. H., Sanguin Gendreau, V., & Muller, W. J. (2017). β-Catenin haploinsufficiency promotes mammary tumorigenesis in an ErbB2-positive basal breast cancer model. *Proceedings of the National Academy of Sciences of the United States of America*, *114*, E707–E716.

Canadian Cancer Statistics Advisory Committee. (2018). *Canadian cancer statistics. 2018*. Toronto, ON: Canadian Cancer Society.

Cardiff, R., & Muller, W. J. (1993). Transgenic mouse models of mammary tumorigenesis. *Cancer Surveys*, *16*, 97–113.

Carrano, A. C., Eytan, E., Hershko, A., & Pagano, M. (1999). SKP2 is required for ubiquitin-mediated degradation of the CDK inhibitor p27. *Nature Cell Biology*, *1*, 193.

Carter, P., Presta, L., Gorman, C. M., Ridgway, J. B., Henner, D., Wong, W. L., et al. (1992). Humanization of an anti-p185HER2 antibody for human cancer therapy. *Proceedings of the National Academy of Sciences of the United States of America*, *89*, 4285–4289.

Chan, R., Hardy, W. R., Dankort, D., Laing, M. A., & Muller, W. J. (2004). Modulation of Erbb2 signaling during development: A threshold level of Erbb2 signaling is required for development. *Development*, *131*, 5551.

Cheng, H., Liu, P., Ohlson, C., Xu, E., Symonds, L., Isabella, A., et al. (2016). PIK3CA H1047R-and Her2-initiated mammary tumors escape PI3K dependency by compensatory activation of MEK-ERK signaling. *Oncogene*, *35*, 2961.

Clarkson, R. W., Boland, M. P., Kritikou, E. A., Lee, J. M., Freeman, T. C., Tiffen, P. G., et al. (2006). The genes induced by signal transducer and activators of transcription (STAT) 3 and STAT5 in mammary epithelial cells define the roles of these STATs in mammary development. *Molecular Endocrinology*, *20*, 675–685.

Clarkson, R. W., & Watson, C. J. (1999). NF-kappaB and apoptosis in mammary epithelial cells. *Journal of Mammary Gland Biology and Neoplasia*, *4*, 165–175.

D'Amico, M., Hulit, J., Amanatullah, D. F., Zafonte, B. T., Albanese, C., Bouzahzah, B., et al. (2000). The integrin-linked kinase regulates the cyclin D1 gene through glycogen synthase kinase 3β and cAMP-responsive element-binding protein-dependent pathways. *The Journal of Biological Chemistry*, *275*, 32649–32657.

Dankort, D., Maslikowski, B., Warner, N., Kanno, N., Kim, H., Wang, Z., et al. (2001). Grb2 and Shc adapter proteins play distinct roles in Neu (ErbB-2)-induced mammary tumorigenesis: Implications for human breast cancer. *Molecular and Cellular Biology*, *21*, 1540–1551. https://doi.org/10.1128/MCB.21.5.1540-1551.2001.

Dankort, D. L., Wang, Z., Blackmore, V., Moran, M. F., & Muller, W. J. (1997). Distinct tyrosine autophosphorylation sites negatively and positively modulate neu-mediated transformation. *Molecular and Cellular Biology*, *17*, 5410–5425.

Davie, S. A., Maglione, J. E., Manner, C. K., Young, D., Cardiff, R. D., MacLeod, C. L., et al. (2007). Effects of FVB/NJ and C57Bl/6J strain backgrounds on mammary tumor phenotype in inducible nitric oxide synthase deficient mice. *Transgenic Research*, *16*, 193–201. https://doi.org/10.1007/s11248-006-9056-9.

Deblois, G., Chahrour, G., Perry, M.-C., Sylvain-Drolet, G., Muller, W. J., & Giguère, V. (2010). Transcriptional control of the ERBB2 amplicon by ERRalpha and PGC-1beta promotes mammary gland tumorigenesis. *Cancer Research*, *70*, 10277–10287. https://doi.org/10.1158/0008-5472.CAN-10-2840.

Delcommenne, M., Tan, C., Gray, V., Rue, L., Woodgett, J., & Dedhar, S. (1998). Phosphoinositide-3-OH kinase-dependent regulation of glycogen synthase kinase 3 and protein kinase B/AKT by the integrin-linked kinase. *Proceedings of the National Academy of Sciences of the United States of America*, *95*, 11211–11216.

Derynck, R., Akhurst, R. J., & Balmain, A. (2001). TGF-β signaling in tumor suppression and cancer progression. *Nature Publishing Group*, *29*, 117–129.

Dillon, R. L., Marcotte, R., Hennessy, B. T., Woodgett, J. R., Mills, G. B., & Muller, W. J. (2009). Akt1 and Akt2 play distinct roles in the initiation and metastatic phases of mammary tumor progression. *Cancer Research*, *69*, 5057–5064. https://doi.org/10.1158/0008-5472.CAN-08-4287.

Drebin, J. A., Link, V. C., Stern, D. F., Weinberg, R. A., & Greene, M. I. (1985). Down-modulation of an oncogene protein product and reversion of the transformed phenotype by monoclonal antibodies. *Cell*, *41*, 695–706.

Drebin, J. A., Link, V. C., Weinberg, R. A., & Greene, M. I. (1986). Inhibition of tumor growth by a monoclonal antibody reactive with an oncogene-encoded tumor antigen. *Proceedings of the National Academy of Sciences of the United States of America*, *83*, 9129–9133.

Drebin, J. A., Stern, D. F., Link, V. C., Weinberg, R. A., & Greene, M. I. (1984). Monoclonal antibodies identify a cell-surface antigen associated with an activated cellular oncogene. *Nature*, *312*, 545.

Dubé, N., & Tremblay, M. L. (2004). Beyond the metabolic function of PTP1B. *Cell Cycle*, *3*, 548–551.

Dupuy, F., Griss, T., Blagih, J., Bridon, G., Avizonis, D., Ling, C., et al. (2013). LKB1 is a central regulator of tumor initiation and pro-growth metabolism in ErbB2-mediated breast cancer. *Cancer & Metabolism, 1*, 1–18. https://doi.org/10.1186/2049-3002-1-18.

Falcioni, R., Antonini, A., Nisticò, P., Di Stefano, S., Crescenzi, M., Natali, P. G., et al. (1997). α6β4 and α6β1 integrins associate with ErbB-2 in human carcinoma cell lines. *Experimental Cell Research, 236*, 76–85.

Fan, H., & Guan, J.-L. (2011). Compensatory function of Pyk2 protein in the promotion of focal adhesion kinase (FAK)-null mammary cancer stem cell tumorigenicity and metastatic activity. *The Journal of Biological Chemistry, 286*, 18573–18582.

Finkle, D., Quan, Z. R., Asghari, V., Kloss, J., Ghaboosi, N., Mai, E., et al. (2004). HER2-targeted therapy reduces incidence and progression of midlife mammary tumors in female murine mammary tumor virus huHER2-transgenic mice. *Clinical Cancer Research, 10*, 2499–2511.

Fowler, K. J., Walker, F., Alexander, W., Hibbs, M. L., Nice, E. C., Bohmer, R. M., et al. (1995). A mutation in the epidermal growth factor receptor in waved-2 mice has a profound effect on receptor biochemistry that results in impaired lactation. *Proceedings of the National Academy of Sciences of the United States of America, 92*, 1465–1469.

Franklin, M. C., Carey, K. D., Vajdos, F. F., Leahy, D. J., De Vos, A. M., & Sliwkowski, M. X. (2004). Insights into ErbB signaling from the structure of the ErbB2-pertuzumab complex. *Cancer Cell, 5*, 317–328.

Fry, D. W., Kraker, A. J., McMichael, A., Ambroso, L. A., Nelson, J. M., Leopold, W. R., et al. (1994). A specific inhibitor of the epidermal growth factor receptor tyrosine kinase. *Science, 265*, 1093–1095. https://doi.org/10.1126/science.8066447.

Geyer, F. C., Lacroix-Triki, M., Savage, K., Arnedos, M., Lambros, M. B., MacKay, A., et al. (2011). β-Catenin pathway activation in breast cancer is associated with triple-negative phenotype but not with CTNNB1 mutation. *Modern Pathology, 24*, 209.

Giancotti, F. G., & Ruoslahti, E. (1999). Integrin signaling. *Science, 285*, 1028–1033.

Gillgrass, A., Cardiff, R. D., Sharan, N., Kannan, S., & Muller, W. J. (2003). Epidermal growth factor receptor-dependent activation of Gab1 is involved in ErbB-2-mediated mammary tumor progression. *Oncogene, 22*, 9151 9155. https://doi.org/10.1038/sj.onc.1206983.

Gjerdrum, C., Tiron, C., Høiby, T., Stefansson, I., Haugen, H., Sandal, T., et al. (2010). Axl is an essential epithelial-to-mesenchymal transition-induced regulator of breast cancer metastasis and patient survival. *Proceedings of the National Academy of Sciences of the United States of America, 107*, 1124–1129.

Gonzalez, M. A., Pinder, S. E., Wencyk, P. M., Bell, J. A., Elston, C. W., Nicholson, R. I., et al. (1999). An immunohistochemical examination of the expression of E-cadherin, α-and β/γ-catenins, and α2-and β1-integrins in invasive breast cancer. *The Journal of Pathology, 187*, 523–529.

Goyette, M.-A., Duhamel, S., Aubert, L., Pelletier, A., Savage, P., Thibault, M.-P., et al. (2018). The receptor tyrosine kinase AXL is required at multiple steps of the metastatic cascade during HER2-positive breast cancer progression. *Cell Reports, 23*, 1476–1490. https://doi.org/10.1016/j.celrep.2018.04.019.

Graus Porta, D., Beerli, R. R., Daly, J. M., & Hynes, N. E. (1997). ErbB-2, the preferred heterodimerization partner of all ErbB receptors, is a mediator of lateral signaling. *The EMBO Journal, 16*, 1647–1655.

Guo, W., Pylayeva, Y., Pepe, A., Yoshioka, T., Muller, W. J., Inghirami, G., et al. (2006). β4 integrin amplifies ErbB2 signaling to promote mammary tumorigenesis. *Cell, 126*, 489–502. https://doi.org/10.1016/j.cell.2006.05.047.

Guy, C. T., Cardiff, R. D., & Muller, W. J. (1996). Activated neu induces rapid tumor progression. *The Journal of Biological Chemistry, 271*, 7673–7678.

Guy, C. T., Muthuswamy, S. K., Cardiff, R. D., Soriano, P., & Muller, W. J. (1994). Activation of the c-Src tyrosine kinase is required for the induction of mammary tumors in transgenic mice. *Genes & Development, 8*, 23–32.

Guy, C. T., Webster, M. A., Schaller, M., Parsons, T. J., Cardiff, R. D., & Muller, W. J. (1992). Expression of the neu protooncogene in the mammary epithelium of transgenic mice induces metastatic disease. *Proceedings of the National Academy of Sciences of the United States of America, 89*, 10578–10582.

Hammond, M. E. H., Hayes, D. F., Dowsett, M., Allred, D. C., Hagerty, K. L., Badve, S., et al. (2010). American Society of Clinical Oncology/College of American Pathologists guideline recommendations for immunohistochemical testing of estrogen and progesterone receptors in breast cancer (unabridged version). *Archives of Pathology & Laboratory Medicine, 134*, e48–e72.

Hannigan, G. E., Leung-Hagesteijn, C., Fitz-Gibbon, L., Coppolino, M. G., Radeva, G., Filmus, J., et al. (1996). Regulation of cell adhesion and anchorage-dependent growth by a new β1-integrin-linked protein kinase. *Nature, 379*, 91.

Harari, D., & Yarden, Y. (2000). Molecular mechanisms underlying ErbB2/HER2 action in breast cancer. *Oncogene, 19*, 6102.

Hartman, Z. C., Yang, X.-Y., Glass, O., Lei, G., Osada, T., Dave, S. S., et al. (2011). HER2 overexpression elicits a proinflammatory IL-6 autocrine signaling loop that is critical for tumorigenesis. *Cancer Research, 71*, 4380–4391.

Hawley, S. A., Boudeau, J., Reid, J. L., Mustard, K. J., Udd, L., Mäkelä, T. P., et al. (2003). Complexes between the LKB1 tumor suppressor, STRADα/β and MO25α/β are upstream kinases in the AMP-activated protein kinase cascade. *Journal of Biology, 2*, 28.

Helbig, G., Christopherson, K. W., Bhat-Nakshatri, P., Kumar, S., Kishimoto, H., Miller, K. D., et al. (2003). NF-κ B promotes breast cancer cell migration and metastasis by inducing the expression of the chemokine receptor CXCR4. *The Journal of Biological Chemistry, 278*, 21631–21638.

Hennipman, A., Van Oirschot, B. A., Smits, J., Rijksen, G., & Staal, G. (1989). Tyrosine kinase activity in breast cancer, benign breast disease, and normal breast tissue. *Cancer Research, 49*, 516–521.

Heuberger, J., & Birchmeier, W. (2010). Interplay of cadherin-mediated cell adhesion and canonical Wnt signaling. *Cold Spring Harbor Perspectives in Biology, 2*, a002915.

Hirukawa, A., Singh, S., Wang, J., Rennhack, J. P., Swiatnicki, M., Sanguin Gendreau, V., et al. (2019). Reduction of global H3K27me3 enhances HER2/ErbB2 targeted therapy. *Cell Reports, 29*, 249–257. https://doi.org/10.1016/j.celrep.2019.08.105.

Holm, K., Grabau, D., Lövgren, K., Aradottir, S., Gruvberger-Saal, S., Howlin, J., et al. (2012). Global H3K27 trimethylation and EZH2 abundance in breast tumor subtypes. *Molecular Oncology, 6*, 494–506.

Hubalek, M., Brunner, C., Matthä, K., & Marth, C. (2010). Resistance to HER2-targeted therapy: Mechanisms of trastuzumab resistance and possible strategies to overcome unresponsiveness to treatment. *Wiener Medizinische Wochenschrift, 160*, 506–512.

Huck, L., Pontier, S. M., Zuo, D. M., & Muller, W. J. (2010). β1-integrin is dispensable for the induction of ErbB2 mammary tumors but plays a critical role in the metastatic phase of tumor progression. *Proceedings of the National Academy of Sciences of the United States of America, 107*, 15559–15564.

Hudziak, R. M., Lewis, G. D., Winget, M., Fendly, B. M., Shepard, H. M., & Ullrich, A. (1989). p185HER2 monoclonal antibody has antiproliferative effects in vitro and sensitizes human breast tumor cells to tumor necrosis factor. *Molecular and Cellular Biology, 9*, 1165–1172.

Hulit, J., Lee, R. J., Li, Z., Wang, C., Katiyar, S., Yang, J., et al. (2006). p27Kip1 repression of ErbB2-induced mammary tumor growth in transgenic mice involves Skp2 and Wnt/β-catenin signaling. *Cancer Research, 66*, 8529–8541.

Hutchinson, J. N., Jin, J., Cardiff, R. D., Woodgett, J. R., & Muller, W. J. (2004). Activation of Akt-1 (PKB-α) can accelerate ErbB-2-mediated mammary tumorigenesis but suppresses tumor invasion. *Cancer Research, 64*, 3171–3178.

Jackson-Fisher, A. J., Bellinger, G., Shum, E., Duong, J. K., Perkins, A. S., Gassmann, M., et al. (2006). Formation of Neu/ErbB2-induced mammary tumors is unaffected by loss of ErbB4. *Oncogene, 25*, 5664–5672. https://doi.org/10.1038/sj.onc.1209574.

Joly, M. M., Hicks, D. J., Jones, B., Sánchez, V., Estrada, M. V., Young, C., et al. (2016). Rictor/mTORC2 drives progression and therapeutic resistance of HER2-amplified breast cancers. *Cancer Research, 76*, 4752–4764.

Joly, M. M., Williams, M. M., Hicks, D. J., Jones, B., Sánchez, V., Young, C. D., et al. (2017). Two distinct mTORC2-dependent pathways converge on Rac1 to drive breast cancer metastasis. *Breast Cancer Research, 19*, 74. https://doi.org/10.1186/s13058-017-0868-8.

Julien, S. G., Dubé, N., Read, M., Penney, J., Paquet, M., Han, Y., et al. (2007). Protein tyrosine phosphatase 1B deficiency or inhibition delays ErbB2-induced mammary tumorigenesis and protects from lung metastasis. *Nature Genetics, 39*, 338–346. https://doi.org/10.1038/ng1963.

Kaplan, D. R., Whitman, M., Schaffhausen, B., Pallas, D. C., White, M., Cantley, L., et al. (1987). Common elements in growth factor stimulation and oncogenic transformation: 85 kd phosphoprotein and phosphatidylinositol kinase activity. *Cell, 50*, 1021–1029.

Katsumata, M., Okudaira, T., Samanta, A., Clark, D. P., Drebin, J. A., Jolicoeur, P., et al. (1995). Prevention of breast tumour development in vivo by downregulation of the p185 neu receptor. *Nature Medicine, 1*, 644.

Kim, H., Chan, R., Dankort, D. L., Zuo, D., Najoukas, M., Park, M., et al. (2005). The c-Src tyrosine kinase associates with the catalytic domain of ErbB-2: Implications for ErbB-2 mediated signaling and transformation. *Oncogene, 24*, 7599.

Kleer, C. G., Cao, Q., Varambally, S., Shen, R., Ota, I., Tomlins, S. A., et al. (2003). EZH2 is a marker of aggressive breast cancer and promotes neoplastic transformation of breast epithelial cells. *Proceedings of the National Academy of Sciences of the United States of America, 100*, 11606–11611.

Kun, Y., How, L. C., Hoon, T. P., Bajic, V. B., Lam, T. S., Aggarwal, A., et al. (2003). Classifying the estrogen receptor status of breast cancers by expression profiles reveals a poor prognosis subpopulation exhibiting high expression of the ERBB2 receptor. *Human Molecular Genetics, 12*, 3245–3258.

Lahlou, H., Muller, T., Sanguin-Gendreau, V., Birchmeier, C., & Muller, W. J. (2012). Uncoupling of PI3K from ErbB3 impairs mammary gland development but does not impact on ErbB2-induced mammary tumorigenesis. *Cancer Research, 72*, 3080–3090. https://doi.org/10.1158/0008-5472.CAN-11-3513.

Lahlou, H., Sanguin Gendreau, V., Frame, M. C., & Muller, W. J. (2012). Focal adhesion kinase contributes to proliferative potential of ErbB2 mammary tumour cells but is dispensable for ErbB2 mammary tumour induction in vivo. *Breast Cancer Research, 14*, R36.

Landis, M. D., Seachrist, D. D., Abdul-Karim, F. W., & Keri, R. A. (2006). Sustained trophism of the mammary gland is sufficient to accelerate and synchronize development of ErbB2/Neu-induced tumors. *Oncogene, 25*, 3325.

Landis, M. D., Seachrist, D. D., Montanez-Wiscovich, M. E., Danielpour, D., & Keri, R. A. (2005). Gene expression profiling of cancer progression reveals intrinsic regulation of transforming growth factor-β signaling in ErbB2/Neu-induced tumors from transgenic mice. *Oncogene, 24*, 5173.

Lark, A. L., Livasy, C. A., Dressler, L., Moore, D. T., Millikan, R. C., Geradts, J., et al. (2005). High focal adhesion kinase expression in invasive breast carcinomas is associated with an aggressive phenotype. *Modern Pathology, 18*, 1289.

Lee, R. J., Albanese, C., Fu, M., D'Amico, M., Lin, B., Watanabe, G., et al. (2000). Cyclin D1 is required for transformation by activated Neu and is induced through an E2F-dependent signaling pathway. *Molecular and Cellular Biology, 20*, 672–683.

Lemke, G., & Rothlin, C. V. (2008). Immunobiology of the TAM receptors. *Nature Reviews. Immunology, 8*, 327.

Letterio, J. J., & Roberts, A. B. (1998). Regulation of immune responses by TGF-β. *Annual Review of Immunology, 16*, 137–161. https://doi.org/10.1146/annurev.immunol.16.1.137.

Lewis, G. D., Figari, I., Fendly, B., Wong, W. L., Carter, P., Gorman, C., et al. (1993). Differential responses of human tumor cell lines to anti-p185 HER2 monoclonal antibodies. *Cancer Immunology, Immunotherapy, 37*, 255–263.

Li, J., Davidson, D., Souza, M. C., Zhong, M. C., Wu, N., Park, M., et al. (2015). Loss of PTPN12 stimulates progression of ErbB2-dependent breast cancer by enhancing cell survival, migration, and epithelial-to-mesenchymal transition. *Molecular and Cellular Biology, 35*, 4069–4082. https://doi.org/10.1128/MCB.00741-15.

Liberti, M. V., & Locasale, J. W. (2016). The Warburg effect: How does it benefit cancer cells? *Trends in Biochemical Sciences, 41*, 211–218.

Lightfoot, H. M., Lark, A., Livasy, C. A., Moore, D. T., Cowan, D., Dressler, L., et al. (2004). Upregulation of focal adhesion kinase (FAK) expression in ductal carcinoma in situ (DCIS) is an early event in breast tumorigenesis. *Breast Cancer Research and Treatment, 88*, 109–116.

Ling, C., Su, V.-M.-T., Zuo, D., & Muller, W. J. (2012). Loss of the 14-3-3σ tumor suppressor is a critical event in ErbB2-mediated tumor progression. *Cancer Discovery, 2*, 68–81. https://doi.org/10.1158/2159-8290.CD-11-0189.

Loibl, S., Von Minckwitz, G., Schneeweiss, A., Paepke, S., Lehmann, A., Rezai, M., et al. (2014). PIK3CA mutations are associated with lower rates of pathologic complete response to anti-human epidermal growth factor receptor 2 (her2) therapy in primary HER2-overexpressing breast cancer. *Journal of Clinical Oncology, 32*, 3212–3220.

Luetteke, N. C., Phillips, H. K., Qiu, T. H., Copeland, N. G., Earp, H. S., Jenkins, N. A., et al. (1994). The mouse waved-2 phenotype results from a point mutation in the EGF receptor tyrosine kinase. *Genes & Development, 8*, 399–413.

Macheda, M. L., Rogers, S., & Best, J. D. (2005). Molecular and cellular regulation of glucose transporter (GLUT) proteins in cancer. *Journal of Cellular Physiology, 202*, 654–662.

Mainiero, F., Murgia, C., Wary, K. K., Curatola, A. M., Pepe, A., Blumemberg, M., et al. (1997). The coupling of α6β4 integrin to Ras–MAP kinase pathways mediated by Shc controls keratinocyte proliferation. *The EMBO Journal, 16*, 2365–2375.

Majewski, I. J., Nuciforo, P., Mittempergher, L., Bosma, A. J., Eidtmann, H., Holmes, E., et al. (2015). PIK3CA mutations are associated with decreased benefit to neoadjuvant human epidermal growth factor receptor 2-targeted therapies in breast cancer. *Journal of Clinical Oncology, 33*, 1334.

Marchini, C., Gabrielli, F., Iezzi, M., Zenobi, S., Montani, M., Pietrella, L., et al. (2011). The human splice variant Δ16HER2 induces rapid tumor onset in a reporter transgenic mouse. *PLoS One, 6*, e18727. https://doi.org/10.1371/journal.pone.0018727.

Marcotte, R., Zhou, L., Kim, H., Roskelly, C. D., & Muller, W. J. (2009). c-Src associates with ErbB2 through an interaction between catalytic domains and confers enhanced transforming potential. *Molecular and Cellular Biology, 29*, 5858–5871.

Mariotti, A., Kedeshian, P. A., Dans, M., Curatola, A. M., Gagnoux-Palacios, L., & Giancotti, F. G. (2001). EGF-R signaling through Fyn kinase disrupts the function of integrin α6β4 at hemidesmosomes: Role in epithelial cell migration and carcinoma invasion. *The Journal of Cell Biology, 155*, 447–458.

Massoumi, R., Chmielarska, K., Hennecke, K., Pfeifer, A., & Fässler, R. (2006). Cyld inhibits tumor cell proliferation by blocking Bcl-3-dependent NF-κB signaling. *Cell, 125*, 665–677.

Mathupala, S. P., Rempel, A., & Pedersen, P. L. (2001). Glucose catabolism in cancer cells identification and characterization of a marked activation response of the type II hexokinase gene to hypoxic conditions. *The Journal of Biological Chemistry*, *276*, 43407–43412.

McDonald, P. C., Oloumi, A., Mills, J., Dobreva, I., Maidan, M., Gray, V., et al. (2008). Rictor and integrin-linked kinase interact and regulate Akt phosphorylation and cancer cell survival. *Cancer Research*, *68*, 1618–1624.

Moody, S. E., Sarkisian, C. J., Hahn, K. T., Gunther, E. J., Pickup, S., Dugan, K. D., et al. (2002). Conditional activation of Neu in the mammary epithelium of transgenic mice results in reversible pulmonary metastasis. *Cancer Cell*, *2*, 451–461.

Muller, W. J., Arteaga, C. L., Muthuswamy, S. K., Siegel, P. M., Webster, M. A., Cardiff, R. D., et al. (1996). Synergistic interaction of the Neu proto-oncogene product and transforming growth factor alpha in the mammary epithelium of transgenic mice. *Molecular and Cellular Biology*, *16*, 5726–5736.

Muller, W. J., Sinn, E., Pattengale, P. K., Wallace, R., & Leder, P. (1988). Single-step induction of mammary adenocarcinoma in transgenic mice bearing the activated c-neu oncogene. *Cell*, *54*, 105–115.

Muthuswamy, S. K., & Muller, W. J. (1994). Activation of the Src family of tyrosine kinases in mammary tumorigenesis. *Advances in Cancer Research*, *64*, 111–123.

Muthuswamy, S. K., & Muller, W. J. (1995a). Direct and specific interaction of c-Src with Neu is involved in signaling by the epidermal growth factor receptor. *Oncogene*, *11*, 271–279.

Muthuswamy, S. K., & Muller, W. J. (1995b). Activation of Src family kinases in Neu-induced mammary tumors correlates with their association with distinct sets of tyrosine phosphorylated proteins in vivo. *Oncogene*, *11*, 1801–1810.

Muthuswamy, S. K., Siegel, P. M., Dankort, D. L., Webster, M. A., & Muller, W. J. (1994). Mammary tumors expressing the neu proto-oncogene possess elevated c-Src tyrosine kinase activity. *Molecular and Cellular Biology*, *14*, 735–743.

Neve, R. M., Chin, K., Fridlyand, J., Yeh, J., Baehner, F. L., Fevr, T., et al. (2006). A collection of breast cancer cell lines for the study of functionally distinct cancer subtypes. *Cancer Cell*, *10*, 515–527.

Nurcombe, V., Smart, C. E., Chipperfield, H., Cool, S. M., Boilly, B., & Hondermarck, H. (2000). The proliferative and migratory activities of breast cancer cells can be differentially regulated by heparan sulfates. *The Journal of Biological Chemistry*, *275*, 30009–30018.

Oktay, M. H., Oktay, K., Hamele-Bena, D., Buyuk, A., & Koss, L. G. (2003). Focal adhesion kinase as a marker of malignant phenotype in breast and cervical carcinomas. *Human Pathology*, *34*, 240–245.

Östman, A., & Böhmer, F.-D. (2001). Regulation of receptor tyrosine kinase signaling by protein tyrosine phosphatases. *Trends in Cell Biology*, *11*, 258–266.

Otten, A. D., Sanders, M. M., & McKnight, G. S. (1988). The MMTV LTR Promoter is induced by progesterone and dihydrotestosterone but not by estrogen. *Molecular Endocrinology*, *2*, 143–147. https://doi.org/10.1210/mend-2-2-143.

Ottenhoff-Kalff, A. E., Rijksen, G., Van Beurden, E., Hennipman, A., Michels, A. A., & Staal, G. (1992). Characterization of protein tyrosine kinases from human breast cancer: Involvement of the c-src oncogene product. *Cancer Research*, *52*, 4773–4778.

Padhy, L. C., Shih, C., Cowing, D., Finkelstein, R., & Weinberg, R. A. (1982). Identification of a phosphoprotein specifically induced by the transforming DNA of rat neuroblastomas. *Cell*, *28*, 865–871.

Park, B. K., Zhang, H., Zeng, Q., Dai, J., Keller, E. T., Giordano, T., et al. (2007). NF-κB in breast cancer cells promotes osteolytic bone metastasis by inducing osteoclastogenesis via GM-CSF. *Nature Medicine*, *13*, 62.

Patra, K. C., Wang, Q., Bhaskar, P. T., Miller, L., Wang, Z., Wheaton, W., et al. (2013). Hexokinase 2 is required for tumor initiation and maintenance and its systemic deletion is therapeutic in mouse models of cancer. *Cancer Cell, 24*, 213–228.

Pérez-Tenorio, G., & Stål, O. (2002). Activation of AKT/PKB in breast cancer predicts a worse outcome among endocrine treated patients. *British Journal of Cancer, 86*, 540.

Perou, C. M., Sørlie, T., Eisen, M. B., Van De Rijn, M., Jeffrey, S. S., Rees, C. A., et al. (2000). Molecular portraits of human breast tumours. *Nature, 406*, 747.

Pinheiro, C., Sousa, B., Albergaria, A., Paredes, J., Dufloth, R., Vieira, D., et al. (2011). GLUT1 and CAIX expression profiles in breast cancer correlate with adverse prognostic factors and MCT1 overexpression. *Histology and Histopathology, 26*, 1279–1286.

Pohlmann, P. R., Mayer, I. A., & Mernaugh, R. (2009). Resistance to trastuzumab in breast cancer. *Clinical Cancer Research, 15*, 7479–7491.

Pontier, S. M., Huck, L., White, D. E., Rayment, J., Sanguin-Gendreau, V., Hennessy, B., et al. (2019). Integrin-linked kinase has a critical role in ErbB2 mammary tumor progression: Implications for human breast cancer. *Oncogene, 29*, 3374–3385. https://doi.org/10.1038/onc.2010.86.

Prasad, G. L., Valverius, E. M., McDuffie, E., & Cooper, H. L. (1992). Complementary DNA cloning of a novel epithelial cell marker protein, HME1, that may be down-regulated in neoplastic mammary cells. *Cell Growth and Differentiation, 3*, 507–513.

Radeva, G., Petrocelli, T., Behrend, E., Leung-Hagesteijn, C., Filmus, J., Slingerland, J., et al. (1997). Overexpression of the integrin-linked kinase promotes anchorage-independent cell cycle progression. *The Journal of Biological Chemistry, 272*, 13937–13944.

Ranger, J. J., Levy, D. E., Shahalizadeh, S., Hallett, M., & Muller, W. J. (2009). Identification of a Stat3-dependent transcription regulatory network involved in metastatic progression. *Cancer Research, 69*, 6823–6830. https://doi.org/10.1158/0008-5472.CAN-09-1684.

Ren, Z., & Schaefer, T. S. (2002). ErbB-2 activates Stat3α in a Src-and JAK2-dependent manner. *The Journal of Biological Chemistry, 277*, 38486–38493.

Rimawi, M. F., De Angelis, C., Contreras, A., Pareja, F., Geyer, F. C., Burke, K. A., et al. (2017). Low PTEN levels and PIK3CA mutations predict resistance to neoadjuvant lapatinib and trastuzumab without chemotherapy in patients with HER2 over-expressing breast cancer. *Breast Cancer Research and Treatment, 167*, 731–740. https://doi.org/10.1007/s10549-017-4533-9.

Robey, R. A., & Hay, N. (2006). Mitochondrial hexokinases, novel mediators of the anti-apoptotic effects of growth factors and Akt. *Oncogene, 25*, 4683.

Roe, C. R., Millington, D. S., Norwood, D. L., Kodo, N., Sprecher, H., Mohammed, B. S., et al. (1990). 2,4-Dienoyl-coenzyme A reductase deficiency: A possible new disorder of fatty acid oxidation. *The Journal of Clinical Investigation, 85*, 1703–1707.

Ross, J. S., Slodkowska, E. A., Symmans, W. F., Pusztai, L., Ravdin, P. M., & Hortobagyi, G. N. (2009). The HER-2 receptor and breast cancer: Ten years of targeted anti-HER-2 therapy and personalized medicine. *The Oncologist, 14*, 320–368.

Rothlin, C. V., Ghosh, S., Zuniga, E. I., Oldstone, M. B., & Lemke, G. (2007). TAM receptors are pleiotropic inhibitors of the innate immune response. *Cell, 131*, 1124–1136.

Rusnak, D. W., Affleck, K., Cockerill, S. G., Stubberfield, C., Harris, R., Page, M., et al. (2001). The characterization of novel, dual ErbB-2/EGFR, tyrosine kinase inhibitors. *Cancer Research, 61*, 7196–7203.

Schade, B., Lam, S. H., Cernea, D., Sanguin Gendreau, V., Cardiff, R. D., Jung, B. L., et al. (2007). Distinct ErbB-2-coupled signaling pathways promote mammary tumors with unique pathologic and transcriptional profiles. *Cancer Research, 67*, 7579–7588.

Schade, B., Lesurf, R., Sanguin Gendreau, V., Bui, T., Deblois, G., O'Toole, S. A., et al. (2013). β-Catenin signaling is a critical event in ErbB2-mediated mammary tumor progression. *Cancer Research, 73*, 4474–4487.

Schade, B., Rao, T., Dourdin, N., Lesurf, R., Hallett, M., Cardiff, R. D., et al. (2009). PTEN deficiency in a luminal ErbB-2 mouse model results in dramatic acceleration of mammary tumorigenesis and metastasis. *The Journal of Biological Chemistry, 284*, 19018–19026.

Schechter, A. L., Stern, D. F., Vaidyanathan, L., Decker, S. J., Drebin, J. A., Greene, M. I., et al. (1984). The neu oncogene: An erb-B-related gene encoding a 185,000-Mr tumour antigen. *Nature, 312*, 513–516.

Schulze, W. X., Deng, L., & Mann, M. (2005). Phosphotyrosine interactome of the ErbB-receptor kinase family. *Molecular Systems Biology, 1*. https://doi.org/10.1038/msb4100012, 2005.0008.

Shaw, R. J., Kosmatka, M., Bardeesy, N., Hurley, R. L., Witters, L. A., DePinho, R. A., et al. (2004). The tumor suppressor LKB1 kinase directly activates AMP-activated kinase and regulates apoptosis in response to energy stress. *Proceedings of the National Academy of Sciences of the United States of America, 101*, 3329–3335.

Shaw, L. M., Rabinovitz, I., Wang, H. H.-F., Toker, A., & Mercurio, A. M. (1997). Activation of phosphoinositide 3-OH kinase by the α6β4 integrin promotes carcinoma invasion. *Cell, 91*, 949–960.

Sherr, C. J., & Roberts, J. M. (1999). CDK inhibitors: Positive and negative regulators of G1-phase progression. *Genes & Development, 13*, 1501–1512.

Shih, C., Padhy, L. C., Murray, M., & Weinberg, R. A. (1981). Transforming genes of carcinomas and neuroblastomas introduced into mouse fibroblasts. *Nature, 290*, 261.

Shinohara, Y., Yamamoto, K., Kogure, K., Ichihara, J., & Terada, H. (1994). Steady state transcript levels of the type II hexokinase and type 1 glucose transporter in human tumor cell lines. *Cancer Letters, 82*, 27–32.

Shou, J., Massarweh, S., Osborne, C. K., Wakeling, A. E., Ali, S., Weiss, H., et al. (2004). Mechanisms of tamoxifen resistance: Increased estrogen receptor-HER2/neu cross-talk in ER/HER2-positive breast cancer. *Journal of the National Cancer Institute, 96*, 926–935.

Sieg, D. J., Hauck, C. R., Ilic, D., Klingbeil, C. K., Schaefer, E., Damsky, C. H., et al. (2000). FAK integrates growth-factor and integrin signals to promote cell migration. *Nature Cell Biology, 2*, 249.

Siegel, P. M., Dankort, D. L., Hardy, W. R., & Muller, W. J. (1994). Novel activating mutations in the neu proto-oncogene involved in induction of mammary tumors. *Molecular and Cellular Biology, 14*, 7068–7077.

Siegel, P. M., Ryan, E. D., Cardiff, R. D., & Muller, W. J. (1999). Elevated expression of activated forms of Neu/ErbB-2 and ErbB-3 are involved in the induction of mammary tumors in transgenic mice: Implications for human breast cancer. *The EMBO Journal, 18*, 2149–2164.

Siegel, P. M., Shu, W., Cardiff, R. D., Muller, W. J., & Massagué, J. (2003). Transforming growth factor β signaling impairs Neu-induced mammary tumorigenesis while promoting pulmonary metastasis. *Proceedings of the National Academy of Sciences of the United States of America, 100*, 8430–8435.

Simond, A. M., Rao, T., Zuo, D., Zhao, J. J., & Muller, W. J. (2017). ErbB2-positive mammary tumors can escape PI3K-p110α loss through downregulation of the Pten tumor suppressor. *Oncogene, 36*, 6059–6066. https://doi.org/10.1038/onc.2017.264.

Smith, H. W., Hirukawa, A., Sanguin Gendreau, V., Nandi, I., Dufour, C. R., Zuo, D., et al. (2019). An ErbB2/c-Src axis links bioenergetics with PRC2 translation to drive epigenetic reprogramming and mammary tumorigenesis. *Nature Communications, 10*, 2901. https://doi.org/10.1038/s41467-019-10681-4.

Sørlie, T., Perou, C. M., Tibshirani, R., Aas, T., Geisler, S., Johnsen, H., et al. (2001). Gene expression patterns of breast carcinomas distinguish tumor subclasses with clinical implications. *Proceedings of the National Academy of Sciences of the United States of America, 98*, 10869–10874.

Stöcklin, E., Botteri, F., & Groner, B. (1993). An activated allele of the c-erbB-2 oncogene impairs kidney and lung function and causes early death of transgenic mice. *The Journal of Cell Biology*, *122*, 199–208.

Sun, T., Aceto, N., Meerbrey, K. L., Kessler, J. D., Zhou, C., Migliaccio, I., et al. (2011). Activation of multiple proto-oncogenic tyrosine kinases in breast cancer via loss of the PTPN12 phosphatase. *Cell*, *144*, 703–718.

Sun, M., Wang, G., Paciga, J. E., Feldman, R. I., Yuan, Z.-Q., Ma, X.-L., et al. (2001). AKT1/PKBα kinase is frequently elevated in human cancers and its constitutive activation is required for oncogenic transformation in NIH3T3 cells. *The American Journal of Pathology*, *159*, 431–437.

Suzuki, T., Miki, Y., Moriya, T., Shimada, N., Ishida, T., Hirakawa, H., et al. (2004). Estrogen-related receptor α in human breast carcinoma as a potent prognostic factor. *Cancer Research*, *64*, 4670–4676.

Troussard, A. A., McDonald, P. C., Wederell, E. D., Mawji, N. M., Filipenko, N. R., Gelmon, K. A., et al. (2006). Preferential dependence of breast cancer cells versus normal cells on integrin-linked kinase for protein kinase B/Akt activation and cell survival. *Cancer Research*, *66*, 393–403.

Trusolino, L., Bertotti, A., & Comoglio, P. M. (2001). A signaling adapter function for α6β4 integrin in the control of HGF-dependent invasive growth. *Cell*, *107*, 643–654.

Turpin, J., Ling, C., Crosby, E. J., Hartman, Z. C., Simond, A. M., Chodosh, L. A., et al. (2016). The ErbB2ΔEx16 splice variant is a major oncogenic driver in breast cancer that promotes a pro-metastatic tumor microenvironment. *Oncogene*, *35*, 6053–6064. https://doi.org/10.1038/onc.2016.129.

Tzahar, E., Waterman, H., Chen, X., Levkowitz, G., Karunagaran, D., Lavi, S., et al. (1996). A hierarchical network of interreceptor interactions determines signal transduction by Neu differentiation factor/neuregulin and epidermal growth factor. *Molecular and Cellular Biology*, *16*, 5276–5287.

Ursini Siegel, J., Hardy, W. R., Zuo, D., Lam, S. H., Sanguin Gendreau, V., Cardiff, R. D., et al. (2008). ShcA signalling is essential for tumour progression in mouse models of human breast cancer. *The EMBO Journal*, *27*, 910–920.

Ursini Siegel, J., Rajput, A. B., Lu, H., Sanguin Gendreau, V., Zuo, D., Papavasiliou, V., et al. (2007). Elevated expression of DecR1 impairs ErbB2/Neu-induced mammary tumor development. *Molecular and Cellular Biology*, *27*, 6361–6371. https://doi.org/10.1128/MCB.00686-07.

Ursini Siegel, J., Schade, B., Cardiff, R. D., & Muller, W. J. (2007). Insights from transgenic mouse models of ERBB2-induced breast cancer. *Nature Reviews. Cancer*, *7*, 389.

Utermark, T., Rao, T., Cheng, H., Wang, Q., Lee, S. H., Wang, Z. C., et al. (2012). The p110α and p110β isoforms of PI3K play divergent roles in mammary gland development and tumorigenesis. *Genes & Development*, *26*, 1573–1586.

Vaught, D. B., Stanford, J. C., Young, C., Hicks, D. J., Wheeler, F., Rinehart, C., et al. (2012). HER3 is required for HER2-induced preneoplastic changes to the breast epithelium and tumor formation. *Cancer Research*, *72*, 2672–2682. https://doi.org/10.1158/0008-5472.CAN-11-3594.

Vivanco, I., & Sawyers, C. L. (2002). The phosphatidylinositol 3-kinase–AKT pathway in human cancer. *Nature Reviews. Cancer*, *2*, 489.

Vuoriluoto, K., Haugen, H., Kiviluoto, S., Mpindi, J. P., Nevo, J., Gjerdrum, C., et al. (2011). Vimentin regulates EMT induction by Slug and oncogenic H-Ras and migration by governing Axl expression in breast cancer. *Oncogene*, *30*, 1436.

Wagner, K.-U., Ward, T., Davis, B., Wiseman, R., & Hennighausen, L. (2001). Spatial and temporal expression of the Cre gene under the control of the MMTV-LTR in different lines of transgenic mice. *Transgenic Research*, *10*, 545–553. https://doi.org/10.1023/A:1013063514007.

Wakefield, A., Soukupova, J., Montagne, A., Ranger, J., French, R., Muller, W. J., et al. (2013). Bcl3 selectively promotes metastasis of ERBB2-driven mammary tumors. *Cancer Research*, *73*, 745–755. https://doi.org/10.1158/0008-5472.CAN-12-1321.

Ward, W. H. J., Cook, P. N., Slater, A. M., Davies, D. H., Holdgate, G. A., & Green, L. R. (1994). Epidermal growth factor receptor tyrosine kinase: Investigation of catalytic mechanism, structure-based searching and discovery of a potent inhibitor. *Biochemical Pharmacology*, *48*, 659–666.

Weiner, D. B., Liu, J., Cohen, J. A., Williams, W. V., & Greene, M. I. (1989). A point mutation in the neu oncogene mimics ligand induction of receptor aggregation. *Nature*, *339*, 230–231.

Wellberg, E. A., Johnson, S., Finlay-Schultz, J., Lewis, A. S., Terrell, K. L., Sartorius, C. A., et al. (2016). The glucose transporter GLUT1 is required for ErbB2-induced mammary tumorigenesis. *Breast Cancer Research*, *18*, 131. https://doi.org/10.1186/s13058-016-0795-0.

Wick, M. J., Wick, K. R., Chen, H., He, H., Dong, L. Q., Quon, M. J., et al. (2002). Substitution of the autophosphorylation site Thr516 with a negatively charged residue confers constitutive activity to mouse 3-phosphoinositide-dependent protein kinase-1 in cells. *The Journal of Biological Chemistry*, *277*, 16632–16638.

Wiener, J. R., Hurteau, J. A., Kerns, B.-J. M., Whitaker, R. S., Conaway, M. R., Berchuck, A., et al. (1994). Overexpression of the tyrosine phosphatase PTP1B is associated with human ovarian carcinomas. *American Journal of Obstetrics and Gynecology*, *170*, 1177–1183.

Wilken, J. A., & Maihle, N. J. (2010). Primary trastuzumab resistance: New tricks for an old drug. *Annals of the New York Academy of Sciences*, *1210*, 53.

Wilson, C., Ye, X., Pham, T., Lin, E., Chan, S., McNamara, E., et al. (2014). AXL inhibition sensitizes mesenchymal cancer cells to antimitotic drugs. *Cancer Research*, *74*, 5878–5890.

Wolff, A. C., Hammond, M. E. H., Allison, K. H., Harvey, B. E., Mangu, P. B., Bartlett, J. M., et al. (2018). Human epidermal growth factor receptor 2 testing in breast cancer: American Society of Clinical Oncology/College of American Pathologists clinical practice guideline focused update. *Archives of Pathology & Laboratory Medicine*, *142*, 1364–1382.

Wrana, J. L., Attisano, L., Wieser, R., Ventura, F., & Massagué, J. (1994). Mechanism of activation of the TGF-β receptor. *Nature*, *370*, 341.

Wu, M.-Q., Hu, P., Gao, J., Wei, W.-D., Xiao, X.-S., Tang, H.-L., et al. (2013). Low expression of tyrosine-protein phosphatase nonreceptor type 12 is associated with lymph node metastasis and poor prognosis in operable triple-negative breast cancer. *Asian Pacific Journal of Cancer Prevention*, *14*, 287–292.

Xia, W., Husain, I., Liu, L., Bacus, S., Saini, S., Spohn, J., et al. (2007). Lapatinib antitumor activity is not dependent upon phosphatase and tensin homologue deleted on chromosome 10 in ErbB2-overexpressing breast cancers. *Cancer Research*, *67*, 1170–1175. https://doi.org/10.1158/0008-5472.CAN-06-2101.

Xu, L.-H., Yang, X., Bradham, C. A., Brenner, D. A., Baldwin, A. S., Craven, R. J., et al. (2000). The focal adhesion kinase suppresses transformation-associated, anchorage-independent apoptosis in human breast cancer cells involvement of death receptor-related signaling pathways. *The Journal of Biological Chemistry*, *275*, 30597–30604.

Xunyi, Y., Zhentao, Y., Dandan, J., & Funian, L. (2012). Clinicopathological significance of PTPN12 expression in human breast cancer. *Brazilian Journal of Medical and Biological Research*, *45*, 1334–1340.

Yang, H.-Y., Zhou, B. P., Hung, M.-C., & Lee, M.-H. (2000). Oncogenic signals of HER-2/neu in regulating the stability of the cyclin-dependent kinase inhibitor p27. *The Journal of Biological Chemistry*, *275*, 24735–24739.

Yao, E. S., Zhang, H., Chen, Y.-Y., Lee, B., Chew, K., Moore, D., et al. (2007). Increased β1 integrin is associated with decreased survival in invasive breast cancer. *Cancer Research*, *67*, 659–664.

Yu, Q., Geng, Y., & Sicinski, P. (2001). Specific protection against breast cancers by cyclin D1 ablation. *Nature*, *411*, 1017.

Zhang, D., Tai, L. K., Wong, L. L., Chiu, L.-L., Sethi, S. K., & Koay, E. S. (2005). Proteomic study reveals that proteins involved in metabolic and detoxification pathways are highly expressed in HER-2/neu-positive breast cancer. *Molecular & Cellular Proteomics*, *4*, 1686–1696.

CHAPTER SIX

Advances in insulin-like growth factor biology and -directed cancer therapeutics

Lynsey M. Fettig, Douglas Yee*
Masonic Cancer Center, University of Minnesota, Minneapolis, MN, United States
*Corresponding author: e-mail address: yeexx006@umn.edu

Contents

Abstract

The insulin and insulin-like growth factor (IGF) family of proteins are part of a complex network that regulates cell proliferation and survival. While this system is undoubtedly important in prenatal development and postnatal cell growth, members of this family have been implicated in several different cancer types. Increased circulating insulin and IGF ligands have been linked to increased risk of cancer incidence. This observation has led to targeting the IGF system as a therapeutic strategy in a number of cancers. This chapter aims to describe the well-characterized biology of the IGF1R system, outline the rationale for targeting this system in cancer, summarize the clinical data as it stands, and discuss where we can go from here.

Advances in Cancer Research, Volume 147
ISSN 0065-230X
https://doi.org/10.1016/bs.acr.2020.04.005
© 2020 Elsevier Inc.
All rights reserved.

1. Introduction

The insulin and insulin-like growth factor (IGF) family of proteins was first identified in the middle of the 20th century. Over the following decades, it was discovered how critical the IGF family is for normal development, cell metabolism, and cell survival. As our understanding of the expansive network of proteins and processes involved in and controlled by the IGF family expanded, so too did our understanding of how the downstream pathways could contribute to development and survival of cancer.

Epidemiological and promising preclinical data led to much enthusiasm for the targeting of IGF family members in several cancer types in the 1990s and early 2000s. Early phase clinical trials showed promise as well, however, larger phase III clinical trials failed to show clinical benefit and most therapies directed against the type I IGF receptor (IGF1R) were discontinued. Since then there has been much debate as to whether these trials were abandoned too early as data collected from these same trials has given us better insight into IGF1R biology.

This chapter will review normal IGF family protein biology and signaling, as well as examine the rationale for developing therapeutics against this system in cancer treatment, how they have been targeted in the clinic, and discuss where we go from here.

2. The insulin-like growth factor (IGF) family

2.1 Insulin and IGF ligands

The IGF family of proteins includes three ligands: IGF1, IGF2, and insulin. All three ligands are derived from unique genes on chromosomes 12 (*IGF1*) and 11 (*IGF2* and *INS*). Both IGF1 and IGF2 are primarily synthesized in the liver, however, they are also produced by several other tissues and can act in an endocrine, autocrine, or paracrine manner (Livingstone & Borai, 2014; Yakar et al., 1999). While hepatic IGF1 is produced in response to pituitary-derived growth hormone (GH) and changes with age, circulating levels of both IGF1 and IGF2 can be affected equally by hormones, genetic factors, and lifestyle factors (Frystyk, Skjaerbaek, Dinesen, & Orskov, 1994; Harrela et al., 1996; Miller, Schalch, & Draznin, 1981; Schwander, Hauri, Zapf, & Froesch, 1983). Insulin expression is regulated by many factors, however, glucose is the most important as it has been shown to play a role in insulin transcription, translation, and secretion from pancreatic β-cells to act in an endocrine fashion (Giddings, Chirgwin, & Permutt, 1982; Kulkarni et al., 2011; Layden, Durai, & Lowe, 2010; Poitout et al., 2006).

IGF1 and IGF2 are single chain polypeptides. After synthesis, they are processed in the endoplasmic reticulum where the E-peptide is cleaved, leaving the mature form containing A and B domains. Insulin is first synthesized as a single polypeptide (preproinsulin) in β-cells in the pancreas and is processed into proinsulin in the endoplasmic reticulum. However, unlike the IGF ligands, insulin is then processed into three peptides—the A and B chain peptides, which are covalently linked via two disulfide bonds and form the mature form of insulin, and the C-peptide. As indicated by their names, insulin and IGFs have substantial homology. After processing, IGF1 and IGF2 share about 70% homology in their amino acid sequence and have about 50% homology with the A and B chains in mature insulin (Bach et al., 1993). Further, the 3D structures of both IGF1 and IGF2 closely resemble that of proinsulin (Blundell, Bedarkar, Rinderknecht, & Humbel, 1978). Despite this close homology, the ligands have substantially different affinities for each of the receptors in the IGF family (discussed below).

2.2 IGF binding proteins

Though insulin is found only in a "free" state in circulation, only about 0.4% of IGF1 and 0.2% of IGF2 exists in an unbound state (Frystyk et al., 1994). Instead, IGFs in extracellular fluids are primarily found bound to six structurally related binding proteins, IGFBP-1 through IGFBP-6, which bind IGFs with equal or higher affinity to the IGF receptors, indicating that this binding is an important regulator of IGF signaling (Sitar, Popowicz, Siwanowicz, Huber, & Holak, 2006). In a free state, both insulin and IGFs have a half-life of about 10 min; however, when IGFs are bound to IGFBPs this time is increased to between 25 min and 15 h, depending on the IGFBP complex (Guler, Zapf, Schmid, & Froesch, 1989). IGFBPs share about 50% of their amino acid sequence and contain three distinct domains: an N-terminal domain, a cysteine-rich C-terminal domain, and a central linker domain. The N- and C-terminal domains are highly conserved among family members and facilitate IGF binding. The central linker domain is less conserved and is the site of the majority of the post-translational modifications found on IGFBPs (Baxter, 2000). All six members of the family have high affinity for IGF1 and IGF2, with most K_d values in the 10^{-10} to 10^{-11} M range (Baxter, 2000). IGFBPs are synthesized in a number of different cell types and their genes, which are spread across four chromosomes, are regulated independently.

The gene encoding IGFBP-1 is found on chromosome 7 and its expression is rapidly inhibited by insulin via suppression of transcription (Chahal et al., 2008). IGFBP-1 only binds a small fraction of circulating IGF1, however, this binding reduces bioavailability of IGF1 thereby inhibiting its activity (Brismar, Fernqvist-Forbes, Wahren, & Hall, 1994). Additionally, IGFBP-1 contains an arginine-glycine-aspartate (RGD) domain, which was found to interact with integrins, a family of cell adhesion receptors involved in cell-cell and cell-extracellular matrix interactions (Jones, Gockerman, Busby, Wright, & Clemmons, 1993). The *IGFBP2* gene is located on chromosome 2 and, similar to IGFBP-1, it contains an RGD motif (Jones, Doerr, & Clemmons, 1995). IGFBP-2 protein expression in human plasma inversely correlates with fasting insulin levels and body mass index (BMI), indicating it could be used as an early marker of insulin resistance (Blum et al., 1993; Ko et al., 2012). IGFBP-3, whose gene is also found on chromosome 2, is the most abundant family member. IGFBP-3 does not contain an RGD motif, however, it does contain a basic heparin-binding motif in its C-terminus that also allows interaction with the extracellular matrix (ECM) (Fowlkes, Thrailkill, George-Nascimento, Rosenberg, & Serra, 1997). While all IGFBPs can form binary complexes with IGFs, the majority of circulating IGFs (approximately 75%–80%) form a ternary complex with IGFBP-3 and acid-labile subunit (ALS) (Leong, Baxter, Camerato, Dai, & Wood, 1992). These 150-kDa complexes are highly stable with a half-life of 12–15 h, however, they are unable to cross capillary endothelium (Guler et al., 1989; Lewitt, Saunders, Phuyal, & Baxter, 1994). Similar to IGF1, IGFBP-3 is also produced in the liver in response to GH, likely to ensure there are sufficient levels to bind and stabilize hepatically-derived IGF1 (Blum et al., 1993). *IGFBP4* is found on chromosome 17, though not much is known about its regulation. Its protein product binds to IGF1 and IGF2 with similar affinities, however, unlike other IGFBPs, there is no evidence that IGFBP-4 is able to tether to cell surfaces as it lacks both RGD and heparin-binding motifs (Zhou, Diehl, Hoeflich, Lahm, & Wolf, 2003). *IGFBP5* is found on chromosome 2, and its expression is regulated by glucocorticoids. IGFBP-5 is the most evolutionarily conserved binding protein and, similar to IGFBP-3, it has a heparin-binding motif and can form ternary complexes with IGFs and ALS though at much lower frequency (Andress, 1995; Twigg & Baxter, 1998). IGFBP-6 has the highest affinity for IGF2 of all the IGFBPs and also contains a heparin-binding motif (Bach et al., 1993; Fowlkes et al., 1997). Its gene resides on chromosome 12 and not much is known about the regulation of its expression.

IGFs are released from IGFBPs through methods that reduce their mutual affinity such as by binding of IGFBPS to molecules in the ECM, phosphorylation of IGFBPs, or following proteolytic cleavage of the IGFBP central linker domain by one of several proteases including pregnancy-associated plasma protein A (PAPP-A) and prostate specific antigen (PSA) (reviewed in Bunn & Fowlkes, 2003). The resulting C-terminal and N-terminal domains have substantially lower affinity for IGFs compared to the IGF receptors, but these peptides are still able to partially inhibit IGF activity (Payet, Wang, Baxter, & Firth, 2003). IGFBPs are protected from proteolytic cleavage via post-translational modifications in their linker domain (Coverley, Martin, & Baxter, 2000; Gibson, Aplin, White, & Westwood, 2001; Neumann, Marinaro, & Bach, 1998). Though IGFBPs are typically thought of as inhibitory of IGF actions, their ability to form complexes with IGFs and accumulate in the extracellular space via interactions between the heparin binding or RGD domains and the ECM allows them to act as a pericellular reservoir of IGFs, which can be released via activation of IGFBP-targeting proteases. In this way, IGFBPs can, in fact, promote IGF action by stabilizing the ligands and allowing them to be released in close proximity to their receptors, through which they exert their cellular effects.

2.3 The IGF family of receptors

IGF1, IGF2, and insulin bind to and elicit their effects through the cell surface transmembrane receptors, IGF1R, IGF2R, and insulin receptor (IR) (Fig. 1). *IGF1R* is found on chromosome 15 and its expression is regulated by several growth factors including GH, thyroxine, platelet-derived growth factor, and fibroblast growth factor. The mature receptor is comprised of two dimerized hemi-receptors, each made of an extracellular α-subunit and a membrane spanning β-subunit (De Meyts et al., 1994). Each α- and β-subunit are encoded by a single gene and are synthesized as a single polypeptide, known as a preproreceptor, which is processed into a proreceptor in the endoplasmic reticulum. Additional processing includes addition of a disulfide bond between the α- and β-subunits, dimerization of two proreceptors via disulfide bonds in their α-subunits, and proteolytic cleavage in order to create a mature $\alpha_2\beta_2$ tetrameric receptor. IGF1R has

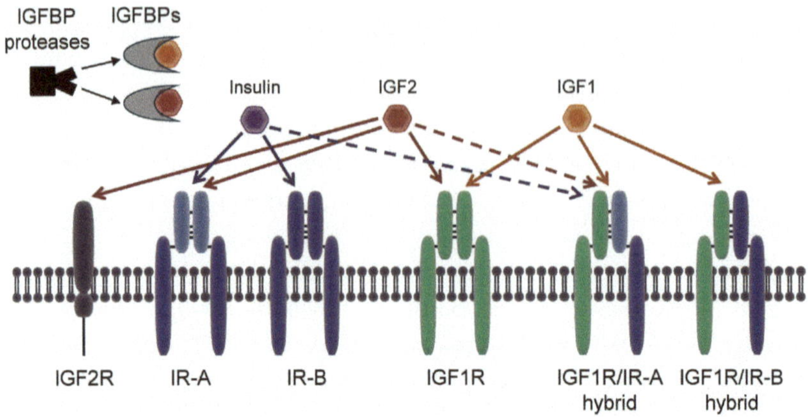

Fig. 1 Insulin and IGF system components. Components of the IGF system are discussed extensively in the text. Briefly, the IGF system is comprised of IR and IGF1R homo- and hetero-dimers which are stimulated via binding of their ligands, insulin, IGF1, and IGF2. Each ligand has differing affinities for each of the receptors—solid lines denote high affinity interactions while dashed lines denote low affinity interactions. IGF1 and IGF2 are both primarily found in circulation bound to IGF binding proteins (IGFBP). IGFs are released from IGFBPs upon proteolytic cleavage by an IGFBP protease. *Adapted from Sachdev, D. & Yee, D. (2007). Disrupting insulin-like growth factor signaling as a potential cancer therapy.* Molecular Cancer Therapeutics, *6(1), 1–12.*

two splice isoforms differing by three nucleotides that ultimately code for either an Arg or Thr-Gly in the extracellular portion of the β-subunit of the receptor (Condorelli, Bueno, & Smith, 1994; Tobin, Yee, Brunner, & Rotwein, 1990).

IGF2R, also called the mannose 6-phosphate receptor, is found on chromosome 6 in humans and is structurally and functionally unrelated to IGF1R and IR. Comprised of an N-terminal extracellular domain, a single membrane-spanning region, and a small cytoplasmic tail, IGF2R lacks the intracellular tyrosine kinase activity of the other two receptors in this family. It does, however, play a role in many IGF-independent biological processes including trafficking of lysosomal enzymes (Ghosh, Dahms, & Kornfeld, 2003). IGF2R expression is ubiquitous across human tissues and the receptor is also found in a soluble form in circulation (Lobel, Dahms, & Kornfeld, 1988; Oshima, Nolan, Kyle, Grubb, & Sly, 1988). In regard to IGF signaling, IGF2R predominantly acts as a decoy receptor for IGF2, as binding of IGF2 to IGF2R leads to its endocytosis and lysosomal degradation without the activation of signaling cascades associated with IGF1R and IR activation (Kornfeld, 1992).

InsR, the gene encoding IR, is located on chromosome 19. Similar to IGF1R, a mature IR is composed of two α-subunits and β-subunits, which are processed and dimerized in the endoplasmic reticulum (Knutson, 1991). Two splice variants of IR exist—full length IR-B and the short form IR-A, which lacks the 36 base pair exon 11 and is primarily expressed in fetal tissues (Frasca et al., 1999). The presence of the 12 amino acid domain encoded by exon 11 in the C-terminus of the α-subunit decreases insulin binding affinity by two to threefold, and reduces receptor internalization and downregulation (Yamaguchi et al., 1991).

The existence of these three unique receptors for IGF1, IGF2, and insulin was originally suggested in 1982, however, the relationship among these receptors has since been found to be rather complex (Sara et al., 1982). IGF1R and IR share a high degree of homology with 84% overlapping amino acid sequence in the tyrosine kinase domain, 45%–65% in the ligand-binding domain, 100% in the ATP binding domain, and over 50% in overall sequence. IGF1R and IR also share structural similarities. Both receptors have six domains in their extracellular regions—L1, CR, L2, Fn1, Fn2, and Fn3 domains—followed by a transmembrane region, an intracellular juxtamembrane region, a tyrosine kinase domain, and a C-terminal tail (De Meyts & Whittaker, 2002; Ullrich et al., 1985). Further, IGF1R and both IR variants can form functional hybrid receptors in all tissues in which the receptors are expressed (Bailyes et al., 1997; Entingh-Pearsall & Kahn, 2004; Moxham, Duronio, & Jacobs, 1989; Valensise et al., 1996). IGF2R, however, is not able to participate in these heterodimers.

Each of the insulin and IGF ligands has vastly different binding affinities for IGF1R, IR, and hybrid receptors (Fig. 1). IGF1 has greater affinity for IGF1R compared to the hybrid receptor, and almost no affinity for IR (Slaaby et al., 2006). Similarly, insulin binds to IR with far higher affinity than the hybrid receptor and has very low affinity for IGF1R (Soos, Field, & Siddle, 1993). IGF2 has equal affinity for IGF1R and IGF2R, and can also bind the IR-A/IGF1R hybrid receptor and IR-A, however, it has lower affinity for IR-B and IR-B/IGF1R hybrid receptors.

In the absence of ligand, IGF1R, IR, and hybrid receptors on the cell surface are held in an auto-inhibitory inverted "v" conformation via reciprocal interaction between the L2-Fn1 domain pairs from opposing α-subunits (Xu et al., 2018). In this state the intracellular components of the β-subunits are spatially separated preventing autophosphorylation. Binding of one molecule of ligand induces an asymmetrical conformational change that both reduces affinity for ligand in the second ligand binding pocket and induces a shift in the intracellular domain such that the kinase domains in the β-subunits come in close enough proximity to trans-phosphorylate tyrosine residues in the kinase domains, leading to intracellular signaling (Kavran et al., 2014; Pang & Shafer, 1984; Ward, Menting, & Lawrence, 2013).

2.4 Signaling

Binding of ligand to IGF1R and IR leads to autophosphorylation on tyrosine residues Y1131, Y1135, and Y1136 on IGF1R or Y1158, Y1162, and Y1163 on IR in their respective kinase domains, followed by phosphorylation of tyrosine residues in the juxtamembrane region (Fig. 2). These residues allow subsequent docking and phosphorylation of adaptor molecules such as insulin receptor substrate (IRS)-1, IRS-2, and SHC (Wei, Hubbard, Hendrickson, & Ellis, 1995). Additional residues allow for alternate adaptor interactions. The downstream consequences of activation of these adaptors have been well characterized. In normal tissues, IRS-1 and IRS-2 are ubiquitously expressed and are critical mediators of insulin and IGF-dependent mitogenesis and metabolic regulation (Bruning, Winnay, Cheatham, & Kahn, 1997; White, 2002; Yamauchi et al., 1996). It is important to note that different normal tissues use these signals in different ways. In the liver, activation of this pathway leads to activation of glycogen storage (Clemmons, 2004). In epithelial cells, however, phosphorylation of IRS-1 leads to activation of the PI3K/Akt/mTOR signaling cascade, ultimately promoting proliferation and pro-survival pathways (Jackson, White, & Yee, 1998). IRS-2 also leads to signaling through PI3K, however this signaling leads to alterations in cell adhesion and motility (Jackson, Zhang, Yoneda, & Yee, 2001). Activation of SHC promotes activation of proliferation via the Ras/Raf/MAPK signaling cascade. While these pathways downstream of IGF and insulin receptors are essential during both prenatal and postnatal growth, aberrant activation of these pathways has also been strongly associated with cancer development and prognosis (Pollak, 2012).

3. The IGF family in cancer

3.1 Epidemiological evidence

Members of the IGF family of proteins are widely expressed in normal and cancerous tissues (Badzio et al., 2010; Cox et al., 2009; Kim et al., 2012;

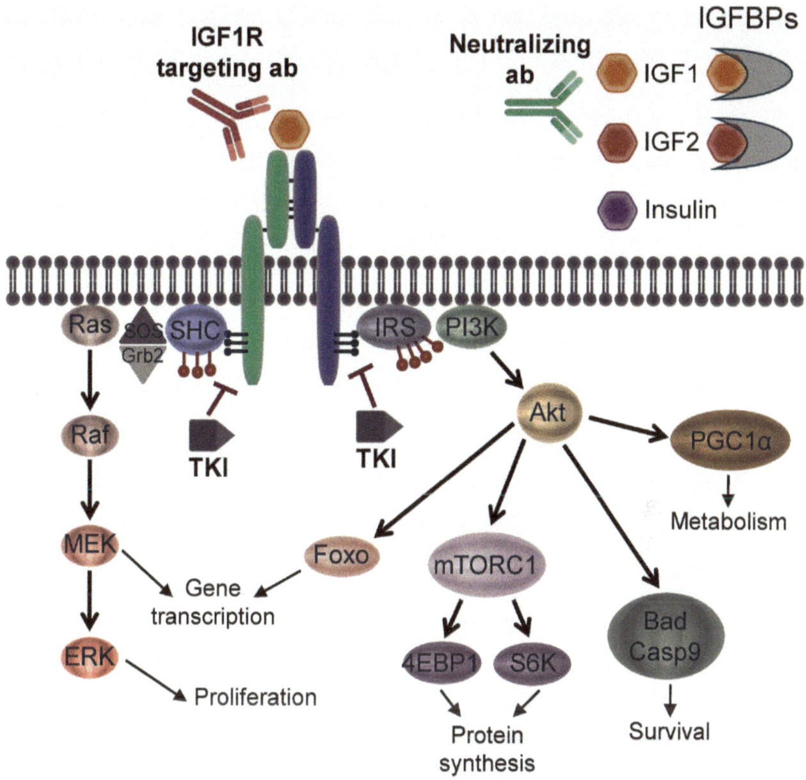

Fig. 2 Downstream pathways activated by insulin and IGF receptors. Receptor activation by insulin and IGF ligands primarily activates two major pathways—Akt and Ras/ERK. Actions of these downstream mediators induce numerous responses important in normal tissue and cancer including proliferation, survival, metabolic changes, protein synthesis, and gene transcription. IGF system-targeting methods include ligand-neutralizing antibodies, IGF1R targeting antibodies, and tyrosine kinase inhibitors (TKIs) which all target important nodes to disrupt the system. *Adapted from Sachdev, D. & Yee, D. (2007). Disrupting insulin-like growth factor signaling as a potential cancer therapy.* Molecular Cancer Therapeutics, *6(1), 1–12.*

Law et al., 2008). Several epidemiological studies have established a link between the IGF family of proteins and an increased risk of cancer incidence or worse prognosis. In normal development, IGF1 produced in response to GH released during puberty is responsible for skeletal development and height. Analysis of data from the Million Women Study found that increased height was significantly associated with an increased risk of cancer incidence in 10 of the 17 cancers assessed (Green et al., 2011). Conversely, patients that have IGF1 deficiency have a significantly lower incidence of cancer (Guevara-Aguirre et al., 2011; Shevah & Laron, 2007).

More directly, data from the Nurses' Health Study cohort showed that, in premenopausal women, there was a very strong positive correlation between circulating IGF1 levels and breast cancer risk (Hankinson et al., 1998). This finding has been further confirmed in breast cancer, as well as extended to other cancer types (Shi, Yu, McLarty, & Glass, 2004). High plasma levels of IGF1 were also associated with an increased risk of lung cancer (OR = 2.06, 95% confidence interval = 1.19–3.56), and methylation of the IGFBP-3 promoter was associated with decreased survival (Chang et al., 2002; Yu et al., 1999). Even more strikingly, the Physicians' Health Study found that men in the highest quartile of plasma IGF1 levels had a risk of developing prostate cancer that was 4.3 times higher than men in the lowest quartile, independent of PSA levels (Chan et al., 1998). Subsequent studies failed to replicate these results, as did studies looking at the correlation between IGFBP-3 levels and prostate cancer development (Cutting et al., 1999; Koistinen et al., 2002). In colorectal cancer, men and women with the highest levels of circulating IGF1 had a significantly greater risk of cancer development compared to the lowest levels and, similar to lung cancer, higher expression of IGFBP-3 was associated with a reduced relative risk (Giovannucci et al., 2000; Ma et al., 1999).

It is important to note that other studies have provided conflicting results to this relationship between IGFBP-3 and lung cancer risk, likely due to the complex relationship between IGFBPs and receptor activation (Hankinson et al., 1998; Harman et al., 2000; Shi et al., 2004). Another consideration when evaluating IGFBPs is the availability of the proteases that allow release of the IGF ligands from their binding proteins. Circulating PAPP-A levels, for example, are significantly higher in lung cancer patients compared to healthy subjects (Bulut et al., 2009). Further, another study showed that, in advanced lung cancer, pleural fluid had a 47-fold increased in PAPP-A compared to serum levels (Espelund et al., 2017). PAPP-A was also shown to be overexpressed in ovarian cancer, Ewing sarcoma, and several other cancers (Alexiadis, Mamers, Chu, & Fuller, 2006; Guo, Bao, Guo, & Yang, 2018; Kalli et al., 2004; Kirschner et al., 2017). In breast cancer, not only is PAPP-A overexpressed, particularly in the luminal B subtype, but is also an independent predictor of early recurrence for patients with stage 1 and 2 disease (Kuhajda, Abeloff, & Eggleston, 1985; Kuhajda & Eggleston, 1985; Mansfield et al., 2014). Similarly, elevated PSA levels strongly correlate with prostate cancer development, despite the lack of consistent association between prostate cancer and IGF1 or IGFBP-3 levels (Lojanapiwat, Anutrakulchai, Chongruksut, & Udomphot, 2014). Together these data indicate that, in some instances, circulating levels of IGFs or their binding proteins may be less important than the ability to release ligand in proximity to the tumor.

Additional studies looking at the effects of high serum insulin levels have uncovered similar results to elevated IGF1. A meta-analysis of several population studies that found that hyperinsulinemia and high C-peptide levels are correlated with increased risk of breast, colorectal, and pancreatic cancers (Giovannucci, 2001; Pollak, 2012). Hyperinsulinemia often occurs in the context of diabetes, obesity, and other metabolic syndrome, and these conditions all promote several additional factors that have been shown to support tumor growth including hyperglycemia, hyperlipidemia, chronic inflammation, and expression of several growth factors and cytokines (Belardi, Gallagher, Novosyadlyy, & LeRoith, 2013). While it is difficult to tease apart the contributions of each of these factors, evidence suggests that the IGF system independently supports cancer initiation and development (reviewed in Stone, McPherson, & Gail Darlington, 2018).

In addition to circulating ligands and binding proteins, increased receptor expression has also been found to correlate with prognosis. While the prognostic value of IGF1R in lung cancer has been controversial, a recent meta-analysis of 22 studies found that high expression of IGF1R in fact does predict shorter disease-free survival in non-small cell lung cancer (NSCLC) patients, however, is does not have predictive value for overall survival in NSCLC, or disease free survival or overall survival in small cell lung cancer (SCLC) (Xu et al., 2019). In prostate cancer samples, high staining for IGF1R, but not IR, was associated with increased proliferation, decreased apoptosis, and an increased risk of lethal disease (Ahearn et al., 2018; Zu et al., 2013). Higher levels of tumor IGF1R expression have also been positively correlated with worse disease outcome in gastric cancer and renal cell carcinoma (Matsubara et al., 2008; Sichani et al., 2010). In breast cancer, reports on IGF1R as a prognostic factor have been mixed. Multiple studies found that IGF1R expression correlates with good prognostic markers in patients with early breast cancer and with the luminal B subtype of breast cancer, however, its expression was associated with worse outcome in patients with the HER2-enriched subtype and shorter disease free survival in patients with triple negative disease (Fu et al., 2011; Hartog et al., 2011; Yerushalmi et al., 2012).

Based on the cumulative weight of these epidemiological studies, several methods for targeting the IGF system have been developed for use in preclinical and clinical settings. These include antibodies to neutralize IGF1 and IGF2, antibodies targeting IGF1R, and tyrosine kinase inhibitors (TKI) to prevent activation of downstream targets.

3.2 Targeting the IGF family

3.2.1 Ligand neutralization

Unlike insulin, many cell types express IGF ligands and, in fact, autocrine production by transformed cells is common. The pro-proliferative properties of IGF1 and IGF2 in mammalian epithelial cells were first identified in 1982, and not long after their effects in driving proliferation in other tissue and cancers types followed (Reddan & Dziedzic, 1982). Elevated IGF2 levels were shown to increase the number of colorectal lesions in Apc^{Min} mice and nearly 70% of transgenic mice overexpressing IGF2 developed lung adenocarcinomas by 18 months of age (Hassan & Howell, 2000; Moorehead, Sanchez, Baldwin, & Khokha, 2003). Further, the structure of IGF1R makes it difficult to activate kinase signaling pathways without ligand binding to the receptor (Kaleko, Rutter, & Miller, 1990).

These and other preclinical observations led to the development of several antibodies to neutralize IGF ligands (Dransfield et al., 2010; Feng et al., 2006; Gao et al., 2011; Zhao, Feng, Zhu, & Dimitrov, 2011). One such antibody, MEDI-573 (dusigitumab), has higher affinity for IGF2 over IGF1, but has no affinity for insulin, and was shown to suppress growth of mouse embryonic fibroblast cell line xenografts engineered to overexpress either human IGF2 and IGF1R, or human IGF1 and IGF1R (Gao et al., 2011). Further, MEDI-573 was shown to inhibit sarcoma cell line growth in vitro and in xenograft models (Zhong et al., 2014). With these promising data, MEDI-573 was taken into clinical trials for breast cancer and hepatocellular carcinoma (HCC). The breast cancer study looked at the efficacy of MEDI-573 plus aromatase inhibitor (AI) versus AI alone in women with metastatic hormone-sensitive breast cancer. Unfortunately no clinical benefit was seen in this patient population (NCT01446159). Phase II of the HCC study was not launched at the discretion of the sponsor (NCT01498952). A second IGF neutralizing antibody, xentuzumab (BI 836845), has higher affinity for IGF1 over IGF2 and also showed great promise in preclinical models (Friedbichler et al., 2014). Xentuzumab is also being tested in a phase I study in patients with NSCLC in combination with the EGFR inhibitor afatinib (completed not published, NCT02191891) and in a phase Ib/II study in patients with metastatic or locally advanced ER+ breast cancer in combination with a abemaciclib, a CDK4/6 inhibitor, and endocrine therapy (NCT03099174). An additional study combining xentuzumab with an AI (exemestane) and the mTOR inhibitor everolimus in metastatic breast cancer to bone is ongoing (NCT03659136).

3.2.2 IGF1R targeted monoclonal antibodies

Much preclinical data has supported the targeting of IGF1R. One striking example comes from a study where mice engineered to express constitutively active IGF1R developed salivary and mammary adenocarcinomas as early as 8 weeks of age (Carboni et al., 2005). The success of monoclonal antibodies targeting other receptors, such as trastuzumab for HER2 expressing cancers, led to their use becoming one of the dominant strategies for targeting the IGF system in clinical trials. These IGF1R-directed monoclonal antibodies most often elicit their effects by binding the extracellular region of the receptor and causing its internalization and degradation (Cohen et al., 2005; Sachdev, Singh, Fujita-Yamaguchi, & Yee, 2006). While these antibodies do not cross-react with IR, they can bind and inhibit the activity of hybrid receptors (Pandini et al., 2007).

As early as 1989, antibodies targeting IGF1R in breast cancer showed promise in preclinical studies (Arteaga et al., 1989). The monoclonal antibody AVE1642 inhibited breast cancer xenograft growth in preclinical studies, and even exhibited anti-tumor activity in solid tumors in phase I trials (Sachdev, Zhang, Matise, Gaillard-Kelly, & Yee, 2010; Zeng, Sachdev, Zhang, Gaillard-Kelly, & Yee, 2009). AVE1642 treatment led to partial response or stable disease in 25 of 57 patients in one study and 11 of 27 patients in another (Macaulay et al., 2013; Soria et al., 2013). Unfortunately AVE1642 and additional IGF1R antibodies teprotumumab (R1507) and robatumumab all failed in phase II trials in solids tumors (Lin et al., 2014; Pappo et al., 2014; Ramalingam et al., 2011). Phase I clinical trial testing of dalotuzumab as a single agent in solid tumors showed that some patients saw clinical benefit, while in a separate trial in breast cancer, the combination of dalotuzumab plus the mTOR inhibitor ridaforolimus showed partial response or stable disease in 6 of 11 patients with ER+ highly proliferative disease (Atzori et al., 2011; Di Cosimo et al., 2015). Similar to other phase II trials with this type of agent, however, a phase II trial testing blockade of the same pathway components using cixutumumab and the mTOR inhibitor temsirolimus found no efficacy with this combination in patients who had received multiple prior lines of chemotherapy (Ma et al., 2013).

Use of monoclonal antibodies in NSCLC have resulted in equally disappointing results. One phase II study using cisplatin and pemetrexed with or without cixutumumab as first-line therapy in stage IV nonsquamous NSCLC found that patients derived no clinical benefit from addition of cixutumumab but did experience increased incidence of hyperglycemia

(Novello et al., 2017). Further, despite promising phase II trial data, a phase III trial testing the IGF1R-directed monoclonal antibody figitumumab in combination with paclitaxel and carboplatin versus paclitaxel and carboplatin alone was closed early due to lack of clinical benefit and increased rates hyperglycemia and serious adverse events (Goto et al., 2012; Langer et al., 2014). It is important to note, however, that a small subset of patients did respond to treatment and response was associated with higher baseline levels of circulating IGF1 (Goto et al., 2012).

Clinical trials targeting IGF1R in other cancers have shown some promise. In metastatic pancreatic cancer patients, use of the IGF1R-directed antibody ganitumab plus gemcitabine showed no effect on overall survival, however, data from this trial suggest that a subset of patients may derive benefit; further a later trial showed that there may in fact be improved overall survival with this treatment combination (Fuchs et al., 2015; Kindler et al., 2012). In a phase I study in patients with sarcoma or Ewing's sarcoma, almost half of patients had a reduction in tumor volume over baseline, with one patient showing a complete response, one showing a partial response, and eight patients having prolonged stable disease (Olmos et al., 2010). In a separate phase I study in Ewing's sarcoma, cixutumumab combined with an mTOR inhibitor led to tumor regression in 29% (5/17) patients (Naing et al., 2012). Further, a phase II study in refractory Ewing sarcoma showed 10% of patients had a complete or partial response to another IGF1R targeting antibody, teprotumumab (Pappo et al., 2011). A phase III trial combining ganitumab with chemotherapy in metastatic Ewing sarcoma has completed enrollment (NCT02306161).

3.2.3 Tyrosine kinase inhibitors

In contrast to the IGF1R-directed antibodies, the small molecule tyrosine kinase inhibitors (TKI) block downstream IGF1R signaling by competing for the ATP-binding site in the catalytic domain of the receptor, and most often elicit their effect without causing receptor internalization (Ji et al., 2007). Due to the close homology between IGF1R and IR, currently available TKIs not only bind and inhibit IGF1R, but IR as well. Unfortunately, because of this cross-reactivity, some in vivo studies have shown this method of targeting the IGF system leads to insulin resistance and hyperglycemia (King, Aleksic, Haluska, & Macaulay, 2014; Pollak, 2008). Additionally, off target effects on other kinases have been seen with use of some of these TKIs, which may influence their therapeutic potential (Carboni et al., 2009).

Despite this, the efficacy of TKIs has been shown in preclinical studies in multiple cancer types. In breast, endometrial, and neuroblastoma, NVP-AEW541 showed antiproliferative effects in vitro and in vivo (Attias-Geva, Bentov, Fishman, Werner, & Bruchim, 2011; Fagan, Uselman, Sachdev, & Yee, 2012; Garcia-Echeverria et al., 2004; Tanno et al., 2006). Studies in pancreatic cancer cell lines showed that BMS-754807 enhanced gemcitabine activity and reduced xenograft tumor growth in a dose dependent manner by reducing activation of IGF1R and downstream mediators, and inducing apoptosis (Awasthi, Zhang, Ruan, Schwarz, & Schwarz, 2012). Similar effects on signaling and cell proliferation in vitro, plus efficacy as a single agent and in combination with other treatments on xenograft growth were found in several cancer types (Carboni et al., 2009; Franks, Jones, Briah, Murray, & Moorehead, 2016).

Clinical trials using these agents, however, have had less successful results. A study looking at the efficacy of letrozole in combination with BMS-754807 was discontinued early by the sponsor without reaching the primary endpoint of progression free survival at 24 weeks (NCT01225172). In a phase I dose escalation study using linsitinib (OSI-906) in combination with EGFR inhibitor erlotinib in patients with advanced solid tumors, disease control occurred in 51% (38/75) patients (Macaulay et al., 2016). Unfortunately in a phase III trial in patients with adrenocortical carcinoma, linsitinib showed no benefit over placebo (Fassnacht et al., 2015). It is important to note, however, that 3% of patients had a partial response indicating that there may be a subpopulation of patients that may derive benefit. Unfortunately in advanced NSCLC, patients treated with AXL1717 showed no difference in progression free survival compared to the docetaxel treated group in a phase II trial (Bergqvist et al., 2017).

4. Lessons learned and where we go from here

Epidemiological and preclinical data provide a strong rationale for targeting the IGF system as it has cumulatively shown that every aspect of the IGF system can be altered in cancer. Despite this potential, clinical trial data has been less compelling. Negative phase II and phase III trial data have led some to believe that there is sufficient evidence to halt the investigation of targeting the IGF system in most cancers. However, several studies did show that certain populations of patients derived clinical benefit from treatment, and IGF system-targeting therapies were well tolerated.

Before completely discounting the targeting of this system, it may be worth asking ourselves who benefitted, who did not, and why.

4.1 Biomarkers

Due to the desire for rapid clinical development of new therapeutic strategies, most of the discussed clinical trials were initiated in unselected patient populations as no predictive biomarkers had been identified. In the most simplistic approach, one might hypothesize that IGF1R expression may be sufficient to predict response to these therapies. Unfortunately, while low levels of IGF1R were shown in preclinical studies to be a strong negative predictor of response, IGF1R expression itself is necessary but not sufficient to accurately predict sensitivity (Lee & Yee, 2011; Zha et al., 2009). In some trials, patients were given IGF1R-targeting treatment regardless of the levels of receptor expression in their tumors, indicating that some of the negative results could be due to a lack of target expression.

Data from initial negative trials have been used to identify candidate predictive biomarkers. For example, circulating ligand levels may prove to be useful in stratifying patients who may benefit. In patients with metastatic pancreatic cancer, high baseline levels of IGF1 or IGF2 identified patients who were likely respond to IGF1R targeting therapies, as were low levels of IGFBP-2 or high IGFBP-3 (McCaffery et al., 2013). Pretreatment circulating levels of IGF1 have also shown some promise as a potential marker in NSCLC, however, some of these studies have been retracted due to a lack of reproducibility (Goto et al., 2012; Gualberto et al., 2011). In vitro, Huang et al. examined gene signatures in sarcoma and neuroblastoma cell lines sensitive and resistant to the TKI BMS-536924 (Huang et al., 2009). Interestingly, high IGF1 and IGF2 expression levels were associated with sensitivity, as was IGF1R expression, and IGFBP-3 and IGFBP-6 were both highly expressed in the resistant lines. Similar results were found in colorectal cancer explants, whereby increased levels of IGF2 mRNA associated with better response to the IGF1R-targeting antibody h10H5 (Zha et al., 2009). It has been proposed that high levels of circulating ligands might have predictive value, as tumors that arise in these patients may be pushed to become dependent on these signals and therefore may be more likely to respond to their blockade. Measuring expression profiles to identify IGF driven breast cancers has also been shown in preclinical models (Litzenburger et al., 2011).

Other unexpected candidate predictive biomarkers under investigation are transforming fusion proteins. Preclinical studies in breast cancer show that activation of the IGF system is a requirement for transformation in cells harboring an ETV6-NTRK3 fusion protein, a translocation that is found in virtually all secretory breast cancers (Tognon et al., 2011). Further, in Ewing's sarcoma a common translocation, EWS-FLI1, leads to increased IGF1 expression and requires expression of IGF1R in order to be transforming (Toretsky, Kalebic, Blakesley, LeRoith, & Helman, 1997).

Additional preclinical studies using patient-derived xenografts looking at gene expression and mutational profiles found that KRAS mutational status may also be predictive of response, and highlighted the concept that rational combination therapies could be another useful approach to increase the efficacy of IGF system targeting agents (Pitts et al., 2010).

4.2 Combination therapies

Results from clinical trials suggest that targeting the IGF system alone is likely not sufficient for sustained tumor inhibition. Several targeted therapies have been used successfully in combination with other treatments, and in fact dual inhibition of mTOR and IGF1R in Ewing's sarcoma patients has shown great promise (Naing et al., 2012). There have been several other studies using IGF family inhibitors in combination with other agents, however, many of these combinations have not been selected by examining preclinical evidence. However, preclinical data exists that may help in identifying clinically useful combinations.

In vitro studies have pointed to overexpression of EGFR and its ligands as a possible mechanism of resistance to IGF family-targeted therapies, indicating that, for certain cancers, dual targeting may be beneficial (Huang et al., 2009). Additional data suggest that, in melanoma, addition of BRAF inhibitors may work cooperatively with IGF1R inhibition to induce apoptosis (Villanueva et al., 2010; Yeh, Bohula, & Macaulay, 2006). Beyond Ewing sarcoma, other cancers may also benefit from dual inhibition of mTOR and the IGF system (Friedbichler et al., 2014).

Another important factor to consider when discussing combination therapies is the timing and order of treatment administration. In breast cancer trials, for example, almost all women had previously been treated with endocrine therapy before receiving IGF1R-targeting antibodies. Preclinical data has shown that endocrine therapy resistance is associated with loss of IGF1R expression; this trend was also seen in patients, indicating that from the start these trials were set to fail (Drury et al., 2011; Fagan et al., 2012).

4.3 IR compensation

Some evidence suggests that the failure of IGF1R targeting therapies has been, at least in part, due to continued activation of downstream pathways through IR. Preclinical studies have shown that targeting of IGF1R can in fact lead to an increase in IR expression and may confer inherent resistance to IGF1R-targeting therapies (Ulanet, Ludwig, Kahn, & Hanahan, 2010). Early trials using the IGF1R targeting antibody figitumumab showed that patients taking the drug had increased circulating insulin, indicating this may be a potential mechanism for resistance (Haluska et al., 2007). This observation also led to the concern that tumors in patients with hyperinsulinemia may have an intrinsic resistance to IGF1R targeted therapies due to activation of IR. Similar trends have been seen in vitro and in vivo, as hyperinsulinemic mice have increased xenograft growth compared to their normal counterparts; however blocking this increased insulin signaling using BMS-536924 reduced tumor burden (Novosyadlyy et al., 2010).

While targeting of IR was previously avoided due to the potential for detrimental metabolic side effects, there has recently been renewed interest. Optimism has been driven by data showing that small molecule inhibitors of IGF1R/IR were tolerated far better than expected, possibly due to their differential distribution to insulin target tissues (Dool et al., 2011; Fassnacht et al., 2015). This suggests that targeting IR may be possible without inducing severe hypoglycemia. Further, upregulation of the IR-A isoform in several cancer types may provide a better, more specific cancer target. This method is also appealing as IGF2 has equal affinity for IR-A as insulin. However, isoform specific modulators have yet to be developed.

5. Conclusions

It is clear that single agent targeting of the IGF system does not work. This does not rule out the possible use of these agents for all indications. Thoughtful investigation into combination therapies and biomarker identification are needed in order to illuminate where these therapies will be useful in the clinic.

Acknowledgment

This work was supported by grants from the NIH T32 CA009138 (L.M.F.), 5P30 CA077598 (D.Y.) and P50 CA116201 (D.Y.).

References

Ahearn, T. U., Peisch, S., Pettersson, A., Ebot, E. M., Zhou, C. K., Graff, R. E., et al. (2018). Expression of IGF/insulin receptor in prostate cancer tissue and progression to lethal disease. *Carcinogenesis*, *39*(12), 1431–1437.

Alexiadis, M., Mamers, P., Chu, S., & Fuller, P. J. (2006). Insulin-like growth factor, insulin-like growth factor-binding protein-4, and pregnancy-associated plasma protein-A gene expression in human granulosa cell tumors. *International Journal of Gynecological Cancer*, *16*(6), 1973–1979.

Andress, D. L. (1995). Heparin modulates the binding of insulin-like growth factor (IGF) binding protein-5 to a membrane protein in osteoblastic cells. *The Journal of Biological Chemistry*, *270*(47), 28289–28296.

Arteaga, C. L., Kitten, L. J., Coronado, E. B., Jacobs, S., Kull, F. C., Jr., Allred, D. C., et al. (1989). Blockade of the type I somatomedin receptor inhibits growth of human breast cancer cells in athymic mice. *The Journal of Clinical Investigation*, *84*(5), 1418–1423.

Attias-Geva, Z., Bentov, I., Fishman, A., Werner, H., & Bruchim, I. (2011). Insulin-like growth factor-I receptor inhibition by specific tyrosine kinase inhibitor NVP-AEW541 in endometrioid and serous papillary endometrial cancer cell lines. *Gynecologic Oncology*, *121*(2), 383–389.

Atzori, F., Tabernero, J., Cervantes, A., Prudkin, L., Andreu, J., Rodriguez-Braun, E., et al. (2011). A phase I pharmacokinetic and pharmacodynamic study of dalotuzumab (MK-0646), an anti-insulin-like growth factor-1 receptor monoclonal antibody, in patients with advanced solid tumors. *Clinical Cancer Research*, *17*(19), 6304–6312.

Awasthi, N., Zhang, C., Ruan, W., Schwarz, M. A., & Schwarz, R. E. (2012). BMS-754807, a small-molecule inhibitor of insulin-like growth factor-1 receptor/insulin receptor, enhances gemcitabine response in pancreatic cancer. *Molecular Cancer Therapeutics*, *11*(12), 2644–2653.

Bach, L. A., Hsieh, S., Sakano, K., Fujiwara, H., Perdue, J. F., & Rechler, M. M. (1993). Binding of mutants of human insulin-like growth factor II to insulin-like growth factor binding proteins 1-6. *The Journal of Biological Chemistry*, *268*(13), 9246–9254.

Badzio, A., Wynes, M. W., Dziadziuszko, R., Merrick, D. T., Pardo, M., Rzyman, W., et al. (2010). Increased insulin-like growth factor 1 receptor protein expression and gene copy number in small cell lung cancer. *Journal of Thoracic Oncology*, *5*(12), 1905–1911.

Bailyes, E. M., Nave, B. T., Soos, M. A., Orr, S. R., Hayward, A. C., & Siddle, K. (1997). Insulin receptor/IGF-I receptor hybrids are widely distributed in mammalian tissues: Quantification of individual receptor species by selective immunoprecipitation and immunoblotting. *The Biochemical Journal*, *327*(Pt. 1), 209–215.

Baxter, R. C. (2000). Insulin-like growth factor (IGF)-binding proteins: Interactions with IGFs and intrinsic bioactivities. *American Journal of Physiology Endocrinology and Metabolism*, *278*(6), E967–E976.

Belardi, V., Gallagher, E. J., Novosyadlyy, R., & LeRoith, D. (2013). Insulin and IGFs in obesity-related breast cancer. *Journal of Mammary Gland Biology and Neoplasia*, *18*(3–4), 277–289.

Bergqvist, M., Holgersson, G., Bondarenko, I., Grechanaya, E., Maximovich, A., Andor, G., et al. (2017). Phase II randomized study of the IGF-1R pathway modulator AXL1717 compared to docetaxel in patients with previously treated, locally advanced or metastatic non-small cell lung cancer. *Acta Oncologica*, *56*(3), 441–447.

Blum, W. F., Horn, N., Kratzsch, J., Jorgensen, J. O., Juul, A., Teale, D., et al. (1993). Clinical studies of IGFBP-2 by radioimmunoassay. *Growth Regulation*, *3*(1), 100–104.

Blundell, T. L., Bedarkar, S., Rinderknecht, E., & Humbel, R. E. (1978). Insulin-like growth factor: A model for tertiary structure accounting for immunoreactivity and receptor binding. *Proceedings of the National Academy of Sciences of the United States of America*, *75*(1), 180–184.

Brismar, K., Fernqvist-Forbes, E., Wahren, J., & Hall, K. (1994). Effect of insulin on the hepatic production of insulin-like growth factor-binding protein-1 (IGFBP-1), IGFBP-3, and IGF-I in insulin-dependent diabetes. *The Journal of Clinical Endocrinology and Metabolism*, *79*(3), 872–878.

Bruning, J. C., Winnay, J., Cheatham, B., & Kahn, C. R. (1997). Differential signaling by insulin receptor substrate 1 (IRS-1) and IRS-2 in IRS-1-deficient cells. *Molecular and Cellular Biology*, *17*(3), 1513–1521.

Bulut, I., Coskun, A., Ciftci, A., Cetinkaya, E., Altiay, G., Caglar, T., et al. (2009). Relationship between pregnancy-associated plasma protein-A and lung cancer. *The American Journal of the Medical Sciences*, *337*(4), 241–244.

Bunn, R. C., & Fowlkes, J. L. (2003). Insulin-like growth factor binding protein proteolysis. *Trends in Endocrinology and Metabolism*, *14*(4), 176–181.

Carboni, J. M., Lee, A. V., Hadsell, D. L., Rowley, B. R., Lee, F. Y., Bol, D. K., et al. (2005). Tumor development by transgenic expression of a constitutively active insulin-like growth factor I receptor. *Cancer Research*, *65*(9), 3781–3787.

Carboni, J. M., Wittman, M., Yang, Z., Lee, F., Greer, A., Hurlburt, W., et al. (2009). BMS-754807, a small molecule inhibitor of insulin-like growth factor-1R/IR. *Molecular Cancer Therapeutics*, *8*(12), 3341–3349.

Chahal, J., Chen, C. C., Rane, M. J., Moore, J. P., Barati, M. T., Song, Y., et al. (2008). Regulation of insulin-response element binding protein-1 in obesity and diabetes: Potential role in impaired insulin-induced gene transcription. *Endocrinology*, *149*(10), 4829–4836.

Chan, J. M., Stampfer, M. J., Giovannucci, E., Gann, P. H., Ma, J., Wilkinson, P., et al. (1998). Plasma insulin-like growth factor-I and prostate cancer risk: A prospective study. *Science*, *279*(5350), 563–566.

Chang, Y. S., Wang, L., Liu, D., Mao, L., Hong, W. K., Khuri, F. R., et al. (2002). Correlation between insulin-like growth factor-binding protein-3 promoter methylation and prognosis of patients with stage I non-small cell lung cancer. *Clinical Cancer Research*, *8*(12), 3669–3675.

Clemmons, D. R. (2004). Role of insulin-like growth factor in maintaining normal glucose homeostasis. *Hormone Research*, *62*(Suppl. 1), 77–82.

Cohen, B. D., Baker, D. A., Soderstrom, C., Tkalcevic, G., Rossi, A. M., Miller, P. E., et al. (2005). Combination therapy enhances the inhibition of tumor growth with the fully human anti-type 1 insulin-like growth factor receptor monoclonal antibody CP-751,871. *Clinical Cancer Research*, *11*(5), 2063–2073.

Condorelli, G., Bueno, R., & Smith, R. J. (1994). Two alternatively spliced forms of the human insulin-like growth factor I receptor have distinct biological activities and internalization kinetics. *The Journal of Biological Chemistry*, *269*(11), 8510–8516.

Coverley, J. A., Martin, J. L., & Baxter, R. C. (2000). The effect of phosphorylation by casein kinase 2 on the activity of insulin-like growth factor-binding protein-3. *Endocrinology*, *141*(2), 564–570.

Cox, M. E., Gleave, M. E., Zakikhani, M., Bell, R. H., Piura, E., Vickers, E., et al. (2009). Insulin receptor expression by human prostate cancers. *Prostate*, *69*(1), 33–40.

Cutting, C. W., Hunt, C., Nisbet, J. A., Bland, J. M., Dalgleish, A. G., & Kirby, R. S. (1999). Serum insulin-like growth factor-1 is not a useful marker of prostate cancer. *BJU International, 83*(9), 996–999.

De Meyts, P., Wallach, B., Christoffersen, C. T., Urso, B., Gronskov, K., Latus, L. J., et al. (1994). The insulin-like growth factor-I receptor. Structure, ligand-binding mechanism and signal transduction. *Hormone Research, 42*(4–5), 152–169.

De Meyts, P., & Whittaker, J. (2002). Structural biology of insulin and IGF1 receptors: Implications for drug design. *Nature Reviews Drug Discovery, 1*(10), 769–783.

Di Cosimo, S., Sathyanarayanan, S., Bendell, J. C., Cervantes, A., Stein, M. N., Brana, I., et al. (2015). Combination of the mTOR inhibitor ridaforolimus and the anti-IGF1R monoclonal antibody dalotuzumab: Preclinical characterization and phase I clinical trial. *Clinical Cancer Research, 21*(1), 49–59.

Dool, C. J., Mashhedi, H., Zakikhani, M., David, S., Zhao, Y., Birman, E., et al. (2011). IGF1/insulin receptor kinase inhibition by BMS-536924 is better tolerated than alloxan-induced hypoinsulinemia and more effective than metformin in the treatment of experimental insulin-responsive breast cancer. *Endocrine-Related Cancer, 18*(6), 699–709.

Dransfield, D. T., Cohen, E. H., Chang, Q., Sparrow, L. G., Bentley, J. D., Dolezal, O., et al. (2010). A human monoclonal antibody against insulin-like growth factor-II blocks the growth of human hepatocellular carcinoma cell lines in vitro and in vivo. *Molecular Cancer Therapeutics, 9*(6), 1809–1819.

Drury, S. C., Detre, S., Leary, A., Salter, J., Reis-Filho, J., Barbashina, V., et al. (2011). Changes in breast cancer biomarkers in the IGF1R/PI3K pathway in recurrent breast cancer after tamoxifen treatment. *Endocrine-Related Cancer, 18*(5), 565–577.

Entingh-Pearsall, A., & Kahn, C. R. (2004). Differential roles of the insulin and insulin-like growth factor-I (IGF-I) receptors in response to insulin and IGF-I. *The Journal of Biological Chemistry, 279*(36), 38016–38024.

Espelund, U. S., Bjerre, M., Hjortebjerg, R., Rasmussen, T. R., Lundby, A., Hoeflich, A., et al. (2017). Insulin-like growth factor bioactivity, stanniocalcin-2, pregnancy-associated plasma protein-A, and IGF-binding protein-4 in pleural fluid and serum from patients with pulmonary disease. *The Journal of Clinical Endocrinology and Metabolism, 102*(9), 3526–3534.

Fagan, D. H., Uselman, R. R., Sachdev, D., & Yee, D. (2012). Acquired resistance to tamoxifen is associated with loss of the type I insulin-like growth factor receptor: Implications for breast cancer treatment. *Cancer Research, 72*(13), 3372–3380.

Fassnacht, M., Berruti, A., Baudin, E., Demeure, M. J., Gilbert, J., Haak, H., et al. (2015). Linsitinib (OSI-906) versus placebo for patients with locally advanced or metastatic adrenocortical carcinoma: A double-blind, randomised, phase 3 study. *The Lancet Oncology, 16*(4), 426–435.

Feng, Y., Zhu, Z., Xiao, X., Choudhry, V., Barrett, J. C., & Dimitrov, D. S. (2006). Novel human monoclonal antibodies to insulin-like growth factor (IGF)-II that potently inhibit the IGF receptor type I signal transduction function. *Molecular Cancer Therapeutics, 5*(1), 114–120.

Fowlkes, J. L., Thrailkill, K. M., George-Nascimento, C., Rosenberg, C. K., & Serra, D. M. (1997). Heparin-binding, highly basic regions within the thyroglobulin type-1 repeat of insulin-like growth factor (IGF)-binding proteins (IGFBPs) -3, -5, and -6 inhibit IGFBP-4 degradation. *Endocrinology, 138*(6), 2280–2285.

Franks, S. E., Jones, R. A., Briah, R., Murray, P., & Moorehead, R. A. (2016). BMS-754807 is cytotoxic to non-small cell lung cancer cells and enhances the effects of platinum chemotherapeutics in the human lung cancer cell line A549. *BMC Research Notes, 9*, 134.

Frasca, F., Pandini, G., Scalia, P., Sciacca, L., Mineo, R., Costantino, A., et al. (1999). Insulin receptor isoform A, a newly recognized, high-affinity insulin-like growth factor II receptor in fetal and cancer cells. *Molecular and Cellular Biology*, *19*(5), 3278–3288.

Friedbichler, K., Hofmann, M. H., Kroez, M., Ostermann, E., Lamche, H. R., Koessl, C., et al. (2014). Pharmacodynamic and antineoplastic activity of BI 836845, a fully human IGF ligand-neutralizing antibody, and mechanistic rationale for combination with rapamycin. *Molecular Cancer Therapeutics*, *13*(2), 399–409.

Frystyk, J., Skjaerbaek, C., Dinesen, B., & Orskov, H. (1994). Free insulin-like growth factors (IGF-I and IGF-II) in human serum. *FEBS Letters*, *348*(2), 185–191.

Fu, P., Ibusuki, M., Yamamoto, Y., Hayashi, M., Murakami, K., Zheng, S., et al. (2011). Insulin-like growth factor-1 receptor gene expression is associated with survival in breast cancer: A comprehensive analysis of gene copy number, mRNA and protein expression. *Breast Cancer Research and Treatment*, *130*(1), 307–317.

Fuchs, C. S., Azevedo, S., Okusaka, T., Van Laethem, J. L., Lipton, L. R., Riess, H., et al. (2015). A phase 3 randomized, double-blind, placebo-controlled trial of ganitumab or placebo in combination with gemcitabine as first-line therapy for metastatic adenocarcinoma of the pancreas: The GAMMA trial. *Annals of Oncology*, *26*(5), 921–927.

Gao, J., Chesebrough, J. W., Cartlidge, S. A., Ricketts, S. A., Incognito, L., Veldman-Jones, M., et al. (2011). Dual IGF-I/II-neutralizing antibody MEDI-573 potently inhibits IGF signaling and tumor growth. *Cancer Research*, *71*(3), 1029–1040.

Garcia-Echeverria, C., Pearson, M. A., Marti, A., Meyer, T., Mestan, J., Zimmermann, J., et al. (2004). In vivo antitumor activity of NVP-AEW541-A novel, potent, and selective inhibitor of the IGF-IR kinase. *Cancer Cell*, *5*(3), 231–239.

Ghosh, P., Dahms, N. M., & Kornfeld, S. (2003). Mannose 6-phosphate receptors: New twists in the tale. *Nature Reviews Molecular Cell Biology*, *4*(3), 202–212.

Gibson, J. M., Aplin, J. D., White, A., & Westwood, M. (2001). Regulation of IGF bioavailability in pregnancy. *Molecular Human Reproduction*, 7(1), 79–87.

Giddings, S. J., Chirgwin, J., & Permutt, M. A. (1982). Effects of glucose on proinsulin messenger RNA in rats in vivo. *Diabetes*, *31*(7), 624–629.

Giovannucci, E. (2001). Insulin, insulin-like growth factors and colon cancer: A review of the evidence. *The Journal of Nutrition*, *131*(11 Suppl), 3109S–3120S.

Giovannucci, E., Pollak, M. N., Platz, E. A., Willett, W. C., Stampfer, M. J., Majeed, N., et al. (2000). A prospective study of plasma insulin-like growth factor-1 and binding protein-3 and risk of colorectal neoplasia in women. *Cancer Epidemiology, Biomarkers & Prevention*, *9*(4), 345–349.

Goto, Y., Sekine, I., Tanioka, M., Shibata, T., Tanai, C., Asahina, H., et al. (2012). Figitumumab combined with carboplatin and paclitaxel in treatment-naive Japanese patients with advanced non-small cell lung cancer. *Investigational New Drugs*, *30*(4), 1548–1556.

Green, J., Cairns, B. J., Casabonne, D., Wright, F. L., Reeves, G., Beral, V., et al. (2011). Height and cancer incidence in the Million Women Study: Prospective cohort, and meta-analysis of prospective studies of height and total cancer risk. *The Lancet Oncology*, *12*(8), 785–794.

Gualberto, A., Hixon, M. L., Karp, D. D., Li, D., Green, S., Dolled-Filhart, M., et al. (2011). Pre-treatment levels of circulating free IGF-1 identify NSCLC patients who derive clinical benefit from figitumumab. *British Journal of Cancer*, *104*(1), 68–74.

Guevara-Aguirre, J., Balasubramanian, P., Guevara-Aguirre, M., Wei, M., Madia, F., Cheng, C. W., et al. (2011). Growth hormone receptor deficiency is associated with a major reduction in pro-aging signaling, cancer, and diabetes in humans. *Science Translational Medicine*, *3*(70), 70ra13.

Guler, H. P., Zapf, J., Schmid, C., & Froesch, E. R. (1989). Insulin-like growth factors I and II in healthy man. Estimations of half-lives and production rates. *Acta Endocrinologica*, *121*(6), 753–758.

Guo, Y., Bao, Y., Guo, D., & Yang, W. (2018). Pregnancy-associated plasma protein a in cancer: Expression, oncogenic functions and regulation. *American Journal of Cancer Research*, *8*(6), 955–963.

Haluska, P., Shaw, H. M., Batzel, G. N., Yin, D., Molina, J. R., Molife, L. R., et al. (2007). Phase I dose escalation study of the anti insulin-like growth factor-I receptor monoclonal antibody CP-751,871 in patients with refractory solid tumors. *Clinical Cancer Research*, *13*(19), 5834–5840.

Hankinson, S. E., Willett, W. C., Colditz, G. A., Hunter, D. J., Michaud, D. S., Deroo, B., et al. (1998). Circulating concentrations of insulin-like growth factor-I and risk of breast cancer. *Lancet*, *351*(9113), 1393–1396.

Harman, S. M., Metter, E. J., Blackman, M. R., Landis, P. K., Carter, H. B., & Baltimore Longitudinal Study on Aging (2000). Serum levels of insulin-like growth factor I (IGF-I), IGF-II, IGF-binding protein-3, and prostate-specific antigen as predictors of clinical prostate cancer. *The Journal of Clinical Endocrinology and Metabolism*, *85*(11), 4258–4265.

Harrela, M., Koistinen, H., Kaprio, J., Lehtovirta, M., Tuomilehto, J., Eriksson, J., et al. (1996). Genetic and environmental components of interindividual variation in circulating levels of IGF-I, IGF-II, IGFBP-1, and IGFBP-3. *The Journal of Clinical Investigation*, *98*(11), 2612–2615.

Hartog, H., Horlings, H. M., van der Vegt, B., Kreike, B., Ajouaou, A., van de Vijver, M. J., et al. (2011). Divergent effects of insulin-like growth factor-1 receptor expression on prognosis of estrogen receptor positive versus triple negative invasive ductal breast carcinoma. *Breast Cancer Research and Treatment*, *129*(3), 725–736.

Hassan, A. B., & Howell, J. A. (2000). Insulin-like growth factor II supply modifies growth of intestinal adenoma in Apc(Min/+) mice. *Cancer Research*, *60*(4), 1070–1076.

Huang, F., Greer, A., Hurlburt, W., Han, X., Hafezi, R., Wittenberg, G. M., et al. (2009). The mechanisms of differential sensitivity to an insulin-like growth factor-1 receptor inhibitor (BMS-536924) and rationale for combining with EGFR/HER2 inhibitors. *Cancer Research*, *69*(1), 161–170.

Jackson, J. G., White, M. F., & Yee, D. (1998). Insulin receptor substrate-1 is the predominant signaling molecule activated by insulin-like growth factor-I, insulin, and interleukin-4 in estrogen receptor-positive human breast cancer cells. *The Journal of Biological Chemistry*, *273*(16), 9994–10003.

Jackson, J. G., Zhang, X., Yoneda, T., & Yee, D. (2001). Regulation of breast cancer cell motility by insulin receptor substrate-2 (IRS-2) in metastatic variants of human breast cancer cell lines. *Oncogene*, *20*(50), 7318–7325.

Ji, Q. S., Mulvihill, M. J., Rosenfeld-Franklin, M., Cooke, A., Feng, L., Mak, G., et al. (2007). A novel, potent, and selective insulin-like growth factor-I receptor kinase inhibitor blocks insulin-like growth factor-I receptor signaling in vitro and inhibits insulin-like growth factor-I receptor dependent tumor growth in vivo. *Molecular Cancer Therapeutics*, *6*(8), 2158–2167.

Jones, J. I., Doerr, M. E., & Clemmons, D. R. (1995). Cell migration: Interactions among integrins, IGFs and IGFBPs. *Progress in Growth Factor Research*, *6*(2–4), 319–327.

Jones, J. I., Gockerman, A., Busby, W. H., Jr., Wright, G., & Clemmons, D. R. (1993). Insulin-like growth factor binding protein 1 stimulates cell migration and binds to the alpha 5 beta 1 integrin by means of its Arg-Gly-Asp sequence. *Proceedings of the National Academy of Sciences of the United States of America*, *90*(22), 10553–10557.

Kaleko, M., Rutter, W. J., & Miller, A. D. (1990). Overexpression of the human insulin like growth factor I receptor promotes ligand-dependent neoplastic transformation. *Molecular and Cellular Biology*, *10*(2), 464–473.

Kalli, K. R., Chen, B. K., Bale, L. K., Gernand, E., Overgaard, M. T., Oxvig, C., et al. (2004). Pregnancy-associated plasma protein-A (PAPP-A) expression and insulin-like growth factor binding protein-4 protease activity in normal and malignant ovarian surface epithelial cells. *International Journal of Cancer*, *110*(5), 633–640.

Kavran, J. M., McCabe, J. M., Byrne, P. O., Connacher, M. K., Wang, Z., Ramek, A., et al. (2014). How IGF-1 activates its receptor. *eLife*, *3*, e03772.
Kim, J. S., Kim, E. S., Liu, D., Lee, J. J., Solis, L., Behrens, C., et al. (2012). Prognostic impact of insulin receptor expression on survival of patients with nonsmall cell lung cancer. *Cancer*, *118*(9), 2454–2465.
Kindler, H. L., Richards, D. A., Garbo, L. E., Garon, E. B., Stephenson, J. J., Jr., Rocha-Lima, C. M., et al. (2012). A randomized, placebo-controlled phase 2 study of ganitumab (AMG 479) or conatumumab (AMG 655) in combination with gemcitabine in patients with metastatic pancreatic cancer. *Annals of Oncology*, *23*(11), 2834–2842.
King, H., Aleksic, T., Haluska, P., & Macaulay, V. M. (2014). Can we unlock the potential of IGF-1R inhibition in cancer therapy? *Cancer Treatment Reviews*, *40*(9), 1096–1105.
Kirschner, A., Thiede, M., Grunewald, T. G., Alba Rubio, R., Richter, G. H., Kirchner, T., et al. (2017). Pappalysin-1 T cell receptor transgenic allo-restricted T cells kill Ewing sarcoma in vitro and in vivo. *Oncoimmunology*, *6*(2), e1273301.
Knutson, V. P. (1991). Cellular trafficking and processing of the insulin receptor. *The FASEB Journal*, *5*(8), 2130–2138.
Ko, J. M., Park, H. K., Yang, S., Kim, E. Y., Chung, S. C., & Hwang, I. T. (2012). Association between insulin-like growth factor binding protein-2 levels and cardiovascular risk factors in Korean children. *Endocrine Journal*, *59*(4), 335–343.
Koistinen, H., Paju, A., Koistinen, R., Finne, P., Lovgren, J., Wu, P., et al. (2002). Prostate-specific antigen and other prostate-derived proteases cleave IGFBP-3, but prostate cancer is not associated with proteolytically cleaved circulating IGFBP-3. *Prostate*, *50*(2), 112–118.
Kornfeld, S. (1992). Structure and function of the mannose 6-phosphate/insulin like growth factor II receptors. *Annual Review of Biochemistry*, *61*, 307–330.
Kuhajda, F. P., Abeloff, M. D., & Eggleston, J. C. (1985). Pregnancy-associated plasma protein A: A clinically significant predictor of early recurrence in stage II breast carcinoma. *Human Pathology*, *16*(3), 228–235.
Kuhajda, F. P., & Eggleston, J. C. (1985). Pregnancy-associated plasma protein A and extensive necrosis. Clinically significant predictors of early recurrence in stage I estrogen receptor-negative breast carcinoma. *Laboratory Investigation*, *53*(1), 101–107.
Kulkarni, S. D., Muralidharan, B., Panda, A. C., Bakthavachalu, B., Vindu, A., & Seshadri, V. (2011). Glucose-stimulated translation regulation of insulin by the 5' UTR-binding proteins. *The Journal of Biological Chemistry*, *286*(16), 14146–14156.
Langer, C. J., Novello, S., Park, K., Krzakowski, M., Karp, D. D., Mok, T., et al. (2014). Randomized, phase III trial of first-line figitumumab in combination with paclitaxel and carboplatin versus paclitaxel and carboplatin alone in patients with advanced non-small-cell lung cancer. *Journal of Clinical Oncology*, *32*(19), 2059–2066.
Law, J. H., Habibi, G., Hu, K., Masoudi, H., Wang, M. Y., Stratford, A. L., et al. (2008). Phosphorylated insulin-like growth factor-i/insulin receptor is present in all breast cancer subtypes and is related to poor survival. *Cancer Research*, *68*(24), 10238–10246.
Layden, B. T., Durai, V., & Lowe, W. L., Jr. (2010). G-protein-coupled receptors, pancreatic islets, and diabetes. *Nature Education*, *3*(9), 13.
Lee, A. V., & Yee, D. (2011). Targeting IGF-1R: At a crossroad. *Oncology (Williston Park)*, *25*(6), 535–536, Discussion 551.
Leong, S. R., Baxter, R. C., Camerato, T., Dai, J., & Wood, W. I. (1992). Structure and functional expression of the acid-labile subunit of the insulin-like growth factor-binding protein complex. *Molecular Endocrinology*, *6*(6), 870–876.
Lewitt, M. S., Saunders, H., Phuyal, J. L., & Baxter, R. C. (1994). Complex formation by human insulin-like growth factor-binding protein-3 and human acid-labile subunit in growth hormone-deficient rats. *Endocrinology*, *134*(6), 2404–2409.

Lin, E. H., Lenz, H. J., Saleh, M. N., Mackenzie, M. J., Knost, J. A., Pathiraja, K., et al. (2014). A randomized, phase II study of the anti-insulin-like growth factor receptor type 1 (IGF-1R) monoclonal antibody robatumumab (SCH 717454) in patients with advanced colorectal cancer. *Cancer Medicine*, *3*(4), 988–997.

Litzenburger, B. C., Creighton, C. J., Tsimelzon, A., Chan, B. T., Hilsenbeck, S. G., Wang, T., et al. (2011). High IGF-IR activity in triple-negative breast cancer cell lines and tumor grafts correlates with sensitivity to anti-IGF-IR therapy. *Clinical Cancer Research*, *17*(8), 2314–2327.

Livingstone, C., & Borai, A. (2014). Insulin-like growth factor-II: Its role in metabolic and endocrine disease. *Clinical Endocrinology*, *80*(6), 773–781.

Lobel, P., Dahms, N. M., & Kornfeld, S. (1988). Cloning and sequence analysis of the cation-independent mannose 6-phosphate receptor. *The Journal of Biological Chemistry*, *263*(5), 2563–2570.

Lojanapiwat, B., Anutrakulchai, W., Chongruksut, W., & Udomphot, C. (2014). Correlation and diagnostic performance of the prostate-specific antigen level with the diagnosis, aggressiveness, and bone metastasis of prostate cancer in clinical practice. *Prostate International*, *2*(3), 133–139.

Ma, J., Pollak, M. N., Giovannucci, E., Chan, J. M., Tao, Y., Hennekens, C. H., et al. (1999). Prospective study of colorectal cancer risk in men and plasma levels of insulin-like growth factor (IGF)-I and IGF-binding protein-3. *Journal of the National Cancer Institute*, *91*(7), 620–625.

Ma, C. X., Suman, V. J., Goetz, M., Haluska, P., Moynihan, T., Nanda, R., et al. (2013). A phase I trial of the IGF-1R antibody Cixutumumab in combination with temsirolimus in patients with metastatic breast cancer. *Breast Cancer Research and Treatment*, *139*(1), 145–153.

Macaulay, V. M., Middleton, M. R., Eckhardt, S. G., Rudin, C. M., Juergens, R. A., Gedrich, R., et al. (2016). Phase I dose-escalation study of linsitinib (OSI-906) and erlotinib in patients with advanced solid tumors. *Clinical Cancer Research*, *22*(12), 2897–2907.

Macaulay, V. M., Middleton, M. R., Protheroe, A. S., Tolcher, A., Dieras, V., Sessa, C., et al. (2013). Phase I study of humanized monoclonal antibody AVE1642 directed against the type 1 insulin-like growth factor receptor (IGF-1R), administered in combination with anticancer therapies to patients with advanced solid tumors. *Annals of Oncology*, *24*(3), 784–791.

Mansfield, A. S., Visscher, D. W., Hart, S. N., Wang, C., Goetz, M. P., Oxvig, C., et al. (2014). Pregnancy-associated plasma protein-A expression in human breast cancer. *Growth Hormone & IGF Research*, *24*(6), 264–267.

Matsubara, J., Yamada, Y., Hirashima, Y., Takahari, D., Okita, N. T., Kato, K., et al. (2008). Impact of insulin-like growth factor type 1 receptor, epidermal growth factor receptor, and HER2 expressions on outcomes of patients with gastric cancer. *Clinical Cancer Research*, *14*(10), 3022–3029.

McCaffery, I., Tudor, Y., Deng, H., Tang, R., Suzuki, S., Badola, S., et al. (2013). Putative predictive biomarkers of survival in patients with metastatic pancreatic adenocarcinoma treated with gemcitabine and ganitumab, an IGF1R inhibitor. *Clinical Cancer Research*, *19*(15), 4282–4289.

Miller, L. L., Schalch, D. S., & Draznin, B. (1981). Role of the liver in regulating somatomedin activity: Effects of streptozotocin diabetes and starvation on the synthesis and release of insulin-like growth factor and its carrier protein by the isolated perfused rat liver. *Endocrinology*, *108*(4), 1265–1271.

Moorehead, R. A., Sanchez, O. H., Baldwin, R. M., & Khokha, R. (2003). Transgenic overexpression of IGF-II induces spontaneous lung tumors: A model for human lung adenocarcinoma. *Oncogene*, *22*(6), 853–857.

Moxham, C. P., Duronio, V., & Jacobs, S. (1989). Insulin-like growth factor I receptor beta-subunit heterogeneity. Evidence for hybrid tetramers composed of insulin-like growth factor I and insulin receptor heterodimers. *The Journal of Biological Chemistry*, *264*(22), 13238–13244.

Naing, A., LoRusso, P., Fu, S., Hong, D. S., Anderson, P., Benjamin, R. S., et al. (2012). Insulin growth factor-receptor (IGF-1R) antibody cixutumumab combined with the mTOR inhibitor temsirolimus in patients with refractory Ewing's sarcoma family tumors. *Clinical Cancer Research*, *18*(9), 2625–2631.

Neumann, G. M., Marinaro, J. A., & Bach, L. A. (1998). Identification of O-glycosylation sites and partial characterization of carbohydrate structure and disulfide linkages of human insulin-like growth factor binding protein 6. *Biochemistry*, *37*(18), 6572–6585.

Novello, S., Scagliotti, G., de Castro, G., Jr., Kiyik, M., Kowalyszyn, R., Deppermann, K. M., et al. (2017). An open-label, multicenter, randomized, phase II study of cisplatin and pemetrexed with or without cixutumumab (IMC-A12) as a first-line therapy in patients with advanced nonsquamous non-small cell lung cancer. *Journal of Thoracic Oncology*, *12*(2), 383–389.

Novosyadlyy, R., Lann, D. E., Vijayakumar, A., Rowzee, A., Lazzarino, D. A., Fierz, Y., et al. (2010). Insulin-mediated acceleration of breast cancer development and progression in a nonobese model of type 2 diabetes. *Cancer Research*, *70*(2), 741–751.

Olmos, D., Postel-Vinay, S., Molife, L. R., Okuno, S. H., Schuetze, S. M., Paccagnella, M. L., et al. (2010). Safety, pharmacokinetics, and preliminary activity of the anti-IGF-1R antibody figitumumab (CP-751,871) in patients with sarcoma and Ewing's sarcoma: A phase 1 expansion cohort study. *The Lancet Oncology*, *11*(2), 129–135.

Oshima, A., Nolan, C. M., Kyle, J. W., Grubb, J. H., & Sly, W. S. (1988). The human cation-independent mannose 6-phosphate receptor. Cloning and sequence of the full-length cDNA and expression of functional receptor in COS cells. *The Journal of Biological Chemistry*, *263*(5), 2553–2562.

Pandini, G., Wurch, T., Akla, B., Corvaia, N., Belfiore, A., & Goetsch, L. (2007). Functional responses and in vivo anti-tumour activity of h7C10: A humanised monoclonal antibody with neutralising activity against the insulin-like growth factor-1 (IGF-1) receptor and insulin/IGF-1 hybrid receptors. *European Journal of Cancer*, *43*(8), 1318–1327.

Pang, D. T., & Shafer, J. A. (1984). Evidence that insulin receptor from human placenta has a high affinity for only one molecule of insulin. *The Journal of Biological Chemistry*, *259*(13), 8589–8596.

Pappo, A. S., Patel, S. R., Crowley, J., Reinke, D. K., Kuenkele, K. P., Chawla, S. P., et al. (2011). R1507, a monoclonal antibody to the insulin-like growth factor 1 receptor, in patients with recurrent or refractory Ewing sarcoma family of tumors: Results of a phase II Sarcoma Alliance for Research Through Collaboration Study. *Journal of Clinical Oncology*, *29*(34), 4541–4547.

Pappo, A. S., Vassal, G., Crowley, J. J., Bolejack, V., Hogendoorn, P. C., Chugh, R., et al. (2014). A phase 2 trial of R1507, a monoclonal antibody to the insulin-like growth factor-1 receptor (IGF-1R), in patients with recurrent or refractory rhabdomyosarcoma, osteosarcoma, synovial sarcoma, and other soft tissue sarcomas: Results of a Sarcoma Alliance for Research Through Collaboration Study. *Cancer*, *120*(16), 2448–2456.

Payet, L. D., Wang, X. H., Baxter, R. C., & Firth, S. M. (2003). Amino- and carboxyl-terminal fragments of insulin-like growth factor (IGF) binding protein-3 cooperate to bind IGFs with high affinity and inhibit IGF receptor interactions. *Endocrinology*, *144*(7), 2797–2806.

Pitts, T. M., Tan, A. C., Kulikowski, G. N., Tentler, J. J., Brown, A. M., Flanigan, S. A., et al. (2010). Development of an integrated genomic classifier for a novel agent in colorectal cancer: Approach to individualized therapy in early development. *Clinical Cancer Research*, *16*(12), 3193–3204.

Poitout, V., Hagman, D., Stein, R., Artner, I., Robertson, R. P., & Harmon, J. S. (2006). Regulation of the insulin gene by glucose and fatty acids. *The Journal of Nutrition, 136*(4), 873–876.

Pollak, M. (2008). Insulin and insulin-like growth factor signalling in neoplasia. *Nature Reviews Cancer, 8*(12), 915–928.

Pollak, M. (2012). The insulin and insulin-like growth factor receptor family in neoplasia: An update. *Nature Reviews Cancer, 12*(3), 159–169.

Ramalingam, S. S., Spigel, D. R., Chen, D., Steins, M. B., Engelman, J. A., Schneider, C. P., et al. (2011). Randomized phase II study of erlotinib in combination with placebo or R1507, a monoclonal antibody to insulin-like growth factor-1 receptor, for advanced-stage non-small-cell lung cancer. *Journal of Clinical Oncology, 29*(34), 4574–4580.

Reddan, J. R., & Dziedzic, D. C. (1982). Insulin-like growth factors, IGF-1, IGF-2 and somatomedin C trigger cell proliferation in mammalian epithelial cells cultured in a serum-free medium. *Experimental Cell Research, 142*(2), 293–300.

Sachdev, D., Singh, R., Fujita-Yamaguchi, Y., & Yee, D. (2006). Down-regulation of insulin receptor by antibodies against the type I insulin-like growth factor receptor: Implications for anti-insulin-like growth factor therapy in breast cancer. *Cancer Research, 66*(4), 2391–2402.

Sachdev, D., Zhang, X., Matise, I., Gaillard-Kelly, M., & Yee, D. (2010). The type I insulin-like growth factor receptor regulates cancer metastasis independently of primary tumor growth by promoting invasion and survival. *Oncogene, 29*(2), 251–262.

Sara, V. R., Hall, K., Von Holtz, H., Humbel, R., Sjogren, B., & Wetterberg, L. (1982). Evidence for the presence of specific receptors for insulin-like growth factors 1 (IGE-1) and 2 (IGF-2) and insulin throughout the adult human brain. *Neuroscience Letters, 34*(1), 39–44.

Schwander, J. C., Hauri, C., Zapf, J., & Froesch, E. R. (1983). Synthesis and secretion of insulin-like growth factor and its binding protein by the perfused rat liver: Dependence on growth hormone status. *Endocrinology, 113*(1), 297–305.

Shevah, O., & Laron, Z. (2007). Patients with congenital deficiency of IGF-I seem protected from the development of malignancies: A preliminary report. *Growth Hormone & IGF Research, 17*(1), 54–57.

Shi, R., Yu, H., McLarty, J., & Glass, J. (2004). IGF-I and breast cancer: A meta-analysis. *International Journal of Cancer, 111*(3), 418–423.

Sichani, M. M., Yazdi, F. S., Moghaddam, N. A., Chehrei, A., Kabiri, M., Naeimi, A., et al. (2010). Prognostic value of insulin- like growth factor-I receptor expression in renal cell carcinoma. *Saudi Journal of Kidney Diseases and Transplantation, 21*(1), 69–74.

Sitar, T., Popowicz, G. M., Siwanowicz, I., Huber, R., & Holak, T. A. (2006). Structural basis for the inhibition of insulin-like growth factors by insulin-like growth factor-binding proteins. *Proceedings of the National Academy of Sciences of the United States of America, 103*(35), 13028–13033.

Slaaby, R., Schaffer, L., Lautrup-Larsen, I., Andersen, A. S., Shaw, A. C., Mathiasen, I. S., et al. (2006). Hybrid receptors formed by insulin receptor (IR) and insulin-like growth factor I receptor (IGF-IR) have low insulin and high IGF-1 affinity irrespective of the IR splice variant. *The Journal of Biological Chemistry, 281*(36), 25869–25874.

Soos, M. A., Field, C. E., & Siddle, K. (1993). Purified hybrid insulin/insulin-like growth factor-I receptors bind insulin-like growth factor-I, but not insulin, with high affinity. *The Biochemical Journal, 290*(Pt. 2), 419–426.

Soria, J. C., Massard, C., Lazar, V., Ozoux, M. L., Mery-Mignard, D., Deslandes, A., et al. (2013). A dose finding, safety and pharmacokinetic study of AVE1642, an anti-insulin-like growth factor-1 receptor (IGF-1R/CD221) monoclonal antibody, administered as a single agent and in combination with docetaxel in patients with advanced solid tumours. *European Journal of Cancer, 49*(8), 1799–1807.

Stone, T. W., McPherson, M., & Gail Darlington, L. (2018). Obesity and cancer: Existing and new hypotheses for a causal connection. *eBioMedicine*, *30*, 14–28.

Tanno, B., Mancini, C., Vitali, R., Mancuso, M., McDowell, H. P., Dominici, C., et al. (2006). Down-regulation of insulin-like growth factor I receptor activity by NVP-AEW541 has an antitumor effect on neuroblastoma cells in vitro and in vivo. *Clinical Cancer Research*, *12*(22), 6772–6780.

Tobin, G., Yee, D., Brunner, N., & Rotwein, P. (1990). A novel human insulin-like growth factor I messenger RNA is expressed in normal and tumor cells. *Molecular Endocrinology*, *4*(12), 1914–1920.

Tognon, C. E., Somasiri, A. M., Evdokimova, V. E., Trigo, G., Uy, E. E., Melnyk, N., et al. (2011). ETV6-NTRK3-mediated breast epithelial cell transformation is blocked by targeting the IGF1R signaling pathway. *Cancer Research*, *71*(3), 1060–1070.

Toretsky, J. A., Kalebic, T., Blakesley, V., LeRoith, D., & Helman, L. J. (1997). The insulin-like growth factor-I receptor is required for EWS/FLI-1 transformation of fibroblasts. *The Journal of Biological Chemistry*, *272*(49), 30822–30827.

Twigg, S. M., & Baxter, R. C. (1998). Insulin-like growth factor (IGF)-binding protein 5 forms an alternative ternary complex with IGFs and the acid-labile subunit. *The Journal of Biological Chemistry*, *273*(11), 6074–6079.

Ulanet, D. B., Ludwig, D. L., Kahn, C. R., & Hanahan, D. (2010). Insulin receptor functionally enhances multistage tumor progression and conveys intrinsic resistance to IGF-1R targeted therapy. *Proceedings of the National Academy of Sciences of the United States of America*, *107*(24), 10791–10798.

Ullrich, A., Bell, J. R., Chen, E. Y., Herrera, R., Petruzzelli, L. M., Dull, T. J., et al. (1985). Human insulin receptor and its relationship to the tyrosine kinase family of oncogenes. *Nature*, *313*(6005), 756–761.

Valensise, H., Liu, Y. Y., Federici, M., Lauro, D., Dell'anna, D., Romanini, C., et al. (1996). Increased expression of low-affinity insulin receptor isoform and insulin/insulin-like growth factor-I hybrid receptors in term placenta from insulin-resistant women with gestational hypertension. *Diabetologia*, *39*(8), 952–960.

Villanueva, J., Vultur, A., Lee, J. T., Somasundaram, R., Fukunaga-Kalabis, M., Cipolla, A. K., et al. (2010). Acquired resistance to BRAF inhibitors mediated by a RAF kinase switch in melanoma can be overcome by cotargeting MEK and IGF-1R/PI3K. *Cancer Cell*, *18*(6), 683–695.

Ward, C. W., Menting, J. G., & Lawrence, M. C. (2013). The insulin receptor changes conformation in unforeseen ways on ligand binding: Sharpening the picture of insulin receptor activation. *Bioessays*, *35*(11), 945–954. https://doi.org/10.1002/bies.201370111.

Wei, L., Hubbard, S. R., Hendrickson, W. A., & Ellis, L. (1995). Expression, characterization, and crystallization of the catalytic core of the human insulin receptor protein-tyrosine kinase domain. *The Journal of Biological Chemistry*, *270*(14), 8122–8130.

White, M. F. (2002). IRS proteins and the common path to diabetes. *American Journal of Physiology Endocrinology and Metabolism*, *283*(3), E413–E422.

Xu, J., Bie, F., Wang, Y., Chen, X., Yan, T., & Du, J. (2019). Prognostic value of IGF-1R in lung cancer: A PRISMA-compliant meta-analysis. *Medicine (Baltimore)*, *98*(19), e15467.

Xu, Y., Kong, G. K., Menting, J. G., Margetts, M. B., Delaine, C. A., Jenkin, L. M., et al. (2018). How ligand binds to the type 1 insulin-like growth factor receptor. *Nature Communications*, *9*(1), 821.

Yakar, S., Liu, J. L., Stannard, B., Butler, A., Accili, D., Sauer, B., et al. (1999). Normal growth and development in the absence of hepatic insulin-like growth factor I. *Proceedings of the National Academy of Sciences of the United States of America*, *96*(13), 7324–7329.

Yamaguchi, Y., Flier, J. S., Yokota, A., Benecke, H., Backer, J. M., & Moller, D. E. (1991). Functional properties of two naturally occurring isoforms of the human insulin receptor in Chinese hamster ovary cells. *Endocrinology, 129*(4), 2058–2066.

Yamauchi, T., Tobe, K., Tamemoto, H., Ueki, K., Kaburagi, Y., Yamamoto-Honda, R., et al. (1996). Insulin signalling and insulin actions in the muscles and livers of insulin-resistant, insulin receptor substrate 1-deficient mice. *Molecular and Cellular Biology, 16*(6), 3074–3084.

Yeh, A. H., Bohula, E. A., & Macaulay, V. M. (2006). Human melanoma cells expressing V600E B-RAF are susceptible to IGF1R targeting by small interfering RNAs. *Oncogene, 25*(50), 6574–6581.

Yerushalmi, R., Gelmon, K. A., Leung, S., Gao, D., Cheang, M., Pollak, M., et al. (2012). Insulin-like growth factor receptor (IGF-1R) in breast cancer subtypes. *Breast Cancer Research and Treatment, 132*(1), 131–142.

Yu, H., Spitz, M. R., Mistry, J., Gu, J., Hong, W. K., & Wu, X. (1999). Plasma levels of insulin-like growth factor-I and lung cancer risk: A case-control analysis. *Journal of the National Cancer Institute, 91*(2), 151–156.

Zeng, X., Sachdev, D., Zhang, H., Gaillard-Kelly, M., & Yee, D. (2009). Sequencing of type I insulin-like growth factor receptor inhibition affects chemotherapy response in vitro and in vivo. *Clinical Cancer Research, 15*(8), 2840–2849.

Zha, J., O'Brien, C., Savage, H., Huw, L. Y., Zhong, F., Berry, L., et al. (2009). Molecular predictors of response to a humanized anti-insulin-like growth factor-I receptor monoclonal antibody in breast and colorectal cancer. *Molecular Cancer Therapeutics, 8*(8), 2110–2121.

Zhao, Q., Feng, Y., Zhu, Z., & Dimitrov, D. S. (2011). Human monoclonal antibody fragments binding to insulin-like growth factors I and II with picomolar affinity. *Molecular Cancer Therapeutics, 10*(9), 1677–1685.

Zhong, H., Fazenbaker, C., Breen, S., Chen, C., Huang, J., Morehouse, C., et al. (2014). MEDI-573, alone or in combination with mammalian target of rapamycin inhibitors, targets the insulin-like growth factor pathway in sarcomas. *Molecular Cancer Therapeutics, 13*(11), 2662–2673.

Zhou, R., Diehl, D., Hoeflich, A., Lahm, H., & Wolf, E. (2003). IGF-binding protein-4: Biochemical characteristics and functional consequences. *The Journal of Endocrinology, 178*(2), 177–193.

Zu, K., Martin, N. E., Fiorentino, M., Flavin, R., Lis, R. T., Sinnott, J. A., et al. (2013). Protein expression of PTEN, insulin-like growth factor I receptor (IGF-IR), and lethal prostate cancer: A prospective study. *Cancer Epidemiology, Biomarkers & Prevention, 22*(11), 1984–1993.

CHAPTER SEVEN

MET receptor in oncology: From biomarker to therapeutic target

Raeva Malik[a,†], Isa Mambetsariev[b,†], Jeremy Fricke[b,†], Neal Chawla[c], Arin Nam[b], Rebecca Pharaon[b], Ravi Salgia[b,*,†]

[a]George Washington University Hospital, Washington, DC, United States
[b]Department of Medical Oncology and Therapeutics Research, City of Hope National Medical Center, Duarte, CA, United States
[c]Department of Medicine, Advocate Illinois Masonic Medical Center, Chicago, IL, United States
*Corresponding author: e-mail address: rsalgia@coh.org

Contents

† These authors contributed equally to this work and should be considered co-first authors.

Advances in Cancer Research, Volume 147
ISSN 0065-230X
https://doi.org/10.1016/bs.acr.2020.04.006

© 2020 Elsevier Inc.
All rights reserved.

Abstract

First discovered in the 1984, the MET receptor tyrosine kinase (RTK) and its ligand hepatocyte growth factor or HGF (also known as scatter factor or SF) are implicated as key players in tumor cell migration, proliferation, and invasion in a variety of cancers. This pathway also plays a key role during embryogenesis in the development of muscular and nervous structures. High expression of the MET receptor has been shown to correlate with poor prognosis and resistance to therapy. MET exon 14 splicing variants, initially identified by us in lung cancer, is actionable through various tyrosine kinase inhibitors (TKIs). For this reason, this pathway is of interest as a therapeutic target. In this chapter we will be discussing the history of MET, the genetics of this RTK, and give some background on the receptor biology. Furthermore, we will discuss directed therapeutics, mechanisms of resistance, and the future of MET as a therapeutic target.

1. History of MET

1.1 Structure

MET was first discovered in 1984 in a human osteosarcoma cell line, which was treated with methyl-N′ nitor-N-nitrosoguanidine, a known carcinogen. The resulting fusion protein involved a rearrangement between TPR gene and MET gene. The TPR gene confers activation of MET kinase through constitutive dimerization (Cooper et al., 1984). This kinase activation confers transforming activity of MET. Concurrently, hepatocyte growth factor (HGF) was discovered in mice plasma, and was found to stimulate mitosis in hepatocytes (Nakamura, Nawa, & Ichihara, 1984). A few years later, a molecule termed as "Scatter Factor" was found to cause epithelial cells to scatter in media conditioned with fibroblasts (Stoker, Gherardi, Perryman, & Gray, 1987). In 1991, HGF and SF came to be considered the same protein following the discovery of a common encoding gene and similar action on the MET receptor (Weidner et al., 1991).

In the mid-1990s the structure of MET, its ligand HgF came to be better understood. In 1994, a tyrosine kinase domain was found to be in the intracellular domain. This TK domain has a C-terminal site which possess a multi substrate binding site, that recruits signaling proteins as a result of HGF activation of MET (Furlan, Kherrouche, Montagne, Copin, & Tulasne, 2014;

Gherardi et al., 2006; Schiering et al., 2003). 2003 MET structure was further described, pertaining to the extracellular component of MET receptor, found to have a seven-bladed propeller like component, called the SEMA domain. The SEMA domain is attached to PSI and IPT domains (Furlan et al., 2014; Gherardi et al., 2006; Schiering et al., 2003).

1.1.1 Function and regulation

By the early 1990s the role of MET and HGF in development and homeostasis of various tissues was beginning to be understood. In the 1995, HGF knockout mice found to have malformation of placental labyrinth, resulting in intra-uterine death. The following year, MET was found to influence survival, morphogenesis, angiogenesis, and neural development (Furlan et al., 2014). MET and HGF are found in both endodermal and mesodermal tissues. Furthermore, the presence of HGF in mesenchymal cells was noted, while MET was discovered in organs such as lung, kidney, liver, and skin, which suggested that this receptor and its ligand were pivotal in the development of epithelial organs (Andermarcher, Surani, & Gherardi, 1996).

In 2001, ubiquitination of MET was found to occur following autophosphorylation on y1003 which results in CBL E3 Ubiquitin ligase recruitment (Drilon, Cappuzzo, Ou, & Camidge, 2017). Ubiquitination of MET results in internalization and degradation, which seems to be an important negative regulator of MET. In the absence of HGF, MET was found to promote apoptosis through caspase cleavage (Tulasne et al., 2004). The same year, found that MET inhibition impedes liver, kidney, and skin regeneration. Furthermore, the addition of HGF was observed to augment regeneration of these tissues (Huh et al., 2004). In 2003, we demonstrated that MET exon 14 splicing variants occur in small cell lung cancer (SCLC) and in 2005 we described them in non-small cell lung cancer (NSCLC) (Ma, Kijima, et al., 2003; Ma, Jagadeeswaran, et al., 2005; Ma, Maulik, Christensen, & Salgia, 2003). In 2008, Kermorgant and authors demonstrated that STAT3 and MET co-localize on endosomes (Kermorgant & Parker, 2008). STAT3 acts as transcription factor and oncogene signal transducer (Kermorgant & Parker, 2008).

1.2 Pathway

In the mid-1990s, MET-ligand interaction was found to trigger dimerization and autophosphorylation of the T2 tyrosine residues. As a result, other tyrosine residues outside the kinase domain, in particular those found in the juxtamembrane, and C-terminal tail undergo phosphorylation.

These enable interacting and binding with effectors such as Pi3k, Srv, STAT3, PLCgamma, Gabl1, grb2, Shc7 (Maina et al., 1996). Mutation of the juxtamembrane tyrosine 1003 residues in mice models, results in similar phenotypic expression as MET-deficient mice (Maina et al., 1996). Royal et al. elucidate downstream activation of several signaling cascades via small g proteins such as RAS, Rac, PAK. In 2007, role of MET in cell motility through AKT-Pi3k activation is described (Moumen et al., 2007). A study in 2011, suggested that MET may direct p53 activity toward survival via Abl and p38-MAPK (Furlan et al., 2011).

1.3 Tumorigenesis

MET was found to be a transforming factor through fusion of TPR-MET in 1984 in animal models (Cooper et al., 1984). However, its role in human cancers would not be established until about a decade later. In 1996, Natali et al. found that MET is overexpressed in several cancers and correlates with a poor prognosis (Natali et al., 1996). A year later, tyrosine kinase mutations in MET gene were described in papillary renal carcinomas (Schmidt et al., 1997). In 2004, selective duplication of mutant MET allele resulted in transgenic mice with multiple carcinoma, sarcoma, and lymphoma (Joffre et al., 2011). In 2012, HGF secretion by the tumor microenvironment was described to induce resistance to RAF inhibitors (Straussman et al., 2012).

2. MET genetics

2.1 Chromosomal localization

The *MET* proto-oncogene encodes for receptor kinase c-Met, a member of the receptor tyrosine kinase (RTK) family. It is located on chromosome 7q31.2 and it is more than 120 kilobases long with 21 exons separated by 21 introns. The gene produces tyrosine kinase receptors cleaving the precursor single-chain protein to generate alpha and beta subunits linked through disulfide bonds. This allows the receptor to be dimerized through ligand-receptor binding. MET's ligand, hepatocyte growth factor (HGF), binds to a tyrosine kinase receptor like c-MET or hepatocyte growth factor receptor (HGFR). Binding of the ligand to the receptor induces dimerization and activates the receptor, and recruits proteins involved in signal transduction like Grb2, Shc, Src, and phosphoinositide 3′ kinase. This in turn leads to downstream cascades in pathways involved in cell apoptosis, embryogenesis, and cellular motility and invasion. The extracellular portion of HGFR is composed of a Sema domain, a cysteine-rich Met-related sequence (MRS) domain, and four immunoglobulin-like structure domain while the intracellular portion is composed of a juxtamembrane (JM) domain, a tyrosine kinase domain, and a C-terminal regulatory tail. MET and its ligand, HGF, are expressed in various tissues but are predominantly located in epithelial and mesenchymal cells. Mutations, amplification, and overexpression have been reported in multiple solid tumor cancers.

2.2 Mutations

The MET pathway is involved in numerous cellular functions include cell proliferation, cell growth, survival, apoptosis, and epithelial-mesenchymal transition (EMT). Mutations in *MET* were initially discovered in hereditary papillary renal carcinoma (HRPC). Whole genome sequencing of families with a history of HRPC was performed and evidence was found demonstrated *MET* mutations as oncogenic drivers for HRPC (Schmidt et al., 1997). Interestingly, mutations in the same region of the *MET* gene were also discovered in sporadic renal carcinomas (Schmidt et al., 1997). Germline *MET* mutations occur in 100% of HRPC. Since then, various studies revealed evidence of oncogenic *MET* mutations in breast cancer, non-small cell lung cancer (NSCLC), head and neck squamous cell carcinoma, gastric cancer, colorectal cancer, and other solid tumors (de Melo Gagliato et al., 2014; Di Renzo et al., 2000; Ghadjar et al., 2009; Lee et al., 2000; Liu et al., 2015; Maulik et al., 2002; Neklason et al., 2011; Puri et al., 2007; Seiwert et al., 2009). The discovery of *MET* in so many different cancers suggests the significance of this gene in oncogenesis and tumorigenesis.

MET mutations are normally identified using next-generation sequencing (NGS) platforms. In lung cancer, *MET* mutations are estimated to be found in roughly 6% of patients diagnosed with NSCLC and 3% of patients diagnosed with squamous cell lung cancer (Cancer Genome Atlas Research Network, 2012, 2014). In gynecological cancers, *MET* mutations have been identified in 15–51% of breast cancers and 7.4% of ovarian cancers (Bieche, Champeme, & Lidereau, 1999; Ma et al., 2009; Tang et al., 2014). Common mutations that are found include *MET* missense mutations and *MET* exon 14 skipping mutation (Fig. 1). *MET* missense mutations can occur in the JM domain or the Sema domain. JM domain mutations have been shown to promote ubiquitination of the MET

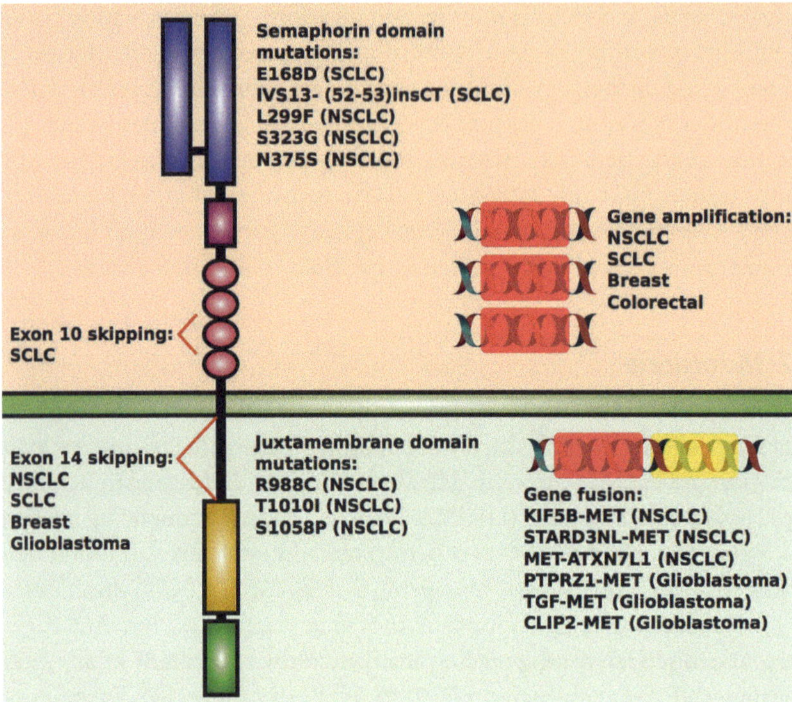

Fig. 1 Genetic alterations found in MET.

receptor to stimulate continued MET signaling while Sema domain mutations are likely to influence the ligand-receptor interaction.

MET missense mutations have been identified in various solid tumors. Early studies in lung cancer were the first to establish the presence and significance of MET mutations in oncogenesis (Ma, Jagadeeswaran, et al., 2005; Ma, Kijima, et al., 2003). Ma et al. reported several JM domain mutations (R988C in H1437 NSCLC cell line, and T1010I and S1058P in NSCLC tumor tissue) and Sema domain mutations (E168D) and two-base-pair insertional mutations (IVS13- (52–53)insCT in SCLC tumor tissue, and L299F, S323G, and N375S in NSCLC tumor tissue) in lung cancer, including *MET* exon 14 skipping mutation (Ma, Jagadeeswaran, et al., 2005; Ma, Kijima, et al., 2003; Maulik et al., 2002; Sattler & Salgia, 2016). The JM domain mutations altered RTK signaling, leading to prolonged signal transduction and increased cellular functions associated with *MET*, specifically cancer invasion and metastasis.

MET exon 14 skipping mutation, a JM domain mutation that has been identified in 4% of lung cancers and other solid tumors like breast cancer and glioblastoma multiforme, deletes the JM domain containing the casitas B-lineage lymphoma (CBL) E3-ubiquitin ligase binding site and decreases the turnover of the resultant aberrant MET protein (Cancer Genome Atlas Research Network, 2014; Tan et al., 2010). Loss of *MET* exon 14 leads to prolonged signal activation once the HGF binds to c-MET. In the early 2000s, we discovered novel mutations involving *MET* exon 14 in both NSCLC and small cell lung cancer (SCLC) (Kong-Beltran et al., 2006; Ma, Jagadeeswaran, et al., 2005; Ma, Kijima, et al., 2003). A landmark study later on demonstrated clinical efficacy of treating NSCLC patients with a *MET* exon 14 skipping mutation with small molecular MET inhibitors crizotinib or cabozantinib (Paik et al., 2015). Since then, multiple clinical trials have evaluated other small molecule inhibitors including capmantinib from the GEOMETRY mono-1 trial or tepotinib from the VISION clinical trial that have exhibited treatment efficacy in targeting *MET* exon 14 skipping mutation. Results from the phase II GEOMETRY mono-1 clinical trial in advanced NSCLC patients with *MET* exon 14 skipping mutation or *MET* amplification reported good overall response rates (ORR), and a median progression-free survival (PFS) of 5.42 months in the pretreated cohort and 9.69 months in the treatment-naïve cohort (Wolf et al., 2019). The data reported in this trial on ORR and PFS suggests the important of identifying these mutations at initial diagnosis to maximize response and survival by treating patients with a therapy that targets their oncogenic driver. Furthermore, results from a phase II trial of tepotinib was presented this year and also demonstrated promising results with a favorable duration of response (DoR) in MET-mutated patients (Paik et al., 2019). Both small molecule MET inhibitors have been granted by the Food and Drug Administration (FDA) a fast track designation for this group of patients.

2.3 MET amplification

Amplification of the *MET* gene can be identified through fluorescence in situ hybridization (FISH), reverse transcriptase-PCR, or NGS testing. *MET* amplification has been identified in approximately 8% of breast cancers and was found to be linked to triple negative breast cancers (Gonzalez-Angulo et al., 2013). Moreover, there was a correlation with high copy numbers of *MET* and poor prognostic features in breast and lung cancer (Gonzalez-Angulo et al., 2013; Kanteti, Yala, Ferguson, & Salgia, 2009). This correlation has been observed in lung cancer as well, shown by the average overall survival (OS) in surgically resection NSCLC patients with positive MET by FISH was 25.8 months vs 47.5 months in patients with negative MET by FISH ($P=0.0045$) (Cappuzzo et al., 2009; Okuda, Sasaki, Yukiue, Yano, & Fujii, 2008). This correlation was further established in other NSCLC studies, confirming the role of *MET* amplification as a poor prognosis of survival (Dimou, Non, Chae, Tester, & Syrigos, 2014; Go et al., 2010). Previous studies have also established *MET* amplification as a mechanism of acquired resistance through ERBB3 signaling in NSCLC and colorectal patients who undergo treatment with EGFR tyrosine kinase inhibitors (TKIs) (Engelman et al., 2007; Raghav et al., 2016). In lung cancer, *MET* amplification makes up 1–4% of patients who have not undergone treatment with EGF TKIs and 5–20% of patients who develop it in response to treatment with EGFR TKIs (Cappuzzo et al., 2009).

2.4 MET rearrangement translocation

Rearrangement of the *MET* gene is not frequently found in cancers, however, novel fusions with *MET* have been reported in lung cancer. The overall incidence of *MET* fusions in lung adenocarcinoma was reported to be 0.5% (Plenker et al., 2018). Plenker et al. identified two patients with a *KIF5B-MET* and *STARD3NL-MET* gene fusions who were both treated with criztonib, a MET inhibitor (Plenker et al., 2018). A multicenter study in China discovered one patient with a novel *MET-ATXN7L1* fusion from a cohort of 2410 NSCLC patient samples with NGS testing (Wang et al., 2018). The patient was documented to have a partial response to crizotinib, similar to the two patients in the study by Plenker et al. In pediatric glioblastoma, a common and aggressive brain tumor, an estimated 10% of patients were identified to have *MET* fusions by NGS testing, including *PTPRZ1-MET*, *TGF-MET*, and *CLIP2-MET* (Bender et al., 2016). In the same study that identified these fusions, the patients were treated with crizotinib and showed considerable treatment response in the tumor but the treatment unfortunately led to new resistant lesions (Bender et al., 2016). Although rare, these *MET* rearrangements represent a small percentage of patients with evidence demonstrating treatment efficacy with MET inhibitors.

3. MET receptor biology

The 150-kDa MET precursor protein is formed into a disulfide bond linked 50-kDa extracellular α-chain and a 140-kDa transmembrane β-chain once it is post-translationally digested and glycosylated. The β-chain portion includes the Semaphorin (Sema) domain, Plexin-Semaphorin-Integrin (PSI) domain, four Immunoglobin-like fold shared by Plexins and Transcription factors (IPT) domains, a transmembrane domain, tyrosine kinase domain (TK), and juxtamembraine (JM) domain, which are homologous structurally shared with other proteins (Sadiq & Salgia, 2013). The Sema domain has a seven-bladed β-propellor at the N-terminus of the MET receptor, contains the ligand-binding domain, and is required for receptor dimerization (Kong-Beltran, Stamos, & Wickramasinghe, 2004; Stamos, Lazarus, Yao, Kirchhofer, & Wiesmann, 2004). The Sema and four IPT domains are connected through the PSI domain which correctly positions the ligand-binding domain (Kozlov et al., 2004). Two phosphorylation sites located in the JM domain, Tyr1003 and Ser985, negatively regulates MET-dependent signal transduction (Fig. 2) (Matsumoto, Umitsu, De Silva, Roy, & Bottaro, 2017). The adenosine triphosphate (ATP)-binding pocket consists of two lobes in the TK domain that remain blocked

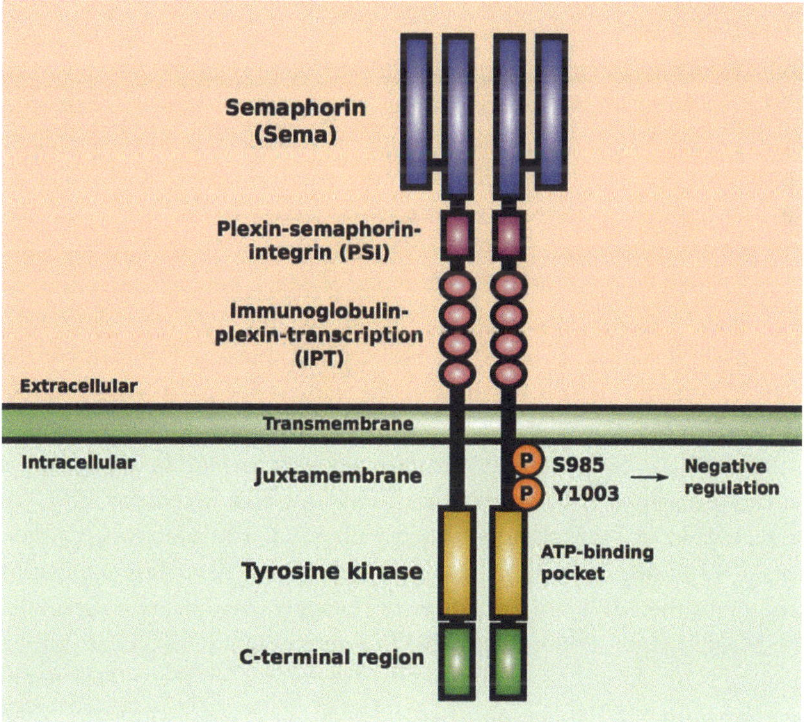

Fig. 2 Structure of MET.

by the activation loop while the receptor is in its inactive state (Schiering et al., 2003). A number of organs including the liver, pancreas, prostate, kidney, muscle and bone marrow express MET on the cell surface during embryogenesis through adulthood (Comoglio, Giordano, & Trusolino, 2008). HGF, the only ligand for MET activation, was first recognized as a growth factor for hepatocytes and also as a fibroblast-derived cell motility factor (Gohda et al., 1986; Nakamura, Nawa, Ichihara, Kaise, & Nishino, 1987). Mesenchymal cells secrete HGF precursor as a single-chain protein before it is cleaved and linked with a disulfide bond into the mature protein (Sonnenberg, Meyer, Weidner, & Birchmeier, 1993). HGF consists of six domains including an N-terminal domain, four kringle domains, and a C-terminal domain (Sadiq & Salgia, 2013). There are two binding sites for MET and HGF. Independent of HGF maturation, the N-terminal domain of HGF binds to the IPT3 and IPT4 domains of MET with high affinity, while only the active form of HGF will bind to the MET Sema domain with low affinity (Basilico, Arnesano, Galluzzo, Comoglio, & Michieli, 2008; Stamos et al., 2004).

3.1 Mitogenesis

HGF up-regulates genes encoding for proteases required for HGF and MET processing leading to constant ligand stimulation and overexpression of MET (Ma et al., 2008; Zhang & Vande Woude, 2003). Several epithelial cell-derived cancers share MET overexpression as common characteristic and contributes to adverse outcomes (Rong et al., 1995). Abnormal paracrine or autocrine ligand production leads to the continuous kinase activation regardless of the presence or absence of MET gene amplification and MET gene mutation acting as another oncogenic pathway (Peschard & Park, 2007; Rong et al., 1995). Hereditary and sporadic forms of papillary renal cell carcinoma (PRC) showed the earliest reported MET missense mutations located in the TK domain (Dharmawardana, Giubellino, & Bottaro, 2004). Some of the strongest support of MET oncogenity in humans are between these missense mutations and hereditary PRC and the regression of PRC tumors treated with MET TK inhibitors (Choueiri et al., 2013). Lung cancer and head and neck cancers have shown mutations throughout the MET coding sequence (Lengyel, Sawada, & Salgia, 2007; Ma, Maulik, et al., 2003; Salgia, 2017; Szturz et al., 2017). HGF-binding antagonists block anchorage-independent growth in epithelial cells requiring HGF with MET mutations (Bardelli et al., 1998; Giordano et al., 2000;

Michieli et al., 1999). Murine engineered PRC-associated MET mutations did not develop PRC but did grow sarcomas and some lymphomas (Graveel, London, & Vande Woude, 2005). Most tumors containing the duplication of the mutant MET allele and trisomy of chromosome 6 comprising the murine MET locus, similarly seen in the trisomy of chromosome 7 in human PRC tumors (Graveel et al., 2005). Defects in MET internalization resulted in insistent signaling in two PRC-associated MET mutants similarly seen in the MET JM domain (Joffre et al., 2011). In non-small cell lung carcinoma (NSCLC)-derived cell lines, 12.5% of small lung cell cancer (SCLC) cases, and in 8% of samples of lung adenocarcinoma tissues there are additional alterations identified in MET encoding the extracellular Sema domain (E168D, L229F, S323G and N375S) and the intracellular JM domain (R970C, T992I, S1040P and exon 14 deletions) (Kong-Beltran et al., 2006; Ma, Jagadeeswaran, et al., 2005; Ma, Kijima, et al., 2003). Cultured cells show that these mutations activate proliferation, motility and invasiveness (Lengyel et al., 2007). Regulation of MET activation in response to liver injury and regeneration is produced by the phosphorylation at S985 to negatively regulate MET by protein kinase C and intracellular calcium (Gandino et al., 1991; Gandino, Di Renzo, Giordano, Bussolino, & Comoglio, 1990; Hashigasako, Machide, Nakamura, Matsumoto, & Nakamura, 2004; Nakayama et al., 2013). Persistent HGF-stimulated signaling directs increasing transformation and tumorigenicity of cells with the loss of exon 14 and Y1003 in the MET JM domain (Kong-Beltran et al., 2006; Ma, Jagadeeswaran, et al., 2005). The JM mutations R970C and T992I may increase the risk of cancer but can be found in individuals without cancer, thus, the oncogenic potential has yet to be determined (Ma, Kijima, et al., 2003; Tyner et al., 2010). Approximately 6% of high-grade gliomas reported a deletion in MET extracellular IPT domains (MET exon 7–8 skipping variant) generating a single pseudo-IPT domain which could impair biosynthetic processing and translocation to the cell membrane (Navis et al., 2015). Oncogenic MET signaling is similar to normal cellular organogenesis with the developmental transitions between epithelial and mesenchymal cell types being regulated by HGF, while tumor invasiveness and metastasis is facilitated with an increase in protease production and cell dissociation and motility (Fig. 3) (Birchmeier, Birchmeier, Gherardi, & Vande Woude, 2003; Jagadeeswaran, Jagadeeswaran, Bindokas, & Salgia, 2007; Rosario & Birchmeier, 2003; Zhang & Vande Woude, 2003). Mouse models of tumor cells with an overexpressed MET gene showed suppression of tumor growth and metastasis when silenced (Corso et al., 2008).

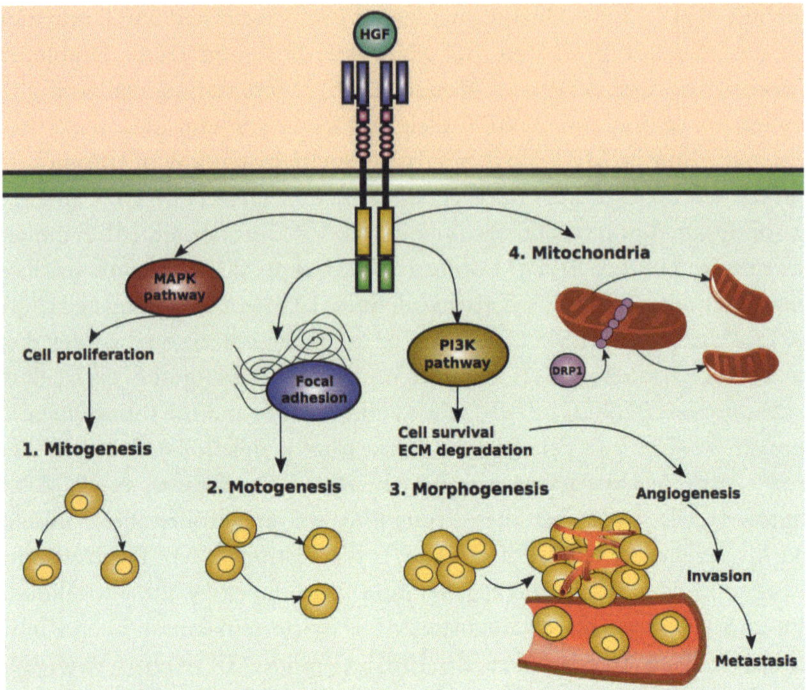

Fig. 3 Effects of MET on mitogenesis, motogenesis, morphogenesis, and mitochondria.

Additionally, hypoxic growth-limited cancers promote tumor metastasis by stimulating angiogenesis from HGF/MET signaling in vascular endothelial cells, while hypoxic conditions boosts MET expression and increases HGF signaling in cultured and mouse tumor cell models (Pennacchietti et al., 2003; Peruzzi, Athauda, & Bottaro, 2006). HGF/MET signaling triggers tumor cell proliferation in human breast cancers in the mammary fat pads of nude mice (Rosen et al., 1997).

3.2 Motogenesis

Cell scattering consists of three phases: cell spreading, cell-to-cell dissociation, and cell migration. The weakening or dissolution of cell-to-cell adhesions is necessary for epithelial cells to scatter. A diversity of different cell types is capable at scattering adherent cells. Intercellular junctions are constantly closely monitored and maintained under normal conditions. Specific physiologic and pathologic conditions like during normal development and tumor cell invasion and/or metastasis are capable or disassembling these junctions (Warburton et al., 2000). HGF, also known as "scatter factor," is an effective stimulator of cell scattering. The identification of HGF as a product secreted by fibroblasts and smooth muscle cells occurred in observations that HGF encourages cell dissociation and motility of epithelial cells similarly to semaphorins (Boccaccio et al., 1998). MDCK epithelial cells can be induced to scatter by HGF when the proteasome is blocked and inhibited. Proteins at the intercellular junctions stabilize the scattering of cells, but proteasome inhibition has no effect on cell migration. Cell-to-cell dissociation facilitated by the proteasome seems to play an integral role (Hammond, Urbe, Vande Woude, & Clague, 2001; Jeffers, Taylor, Weidner, Omura, & Vande Woude, 1997). Cell motility is necessary in normal cellular physiology but also in various disease processes. Chemotactic gradients or randomly, if no gradient exists, are utilized in animals cell to move through or over a substrate. The basic propulsion for crawling/gliding is the actin cytoskeleton. Dense arrays of actin filaments are located in filopodia and lamellipodia, which make up the initial protrusive structure oriented with their barbed ends in the direction of protrusion (Mitchison & Cramer, 1996). The polymerization of actin, formation of actin stress fibers, and focal adhesion formation are the important actions regulating cell motility (Weisberg, Sattler, Ewaniuk, & Salgia, 1997). The motility of epithelial cells has been shown to occur through HGF and MET signaling. The motility of Madin-Darby canine kidney cells were prompted by mutationally active MET (Jeffers et al., 1998).

HGF induces the in vitro invasiveness of poorly aggressive Caco-2 colon cancer epithelial cells by the overproduction of proteases and stimulation of cell motility (Kermorgant, Aparicio, Dessirier, Lewin, & Lehy, 2001). Stromal fibroblasts support cancer cell invasion into 3D collagen matrices utilizing human oral squamous cell carcinoma cells and stromal fibroblasts (Matsumoto, Horikoshi, Rikimaru, & Enomoto, 1989). Stromal fibroblast prompted cancer cell invasion is countered by inhibition of HGF (Matsumoto, Matsumoto, Nakamura, & Kramer, 1994). In experimental animal models, suppression of invasion and metastasis in several kinds of cancers can be achieved by the selective inhibition of the HGF/MET signaling pathway (Cecchi, Rabe, & Bottaro, 2012; Gherardi, Birchmeier, Birchmeier, & Vande, 2012; Sakai, Aoki, & Matsumoto, 2015). Materializing as a critical actor in metastatic colonization and growth, the metastatic tumor microenvironment has an important impact in cancer (Joyce & Pollard, 2009). A large diversity of resident stromal cells or enlisted cells from the bone marrow provide aid to the development of the metastatic microenvironment including macrophages, inflammatory cells, endothelial cells, and cancer-associated fibroblasts (Cirri & Chiarugi, 2012; Joyce & Pollard, 2009). The interactions between cancer cells and stromal cells act as vital players in developing the metastatic microenvironment, and are facilitated by chemokines and growth factors (such as HGF, epidermal growth factor, endothelial cell growth factor, and transforming growth factor-β) (Cirri & Chiarugi, 2012; Joyce & Pollard, 2009). In the tumor microenvironment, cell invasiveness in heavily influenced HGF which performs as a stromal cell-derived factor (Cirri & Chiarugi, 2012; Joyce & Pollard, 2009). Metastatic microenvironment formation is promoted by MET in exosomes of metastatic melanoma (Adachi et al., 2016; Peinado et al., 2012). There are high levels of MET in the exosomes of highly metastatic mouse and human melanoma cells, and the migration of tumor cells utilizing these exosomes in the circulation can be localized to sites of metastatic tissues with increased vascular permeability (Adachi et al., 2016; Peinado et al., 2012). In bone marrow-derived cells, circulating exosomes amplified activated MET altering the cells to a proangiogenic phenotype allowing the cells to mobilize to the lungs enabling angiogenesis, invasion, and metastasis (Peinado et al., 2012). Melanoma cells with lower metastatic aptitude will increase their ability when directed with dosed with exomomes containing high levels of MET (Adachi et al., 2016). Increased MET transcript levels and phosphorylated MET protein were found in the circulating bone-marrow-derived cells of patients with advance melanoma, and melanoma progression can be correctly anticipated with the amount of MET in circulating in serum exosomes (Peinado et al., 2012). The adaptor protein GAB1 is partially responsible for the biochemical foundation of the strength that HGF advances tumor cell survival and motility (Sachs et al., 2000; Weidner et al., 1996). There are common genetic defects illustrating phenotypic similarities between mice that harbor deletions of HGF, MET, and GAB1 suggesting that GAB1 is an important signaling intermediary in the HGF/MET pathway (Sachs et al., 2000).

3.3 Morphogenesis

One of the central mechanisms for tumor cell proliferation and metastasis is the formation of new blood vessels, also known as angiogenesis. Tumor cells can quickly enter into circulation through the growth of new capillaries with fragmented basement membranes based on tumor-derived angiogenic factors. Localized factors that promote or inhibit neovascularization are in constant demand to regulate angiogenesis equilibrium (Folkman, Browder, & Palmblad, 2001). When stimulated, HGF, a powerful angiogenic molecule that performs on vascular endothelial cells, promotes migration, proliferation, protease production, invasion, and organization into capillary-like tubes (Rosen et al., 1997). In head and neck squamous cell carcinoma cells, both the MEK- and phosphotidylinositol-3-OH kinase (PI3K)-dependent pathways are responsible for the dose-dependent increase in cytokines IL-8 and vascular endothelial growth factor (VEGF) when prompted by HGF to support angiogenesis (Dong et al., 2001; Rolle et al., 2017). To enter circulation and begin metastasis, tumor cells must breakdown the barrier of the extracellular matrix (ECM). The interaction of tumor cells with the basement membrane (BM) is the crucial event of tumor invasion. The majority of the BM consists of type IV collagen, laminin, and heparan sulfate proteoglycan, while the ECM is comprised of the basement membrane, fibronectin, and vitronectin. The proteinases of plasminogen activator (PA) and matrix metalloproteinase (MMP) are required for the destruction of the ECM (Curran & Murray, 2000). The HGF/MET pathway is heavily involved in the invasive tumor phenotype. The increased expression of the serine protease urokinase (uPA) and its receptor (uPAR) through HGF and MAPK1/2 signaling leads to the accumulation of uPA at the cell surface where uPA can degrade components of the ECM and cleave plasminogen to the protease plasmin. Plasmin itself is effective at degrading ECM and BM components, but it also activates metalloproteinases which their own effective means to breakdown the ECM and BM (Hotary, Allen, Punturieri, Yana, & Weiss, 2000).

3.4 Mitochondria

Mitochondria are high powered organelles performing important tasks in cellular physiology with a diverse number of cancers resulting in alterations to mitochondrial structures and functions (Cassidy-Stone et al., 2008; Chang & Blackstone, 2010; Frezza et al., 2006; Lee, Jeong, Karbowski, Smith, & Youle, 2004). Cancer cells support their proliferation, migration and resistance to therapeutics by altering the dynamics, bioenergetics, and biosynthetic properties of mitochondria (Brenner & Mak, 2009; Brooks, Wei, Cho, & Dong, 2009). Mitochondria structure and function depends on the regulation of mitochondrial fission through dynamin related protein 1 (DRP1), a mechanochemical GTPase. DRP1 is an encouraging target for cancer therapy because alterations in DRP1 expression ore activation are associated to several cancers (Rehman et al., 2012). Mdivi-1 (mitochondrial division inhibitor) inhibits mitochondrial fission by selectively inhibiting Drp1 over other dynamin family members (Cassidy-Stone et al., 2008).

MET pathway activation prompts alterations in mitochondrial morphology and affects mitochondrial staining intensity in HEK-293 cells, indicating that MET signaling may regulate mitochondrial morphology and functions (Kawada et al., 2013). In NSCLC and MPM, MET does regulate mitochondrial organization and function through the mitochondrial fission proteins DRP1 and to a smaller degree Fis1 uncovering a novel mechanism and suggesting that a combination therapy inhibiting MET and DRP1 could be a treatment plan (Wang et al., 2018).

3.5 Ligand binding with HGF

MET signaling initiates when HGF binds to a MET receptor at the plasma membrane causing the dimerization of two MET receptors. MET is activated through trans-phosphorylation of two tyrosine residues in the tyrosine kinase domain, Tyr1234 and Tyr1235 (Longati, Bardelli, Ponzetto, Naldini, & Comoglio, 1994). Two more tyrosines in the C-terminal region, Tyr1349 and Tyr1356, are trans-phosphorylated to form the unique MET subfamily multifunctional docking site (Garajová, Giovannetti, Biasco, & Peters, 2015). These sites are essential for MET signaling because downstream signaling is activated with the binding of the adapter proteins growth factor receptor bound protein 2 (Grb2), Grb2 associated binding protein 1 (Gab1) and Src homology 2 domain containing (SHC) will bind to the receptor, thus activating downstream signaling (Fixman, Fournier, Kamikura, Naujokas, & Park, 1996; Ponzetto et al., 1994; Weidner et al., 1996). A few of the activated signaling pathways include: mitogen-activated protein kinase (MAPK), phosphoinositide 3-kinase (PI3K)/AKT (protein kinase B), signal transducer and activator of transcription (STAT) proteins, and nuclear factor-κB (NF-κB) (Maulik et al., 2002; Sipeki et al., 1999; Van Der Steen et al., 2015; Zhang, Wang, Jove, & Vande Woude, 2002). MET is negatively regulated when phosphorylated at residue Ser985 of the JM domain (Gandino, Longati, Medico, Prat, & Comoglio, 1994). When phosphorylated, the MET receptor is regulated in MET at Tyr1003 within the JM domain by recruiting CBL, an E3- ubiquitin ligase. The MET receptor is internalized, trafficked to late endosomes, and degraded once ubiquitination occurs (Abella et al., 2005). Endophilin, an enzyme involved in membrane curvature, acts as an adaptor for CBL to regulate internalization of MET (Petrelli et al., 2002; Soubeyran, Kowanetz, Szymkiewicz, Langdon, & Dikic, 2002). The ubiquitin-associated (UBA) domain is responsible for the dimerization and activation of CBL to phosphorylate

MET (Kozlov et al., 2007; Tan et al., 2017). Internalization and retention of the ubiquitinated receptor within the bilayered clathrin coat is facilitated by the interaction of ubiquinated MET and its substrate HGF-regulated tyrosine kinase substrate (Hrs) (Bache, Raiborg, Mehlum, & Stenmark, 2003).

3.6 Biological downstream effects

The MAPK, PI3K/AKT cascades, STAT, and NF-κB signaling pathways drive downstream signaling of MET controlling proliferation, cell survival, migration and invasiveness (Fig. 4) (Fan et al., 2005; Furge, Zhang, & Vande Woude, 2000; Johnson & Lapadat, 2002; Müller, Morotti, & Ponzetto, 2002; Ponzetto et al., 1994). The MAPKs, a collection of serine/threonine protein kinases that are activated by a range of extracellular stimuli to mediate signal transduction from cellular membrane to the nucleus, are responsible for the regulation of normal cell proliferation, survival, and

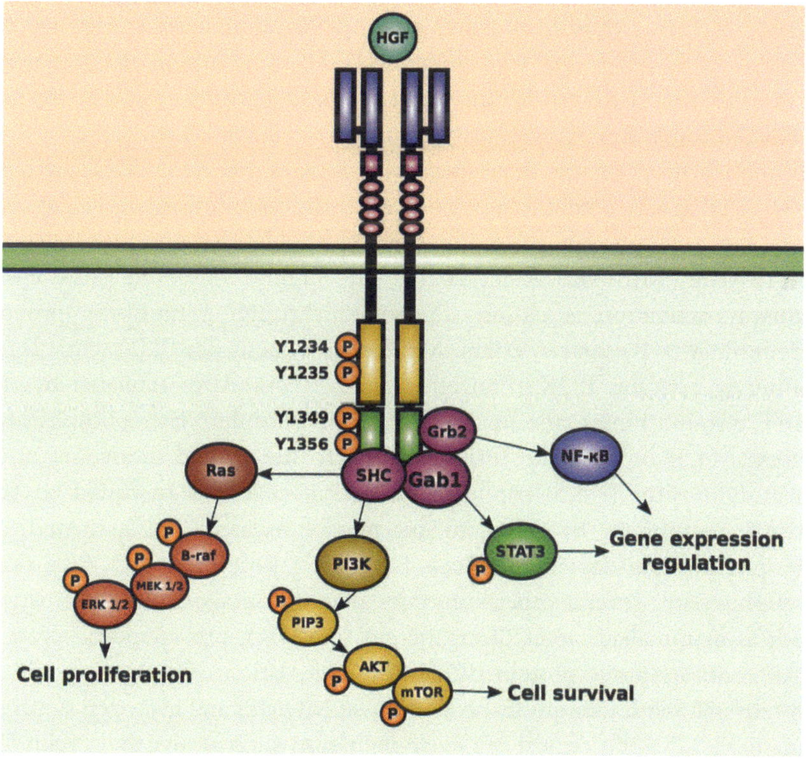

Fig. 4 MET receptor signaling and downstream pathways.

differentiation (Roberts & Der, 2007). The protein kinases are incompletely controlled by protein phosphorylation and the cascade contains three proteins named a MAPK kinase kinase (MAPKKK), a MAPK kinase (MAPKK), and a MAPK (Johnson & Lapadat, 2002; Kesavan et al., 2004). In mammalian cells, there are several groups of MAPKs that act as the final effectors of the cascade including the Extracellular signal-regulated kinases (ERKs), the p38MAPKs, and the c-Jun NH2-terminal kinases (JNKs) (Trusolino, Bertotti, & Comoglio, 2010). One-third of all human cancers have a deregulation of the ERK pathway (Dhillon, Hagan, Rath, & Kolch, 2007). Tyrosine kinase-dependent stimulation of RAS is responsible for the activation of ERKs (Trusolino et al., 2010). RAS activation occurs through the GRB2/SOS complex and is initiated directly with the MET carboxy-terminal multifunctional docking site or indirectly with the SHC adaptor protein (Pelicci et al., 1995; Ponzetto et al., 1994). RAS/ERK activation transpires as the result of the tyrosine phosphatase SHP2 dephosphorylating the binding site on the adaptor protein GAB1 (Maroun, Naujokas, Holgado-Madruga, Wong, & Park, 2000; Montagner et al., 2005). The loading of the RAS GTPase with guanosine triphosphate (GTP), and thus recruiting RAF kinases to the plasma membrane for activation, is prompted by the ligand-mediated activation of receptor tyrosine kinases (Dhillon et al., 2007). The phosphorylation of the duel specificity kinases MEK1/MEK2 happens after RAF is activated as it translocates to the membrane. ERK1/ERK2 is the last to be phosphorylated by the activated MEKs, and thus finishing the cascade (Trusolino et al., 2010). The highly conserved PI3K/AKT pathway activation is closely controlled through a multistep process (Hemmings & Restuccia, 2012). MET directly activates PI3K while RAS indirectly activates PI3K (Ponzetto et al., 1994). After activated by the MET tyrosine kinase, the lipid kinase PI3K phosphorylates phosphatidylinositol (4,5)-biphosphate (PIP2) leading to the second messenger phosphatidylinositol (3,4,5)-triphosphate (PIP3). AKT is recruited to the plasma membrane by PIP3 to be phosphorylated and activated by phosphoinositide-dependent kinase-1 (PDK-1). Once activated, AKT then phosphorylates several other substrates that encourages tumor formation such as mammalian target of rapamycin (mTOR), pro-apoptotic protein BAD, anti-apoptotic protein BCL2, the degradation of the pro-apoptotic protein p53 via E3 ubiquitin-protein ligase MDM2, and glycogen synthase kinase 3β (GSK3β); critical cell cycle regulators such as myc and cyclin D1 are disrupted by the substrates mentioned above (Trusolino et al., 2010).

The Janus kinase (JAK)/(STAT) pathway has been shown to be important for cancer progression and comprises of seven members of the STAT protein family with STAT3 and STAT5 most prevalent in cancer (Haura, Turkson, & Jove, 2005; Yu & Jove, 2004; Yu, Lee, Herrmann, Buettner, & Jove, 2014). The phosphorylation of the MET docking site leads to the phosphorylation of STAT3 causing the separation of STATs from the receptors (Boccaccio et al., 1998). In addition to its main job as a direct transcription factor, STAT3 has other roles such as gene expression regulation through epigenetic mechanisms, mitochondrion function regulation, angiogenesis, and cancer stem cell regulation; emphasizing this pathways function in cancer is its ability for tumor cell proliferation, survival, and tumor invasion (Yu et al., 2014). NF-κB regulates gene expression through binding to the promoters of a mass of genes and is present in the vast majority of cell types and tissues. There are five proteins in the NF-κB family that regulate target genes by forming homo- and heterodimeric complexes with each other and comprise of p65 (RelA), RelB, c-Rel, p105/p50 (NF-κB1) and p100/52 (NF-κB2) (Mukherjee, Houston, Cardenas, & Ghosh, 2015). The inhibitor of κB (IκB) causes the NF-κB system to be inactive in the cytoplasm (Karin, Cao, Greten, & Li, 2002). The classical or alternative pathways are generally responsible for NF-κB activation (Mukherjee et al., 2015). The classical pathway starts with MET activation of the inhibitor of nuclear factor kappa-B kinase (IKK) complex. The subsequent phosphorylation of two N-terminal serine residues of IκB proteins leads to its degradation through the ubiquitin-mediated pathway. The alternative pathway is mediated by the phosphorylation by NF-κB inducing kinase on IKKα. The phosphorylation of p100 by IKKα targets the inhibitory molecules for polyubiquitination and degradation by the proteasome. The breakdown of IκB proteins in both pathways reveals the nucleus localization signal to release NF-κB allowing for it to travel to the nucleus to bind to gene promoter regions for target gene activation (Fan et al., 2005; Müller et al., 2002; Proto et al., 2015; Wang et al., 2014).

4. MET therapeutics

4.1 History

As this chapter has thus far elucidated, the combination of its ligand HGF and MET causes a change in the c-MET receptor protein TK domain which is activated by autophosphorylation. The downstream signaling pathways are then activated. The subsequent cellular responses include the aforementioned mitogenesis, morphogenesis, and proliferative effects. As was stated earlier, it has been shown in prior studies that amplification of MET is shown to be associated with poor prognostic indicators in many types of malignancies. The question is how best to approach developing therapeutics to target this pathway (Mo & Liu, 2017).

The development of the first MET inhibitor, k252a, in 2002 was noted to be weakly selective (Morotti, Mila, Accornero, Tagliabue, & Ponzetto, 2002). A year later PHA-665752, an ATP competitor was found to be more specific (Christensen et al., 2003; Morotti et al., 2002). The mid 2000s brought further clarification of the role of MET in various processes, such as cell proliferation, differentiation, transcription control, G1/s checkpoint, motility, survival, and apoptosis (Ma, Schaefer, Christensen, & Salgia, 2005). The similar role of EGFR in these processes spurred further preclinical study of the synergistic potential of targeting both of these receptor tyrosine kinases.

The discovery of the additive effect of EGF and HGF in cell membrane ruffling further demonstrated its therapeutic potential and we were the first to identify MET as a viable therapeutic target (Puri & Salgia, 2008; Sattler, Ma, & Salgia, 2004). In 2007, Engelman and authors demonstrated that anti-EGFR therapy with gefitinib in lung cancer led to MET amplification through ERBB3 signaling (Engelman et al., 2007). We identified the first generation MET inhibitor, SU11274, and showed synergism in combination with an EGFR inhibitor demonstrating inhibition of cell proliferation and apoptosis (Ma, Jagadeeswaran, et al., 2005; Puri & Salgia, 2008). These preclinical findings established the presence of cross-talk between EGFR and MET pathway (Puri & Salgia, 2008). In 2013, MET amplification was implicated in driving anti-EGFR resistance in colorectal cancer (Bardelli et al., 2013).

There are three main ways for inhibiting the kinase activity of the MET pathway. The first is to prevent the extracellular binding of MET and HGF. This can be done with a biologic antagonist or neutralizing antibody. The second way is to prevent phosphorylation of the TK domain with a small molecule inhibitor. Lastly, one could block downstream signaling in the pathway by stopping the later kinase signaling pathway or signal transducer (Ma et al., 2007; Mo & Liu, 2017). Agents targeting MET or HGF are broadly classified as small molecule inhibitors and monoclonal antibodies. The small molecule inhibitors include TKIs which are either selective or multikinase TKIs. Cabozantinib and crizotinib are examples of multikinase TKIs. Tepotinib and capmatinib are examples of selective MET inhibitors.

The monoclonal antibodies that target HGF include riltumumab and ficlatuzumab. Anti-MET antibodies include onartuzumab and emibetuzumab (Salgia et al., 2014; Yap & Popat, 2017). To date the results of clinical trials targeting MET have been widely variable. There have been some dramatic responses in preselected single arm studies and many reported failures in phase III studies with larger and more molecularly heterogeneous populations. Ultimately, most c-Met inhibitor clinical trials have not shown significant improvement in survival (Hughes & Siemann, 2019). A multitude of studies have failed to reach their primary endpoint (Drilon et al., 2017).

4.2 Clinical application in non-small cell lung cancer (NSCLC)

As was mentioned prior, MET mutations are estimated to be found in roughly 6% of patients diagnosed with NSCLC (Cancer Genome Atlas Research Network, 2014). MET inhibitors combined with EGFR tyrosine kinase inhibitors have been evaluated in several clinical trials. First line therapies aimed at EGFR, which stimulates c-MET activation, have been recommended for NSCLS with targeted mutations (Mo & Liu, 2017). For example, A 2016 randomized controlled phase II trial contrasted cabozantinib alone or in combination with erlotinib vs erlotinib alone in patients with advanced NSCLC with wild-type EGFR and found that both cabozantinib alone (4.3 months) or in combination with erlotinib (4.7 months) had longer progression-free survival than erlotinib alone (1.8 months) (Neal et al., 2016).

There initial great deal of interest in the treatment of MET overexpressing lung cancers was due to a phase II trial of erlotinib and onartuzumab (Spigel et al., 2013). Patients were randomized to erlotinib plus or minus onartuzumab and patients who expressed higher levels of MET did show some progression-free survival benefit. However, subsequently, a phase III trial with NSCLC patient that had MET overexpression with erlotinib plus or minus onartuzumab was ended early due to concerns about potential harm (Spigel et al., 2017). At this point crizotinib, an ATP-competitive small molecule inhibitor, which targets both MET and anaplastic lymphoma kinase (ALK) is FDA approved for treatment of ALK positive NSCLC (Brennan et al., 2013; Clark et al., 2020). In one study of 4672 NSCLC patients 160 ALK-rearranged cases had 16 patients that were identified among them with c-Met overexpression (Chen et al., 2018). Of those patients, 116 ALK-rearranged patients received treatment with crizotinib and the objective response rate (ORR) was 86.7% in ALK-rearranged patients with c-Met overexpression and 59.4% in those without c-Met overexpression this suggesting the advantage in the c-Met group (Chen et al., 2018). This highlights the initial potency of treatment with this drug, however, resistance then develops making this ineffective as a monotherapy (Nanjo et al., 2013).

4.3 Clinical application in central nervous system (CNS) tumors

MET is also often upregulated in Glioblastoma, the most common primary brain tumor and dysregulation of RTKs have been noted in approximately 90% of GBM (Brennan et al., 2013). The previously mentioned crizotinib, which as was noted is FDA approved for treatment of ALK positive NSCLC, is in phase II clinical trials for CNS and brain tumors (Cruickshanks et al., 2017). Cabozantinib has been studied in in GBM tumor models and has shown the inhibition phosphorylation of MET resulting in decreased proliferation and invasion (Yakes et al., 2011). A phase II trial has recently been completed for this drug. Another ATP-competitive MET inhibitor, capmatinib, has completed phase I trials for multiple cancers and is now in a phase I trial in GBM patients (Bauer et al., 2016). The first non-ATP-competitive inhibitor of MET, tivantinib, is currently in phase I trials for patients with metastasis and is showing to be well tolerated (Eder, Vande Woude, Boerner, & LoRusso, 2009).

4.4 Clinical application in gastrointestinal (GI) malignancy

In regards to GI malignancies that show MET amplification approximately 4–10% of upper GI cancers and 50% of advanced gastric cancer show overexpression (Catenacci et al., 2011; Mo & Liu, 2017). There have been a multitude of trials evaluating MET inhibitors as individual treatment or combined with cytotoxic chemotherapy with low success (Mo & Liu, 2017). For example, rilotumumab, a human monoclonal antibody that neutralizes HGF, was studied in a phase two trial of GI cancer patients in combination with epirubicin, cisplatin, and capecitabine vs placebo when combined with the same three drugs and there was some improved overall survival noted in the rilotumumab group (Iveson et al., 2014). However, in phase III trials of rilotumumab vs placebo plus epirubicin, cisplatin, and capecitabine in patients with gastroesophageal adenocarcinoma it did not show to be superior to placebo and proved to be worse (Catenacci et al., 2017). There were increased deaths in the patient population within the rilotumumab arm (Catenacci et al., 2017). This is only an example, but many c-MET inhibitors have been developed and evaluated in patients with GI cancer in various phase II/III clinical trials and there has been a reported a lack of improvement in efficacy with these treatments (Mo & Liu, 2017).

4.5 Role in hepatocellular carcinoma (HCC)

For those patients who cannot undergo surgery, Sorafenib has been the standard of care for unresectable HCC, but approvals of new TKIs and PD-1 antibodies has led to further interest in other drug targets for treatment (Jafarnejad et al., 2019). A phase II randomized controlled trial compared tivantinib vs placebo as second-line therapy in 108 patients with advanced HCC and in subgroup analysis of patients with high c-MET expression, tivantinib lengthened overall survival (Santoro et al., 2013). However, in a phase III trial of inoperable HCC in patients with MET overexpressing disease tivantinib failed to improve outcomes when compared to placebo (Rimassa et al., 2017). A phase III trial is currently underway to compare cabozantinib and placebo as second-line treatment after sorafenib (Mo & Liu, 2017). A phase II trial that included 41 HCC patients with advanced disease evaluated cabozantinib and there appeared to be a partial response (Verslype et al., 2012). Various phase III trials are underway.

4.6 Clinical application in medullary thyroid cancer

In regards to current first line treatment at this point in time, cabozanitib, whose targets also include VEGF, has been approved for the treatment of metastatic medullary thyroid cancer (Choueiri et al., 2015; Kurzrock et al., 2011). It is a small molecule inhibitor that was approved by the FDA in 2012 for metastatic disease (Priya, Dravid, Digumarti, & Dandekar, 2017). In a randomized phase III trial 330 patients with progressive metastatic medullary thyroid carcinoma were assigned to either cabozantinib or placebo and there was a statistically significant prolongation in progression-free survival in the cabozantinib group (Priya et al., 2017). As will be discussed next, when considering the development of MET directed therapeutics and ongoing research, consideration must be given to the development of resistance to drugs and the evolution of the malignancy being treated.

5. MET mechanisms of resistance and tumor evolution

MET kinase mutations while initially identified in hereditary papillary renal cell carcinoma have been described in various human cancers. The hepatocyte growth factor (HGF) receptor MET is a receptor tyrosine kinase (RTK) that has been implicated in numerous biological functions from embryonic development to tissue repair (Graveel, Tolbert, & Vande Woude, 2013). NSCLC patients harboring MET exon 14 mutation have been reported to respond effectively to MET inhibitors such as crizotinib, with several other type I and type II small molecule kinase inhibitors coming to fruition including cabozantinib and crizotinib, with MET inhibitory activity that have been approved by the FDA (Cui et al., 2011; Yakes et al., 2011). Since then two additional MET inhibitors, tepotinib and capmatinib, have both been granted breakthrough therapy designations for patients with MET-mutated advanced non-small cell lung cancer (Merck KGaA, 2019; Novartis, 2019). However, as observed in other RTK inhibitors, resistance eventually develops and the patients invariably have progression of their disease. This acquired resistance to targeted therapy can occur by a variety of mechanisms, including second-site mutations in the kinase domain, bypass signaling pathway activation, copy number changes, and histologic transformation (Lin, Riely, & Shaw, 2017; Ohashi, Maruvka, Michor, & Pao, 2013; Sequist et al., 2011). There have only been a few studies that have characterized this resistance with the majority of which are attributed to different secondary mutations in the MET tyrosine kinase domain and are believed to interfere with the adequate binding of the drug to the kinase (Fig. 5) (Bahcall et al., 2016; Dong et al., 2016; Heist et al., 2016; Ou et al., 2017).

5.1 Resistance through acquired somatic mutations

MET mutations such as D1228 and Y1230 have been identified to cause resistance to type I MET inhibitors, such as crizotinib (Dong et al., 2016; Heist et al., 2016; Ou et al., 2017). It has also been reported that Y1230H and D1228N/H MET mutations can arise alone and simultaneously in resistance to MET inhibition (Schrock, Lai, Ali, Miller, & Raez, 2017). Other MET secondary mutations have also been identified that are associated with resistance to MET targeted therapies (Dong et al., 2016). In a resistance screen of MET inhibitor NVP-BVU972 a number of resistant clones were identified with resistant clones in Y1230 as the most

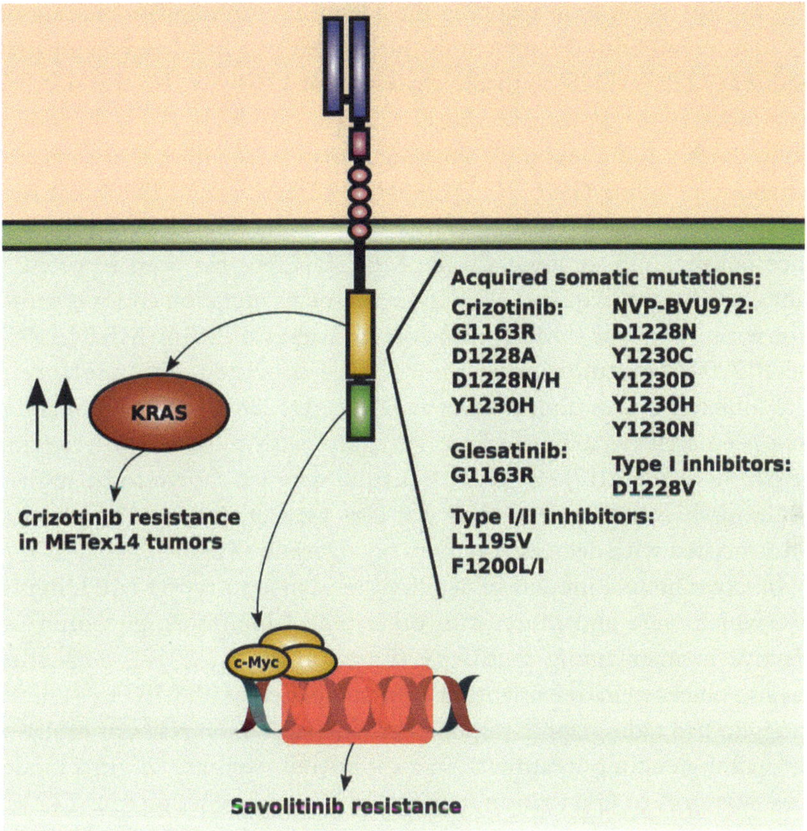

Fig. 5 Mechanisms of resistance to MET inhibitors.

frequent showing highest number of clones of Y1230C, Y1230H, and Y1230N (Tiedt et al., 2011). At much lower rates resistant clones were shown in D1228, F1200, and V1155 (Tiedt et al., 2011). MET D1228N, Y1230C, Y1230H and Y1230D mutations have been identified not only in lung cancer but also in hereditary or sporadic PRCC (Schmidt et al., 1997, 1999) and Y1230C has been reported in metastases of head and neck cancer (Di Renzo et al., 2000; Lorenzato et al., 2002). In one case report of a patient with a MET exon 14 skipping D1010H mutation that was treated with crizotinib, a D1228N mutation in exon 19 of MET emerged within the activation loop of the MET kinase domain (Heist et al., 2016). D1228N/H and Y1230H mutations in the MET kinase domain have been linked to resistance to crizotinib at time of progression (Dong et al., 2016). The MET receptor tyrosine kinase adopts a unique inactive conformation, with the activation loop spanning the adenosine triphosphate binding site through several critical interactions, such as salt pair involving an aspartate residue at 1228 (D1228) with a lysate residue at 1110 (K1110) that stabilizes the position of the tyrosine residue at 1230 (Y1230) (Cui, 2014). In another patient with a MET exon 14 skipping mutation acquired four resistance mutations including G1163R, D1228H, D1228A, and Y1230H that were detected in cell-free circulating-tumor DNA at the time of crizotinib resistance (Zhang, Yin, & Peng, 2017). MET D1228V has been reported by Bahcall et al. as an acquired MET type I resistant mutation and was shown to be sensitive to type II MET inhibitors (Bahcall et al., 2016). MET L1195V and F1200L/I mutations were shown to be associated with resistance to both inhibitor classes (Engstrom et al., 2017). D1228 and Y1230 mutations have been well documented in case reports with resistance to crizotinib (Engstrom et al., 2017). G1163R mutation that was shown to be resistant to glesatinib in one patient case has also been previously characterized and associated with decreased activity of crizotinib (Engstrom et al., 2017).

In experimental models of acquired resistance to type I MET inhibitors, which were also resistant to other type I inhibitors, glesatinib was effective in maintaining sensitivity (Engstrom et al., 2017). This effect was also observed in one crizotinib-resistant patient with METex14 mutation who had radiographic response and loss of A-loop resistant mutation allele after glesatinib treatment. In a capmatinib-resistant in vitro model, there was an A to G mutation (c.3689A > G) corresponding to a Y1230C mutation. Meanwhile in a crizotinib-resistant model, a T to C mutation (c.3688T > C) corresponding to a Y1230H mutation as well as a T to C mutation (c.3598T > C) corresponding to a F1200L mutation were identified. While crizotinib was less potent in the capmatinib-resistant cell line, capmatinib was completely inactive in the crizotinib-resistant cell line. Surprisingly glesatinib was potent in the crizotinib-resistant cell line but showed similar resistance in the capmatinib-resistant model. This suggests that the F1200L variant in cis with Y1230H may be responsible for resistance against crizotinib and glesatinib. In an in vivo model, while Y1230C double mutation maintained glesatinib sensitivity, the D1228N double mutation in the A-loop was not as effective and eventually tumor growth recovered in the mice. This resistance was reconstituted in a single METex14 del mutation patient with MET gene amplification, D1228N, Y1230H, Y1230S, and G1163R mutations. After treatment with glesatinib, liver lesions continued to grow and some increased by 35%. A liquid biopsy after treatment also showed that Y1230H and

Y1230S mutations were no longer detectable, but the patient still secreted D1228N, G1163R and a newly acquired mutation L1195V—with an increase in MET amplification from 50th percentile to 90th percentile (Engstrom et al., 2017). The activity of glesatinib, as a type II inhibitor, against some type I inhibitor resistance mutations suggests for potential sequential strategies to overcome this resistance (Engstrom et al., 2017). Meanwhile other type II MET inhibitors are predicted to have a differential resistance profile of certain active site mutations and the resistance profiles for type II inhibitors such as cabozatinib and tepotinib have not yet been well described (Henry et al., 2016; Qi et al., 2011; Weber et al., 2014). Future resistance mechanisms may be detected in patients utilizing liquid biopsies, however, the limitation to the ctDNA method includes the inability to detect histological changes including epithelial-mesenchymal transition, small cell transformation, or potential upregulation of the MET ligand (Gainor et al., 2016; Pennacchietti et al., 2014).

5.2 Off-target mechanisms of resistance

While most of MET inhibitor resistance profiles have been target-dependent mechanisms involving the acquisition of a secondary mutation in the MET kinase domain, a study by Suzawa et al. identified that activation of KRAS mutation concurrently with METex14 in lung cancer conferred either primary or acquired resistance to crizotinib in patients (Suzawa et al., 2019). MET has been shown to play a role in Ras-mediated tumorigenesis and the activation of the RAS-ERK pathway induces higher MET expression via a transcriptional effect (Furge et al., 2001; Ivan, Bond, Prat, Comoglio, & Wynford-Thomas, 1997; Webb et al., 1998). More so, Suzawa et al. found that METex14 tumors also had a higher degree of RAS pathway activation than tumors with wild-type MET amplification suggesting that METex14 tumors may be more tolerant of KRAS activation (Suzawa et al., 2019). In another preclinical study, Bahcall et al. identified a unique mechanism by which KRAS amplification leads to drug resistance in METex14 patients which was shown to be counter-acted by dual MET/PI3K inhibition against this resistance (Bahcall et al., 2018). Surprisingly, the expression of mutant KRAS prolonged the half-life of METex14 protein suggesting that KRAS regulates METex14 at multiple levels (Suzawa et al., 2019).

MYC expression and particularly MYC overexpression have been implicated in resistance to savolitinib treatment. In an in vitro experiment, it was shown that MYC downregulation attenuated savolitinib efficacy while sustained MYC overexpression was observed in savolitinib resistant clones (Henry et al., 2016). Furthermore, PIM signaling has been implicated in acquired MET inhibitor resistance whereby PIM kinases upregulate the translation of BCL-2 in response to apoptosis induced by MET inhibitors (An, Xiong, LaRue, Kraft, & Cen, 2015). It has been shown that blocking this activity in combination with MET inhibition resensitized the cells to savolitinib in vitro (Henry et al., 2016). It was also reported that lack of MET expression or RAS pathway activation was associated with poor MET TKI outcomes in MET exon-14 altered lung cancers—which suggests that off-target acquired resistance may be mediated by RAS/MDM2/EGFR pathway activation (Guo et al., 2019). Further research into understanding the off-target mechanisms of resistance for MET is required in order to fully understand the required sequential therapy strategies of the currently available MET inhibitors to ensure best outcomes for patients.

6. Future directions

This chapter has discussed the history of MET, the genetics of this TKR, and given some background on the receptor biology. It has explored the development and failure of some directed therapeutics and the mechanisms of resistance that have developed to treatment. It has now been over 3 decades since the discovery of the MET pathway. HGF and MET receptor signaling has been shown to promote tumor cell survival, proliferation, and invasion in a variety of cancers. Inhibition of this pathway has shown to be beneficial in an array of malignancies in prior trials, however, there have been setbacks and disappointments throughout the process.

As has been mentioned earlier in this chapter, MET inhibitors have displayed mixed outcomes in clinical trials. Continued advances in clinical technology and ability to better select for patients in trials is key to further exploring the full potential of targeting this pathway for treatment of various types of cancer. It is clear that a subgroup of patients that exhibit amplification of the MET gene are better served by MET targeted therapies (Cruickshanks et al., 2017). Better definition of targets at a molecular level and specific populations for this drug class benefit will be key.

Research has also shown that autocrine production of HGF is a predictor of response to MET inhibition therapy (Zhang et al., 2013). Given the development of resistance with monotherapy, combining MET inhibitors with other RTKs may also be of benefit in subsequent trials (Lee, Joo, Lee, Yoon, & Nam, 2014). If the inhibitors alone are not sufficient, they may enhance the ability of other drugs on cancer cells (Garajová et al., 2015). Further clarifying how best to enhance this synergy is important moving forward. There is still much to be explored about how best to target and treat cancer in relation to MET-based therapeutics and recent advances have reignited the field.

References

Abella, J. V., Peschard, P., Naujokas, M. A., Lin, T., Saucier, C., Urbé, S., et al. (2005). Met/Hepatocyte growth factor receptor ubiquitination suppresses transformation and is required for Hrs phosphorylation. *Molecular and Cellular Biology*, *25*(21), 9632–9645. PMID: 16227611.

Adachi, E., Sakai, K., Nishiuchi, T., Imamura, R., Sato, H., & Matsumoto, K. (2016). Different growth and metastatic phenotypes associated with a cell-intrinsic change of Met in metastatic melanoma. *Oncotarget*, 7(43), 70779–70793. PMID: 27683122; PMCID: PMC5342589.

An, N., Xiong, Y., LaRue, A. C., Kraft, A. S., & Cen, B. (2015). Activation of pim kinases is sufficient to promote resistance to MET small-molecule inhibitors. *Cancer Research*, *75*(24), 5318–5328. PMID: 26670562; PMCID: PMC4683614.

Andermarcher, E., Surani, M. A., & Gherardi, E. (1996). Co-expression of the HGF/SF and c-met genes during early mouse embryogenesis precedes reciprocal expression in adjacent tissues during organogenesis. *Developmental Genetics*, *18*(3), 254–266. PMID: 8631159.

Bache, K. G., Raiborg, C., Mehlum, A., & Stenmark, H. (2003). STAM and Hrs are subunits of a multivalent ubiquitin-binding complex on early endosomes. *The Journal of Biological Chemistry*, *278*(14), 12513–12521. PMID: 12551915.

Bahcall, M., Awad, M. M., Sholl, L. M., Wilson, F. H., Xu, M., Wang, S., et al. (2018). Amplification of wild-type KRAS Imparts resistance to crizotinib in MET Exon 14 mutant non-small cell lung cancer. *Clinical Cancer Research*, *24*(23), 5963–5976. PMID: 30072474; PMCID: PMC6279568.

Bahcall, M., Sim, T., Paweletz, C. P., Patel, J. D., Alden, R. S., Kuang, Y., et al. (2016). Acquired METD1228V mutation and resistance to MET inhibition in lung cancer. *Cancer Discovery*, *6*(12), 1334–1341. PMID: 27694386; PMCID: PMC5140694.

Bardelli, A., Corso, S., Bertotti, A., Hobor, S., Valtorta, E., Siravegna, G., et al. (2013). Amplification of the MET receptor drives resistance to anti-EGFR therapies in colorectal cancer. *Cancer Discovery*, *3*(6), 658–673. PMID: 23729478; PMCID: PMC4078408.

Bardelli, A., Longati, P., Gramaglia, D., Basilico, C., Tamagnone, L., Giordano, S., et al. (1998). Uncoupling signal transducers from oncogenic MET mutants abrogates cell transformation and inhibits invasive growth. *Proceedings of the National Academy of Sciences of the United States of America*, *95*(24), 14379–14383. PMID: 9826708.

Basilico, C., Arnesano, A., Galluzzo, M., Comoglio, P. M., & Michieli, P. (2008). A high affinity hepatocyte growth factor-binding site in the immunoglobulin-like region of Met. *The Journal of Biological Chemistry*, *283*(30), 21267–21277. PMID: 18495663.

Bauer, T. M., Schuler, M., Berardi, R., Lim, W.-T., Van Geel, R., De Jonge, M., et al. (2016). MINI01.03: Phase (Ph) I study of the safety and efficacy of the cMET inhibitor capmatinib (INC280) in patients with advanced cMET+ NSCLC: Topic: Medical oncology. *Journal of Thoracic Oncology*, *11*(Suppl. 11), S257-S8.

Bender, S., Gronych, J., Warnatz, H.-J., Hutter, B., Gröbner, S., Ryzhova, M., et al. (2016). Recurrent MET fusion genes represent a drug target in pediatric glioblastoma. *Nature Medicine*, *22*(11), 1314–1320.

Bieche, I., Champeme, M. H., & Lidereau, R. (1999). Infrequent mutations of the MET gene in sporadic breast tumours. *International Journal of Cancer, 82*(6), 908–910. PMID: 10446461.

Birchmeier, C., Birchmeier, W., Gherardi, E., & Vande Woude, G. F. (2003). Met, metastasis, motility and more. *Nature Reviews. Molecular Cell Biology, 4*(12), 915–925. PMID: 14685170.

Boccaccio, C., Ando, M., Tamagnone, L., Bardelli, A., Michieli, P., Battistini, C., et al. (1998). Induction of epithelial tubules by growth factor HGF depends on the STAT pathway. *Nature, 391*(6664), 285–288. PMID: 9440692.

Brennan, C. W., Verhaak, R. G., McKenna, A., Campos, B., Noushmehr, H., Salama, S. R., et al. (2013). The somatic genomic landscape of glioblastoma. *Cell, 155*(2), 462–477. PMID: 24120142; PMCID:PMC3910500.

Brenner, D., & Mak, T. W. (2009). Mitochondrial cell death effectors. *Current Opinion in Cell Biology, 21*(6), 871–877. PMID: 19822411.

Brooks, C., Wei, Q., Cho, S.-G., & Dong, Z. (2009). Regulation of mitochondrial dynamics in acute kidney injury in cell culture and rodent models. *The Journal of Clinical Investigation, 119*(5), 1275–1285. PMID: 19349686.

Cancer Genome Atlas Research Network. (2012). Comprehensive genomic characterization of squamous cell lung cancers. *Nature, 489*(7417), 519–525. PMID: 22960745; PMCID: PMC3466113.

Cancer Genome Atlas Research Network. (2014). Comprehensive molecular profiling of lung adenocarcinoma. *Nature, 511*(7511), 543–550. PMID: 25079552; PMCID: PMC4231481.

Cappuzzo, F., Janne, P. A., Skokan, M., Finocchiaro, G., Rossi, E., Ligorio, C., et al. (2009). MET increased gene copy number and primary resistance to gefitinib therapy in non-small-cell lung cancer patients. *Annals of Oncology, 20*(2), 298–304. PMID: 18836087; PMCID: PMC2733067.

Cappuzzo, F., Marchetti, A., Skokan, M., Rossi, E., Gajapathy, S., Felicioni, L., et al. (2009). Increased MET gene copy number negatively affects survival of surgically resected non-small-cell lung cancer patients. *Journal of Clinical Oncology: Official Journal of the American Society of Clinical Oncology, 27*(10), 1667–1674. PMID: 19255323.

Cassidy-Stone, A., Chipuk, J. E., Ingerman, E., Song, C., Yoo, C., Kuwana, T., et al. (2008). Chemical inhibition of the mitochondrial division dynamin reveals its role in Bax/Bak-dependent mitochondrial outer membrane permeabilization. *Developmental Cell, 14*(2), 193–204. PMID: 18267088.

Catenacci, D. V., Henderson, L., Xiao, S. Y., Patel, P., Yauch, R. L., Hegde, P., et al. (2011). Durable complete response of metastatic gastric cancer with anti-Met therapy followed by resistance at recurrence. *Cancer Discovery, 1*(7), 573–579. PMID: 22389872; PMCID: PMC3289149.

Catenacci, D. V. T., Tebbutt, N. C., Davidenko, I., Murad, A. M., Al-Batran, S. E., Ilson, D. H., et al. (2017). Rilotumumab plus epirubicin, cisplatin, and capecitabine as first-line therapy in advanced MET-positive gastric or gastro-oesophageal junction cancer (RILOMET-1): A randomised, double-blind, placebo-controlled, phase 3 trial. *Lancet Oncology, 18*(11), 1467–1482. PMID: 28958504; PMCID: PMC5898242.

Cecchi, F., Rabe, D. C., & Bottaro, D. P. (2012). Targeting the HGF/Met signaling pathway in cancer therapy. *Expert Opinion on Therapeutic Targets, 16*(6), 553–572. PMID: 22530990; PMCID: PMC3711667.

Chang, C.-R., & Blackstone, C. (2010). Dynamic regulation of mitochondrial fission through modification of the dynamin-related protein Drp1. *Annals of the New York Academy of Sciences, 1201*, 34–39. PMID: 20649536.

Chen, R. L., Zhao, J., Zhang, X. C., Lou, N. N., Chen, H. J., Yang, X., et al. (2018). Crizotinib in advanced non-small-cell lung cancer with concomitant ALK rearrangement and c-Met overexpression. *BMC Cancer, 18*(1), 1171. PMID: 30477470; PMCID: PMC6258481.

Choueiri, T. K., Escudier, B., Powles, T., Mainwaring, P. N., Rini, B. I., Donskov, F., et al. (2015). Cabozantinib versus everolimus in advanced renal-cell carcinoma. *The New England Journal of Medicine*, *373*(19), 1814–1823. PMID: 26406150; PMCID: PMC5024539.

Choueiri, T. K., Vaishampayan, U., Rosenberg, J. E., Logan, T. F., Harzstark, A. L., Bukowski, R. M., et al. (2013). Phase II and biomarker study of the dual MET/VEGFR2 inhibitor foretinib in patients with papillary renal cell carcinoma. *Journal of Clinical Oncology: Official Journal of the American Society of Clinical Oncology*, *31*(2), 181–186. PMID: 23213094.

Christensen, J. G., Schreck, R., Burrows, J., Kuruganti, P., Chan, E., Le, P., et al. (2003). A selective small molecule inhibitor of c-Met kinase inhibits c-Met-dependent phenotypes in vitro and exhibits cytoreductive antitumor activity in vivo. *Cancer Research*, *63*(21), 7345–7355. PMID: 14612533.

Cirri, P., & Chiarugi, P. (2012). Cancer-associated-fibroblasts and tumour cells: A diabolic liaison driving cancer progression. *Cancer Metastasis Reviews*, *31*(1–2), 195–208. PMID: 22101652.

Clark, J. W., Camidge, D. R., Kwak, E. L., Maki, R. G., Shapiro, G. I., Chen, I., et al. (2020). Dose-escalation trial of the ALK, MET & ROS1 inhibitor, crizotinib, in patients with advanced cancer. *Future Oncology*, *16*(1), 4289–4301. PMID: 31778074.

Comoglio, P. M., Giordano, S., & Trusolino, L. (2008). Drug development of MET inhibitors: Targeting oncogene addiction and expedience. *Nature Reviews Drug Discovery*, 7(6), 504–516. PMID: 18511928.

Cooper, C. S., Park, M., Blair, D. G., Tainsky, M. A., Huebner, K., Croce, C. M., et al. (1984). Molecular cloning of a new transforming gene from a chemically transformed human cell line. *Nature*, *311*(5981), 29–33. PMID: 6590967.

Corso, S., Migliore, C., Ghiso, E., De Rosa, G., Comoglio, P. M., & Giordano, S. (2008). Silencing the MET oncogene leads to regression of experimental tumors and metastases. *Oncogene*, *27*(5), 684–693. PMID: 17684486.

Cruickshanks, N., Zhang, Y., Yuan, F., Pahuski, M., Gibert, M., & Abounader, R. (2017). Role and therapeutic targeting of the HGF/MET pathway in glioblastoma. *Cancers*, *9*(7), 87. PMID: 28696366; PMCID: PMC5532623.

Cui, J. J. (2014). Targeting receptor tyrosine kinase MET in cancer: Small molecule inhibitors and clinical progress. *Journal of Medicinal Chemistry*, *57*(11), 4427–4453. PMID: 24320965.

Cui, J. J., Tran-Dube, M., Shen, H., Nambu, M., Kung, P. P., Pairish, M., et al. (2011). Structure based drug design of crizotinib (PF-02341066), a potent and selective dual inhibitor of mesenchymal-epithelial transition factor (c-MET) kinase and anaplastic lymphoma kinase (ALK). *Journal of Medicinal Chemistry*, *54*(18), 6342 6363. PMID: 21812414.

Curran, S., & Murray, G. I. (2000). Matrix metalloproteinases: Molecular aspects of their roles in tumour invasion and metastasis. *European Journal of Cancer*, *36*(Spec. 13), 1621–1630. PMID: 10959048.

de Melo Gagliato, D., Jardim, D. L. F., Falchook, G., Tang, C., Zinner, R., Wheler, J. J., et al. (2014). Analysis of MET genetic aberrations in patients with breast cancer at MD Anderson Phase I unit. *Clinical Breast Cancer*, *14*(6), 468–474. PMID: 25065564.

Dharmawardana, P. G., Giubellino, A., & Bottaro, D. P. (2004). Hereditary papillary renal carcinoma type I. *Current Molecular Medicine*, *4*(8), 855–868. PMID: 15579033.

Dhillon, A. S., Hagan, S., Rath, O., & Kolch, W. (2007). MAP kinase signalling pathways in cancer. *Oncogene*, *26*(22), 3279–3290. PMID: 17496922.

Di Renzo, M. F., Olivero, M., Martone, T., Maffe, A., Maggiora, P., Stefani, A. D., et al. (2000). Somatic mutations of the MET oncogene are selected during metastatic spread of human HNSC carcinomas. *Oncogene*, *19*(12), 1547–1555. PMID: 10734314.

Dimou, A., Non, L., Chae, Y. K., Tester, W. J., & Syrigos, K. N. (2014). MET gene copy number predicts worse overall survival in patients with non-small cell lung cancer (NSCLC): A systematic review and meta-analysis. *PLoS One*, *9*(9), e107677.

Dong, G., Chen, Z., Li, Z. Y., Yeh, N. T., Bancroft, C. C., & Van Waes, C. (2001). Hepatocyte growth factor/scatter factor-induced activation of MEK and PI3K signal pathways contributes to expression of proangiogenic cytokines interleukin-8 and vascular endothelial growth factor in head and neck squamous cell carcinoma. *Cancer Research*, *61*(15), 5911–5918. PMID: 11479233.

Dong, H. J., Li, P., Wu, C. L., Zhou, X. Y., Lu, H. J., & Zhou, T. (2016). Response and acquired resistance to crizotinib in Chinese patients with lung adenocarcinomas harboring MET Exon 14 splicing alternations. *Lung Cancer*, *102*, 118–121. PMID: 27987579.

Drilon, A., Cappuzzo, F., Ou, S. I., & Camidge, D. R. (2017). Targeting MET in lung cancer: Will expectations finally be MET? *Journal of Thoracic Oncology*, *12*(1), 15–26. PMID: 27794501; PMCID: PMC5603268.

Eder, J. P., Vande Woude, G. F., Boerner, S. A., & LoRusso, P. M. (2009). Novel therapeutic inhibitors of the c-Met signaling pathway in cancer. *Clinical Cancer Research*, *15*(7), 2207–2214. PMID: 19318488.

Engelman, J. A., Zejnullahu, K., Mitsudomi, T., Song, Y., Hyland, C., Park, J. O., et al. (2007). MET amplification leads to gefitinib resistance in lung cancer by activating ERBB3 signaling. *Science*, *316*(5827), 1039–1043. PMID: 17463250.

Engstrom, L. D., Aranda, R., Lee, M., Tovar, E. A., Essenburg, C. J., Madaj, Z., et al. (2017). Glesatinib exhibits antitumor activity in lung cancer models and patients harboring MET Exon 14 mutations and overcomes mutation-mediated resistance to type I MET inhibitors in nonclinical models. *Clinical Cancer Research*, *23*(21), 6661–6672. PMID: 28765324.

Fan, S., Gao, M., Meng, Q., Laterra, J. J., Symons, M. H., Coniglio, S., et al. (2005). Role of NF-kappaB signaling in hepatocyte growth factor/scatter factor-mediated cell protection. *Oncogene*, *24*(10), 1749–1766. PMID: 15688034.

Fixman, E. D., Fournier, T. M., Kamikura, D. M., Naujokas, M. A., & Park, M. (1996). Pathways downstream of Shc and Grb2 are required for cell transformation by the tpr-Met oncoprotein. *The Journal of Biological Chemistry*, *271*(22), 13116–13122. PMID: 8662733.

Folkman, J., Browder, T., & Palmblad, J. (2001). Angiogenesis research: Guidelines for translation to clinical application. *Thrombosis and Haemostasis*, *86*(1), 23–33. PMID: 11487011.

Frezza, C., Cipolat, S., Martins de Brito, O., Micaroni, M., Beznoussenko, G. V., Rudka, T., et al. (2006). OPA1 controls apoptotic cristae remodeling independently from mitochondrial fusion. *Cell*, *126*(1), 177–189. PMID: 16839885.

Furge, K. A., Kiewlich, D., Le, P., Vo, M. N., Faure, M., Howlett, A. R., et al. (2001). Suppression of Ras-mediated tumorigenicity and metastasis through inhibition of the Met receptor tyrosine kinase. *Proceedings of the National Academy of Sciences of the United States of America*, *98*(19), 10722–10727. PMID: 11535809; PMCID: PMC58533.

Furge, K. A., Zhang, Y. W., & Vande Woude, G. F. (2000). Met receptor tyrosine kinase: Enhanced signaling through adapter proteins. *Oncogene*, *19*(49), 5582–5589. PMID: 11114738.

Furlan, A., Kherrouche, Z., Montagne, R., Copin, M. C., & Tulasne, D. (2014). Thirty years of research on met receptor to move a biomarker from bench to bedside. *Cancer Research*, *74*(23), 6737–6744. PMID: 25411347.

Furlan, A., Stagni, V., Hussain, A., Richelme, S., Conti, F., Prodosmo, A., et al. (2011). Abl interconnects oncogenic Met and p53 core pathways in cancer cells. *Cell Death and Differentiation*, *18*(10), 1608–1616. PMID: 21455220; PMCID: PMC3172114.

Gainor, J. F., Dardaei, L., Yoda, S., Friboulet, L., Leshchiner, I., Katayama, R., et al. (2016). Molecular mechanisms of resistance to first- and second-generation ALK inhibitors in ALK-rearranged lung cancer. *Cancer Discovery*, *6*(10), 1118–1133. PMID: 27432227; PMCID: PMC5050111.

Gandino, L., Di Renzo, M. F., Giordano, S., Bussolino, F., & Comoglio, P. M. (1990). Protein kinase-c activation inhibits tyrosine phosphorylation of the c-met protein. *Oncogene*, *5*(5), 721–725. PMID: 2111905.

Gandino, L., Longati, P., Medico, E., Prat, M., & Comoglio, P. M. (1994). Phosphorylation of serine 985 negatively regulates the hepatocyte growth factor receptor kinase. *The Journal of Biological Chemistry*, *269*(3), 1815–1820. PMID: 8294430.

Gandino, L., Munaron, L., Naldini, L., Ferracini, R., Magni, M., & Comoglio, P. M. (1991). Intracellular calcium regulates the tyrosine kinase receptor encoded by the MET oncogene. *The Journal of Biological Chemistry*, *266*(24), 16098–16104. PMID: 1651934.

Garajová, I., Giovannetti, E., Biasco, G., & Peters, G. J. (2015). c-Met as a target for personalized therapy. *Translational Oncogenomics*, 7(Suppl. 1), 13–31. PMID: 26628860; PMCID: PMC4659440.

Ghadjar, P., Blank-Liss, W., Simcock, M., Hegyi, I., Beer, K. T., Moch, H., et al. (2009). MET Y1253D-activating point mutation and development of distant metastasis in advanced head and neck cancers. *Clinical & Experimental Metastasis*, *26*(7), 809–815. PMID: 19639388.

Gherardi, E., Birchmeier, W., Birchmeier, C., & Vande, W. G. (2012). Targeting MET in cancer: Rationale and progress. *Nature Reviews Cancer*, *12*(2), 89–103. PMID: 22270953; PMCID: PMC4659440.

Gherardi, E., Sandin, S., Petoukhov, M. V., Finch, J., Youles, M. E., Ofverstedt, L. G., et al. (2006). Structural basis of hepatocyte growth factor/scatter factor and MET signalling. *Proceedings of the National Academy of Sciences of the United States of America*, *103*(11), 4046–4051. PMID: 16537482; PMCID: PMC1449643.

Giordano, S., Maffe, A., Williams, T. A., Artigiani, S., Gual, P., Bardelli, A., et al. (2000). Different point mutations in the met oncogene elicit distinct biological properties. *The FASEB Journal*, *14*(2), 399–406. PMID: 10657996.

Go, H., Jeon, Y. K., Park, H. J., Sung, S. W., Seo, J. W., & Chung, D. H. (2010). High MET gene copy number leads to shorter survival in patients with non-small cell lung cancer. *Journal of Thoracic Oncology*, *5*(3), 305–313. PMID: 20107422.

Gohda, E., Tsubouchi, H., Nakayama, H., Hirono, S., Takahashi, K., Koura, M., et al. (1986). Human hepatocyte growth factor in plasma from patients with fulminant hepatic failure. *Experimental Cell Research*, *166*(1), 139–150. PMID: 3527727.

Gonzalez-Angulo, A. M., Chen, H., Karuturi, M. S., Chavez-MacGregor, M., Tsavachidis, S., Meric-Bernstam, F., et al. (2013). Frequency of mesenchymal epithelial transition factor gene (MET) and the catalytic subunit of phosphoinositide-3-kinase (PIK3CA) copy number elevation and correlation with outcome in patients with early stage breast cancer. *Cancer*, *119*(1), 7–15. PMID: 22736407; PMCID: PMC3461089.

Graveel, C. R., London, C. A., & Vande Woude, G. F. (2005). A mouse model of activating Met mutations. *Cell Cycle*, *4*(4), 518–520. PMID: 15738649.

Graveel, C. R., Tolbert, D., & Vande Woude, G. F. (2013). MET: A critical player in tumorigenesis and therapeutic target. *Cold Spring Harbor Perspectives in Biology*, *5*(7), a009209 PMID: 23818496; PMCID: PMC3685898.

Guo, R., Offin, M., Brannon, A. R., Chow, A., Delasos, L., Somwar, R., et al. (2019). MET inhibitor resistance in patients with MET exon 14-altered lung cancers. *Journal of Clinical Oncology*, *37*(15_suppl), 9006.

Hammond, D. E., Urbe, S., Vande Woude, G. F., & Clague, M. J. (2001). Down-regulation of MET, the receptor for hepatocyte growth factor. *Oncogene*, *20*(22), 2761–2770. PMID: 11420688.

Hashigasako, A., Machide, M., Nakamura, T., Matsumoto, K., & Nakamura, T. (2004). Bi-directional regulation of Ser-985 phosphorylation of c-met via protein kinase C and protein phosphatase 2A involves c-Met activation and cellular responsiveness to hepatocyte growth factor. *The Journal of Biological Chemistry*, *279*(25), 26445–26452. PMID: 15075332.

Haura, E. B., Turkson, J., & Jove, R. (2005). Mechanisms of disease: Insights into the emerging role of signal transducers and activators of transcription in cancer. *Nature Clinical Practice Oncology*, *2*(6), 315–324. PMID: 16264989.

Heist, R. S., Sequist, L. V., Borger, D., Gainor, J. F., Arellano, R. S., Le, L. P., et al. (2016). Acquired resistance to crizotinib in NSCLC with MET Exon 14 skipping. *Journal of Thoracic Oncology*, *11*(8), 1242–1245. PMID: 27343442.

Hemmings, B. A., & Restuccia, D. F. (2012). PI3K-PKB/Akt pathway. *Cold Spring Harbor Perspectives in Biology*, *4*(9), a011189 PMID: 22952397; PMCID: PMC3428770.

Henry, R. E., Barry, E. R., Castriotta, L., Ladd, B., Markovets, A., Beran, G., et al. (2016). Acquired savolitinib resistance in non-small cell lung cancer arises via multiple mechanisms that converge on MET-independent mTOR and MYC activation. *Oncotarget*, 7(36), 57651–57670. PMID: 27472392; PMCID: PMC5295379.

Hotary, K., Allen, E., Punturieri, A., Yana, I., & Weiss, S. J. (2000). Regulation of cell invasion and morphogenesis in a three-dimensional type I collagen matrix by membrane-type matrix metalloproteinases 1, 2, and 3. *The Journal of Cell Biology*, *149*(6), 1309–1323. PMID: 10851027; PMCID: PMC2175112.

Hughes, V. S., & Siemann, D. W. (2019). Failures in preclinical and clinical trials of c-Met inhibitors: Evaluation of pathway activity as a promising selection criterion. *Oncotarget*, *10*(2), 184–197. PMID: 30719213; PMCID: PMC6349429.

Huh, C. G., Factor, V. M., Sanchez, A., Uchida, K., Conner, E. A., & Thorgeirsson, S. S. (2004). Hepatocyte growth factor/c-met signaling pathway is required for efficient liver regeneration and repair. *Proceedings of the National Academy of Sciences of the United States of America*, *101*(13), 4477–4482. PMID: 15070743; PMCID: PMC384772.

Ivan, M., Bond, J. A., Prat, M., Comoglio, P. M., & Wynford-Thomas, D. (1997). Activated ras and ret oncogenes induce over-expression of c-met (hepatocyte growth factor receptor) in human thyroid epithelial cells. *Oncogene*, *14*(20), 2417–2423. PMID: 9188856.

Iveson, T., Donehower, R. C., Davidenko, I., Tjulandin, S., Deptala, A., Harrison, M., et al. (2014). Rilotumumab in combination with epirubicin, cisplatin, and capecitabine as first-line treatment for gastric or oesophagogastric junction adenocarcinoma: An open-label, dose de-escalation phase 1b study and a double-blind, randomised phase 2 study. *The Lancet Oncology*, *15*(9), 1007–1018. PMID: 24965569.

Jafarnejad, M., Sové, R. J., Danilova, L., Mirando, A. C., Zhang, Y., Yarchoan, M., et al. (2019). Mechanistically detailed systems biology modeling of the HGF/Met pathway in hepatocellular carcinoma. *NPJ Systems Biology and Applications*, *5*(1), 29.

Jagadeeswaran, R., Jagadeeswaran, S., Bindokas, V. P., & Salgia, R. (2007). Activation of HGF/c-Met pathway contributes to the reactive oxygen species generation and motility of small cell lung cancer cells. *American Journal of Physiology Lung Cellular and Molecular Physiology*, *292*(6), L1488–L1494. PMID: 17322284.

Jeffers, M., Fiscella, M., Webb, C. P., Anver, M., Koochekpour, S., & Vande Woude, G. F. (1998). The mutationally activated Met receptor mediates motility and metastasis. *Proceedings of the National Academy of Sciences of the United States of America*, *95*(24), 14417–14422. PMID: 9826715; PMCID: PMC24388.

Jeffers, M., Taylor, G. A., Weidner, K. M., Omura, S., & Vande Woude, G. F. (1997). Degradation of the Met tyrosine kinase receptor by the ubiquitin-proteasome pathway. *Molecular and Cellular Biology*, *17*(2), 799–808. PMID: 9001234; PMCID: PMC231806.

Joffre, C., Barrow, R., Menard, L., Calleja, V., Hart, I. R., & Kermorgant, S. (2011). A direct role for Met endocytosis in tumorigenesis. *Nature Cell Biology, 13*(7), 827–837. PMID: 21642981.

Johnson, G. L., & Lapadat, R. (2002). Mitogen-activated protein kinase pathways mediated by ERK, JNK, and p38 protein kinases. *Science, 298*(5600), 1911–1912. PMID: 12471242.

Joyce, J. A., & Pollard, J. W. (2009). Microenvironmental regulation of metastasis. *Nature Reviews Cancer, 9*(4), 239–252. PMID: 19279573; PMCID: PMC3251309.

Kanteti, R., Yala, S., Ferguson, M. K., & Salgia, R. (2009). MET, HGF, EGFR, and PXN gene copy number in lung cancer using DNA extracts from FFPE archival samples and prognostic significance. *Journal of Environmental Pathology, Toxicology and Oncology, 28*(2), 89–98. PMID: 19817696; PMCID: PMC2761881.

Karin, M., Cao, Y., Greten, F. R., & Li, Z. W. (2002). NF-kappaB in cancer: From innocent bystander to major culprit. *Nature Reviews Cancer, 2*(4), 301–310. PMID: 12001991.

Kawada, I., Hasina, R., Lennon, F. E., Bindokas, V. P., Usatyuk, P., Tan, Y.-H. C., et al. (2013). Paxillin mutations affect focal adhesions and lead to altered mitochondrial dynamics: Relevance to lung cancer. *Cancer Biology & Therapy, 14*(7), 679–691. PMID: 23792636.

Kermorgant, S., Aparicio, T., Dessirier, V., Lewin, M. J., & Lehy, T. (2001). Hepatocyte growth factor induces colonic cancer cell invasiveness via enhanced motility and protease overproduction. Evidence for PI3 kinase and PKC involvement. *Carcinogenesis, 22*(7), 1035–1042. PMID: 11408346.

Kermorgant, S., & Parker, P. J. (2008). Receptor trafficking controls weak signal delivery: A strategy used by c-Met for STAT3 nuclear accumulation. *The Journal of Cell Biology, 182*(5), 855–863. PMID: 18779368; PMCID: PMC2528569.

Kesavan, K., Lobel-Rice, K., Sun, W., Lapadat, R., Webb, S., Johnson, G. L., et al. (2004). MEKK2 regulates the coordinate activation of ERK5 and JNK in response to FGF-2 in fibroblasts. *Journal of Cellular Physiology, 199*(1), 140–148. PMID: 14978743.

Kong-Beltran, M., Seshagiri, S., Zha, J., Zhu, W., Bhawe, K., Mendoza, N., et al. (2006). Somatic mutations lead to an oncogenic deletion of met in lung cancer. *Cancer Research, 66*(1), 283–289.

Kong-Beltran, M., Stamos, J., & Wickramasinghe, D. (2004). The Sema domain of Met is necessary for receptor dimerization and activation. *Cancer Cell, 6*(1), 75–84. PMID: 15261143.

Kozlov, G., Perreault, A., Schrag, J. D., Park, M., Cygler, M., Gehring, K., et al. (2004). Insights into function of PSI domains from structure of the Met receptor PSI domain. *Biochemical and Biophysical Research Communications, 321*(1), 234–240.

Kozlov, G., Peschard, P., Zimmerman, B., Lin, T., Moldoveanu, T., Mansur-Azzam, N., et al. (2007). Structural basis for UBA-mediated dimerization of c-Cbl ubiquitin ligase. *The Journal of Biological Chemistry, 282*(37), 27547–27555. PMID: 17635922.

Kurzrock, R., Sherman, S. I., Ball, D. W., Forastiere, A. A., Cohen, R. B., Mehra, R., et al. (2011). Activity of XL184 (Cabozantinib), an oral tyrosine kinase inhibitor, in patients with medullary thyroid cancer. *Journal of Clinical Oncology: Official Journal of the American Society of Clinical Oncology, 29*(19), 2660–2666. PMID: 21606412; PMCID: PMC3646303.

Lee, J. H., Han, S. U., Cho, H., Jennings, B., Gerrard, B., Dean, M., et al. (2000). A novel germ line juxtamembrane Met mutation in human gastric cancer. *Oncogene, 19*(43), 4947–4953. PMID: 11042681.

Lee, Y.-j., Jeong, S.-Y., Karbowski, M., Smith, C. L., & Youle, R. J. (2004). Roles of the mammalian mitochondrial fission and fusion mediators Fis1, Drp1, and Opa1 in apoptosis. *Molecular Biology of the Cell, 15*(11), 5001–5011. PMID: 15356267.

Lee, J. K., Joo, K. M., Lee, J., Yoon, Y., & Nam, D. H. (2014). Targeting the epithelial to mesenchymal transition in glioblastoma: The emerging role of MET signaling. *Oncotargets and Therapy*, 7, 1933–1944. PMID: 25364264; PMCID: PMC4211615.

Lengyel, E., Sawada, K., & Salgia, R. (2007). Tyrosine kinase mutations in human cancer. *Current Molecular Medicine*, 7(1), 77–84. PMID: 17311534.

Lin, J. J., Riely, G. J., & Shaw, A. T. (2017). Targeting ALK: Precision medicine takes on drug resistance. *Cancer Discovery*, 7(2), 137–155. PMID: 28122866; PMCID: PMC5296241.

Liu, S., Meric-Bernstam, F., Parinyanitikul, N., Wang, B., Eterovic, A. K., Zheng, X., et al. (2015). Functional consequence of the MET-T1010I polymorphism in breast cancer. *Oncotarget*, *6*(5), 2604–2614. PMID: 25605252; PMCID: PMC4413604.

Longati, P., Bardelli, A., Ponzetto, C., Naldini, L., & Comoglio, P. M. (1994). Tyrosines1234-1235 are critical for activation of the tyrosine kinase encoded by the MET proto-oncogene (HGF receptor). *Oncogene*, *9*(1), 49–57. PMID: 8302603.

Lorenzato, A., Olivero, M., Patane, S., Rosso, E., Oliaro, A., Comoglio, P. M., et al. (2002). Novel somatic mutations of the MET oncogene in human carcinoma metastases activating cell motility and invasion. *Cancer Research*, *62*(23), 7025–7030. PMID: 12460923.

Ma, J., DeFrances, M. C., Zou, C., Johnson, C., Ferrell, R., & Zarnegar, R. (2009). Somatic mutation and functional polymorphism of a novel regulatory element in the HGF gene promoter causes its aberrant expression in human breast cancer. *The Journal of Clinical Investigation*, *119*(3), 478–491. PMID: 19188684; PMCID: PMC2648687.

Ma, P. C., Jagadeeswaran, R., Jagadeesh, S., Tretiakova, M. S., Nallasura, V., Fox, E. A., et al. (2005). Functional expression and mutations of c-Met and its therapeutic inhibition with SU11274 and small interfering RNA in non-small cell lung cancer. *Cancer Research*, *65*(4), 1479–1488. PMID: 15735036.

Ma, P. C., Kijima, T., Maulik, G., Fox, E. A., Sattler, M., Griffin, J. D., et al. (2003). c-MET mutational analysis in small cell lung cancer: Novel juxtamembrane domain mutations regulating cytoskeletal functions. *Cancer Research*, *63*(19), 6272–6281. PMID: 14559814.

Ma, P. C., Maulik, G., Christensen, J., & Salgia, R. (2003). c-Met: Structure, functions and potential for therapeutic inhibition. *Cancer Metastasis Reviews*, *22*(4), 309–325. PMID: 12884908.

Ma, P. C., Schaefer, E., Christensen, J. G., & Salgia, R. (2005). A selective small molecule c-MET inhibitor, PHA665752, cooperates with rapamycin. *Clinical Cancer Research*, *11*(6), 2312–2319. PMID: 15788682.

Ma, P. C., Tretiakova, M. S., MacKinnon, A. C., Ramnath, N., Johnson, C., Dietrich, S., et al. (2008). Expression and mutational analysis of MET in human solid cancers. *Genes Chromosomes Cancer*, *47*(12), 1025–1037. PMID: 18709663; PMCID: PMC2583960.

Ma, P. C., Tretiakova, M. S., Nallasura, V., Jagadeeswaran, R., Husain, A. N., & Salgia, R. (2007). Downstream signalling and specific inhibition of c-MET/HGF pathway in small cell lung cancer: Implications for tumour invasion. *British Journal of Cancer*, *97*(3), 368–377. PMID: 17667909; PMCID: PMC2360323.

Maina, F., Casagranda, F., Audero, E., Simeone, A., Comoglio, P. M., Klein, R., et al. (1996). Uncoupling of Grb2 from the Met receptor in vivo reveals complex roles in muscle development. *Cell*, *87*(3), 531–542. PMID: 8898205.

Maroun, C. R., Naujokas, M. A., Holgado-Madruga, M., Wong, A. J., & Park, M. (2000). The tyrosine phosphatase SHP-2 is required for sustained activation of extracellular signal-regulated kinase and epithelial morphogenesis downstream from the met receptor tyrosine kinase. *Mol Cell Biol*, *20*(22), 8513–8525. PMID: 11046147; PMCID: PMC102157.

Matsumoto, K., Horikoshi, M., Rikimaru, K., & Enomoto, S. (1989). A study of an in vitro model for invasion of oral squamous cell carcinoma. *Journal of Oral Pathology & Medicine*, *18*(9), 498–501. PMID: 2607470.

Matsumoto, K., Matsumoto, K., Nakamura, T., & Kramer, R. H. (1994). Hepatocyte growth factor/scatter factor induces tyrosine phosphorylation of focal adhesion kinase (p125FAK) and promotes migration and invasion by oral squamous cell carcinoma cells. *The Journal of Biological Chemistry*, *269*(50), 31807–31813. PMID: 7527397.

Matsumoto, K., Umitsu, M., De Silva, D. M., Roy, A., & Bottaro, D. P. (2017). Hepatocyte growth factor/MET in cancer progression and biomarker discovery. *Cancer Science*, *108*(3), 296–307. PMID: 28064454.

Maulik, G., Kijima, T., Ma, P. C., Ghosh, S. K., Lin, J., Shapiro, G. I., et al. (2002). Modulation of the c-Met/hepatocyte growth factor pathway in small cell lung cancer. *Clinical Cancer Research*, *8*(2), 620–627. PMID: 11839685.

Maulik, G., Madhiwala, P., Brooks, S., Ma, P. C., Kijima, T., Tibaldi, E. V., et al. (2002). Activated c-Met signals through PI3K with dramatic effects on cytoskeletal functions in small cell lung cancer. *Journal of Cellular and Molecular Medicine*, *6*(4), 539–553. PMID: 12611639.

Maulik, G., Shrikhande, A., Kijima, T., Ma, P. C., Morrison, P. T., & Salgia, R. (2002). Role of the hepatocyte growth factor receptor, c-Met, in oncogenesis and potential for therapeutic inhibition. *Cytokine & Growth Factor Reviews*, *13*(1), 41–59. PMID: 11750879.

Merck KGaA. (2019). *Merck KGaA, Darmstadt, Germany, Announces FDA breakthrough therapy designation for investigational therapy tepotinib in patients with metastatic NSCLC with METex14 skipping alterations.* Merck KGaA. [October 12, 2019]. Available from: https://prn.to/2kuDROO.

Michieli, P., Basilico, C., Pennacchietti, S., Maffe, A., Tamagnone, L., Giordano, S., et al. (1999). Mutant Met-mediated transformation is ligand-dependent and can be inhibited by HGF antagonists. *Oncogene*, *18*(37), 5221–5231. PMID: 10498872.

Mitchison, T. J., & Cramer, L. P. (1996). Actin-based cell motility and cell locomotion. *Cell*, *84*(3), 371–379. PMID: 8608590.

Mo, H. N., & Liu, P. (2017). Targeting MET in cancer therapy. *Chronic Diseases and Translational Medicine*, *3*(3), 148–153. PMID: 29063069; PMCID: PMC5643781.

Montagner, A., Yart, A., Dance, M., Perret, B., Salles, J. P., & Raynal, P. (2005). A novel role for Gab1 and SHP2 in epidermal growth factor-induced Ras activation. *The Journal of Biological Chemistry*, *280*(7), 5350–5360. PMID: 15574420.

Morotti, A., Mila, S., Accornero, P., Tagliabue, E., & Ponzetto, C. (2002). K252a inhibits the oncogenic properties of Met, the HGF receptor. *Oncogene*, *21*(32), 4885–4893. PMID: 12118367.

Moumen, A., Ieraci, A., Patane, S., Sole, C., Comella, J. X., Dono, R., et al. (2007). Met signals hepatocyte survival by preventing Fas-triggered FLIP degradation in a PI3k-Akt-dependent manner. *Hepatology*, *45*(5), 1210–1217. PMID: 17464994.

Mukherjee, N., Houston, T. J., Cardenas, E., & Ghosh, R. (2015). To be an ally or an adversary in bladder cancer: The NF-kappaB story has not unfolded. *Carcinogenesis*, *36*(3), 299–306. PMID: 25543121; PMCID: PMC4425835.

Müller, M., Morotti, A., & Ponzetto, C. (2002). Activation of NF-kappaB is essential for hepatocyte growth factor-mediated proliferation and tubulogenesis. *Molecular and Cellular Biology*, *22*(4), 1060–1072. PMID: 11809798; PMCID: PMC134627.

Nakamura, T., Nawa, K., & Ichihara, A. (1984). Partial purification and characterization of hepatocyte growth factor from serum of hepatectomized rats. *Biochemical and Biophysical Research Communications*, *122*(3), 1450–1459. PMID: 6477569.

Nakamura, T., Nawa, K., Ichihara, A., Kaise, N., & Nishino, T. (1987). Purification and subunit structure of hepatocyte growth factor from rat platelets. *FEBS Letters*, *224*(2), 311–316. PMID: 3319692.

Nakayama, M., Sakai, K., Yamashita, A., Nakamura, T., Suzuki, Y., & Matsumoto, K. (2013). Met/HGF receptor activation is regulated by juxtamembrane Ser985 phosphorylation in hepatocytes. *Cytokine*, *62*(3), 446–452. PMID: 23618918.

Nanjo, S., Yamada, T., Nishihara, H., Takeuchi, S., Sano, T., Nakagawa, T., et al. (2013). Ability of the Met kinase inhibitor crizotinib and new generation EGFR inhibitors to overcome resistance to EGFR inhibitors. *PLoS One*, *8*(12), e84700, PMID: 24386407; PMCID: PMC3873434.

Natali, P. G., Prat, M., Nicotra, M. R., Bigotti, A., Olivero, M., Comoglio, P. M., et al. (1996). Overexpression of the met/HGF receptor in renal cell carcinomas. *International Journal of Cancer*, *69*(3), 212–217. PMID: 8682590.

Navis, A. C., van Lith, S. A., van Duijnhoven, S. M., de Pooter, M., Yetkin-Arik, B., Wesseling, P., et al. (2015). Identification of a novel MET mutation in high-grade glioma resulting in an auto-active intracellular protein. *Acta Neuropathologica*, *130*(1), 131–144. PMID: 25862637; PMCID: PMC4469304.

Neal, J. W., Dahlberg, S. E., Wakelee, H. A., Aisner, S. C., Bowden, M., Huang, Y., et al. (2016). Erlotinib, cabozantinib, or erlotinib plus cabozantinib as second-line or third-line treatment of patients with EGFR wild-type advanced non-small-cell lung cancer (ECOG-ACRIN 1512): A randomised, controlled, open-label, multicentre, phase 2 trial. *Lancet Oncology*, *17*(12), 1661–1671. PMID: 27825638; PMCID: PMC5154681.

Neklason, D. W., Done, M. W., Sargent, N. R., Schwartz, A. G., Anton-Culver, H., Griffin, C. A., et al. (2011). Activating mutation in MET oncogene in familial colorectal cancer. *BMC Cancer*, *11*, 424. PMID: 21970370; PMCID: PMC3202244.

Novartis. (2019). *Novartis investigational lung cancer therapy capmatinib (INC280) granted FDA breakthrough therapy designation for patients with MET-mutated advanced non-small cell lung cancer*. Novartis. [October 11, 2019]. Available from: https://www.novartis.com/news/media-releases/novartis-investigational-lung-cancer-therapy-capmatinib-inc280-granted-fda-breakthrough-therapy-designation-patients-met-mutated-advanced-non-small-cell-lung.

Ohashi, K., Maruvka, Y. E., Michor, F., & Pao, W. (2013). Epidermal growth factor receptor tyrosine kinase inhibitor-resistant disease. *Journal of Clinical Oncology: Official Journal of the American Society of Clinical Oncology*, *31*(8), 1070–1080. PMID: 23401451; PMCID: PMC3589701.

Okuda, K., Sasaki, H., Yukiue, H., Yano, M., & Fujii, Y. (2008). Met gene copy number predicts the prognosis for completely resected non-small cell lung cancer. *Cancer Science*, *99*(11), 2280–2285.

Ou, S.-H. I., Young, L., Schrock, A. B., Johnson, A., Klempner, S. J., Zhu, V. W., et al. (2017). Emergence of preexisting MET Y1230C mutation as a resistance mechanism to crizotinib in NSCLC with MET exon 14 skipping. *Journal of Thoracic Oncology*, *12*(1), 137–140.

Paik, P. K., Drilon, A., Fan, P.-D., Yu, H., Rekhtman, N., Ginsberg, M. S., et al. (2015). Response to MET inhibitors in patients with stage IV lung adenocarcinomas harboring MET mutations causing exon 14 skipping. *Cancer Discovery*, *5*(8), 842–849.

Paik, P. K., Veillon, R., Cortot, A.B., Felip, E., Sakai, H., Mazieres, J., et al. (2019). Phase II study of tepotinib in NSCLC patients with METex14 mutations. *Journal of Clinical Oncology*, *37*(15_suppl), 9005-.

Peinado, H., Aleckovic, M., Lavotshkin, S., Matei, I., Costa-Silva, B., Moreno-Bueno, G., et al. (2012). Melanoma exosomes educate bone marrow progenitor cells toward a pro-metastatic phenotype through MET. *Nature Medicine*, *18*(6), 883–891. PMID: 22635005; PMCID: PMC3645291.

Pelicci, G., Giordano, S., Zhen, Z., Salcini, A. E., Lanfrancone, L., Bardelli, A., et al. (1995). The motogenic and mitogenic responses to HGF are amplified by the Shc adaptor protein. *Oncogene, 10*(8), 1631–1638. PMID: 7731718.

Pennacchietti, S., Cazzanti, M., Bertotti, A., Rideout, W. M., 3rd, Han, M., Gyuris, J., et al. (2014). Microenvironment-derived HGF overcomes genetically determined sensitivity to anti-MET drugs. *Cancer Research, 74*(22), 6598–6609. PMID: 25217525.

Pennacchietti, S., Michieli, P., Galluzzo, M., Mazzone, M., Giordano, S., & Comoglio, P. M. (2003). Hypoxia promotes invasive growth by transcriptional activation of the met protooncogene. *Cancer Cell, 3*(4), 347–361. PMID: 12726861.

Peruzzi, B., Athauda, G., & Bottaro, D. P. (2006). The von Hippel-Lindau tumor suppressor gene product represses oncogenic beta-catenin signaling in renal carcinoma cells. *Proceedings of the National Academy of Sciences of the United States of America, 103*(39), 14531–14536. PMID: 16983094; PMCID: PMC1599994.

Peschard, P., & Park, M. (2007). From Tpr-Met to Met, tumorigenesis and tubes. *Oncogene, 26*(9), 1276–1285. PMID: 17322912.

Petrelli, A., Gilestro, G. F., Lanzardo, S., Comoglio, P. M., Migone, N., & Giordano, S. (2002). The endophilin-CIN85-Cbl complex mediates ligand-dependent down-regulation of c-Met. *Nature, 416*(6877), 187–190. PMID: 11894096.

Plenker, D., Bertrand, M., de Langen, A. J., Riedel, R., Lorenz, C., Scheel, A. H., et al. (2018). Structural alterations of MET trigger response to MET kinase inhibition in lung adenocarcinoma patients. *Clinical Cancer Research, 24*(6), 1337–1343.

Ponzetto, C., Bardelli, A., Zhen, Z., Maina, F., Dalla Zonca, P., Giordano, S., et al. (1994). A multifunctional docking site mediates signaling and transformation by the hepatocyte growth factor/scatter factor receptor family. *Cell, 77*(2), 261–271. PMID: 7513258.

Priya, S. R., Dravid, C. S., Digumarti, R., & Dandekar, M. (2017). Targeted therapy for medullary thyroid cancer: A review. *Frontiers in Oncology, 7*, 238. PMID: 29057215; PMCID: PMC5635342.

Proto, J. D., Tang, Y., Lu, A., Chen, W. C., Stahl, E., Poddar, M., et al. (2015). NF-kappaB inhibition reveals a novel role for HGF during skeletal muscle repair. *Cell Death & Disease, 6*e1730 PMID: 25906153; PMCID: PMC4650539.

Puri, N., Ahmed, S., Janamanchi, V., Tretiakova, M., Zumba, O., Krausz, T., et al. (2007). c-Met is a potentially new therapeutic target for treatment of human melanoma. *Clinical Cancer Research, 13*(7), 2246–2253. PMID: 17404109.

Puri, N., & Salgia, R. (2008). Synergism of EGFR and c-Met pathways, cross-talk and inhibition, in non-small cell lung cancer. *Journal of Carcinogenesis, 7*, 9. PMID: 19240370; PMCID: PMC2669728.

Qi, J., McTigue, M. A., Rogers, A., Lifshits, E., Christensen, J. G., Janne, P. A., et al. (2011). Multiple mutations and bypass mechanisms can contribute to development of acquired resistance to MET inhibitors. *Cancer Research, 71*(3), 1081–1091. PMID: 21266357; PMCID: PMC3208270.

Raghav, K., Morris, V., Tang, C., Morelli, P., Amin, H. M., Chen, K., et al. (2016). MET amplification in metastatic colorectal cancer: An acquired response to EGFR inhibition, not a de novo phenomenon. *Oncotarget, 7*(34), 54627–54631. PMID: 27421137.

Rehman, J., Zhang, H. J., Toth, P. T., Zhang, Y., Marsboom, G., Hong, Z., et al. (2012). Inhibition of mitochondrial fission prevents cell cycle progression in lung cancer. *The FASEB Journal, 26*(5), 2175–2186. PMID: 22321727.

Rimassa, L., Assenat, E., Peck-Radosavljevic, M., Zagonel, V., Pracht, M., Rota Caremoli, E., et al. (2017). Second-line tivantinib (ARQ 197) vs placebo in patients (Pts) with MET-high hepatocellular carcinoma (HCC): Results of the METIV-HCC phase III trial. *Journal of Clinical Oncology, 35*(15_suppl), 4000.

Roberts, P. J., & Der, C. J. (2007). Targeting the Raf-MEK-ERK mitogen-activated protein kinase cascade for the treatment of cancer. *Oncogene, 26*(22), 3291–3310. PMID: 17496923.

Rolle, C. E., Tan, Y. C., Seiwert, T. Y., Vora, S., Kanteti, R., Hasina, R., et al. (2017). Expression and mutational analysis of c-CBL and its relationship to the MET receptor in head and neck squamous cell carcinoma. *Oncotarget*, *8*(12), 18726–18734. PMID: 27244893; PMCID: PMC5386642.

Rong, S., Donehower, L. A., Hansen, M. F., Strong, L., Tainsky, M., Jeffers, M., et al. (1995). Met proto-oncogene product is overexpressed in tumors of p53-deficient mice and tumors of Li-Fraumeni patients. *Cancer Research*, *55*(9), 1963–1970. PMID: 7728766.

Rosario, M., & Birchmeier, W. (2003). How to make tubes: Signaling by the Met receptor tyrosine kinase. *Trends in Cell Biology*, *13*(6), 328–335. PMID: 12791299.

Rosen, E. M., Lamszus, K., Laterra, J., Polverini, P. J., Rubin, J. S., & Goldberg, I. D. (1997). HGF/SF in angiogenesis. *Ciba Foundation Symposium*, *212*, 215–226. discussion 27–29. PMID: 9524773.

Sachs, M., Brohmann, H., Zechner, D., Muller, T., Hulsken, J., Walther, I., et al. (2000). Essential role of Gab1 for signaling by the c-Met receptor in vivo. *The Journal of Cell Biology*, *150*(6), 1375–1384. PMID: 10995442; PMCID: PMC2150711.

Sadiq, A. A., & Salgia, R. (2013). MET as a possible target for non-small-cell lung cancer. *Journal of Clinical Oncology: Official Journal of the American Society of Clinical Oncology*, *31*(8), 1089–1096. PMID: 23401458.

Sakai, K., Aoki, S., & Matsumoto, K. (2015). Hepatocyte growth factor and Met in drug discovery. *Journal of Biochemistry*, *157*(5), 271–284. PMID: 25770121.

Salgia, R. (2017). MET in lung cancer: Biomarker selection based on scientific rationale. *Molecular Cancer Therapeutics*, *16*(4), 555–565. PMID: 28373408.

Salgia, R., Patel, P., Bothos, J., Yu, W., Eppler, S., Hegde, P., et al. (2014). Phase I dose-escalation study of onartuzumab as a single agent and in combination with bevacizumab in patients with advanced solid malignancies. *Clinical Cancer Research*, *20*(6), 1666–1675. PMID: 24493831.

Santoro, A., Rimassa, L., Borbath, I., Daniele, B., Salvagni, S., Van Laethem, J. L., et al. (2013). Tivantinib for second-line treatment of advanced hepatocellular carcinoma: A randomised, placebo-controlled phase 2 study. *The Lancet Oncology*, *14*(1), 55–63. PMID: 23182627.

Sattler, M., Ma, P. C., & Salgia, R. (2004). Therapeutic targeting of the receptor tyrosine kinase Met. *Cancer Treatment and Research*, *119*, 121–138. PMID: 15164876.

Sattler, M., & Salgia, R. (2016). MET in the driver's seat: Exon 14 skipping mutations as actionable targets in lung cancer. *Journal of Thoracic Oncology*, *11*(9), 1381–1383. PMID: 27565402.

Schiering, N., Knapp, S., Marconi, M., Flocco, M. M., Cui, J., Perego, R., et al. (2003). Crystal structure of the tyrosine kinase domain of the hepatocyte growth factor receptor c-Met and its complex with the microbial alkaloid K-252a. *Proceedings of the National Academy of Sciences of the United States of America*, *100*(22), 12654–12659. PMID: 14559966; PMCID: PMC240673.

Schmidt, L., Duh, F. M., Chen, F., Kishida, T., Glenn, G., Choyke, P., et al. (1997). Germline and somatic mutations in the tyrosine kinase domain of the MET proto-oncogene in papillary renal carcinomas. *Nature Genetics*, *16*(1), 68–73. PMID: 9140397.

Schmidt, L., Junker, K., Nakaigawa, N., Kinjerski, T., Weirich, G., Miller, M., et al. (1999). Novel mutations of the MET proto-oncogene in papillary renal carcinomas. *Oncogene*, *18*(14), 2343–2350. PMID: 10327054.

Schrock, A. B., Lai, A., Ali, S. M., Miller, V. A., & Raez, L. E. (2017). Mutation of MET Y1230 as an acquired mechanism of crizotinib resistance in NSCLC with MET exon 14 skipping. *Journal of Thoracic Oncology*, *12*(7), e89–e90.

Seiwert, T. Y., Jagadeeswaran, R., Faoro, L., Janamanchi, V., Nallasura, V., El Dinali, M., et al. (2009). The MET receptor tyrosine kinase is a potential novel therapeutic target for head and neck squamous cell carcinoma. *Cancer Research*, *69*(7), 3021–3031. PMID: 19318576; PMCID: PMC2871252.

Sequist, L. V., Waltman, B. A., Dias-Santagata, D., Digumarthy, S., Turke, A. B., Fidias, P., et al. (2011). Genotypic and histological evolution of lung cancers acquiring resistance to EGFR inhibitors. *Science Translational Medicine*, *3*(75), 75ra26 PMID: 21430269; PMCID: PMC3132801.

Sipeki, S., Bander, E., Buday, L., Farkas, G., Bacsy, E., Ways, D. K., et al. (1999). Phosphatidylinositol 3-kinase contributes to Erk1/Erk2 MAP kinase activation associated with hepatocyte growth factor-induced cell scattering. *Cellular Signalling*, *11*(12), 885–890. PMID: 10659996.

Sonnenberg, E., Meyer, D., Weidner, K. M., & Birchmeier, C. (1993). Scatter factor/hepatocyte growth factor and its receptor, the c-met tyrosine kinase, can mediate a signal exchange between mesenchyme and epithelia during mouse development. *The Journal of Cell Biology*, *123*(1), 223–235. PMID: 8408200.

Soubeyran, P., Kowanetz, K., Szymkiewicz, I., Langdon, W. Y., & Dikic, I. (2002). Cbl-CIN85-endophilin complex mediates ligand-induced downregulation of EGF receptors. *Nature*, *416*(6877), 183–187. PMID: 11894095.

Spigel, D. R., Edelman, M. J., O'Byrne, K., Paz-Ares, L., Mocci, S., Phan, S., et al. (2017). Results from the phase III randomized trial of onartuzumab plus erlotinib versus erlotinib in previously treated stage IIIB or IV non-small-cell lung cancer: METLung. *Journal of Clinical Oncology: Official Journal of the American Society of Clinical Oncology*, *35*(4), 412–420. PMID: 27937096.

Spigel, D. R., Ervin, T. J., Ramlau, R. A., Daniel, D. B., Goldschmidt, J. H., Jr., Blumenschein, G. R., Jr., et al. (2013). Randomized phase II trial of Onartuzumab in combination with erlotinib in patients with advanced non-small-cell lung cancer. *Journal of Clinical Oncology: Official Journal of the American Society of Clinical Oncology*, *31*(32), 4105–4114. PMID: 24101053; PMCID: PMC4878106.

Stamos, J., Lazarus, R. A., Yao, X., Kirchhofer, D., & Wiesmann, C. (2004). Crystal structure of the HGF beta-chain in complex with the Sema domain of the Met receptor. *The EMBO Journal*, *23*(12), 2325–2335. PMID: 15167892.

Stoker, M., Gherardi, E., Perryman, M., & Gray, J. (1987). Scatter factor is a fibroblast-derived modulator of epithelial cell mobility. *Nature*, *327*(6119), 239–242. PMID: 2952888.

Straussman, R., Morikawa, T., Shee, K., Barzily-Rokni, M., Qian, Z. R., Du, J., et al. (2012). Tumour micro-environment elicits innate resistance to RAF inhibitors through HGF secretion. *Nature*, *487*(7408), 500–504. PMID: 22763439; PMCID: PMC3711467.

Suzawa, K., Offin, M., Lu, D., Kurzatkowski, C., Vojnic, M., Smith, R. S., et al. (2019). Activation of KRAS mediates resistance to targeted therapy in MET exon 14-mutant non-small cell lung cancer. *Clinical Cancer Research*, *25*(4), 1248–1260. PMID: 30352902; PMCID: PMC6377821.

Szturz, P., Raymond, E., Abitbol, C., Albert, S., de Gramont, A., & Faivre, S. (2017). Understanding c-MET signalling in squamous cell carcinoma of the head & neck. *Critical Reviews in Oncology/Hematology*, *111*, 39–51. PMID: 28259294.

Tan, Y. H., Krishnaswamy, S., Nandi, S., Kanteti, R., Vora, S., Onel, K., et al. (2010). CBL is frequently altered in lung cancers: Its relationship to mutations in MET and EGFR tyrosine kinases. *PLoS One*, *5*(1), e8972 PMID: 20126411; PMCID: PMC2813301.

Tan, Y. C., Mirzapoiazova, T., Won, B. M., Zhu, L., Srivastava, M. K., Vokes, E. E., et al. (2017). Differential responsiveness of MET inhibition in non-small-cell lung cancer with altered CBL. *Science Reporter*, 7(1), 9192. PMID: 28835699; PMCID: PMC5569108.

Tang, C., Jardim, D. L., Falchook, G. S., Hess, K., Fu, S., Wheler, J. J., et al. (2014). MET nucleotide variations and amplification in advanced ovarian cancer: Characteristics and outcomes with c-Met inhibitors. *Oncoscience*, *1*(1), 5–13. PMID: 25593979; PMCID: PMC4295762.

Tiedt, R., Degenkolbe, E., Furet, P., Appleton, B. A., Wagner, S., Schoepfer, J., et al. (2011). A drug resistance screen using a selective MET inhibitor reveals a spectrum of mutations that partially overlap with activating mutations found in cancer patients. *Cancer Research*, *71*(15), 5255–5264.

Trusolino, L., Bertotti, A., & Comoglio, P. M. (2010). MET signalling: Principles and functions in development, organ regeneration and cancer. *Nature Reviews Molecular Cell Biology*, *11*(12), 834–848. PMID: 21102609.

Tulasne, D., Deheuninck, J., Lourenco, F. C., Lamballe, F., Ji, Z., Leroy, C., et al. (2004). Proapoptotic function of the MET tyrosine kinase receptor through caspase cleavage. *Molecular and Cellular Biology*, *24*(23), 10328–10339. PMID: 15542841; PMCID: PMC529022.

Tyner, J. W., Fletcher, L. B., Wang, E. Q., Yang, W. F., Rutenberg-Schoenberg, M. L., Beadling, C., et al. (2010). MET receptor sequence variants R970C and T992I lack transforming capacity. *Cancer Research*, *70*(15), 6233–6237. PMID: 20670955; PMCID: PMC2913476.

Van Der Steen, N., Pauwels, P., Gil-Bazo, I., Castañon, E., Raez, L., Cappuzzo, F., et al. (2015). cMET in NSCLC: Can we cut off the head of the Hydra? From the pathway to the resistance. *Cancers*, 7(2), 556–573. PMID: 25815459.

Verslype, C., Cohn, A. L., Kelley, R. K., Yang, T.-S., Su, W.-C., Ramies, D. A., et al. (2012). Activity of cabozantinib (XL184) in hepatocellular carcinoma: Results from a phase II randomized discontinuation trial (RDT). *Journal of Clinical Oncology*, *30*(15_suppl), 4007.

Wang, W., Du, Z., Yan, J., Ma, D., Shi, M., Zhang, M., et al. (2014). Mesenchymal stem cells promote liver regeneration and prolong survival in small-for-size liver grafts: Involvement of C-Jun N-terminal kinase, cyclin D1, and NF-kappaB. *PLoS One*, *9*(12), e112532 , PMID: 25479410; PMCID: PMC4257551.

Wang, J., Mirzapoiazova, T., Carol Tan, Y.-H., Pang, K. M., Pozhitkov, A., Wang, Y., et al. (2018). Inhibiting crosstalk between MET signaling and mitochondrial dynamics and morphology: A novel therapeutic approach for lung cancer and mesothelioma. *Cancer Biology & Therapy*, *19*(11), 1023–1032. PMID: 30311833.

Wang, W., Xu, C., Chen, Y., Zhu, Y.-C., Liu, Y., Wang, H., et al. (2018). MET gene fusions in non-small cell lung cancer (NSCLC) in the Chinese population: A multicenter study. *Journal of Clinical Oncology*, *36*(15–suppl), e13539-e.

Warburton, D., Schwarz, M., Tefft, D., Flores-Delgado, G., Anderson, K. D., & Cardoso, W. V. (2000). The molecular basis of lung morphogenesis. *Mechanisms of Development*, *92*(1), 55–81. PMID: 10704888.

Webb, C. P., Taylor, G. A., Jeffers, M., Fiscella, M., Oskarsson, M., Resau, J. H., et al. (1998). Evidence for a role of Met-HGF/SF during Ras-mediated tumorigenesis/metastasis. *Oncogene*, *17*(16), 2019–2025. PMID: 9798673.

Weber, H., Muller, D., Muller, M., Ortiz, A., Birkle, M., Umber, S., et al. (2014). Cell lines expressing recombinant transmembrane domain-activated receptor kinases as tools for drug discovery. *Journal of Biomolecular Screening*, *19*(10), 1350–1361. PMID: 25260782.

Weidner, K. M., Arakaki, N., Hartmann, G., Vandekerckhove, J., Weingart, S., Rieder, H., et al. (1991). Evidence for the identity of human scatter factor and human hepatocyte growth factor. *Proceedings of the National Academy of Sciences of the United States of America*, *88*(16), 7001–7005. PMID: 1831266; PMCID: PMC52221.

Weidner, K. M., Di Cesare, S., Sachs, M., Brinkmann, V., Behrens, J., & Birchmeier, W. (1996). Interaction between Gab1 and the c-Met receptor tyrosine kinase is responsible for epithelial morphogenesis. *Nature*, *384*(6605), 173–176. PMID: 8906793.

Weisberg, E., Sattler, M., Ewaniuk, D. S., & Salgia, R. (1997). Role of focal adhesion proteins in signal transduction and oncogenesis. *Critical Reviews in Oncogenesis*, *8*(4), 343–358. PMID: 9622054.

Wolf, J., Seto, T., Han, J.-Y., Reguart, N., Garon, E.B., Groen, H.J.M., et al. (2019). Capmatinib (INC280) in METΔex14-mutated advanced non-small cell lung cancer (NSCLC): Efficacy data from the phase II GEOMETRY mono-1 study. *Journal of Clinical Oncology*. *37*(15_suppl), 9004-.

Yakes, F. M., Chen, J., Tan, J., Yamaguchi, K., Shi, Y., Yu, P., et al. (2011). Cabozantinib (XL184), a novel MET and VEGFR2 inhibitor, simultaneously suppresses metastasis, angiogenesis, and tumor growth. *Molecular Cancer Therapeutics*, *10*(12), 2298–2308. PMID: 21926191.

Yap, T. A., & Popat, S. (2017). Targeting MET exon 14 skipping alterations: Has lung cancer MET its match? *Journal of Thoracic Oncology*, *12*(1), 12–14. PMID: 27988098.

Yu, H., & Jove, R. (2004). The STATs of cancer—New molecular targets come of age. *Nature Reviews. Cancer*, *4*(2), 97–105. PMID: 14964307.

Yu, H., Lee, H., Herrmann, A., Buettner, R., & Jove, R. (2014). Revisiting STAT3 signalling in cancer: New and unexpected biological functions. *Nature Reviews. Cancer*, *14*(11), 736–746. PMID: 25342631.

Zhang, Y., Farenholtz, K. E., Yang, Y., Guessous, F., Dipierro, C. G., Calvert, V. S., et al. (2013). Hepatocyte growth factor sensitizes brain tumors to c-MET kinase inhibition. *Clinical Cancer Research*, *19*(6), 1433–1444. PMID: 23386689; PMCID: PMC3602223.

Zhang, Y. W., & Vande Woude, G. F. (2003). HGF/SF-met signaling in the control of branching morphogenesis and invasion. *Journal of Cellular Biochemistry*, *88*(2), 408–417. PMID: 12520544.

Zhang, Y. W., Wang, L. M., Jove, R., & Vande Woude, G. F. (2002). Requirement of Stat3 signaling for HGF/SF-Met mediated tumorigenesis. *Oncogene*, *21*(2), 217–226. PMID: 11803465.

Zhang, Y., Yin, J., & Peng, F. (2017). Acquired resistance to crizotinib in advanced lung adenocarcinoma with MET exon 14 skipping. *Lung Cancer*, *113*, 69–71. PMID: 29110851.

CHAPTER EIGHT

Eph receptors as cancer targets for antibody-based therapy

Fenny H.F. Tang[a,b,†], Deodate Davis[a,b,†], Wadih Arap[a,c,*], Renata Pasqualini[a,b,*], Fernanda I. Staquicini[a,b,*]

[a]Rutgers Cancer Institute of New Jersey, Newark, NJ, United States
[b]Division of Cancer Biology, Department of Radiation Oncology, Rutgers New Jersey Medical School, Newark, NJ, United States
[c]Division of Hematology/Oncology, Department of Medicine, Rutgers New Jersey Medical School, Newark, NJ, United States
*Corresponding authors: e-mail address: wadih.arap@rutgers.edu; renata.pasqualini@rutgers.edu; fernanda.staquicini@rutgers.edu

Contents

Abstract

Receptor tyrosine kinases (RTKs) are integral membrane sensors that govern cell differentiation, proliferation and mobility, and enable rapid communication between cells and their environment. Of the 20 RTK subfamilies currently known, Eph receptors are the largest group. Together with their corresponding ephrin ligands, Eph receptors regulate a diverse array of physiologic processes including axonal guidance, bone remodeling, and immune cell development and trafficking. Deregulation of Eph signaling pathways is linked to cancer and other proliferative diseases and, because RTKs play critical roles in cancer development, the specific targeting of these molecules in malignancies provides a promising treatment approach. Monoclonal antibodies targeting RTKs represent a potentially attractive modality for pharmaceutical development due to their

† These authors contributed equally to this work.

Advances in Cancer Research, Volume 147
ISSN 0065-230X
https://doi.org/10.1016/bs.acr.2020.04.007
© 2020 Elsevier Inc.
All rights reserved.

relatively high target specificity and low off-target binding rates. Therefore, new technologies to generate antibodies able to target RTKs in their native *in vivo* context are likely to facilitate pre-clinical and clinical development of antibody-based therapies. Our group has recently reported a platform discovery methodology termed *S*election of *P*hage-displayed *A*ccessible *R*ecombinant *T*argeted *A*ntibodies (SPARTA). SPARTA is a novel and robust stepwise method, which combines the attributes of *in vitro* screenings of a naïve human recombinant antibody library against known tumor targets with those features of *in vivo* selections based on tumor-homing capabilities of a pre-enriched antibody pool. This unique approach overcomes several rate-limiting challenges to generate human monoclonal antibodies amenable to rapid translation into medical applications.

1. Receptor tyrosine kinases: Structure and function

Receptor tyrosine kinases (RTKs) are single-pass transmembrane proteins with similar domain organization: their five fundamental structures include a large N-terminal extracellular portion thereby providing a site for ligand binding, a short helical transmembrane region, a flexible juxtamembrane domain, a cytoplasmic domain with kinase activity and a flexible C-terminal tail region (Critchley et al., 2018; Hubbard, 1999). The transmembrane domain of RTKs is embedded in the membrane lipid bilayer and plays an active role in cell signaling. It contributes to the stability of the full-length receptor dimer and maintains a signaling-competent dimeric receptor conformation (Li & Hristova, 2010). Single point mutations in this domain can dramatically alter the tyrosine kinase activity of the receptor (Bargmann, Hung, & Weinberg, 1986).

The juxtamembrane domain is located in the cytoplasm and is conserved among receptors of the same family. This region is responsible for modulating receptor functions by initiating signaling pathways through the activation of effector molecules, adjacent receptors and transducers. The regulatory mechanisms mediated by the juxtamembrane domain can vary substantially due to modifications in length and primary sequence characteristics of RTK proteins (Wang, Yennawar, & Hankey, 2014).

The intracellular kinase domain distinguishes RTKs from all other transmembrane receptors. It is highly conserved among RTK members and it is responsible for tyrosine kinase activity as it catalyzes the transfer of the γ phosphate of ATP to tyrosine hydroxyl groups on target proteins. Phosphorylation of tyrosine residues modulates enzymatic activity and creates binding sites for the recruitment of downstream signaling proteins

(Lemmon & Schlessinger, 2010). The C-terminal region of kinase domains varies in length and function to render specificity to RTK members and create diversity in downstream signaling cascades. This segment binds to intracellular signaling transducers upon receptor engagement and it initiates a series of events with functional consequences that include cell growth, adhesion, motility and cellular differentiation (Longati, Comoglio, & Bardelli, 2001).

Up to date, the 58 RTKs identified in humans have been divided into 20 subfamilies (Lemmon & Schlessinger, 2010; Manning, Whyte, Martinez, Hunter, & Sudarsanam, 2002; Robinson, Wu, & Lin, 2000) based on the nature of their corresponding ligand (growth factor) and arrangement of structural extracellular domains. Within each RTK subfamily, cognate ligands share sequence similarities, and members are distinguished from each other based on characteristic secondary and tertiary structures, which are primarily composed of immunoglobulin-like (Ig) domains, leucine-rich domains, cysteine-rich domains, and fibronectin repeats (Longati et al., 2001). The vast majority of RTKs exist as a single polypeptide chain and are monomeric in the absence of a ligand. A notable exception is the insulin receptor, which is composed of an α and β subunit that can form dimers independently of ligand activation (Longati et al., 2001).

1.1 Eph receptors and ephrins

Among the 20 RTK subfamilies, the erythropoietin-producing hepatocellular carcinoma (Eph) receptor tyrosine kinase subfamily comprises the largest group of surface receptors. Eph receptors are further subdivided into two classes: the EphA subclass, consisting of nine members (EphA1—A8 and EphA10) along with the EphB subclass, composed of five members (EphB1-B4 and EphB6). Eph receptors are activated upon binding to their interacting ligands—ephrins, which are also subdivided: ephrin-As are attached to the cell surface by a glycosylphosphotidylinositol (GPI) anchor; ephrin-Bs contain a transmembrane segment and a short cytoplasmic region. There are currently 14 known Eph receptors and eight ephrin ligands. Eph activation by ephrins occurs most commonly *via* cell-to-cell interactions, although ephrin-As can also be released to activate EphA receptors at a distance. However, Eph receptors can also be activated by other cell surface receptors such as EGFR (Larsen et al., 2007). Ephrins-A3 and -B3 have also been shown to bind heparan sulfate proteoglycans, with the goal of potentiating EphA receptor activation and signaling.

The Eph receptor family has greatly expanded throughout evolution, and now includes almost one fourth of the 58 aforementioned human RTKs (Lemmon & Schlessinger, 2010). A large number of Eph receptors and ephrins may be required to achieve and maintain the sophisticated tissue organization of higher organisms. Indeed, many are highly expressed in the brain—the most complex organ—particularly during developmental stages (Yamaguchi & Pasquale, 2004). Eph receptors and ephrins are also present in most–if not all–organs and tissues, with characteristic expression patterns that change dramatically in response to environmental stimuli.

Eph receptors and ephrins engage in a multitude of activities. They typically mediate contact-dependent communication between cells (of the same or different types) to control cell morphology, adhesion, movement, proliferation, survival, and differentiation (Pasquale, 2005). In adult organisms, the Eph/ephrin system regulates remodeling of synapses, epithelial differentiation and integrity, bone remodeling, immune function, insulin secretion, and stem cell self-renewal (Genander & Frisen, 2010; Pasquale, 2008). In addition, expression of Eph receptors and ephrins is often up-regulated in injured tissues, where they inhibit some regenerative processes but promote angiogenesis, as well as in cancer cells, where they can promote and/or suppress tumorigenicity depending on the context (Du, Fu, & Sretavan, 2007; McDonell, Kernohan, Boycott, & Sawyer, 2015; Pasquale, 2008, 2010).

2. Mechanisms of receptor tyrosine kinases activation

Different subfamilies of RTKs generally employ a similar strategy to trigger downstream signal transduction: engagement of the RTK extracellular domain *via* ligand binding induces receptor dimerization and/or oligomerization and ultimately results in activation by auto-phosphorylation of tyrosine residues and downstream signal transduction. Activation of RTKs can occur by both membrane-bound and soluble ligands (Hubbard & Miller, 2007; Hubbard & Till, 2000; Volinsky & Kholodenko, 2013), demonstrating the versatility of their activation mechanisms.

Binding of ligands to the extracellular domain of RTKs induces a dynamic conformational change at the monomeric and self-inactivated receptor that causes homo- or hetero-dimer formation and triggers enzyme activation. Dimerization juxtaposes the catalytic domains and induces a conformational change, which allows for tyrosine phosphorylation at multiple sites within the cytoplasmic domain. Tyrosine phosphorylation initiates the recruitment of specific adaptors and enzymes to the cytosolic face of the plasma membrane (Heldin, 1995; Lemmon & Schlessinger, 1994, 2010).

There are four general modes of RTK dimerization and activation of the tyrosine domain. In the first, bound ligands mediate receptor dimerization without any direct contact between the extracellular domains of the two receptors, as the case with nerve growth factor (NGF) receptor, also known as TrkA (Wehrman et al., 2007). The second mode is exemplified by ErbB family members: dimerization is mediated entirely by the receptor without any physical interaction between the two activating ligands (Zhang, Gureasko, Shen, Cole, & Kuriyan, 2006). In the third mode, accessory molecules (such as heparin or heparin sulfate proteoglycans) participate in receptor dimerization in addition to the combination of bivalent ligand and direct receptor-receptor contact, as observed in the FGFR family (Lemmon & Schlessinger, 2010). Finally, ligand homodimers can crosslink two receptor molecules which then interact with each other across the dimer interface, as seen in KIT, the stem cell factor (SCF) receptor (Yuzawa et al., 2007).

Tyrosine phosphorylation involves tyrosine residues both inside and outside the catalytic domain. Once the RTK is activated, it triggers the formation of a multienzymatic-multiadaptor complex containing Src homology (SH2 and SH3), phosphotyrosine binding (PTB), and pleckstrin homology (PH) domains. These domains act as docking sites for phosphotyrosines or phospholipids and thereby trigger the mobilization of these proteins to the cell surface (Pawson, Gish, & Nash, 2001; Pawson & Nash, 2003). The presence of several phosphotyrosines and their association with various docking proteins gives activated RTKs the ability to recruit and regulate a wide range of signaling pathways including RAS/MAPK, PI-3 K/AKT and JAK2/STAT signaling (Du & Lovly, 2018). Such signals ultimately induce diverse biological responses including cell growth, survival, inhibition of apoptosis, promotion of angiogenesis and activation of cell motility (Hubbard, 1999).

Unlike most RTKs, Eph receptors are activated by juxtacrine signaling: Eph receptors interact with ephrins that are more commonly membrane bound, allowing for both "forward signaling" in the receptor-bound cell and "reverse signaling" in the ephrin-bound cell. This bidirectional signaling between Eph-ephrin plays a role in cell segregation and border sharpening during physiological development (Wu, Ashlin, Xu, & Wilkinson, 2019). In addition to forward signaling, Eph receptors can also signal in the absence of ligand binding and kinase activation through direct interaction with proximal RTKs, such as HER2 (Arvanitis & Davy, 2008).

3. Receptor tyrosine kinases in human cancers

RTKs need to be properly localized on the cell surface to function as signal-receiving and signal-transmitting devices. Disturbed RTK signaling can disrupt the balance between cell growth, cell cycle progression and apoptosis, and trigger RTK-induced oncogenesis. Abnormal RTK expression and activation are linked to cellular transformation, cancer development, and metastasis, and have been shown to play an important role in certain hallmarks of cancer, such as angiogenesis, metabolic adaptation, immune evasion and resistance to radiation therapy and chemotherapy (Bhattacharya, Shetake, Pandey, & Kumar, 2018).

Mutations in genes that encode RTKs in both germline and somatic cells have been associated with various types of cancer and other diseases (McDonell et al., 2015). There are four main mechanisms of RTK deregulation in human cancers: overexpression, genomic rearrangements, autocrine activation, and gain-of-function (GOF) mutations (Du & Lovly, 2018). Overexpression of RTKs has been shown to increase the local concentration of receptors on the cell membrane, which results in elevated RTK signaling and consequentially overwhelms antagonizing regulatory cellular functions (Carraway & Sweeney, 2002). While gene amplification is considered the main cause of RTK overexpression, additional mechanisms may include transcriptional/translational enhancement, oncogenic viruses and derailment of normal regulatory functions such as loss of negative regulators and phosphatase activity (Du & Lovly, 2018). Genomic rearrangements including chromosomal translocations can result in oncogenic fusion proteins with dimerizing capabilities and ligand-independent activation of the tyrosine kinase domain (Blume-Jensen & Hunter, 2001). Autocrine signaling refers to the production and secretion of an extracellular mediator (for example, either a growth factor or a cytokine) followed by binding to receptors on the same cell surface. Autocrine binding events can initiate cell signaling cascades with functional consequences including clonal expansion and tumor formation. Perhaps not surprisingly, such events are widely prevalent and well-characterized in diverse cancers (Esposito et al., 2002; Kentsis et al., 2012). GOF mutations and/or small deletions in RTKs lead to aberrant downstream signal transduction that are not subject to the normal cell cycle checkpoints of physiological signaling. The cellular consequences of such molecular events can be devastating as they may lead to propagation of DNA damage and mutations in daughter cells. Of note, GOF mutations are a common cause of cellular resistance to both chemo- and radiation therapies (Blume-Jensen & Hunter, 2001).

3.1 Eph receptors in human cancers

The RTK EphA5 was among the first Eph receptors identified as a tumor-associated antigen (Chiari et al., 2000). Whereas the function of EphA5 in organism development is well recognized (Caras, 1997), its multi-faceted roles in cancer are still emerging. Since its discovery, EphA5 has been described as a prognostic marker for ovarian cancer (Chen, Wang, Wei, & Wang, 2016), and as a tumor suppressor in both hepatocellular carcinoma (Wang, Yeh, Ho, Ng, & Chen, 2016) and gastric cancer (Zhang et al., 2019). EphA5 has also been found to inhibit the migration and proliferation of colon cancer cells (Wang, Chang, Ho, Wu, & Chen, 2012), while in breast cancer, it has been described as a potential target for epigenetic silencing and a valuable molecular marker of carcinogenesis and disease progression (Fu et al., 2010).

Overexpression of EphA5 has also been observed in patients with adenocarcinoma and squamous cell carcinoma of the lung. In normal lung epithelium, EphA5 presence was found to be low and barely detectable whereas high EphA5 expression was detected both in the membrane and cytoplasm of lung cancer cells. Increased EphA5 expression has also been associated with a negative response to radiation therapy and is accompanied by an increase in locoregional recurrence and a sharp decrease in overall survival. These findings were functionally explained by a previously unrecognized role of EphA5 as a regulator of DNA damage repair (Staquicini et al., 2015).

Functions of EphA5 in lung cancer were first investigated by genetically knocking down EphA5 expression in cultured lung cancer cells and observing the effect on response to DNA damage caused by ionizing radiation (IR). EphA5 knockdown caused radiosensitization of cells due to their inability to resolve IR-induced DNA damage. In normal cells and in cells with intact DNA damage machinery, exposure to increasing doses of IR typically results in the activation of cell cycle checkpoints (G1, S and G2 phases) and repair of DNA damage; only repaired cells are then allowed to continue to cell division. However, in EphA5-negative cells, defective activation of cell cycle checkpoints and insufficient repair of DNA damage most likely results in severe genomic instability, mitotic catastrophe and ultimately cell death (Staquicini et al., 2015).

Although EphA5 was first identified on the cell surface, it has also been detected in the nucleus of irradiated lung cancer cells where it interacts directly with phosphorylated ATM at sites of DNA damage repair. ATM is a protein kinase that is recruited to the site of DNA double- stranded breaks to mediate the phosphorylation of effector proteins like p53 and Mdm2. Such events are followed by checkpoint activation and cell cycle arrest of irradiated cells. Genetic knockdown of both p53 and EphA5 in cultured cells results in cell death after exposure to IR. However, if only EphA5 is absent, the presence of p53 causes the cells to become senescent (Staquicini et al., 2015).

In an effort to explore EphA5 as a potential target for lung cancer treatment, our group generated and validated a murine monoclonal antibody termed 11C12. Pre-clinical testing of 11C12 in combination with radiation therapy in xenograft models of lung cancer showed a marked decrease in tumor growth and an increase in overall survival. The biodistribution of 11C12 was also investigated by using SPECT/CT and showed exquisite localization and time-dependent uptake of the antibody in tumors (Staquicini et al., 2015).

4. Receptor tyrosine kinases as cancer targets

Due to the critical roles played by RTKs in human cancers, the specific targeting of these molecules in malignancies may be a promising approach to cancer treatment. Conventional cancer therapies (including radiation, hormone therapy, and chemotherapy) are designed to target cellular features that are constantly altered in cancer such as cellular metabolism, DNA replication and signaling pathways (Herst & Berridge, 2013). While these treatments can be effective against malignant tumor cells, they can also negatively affect non-malignant cells, something that could perhaps be avoided with therapies based on small-molecule inhibitors or monoclonal antibodies to target RTKs (Yamaoka, Kusumoto, Ando, Ohba, & Ohmori, 2018). Because of the high target specificity and low off-target binding, such monoclonal antibodies may represent a more attractive modality for pharmaceutical drug development, especially in scenarios where cell-cell interactions or ligand-receptor interactions play a crucial role in disease development and progression.

4.1 New technologies to develop RTK-targeted cancer therapies

The development of targeted therapy tools like cancer-specific monoclonal antibodies is a major focus in oncology. Monoclonal antibodies were first generated in mice in 1975 by using the hybridoma technology (Kohler & Milstein, 1975). Classically, the production of a hybridoma begins with the fusion of spleen cells from an immunized mouse with human or mouse myeloma cells (chimeric mouse-human monoclonal antibody). As a result, selected immortalized B-cells are able to proliferate rapidly and produce large amounts of monoclonal antibodies (Peterson & Seed, 1987). The introduction of this transformative technology has enabled the generation of target-specific monoclonal antibodies and was responsible for the explosive growth of antibody research and commercialization over the past 20 years. However, hybridoma technology has some limitations: because of the presence of foreign mouse amino acid stretches, murine monoclonal antibodies are highly immunogenic and elicit a human anti-mouse antibody (HAMA) response when administered to human patients. Moreover, generation and production of monoclonal antibodies through traditional hybridoma protocols is expensive as very large cultures of mammalian cells and extensive purification processes are required (Chames, Van Regenmortel, Weiss, & Baty, 2009; Pasqualini & Arap, 2004).

With the emergence of molecular biology technologies, *in vitro* methods were developed to address these challenges and facilitate the generation of therapeutic human monoclonal antibodies (Hoogenboom & Chames, 2000; Marks et al., 1991). The antibody phage display technology was the first and remains the most widely method for *in vitro* antibody selection. Based on the pioneering research of Dr. George P. Smith and of Dr. Gregory P. Winter, the technology (2018 Nobel Prize in Chemistry) involves the display of antibody fragments or small peptides fused to engineered envelope proteins of a filamentous bacteriophage (Smith, 1985; Winter, 2019). Most phage antibody libraries have been produced in phagemid vectors by cloning antibody genes upstream of the g3p coat protein gene, which results in an antibody-g3p fusion protein (Sblattero & Bradbury, 2000). Advantages of using *in vitro* technologies include the speed with which antibody clones undergo selection and isolation, the direct selection of human recombinant monoclonal antibodies, the absence of immunization steps, and the fact that antibody libraries can be used to generate antibodies to toxic antigens that would otherwise kill the host (Liu, 2014).

More recently, our group has developed a new technology for antibody selection termed *S*election of *P*hage-displayed *A*ccessible *R*ecombinant *T*argeted *A*ntibodies" (SPARTA) (D'Angelo et al., 2018). SPARTA is a 2-step *in-tandem* methodology based on *in vitro* screening and *in vivo* selection to discover target-specific and biologically active human recombinant monoclonal antibodies against cancer cell-surface targets (Fig. 1).

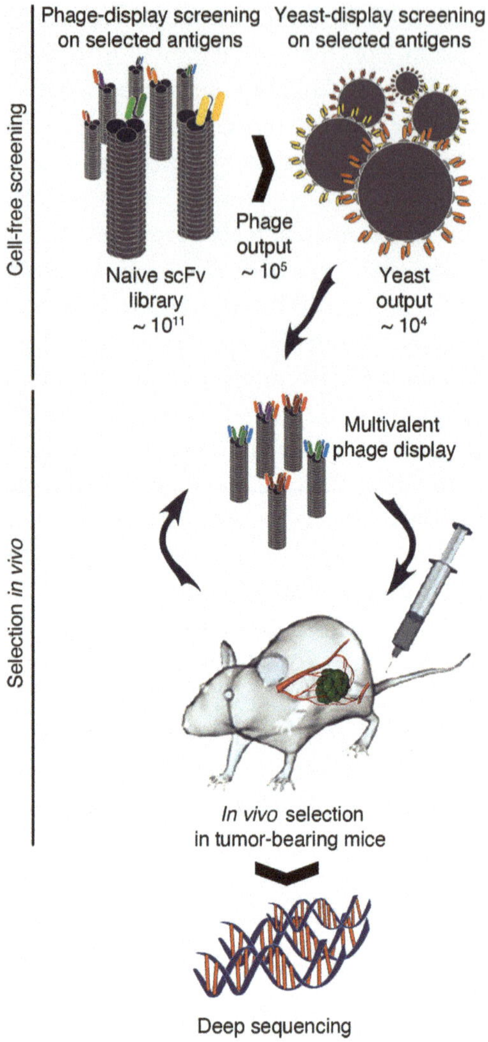

Fig. 1 SPARTA methodology (adapted from Staquicini et al., 2015). A naive human scFv library ($\sim 10^{11}$ TU) is screened *in vitro* against immobilized recombinant antigens. The phage output pool ($\sim 10^6$ TU) is subsequently transferred to a yeast display vector for two additional screening steps. After rounds of positive and negative sorting, the yeast output pool ($\sim 10^4$ TU) is expressed multivalently in phage particles and administered i.v. into tumor-bearing mice. Tumor-homing phage particles are recovered, amplified by PCR and re-expressed in multivalent format for two additional rounds of *in vivo* selection. After NGS, clonal diversity and ranking are determined.

In vitro screening with antibody phage display allows for the isolation of hundreds of human recombinant antibodies against a target (applying a selective pressure), whereas the *in vivo* selection step allows for the identification of antibodies that recognize the cell-surface associated targets within their native *in vivo* context. Unlike an unbiased approach where the receptor is not as yet known, SPARTA may begin with a previously identified tumor cell-surface receptor target. An enriched pool of recombinant human antibodies against the target is generated by using a high-throughput combination of phage and yeast display selections. First, a naive human single-chain variable fragment (scFv) library is screened *in vitro* against immobilized recombinant antigens, and the resulting phage output pool is transferred to a yeast display vector which allows precise fluorescence-activated cell sorting (FACS) to restrict reactivity to those clones recognizing the target (Ferrara et al., 2015). After FACS, the yeast output pool is expressed multivalently in phage particles and administered intravenously into tumor-bearing mice to select tumor-homing phage clones that can be recovered and sequenced to determine clonal diversity (D'Angelo et al., 2018). The combination of phage and yeast antibody display platforms in a single selection strategy provides potential advantages over either platform alone: Screening of a large number of clones in a single experiment and tailored selection for precise functional attributes (D'Angelo et al., 2018).

SPARTA has been proven as an efficient technique to isolate recombinant human antibodies against a selected target. By using this methodology we have been able to isolate several monoclonal antibodies that target EphA5, thus minimizing the risk of failure in later steps of drug development (D'Angelo et al., 2018). *In vivo*, selected anti-EphA5 antibodies recognized and localized to lung cancer xenografts. When antibodies were tested to deliver cytotoxic drugs, they showed effective killing activity against EphA5-expressing lung cancer cells but not control cells, suggesting their potential use as antibody-drug conjugates (ADCs). Antibodies generated by the SPARTA methodology can be tailored for several applications and will greatly benefit pre-screened patients suffering from significant side effects of current treatment options.

5. Conclusion

Cell-cell and cell-environment interactions, largely mediated by RTKs, control most cellular functions in higher organisms. It therefore follows that deregulation of RTK signaling pathways leads to many diseases, including cancers and developmental disorders. A better understanding of the participation of RTKs in cancer development and progression is essential for the generation of new treatment options, in particular antibody-based therapies. Due to their high specificity and affinity toward target antigens, monoclonal antibodies have been increasingly used as therapeutic agents for the treatment of various types of malignancies. Hybridoma and *in vitro* technologies have allowed for rapid development of antibody-based therapies: the monoclonal antibodies cetuximab (anti-EGFR), bevacizumab (anti-VEGF), nivolumab and pembrolizumab (both checkpoint inhibitors anti PD-1) are currently used in clinical settings for the treatment of lung cancer. Several clinical trials are also testing monoclonal antibodies against EGFR, IGF-1R, RANKL and CTLA-4 (Silva, Coelho, Anazetti, & Simioni, 2017). The introduction of novel targets (such as EphA5) and selection methodologies (such as SPARTA) will hopefully enable effective isolation and testing of new drugs to offer new therapeutic opportunities to cancer patients.

Acknowledgments

We thank Dr. Helen Pickersgill and her colleagues (Life Science Editors) for professional editorial services. This work was supported in part by the Cancer Center Support Grants (CCSG/P30) to the Rutgers Cancer Institute of New Jersey (CA072720), NCI R01 Grant (CA218853 to F.I.S.), and by research awards from the Gillson-Longenbaugh Foundation (to R.P. and W.A.).

Conflict of interest

F.I.S., R.P. and W.A. are inventors on patent applications (methods and compositions of matter) related to this work. R.P and W.A. are founders of PhageNova Bio. R.P. is the Chief Scientific Officer and a paid consultant for PhageNova Bio. F.I.S., R.P. and W.A. are entitled to royalties if licensing or commercialization occurs. These arrangements are managed in accordance with the established institutional conflict of interest policies of Rutgers, The State University of New Jersey. The other authors declare that they have no competing interests.

References

Arvanitis, D., & Davy, A. (2008). Eph/ephrin signaling: Networks. *Genes & Development*, *22*(4), 416–429.

Bargmann, C. I., Hung, M. C., & Weinberg, R. A. (1986). Multiple independent activations of the neu oncogene by a point mutation altering the transmembrane domain of p185. *Cell*, *45*(5), 649–657.

Bhattacharya, P., Shetake, N. G., Pandey, B. N., & Kumar, A. (2018). Receptor tyrosine kinase signaling in cancer radiotherapy and its targeting for tumor radiosensitization. *International Journal of Radiation Biology*, *94*(7), 628–644.

Blume-Jensen, P., & Hunter, T. (2001). Oncogenic kinase signalling. *Nature, 411*(6835), 355–365.
Caras, I. W. (1997). A link between axon guidance and axon fasciculation suggested by studies of the tyrosine kinase receptor EphA5/REK7 and its ligand ephrin-A5/AL-1. *Cell and Tissue Research, 290*(2), 261–264.
Carraway, K. L., 3rd, & Sweeney, C. (2002). EGF receptor activation by heterologous mechanisms. *Cancer Cell, 1*(5), 405–406.
Chames, P., Van Regenmortel, M., Weiss, E., & Baty, D. (2009). Therapeutic antibodies: Successes, limitations and hopes for the future. *British Journal of Pharmacology, 157*(2), 220–233.
Chen, X., Wang, X., Wei, X., & Wang, J. (2016). EphA5 protein, a potential marker for distinguishing histological grade and prognosis in ovarian serous carcinoma. *Journal of Ovarian Research, 9*(1), 83.
Chiari, R., Hames, G., Stroobant, V., Texier, C., Maillere, B., Boon, T., et al. (2000). Identification of a tumor-specific shared antigen derived from an Eph receptor and presented to CD4 T cells on HLA class II molecules. *Cancer Research, 60*(17), 4855–4863.
Critchley, W. R., Pellet-Many, C., Ringham-Terry, B., Harrison, M. A., Zachary, I. C., & Ponnambalam, S. (2018). Receptor tyrosine kinase ubiquitination and de-ubiquitination in signal transduction and receptor trafficking. *Cell, 7*(3), 22.
D'Angelo, S., Staquicini, F. I., Ferrara, F., Staquicini, D. I., Sharma, G., Tarleton, C. A., et al. (2018). Selection of phage-displayed accessible recombinant targeted antibodies (SPARTA): Methodology and applications. *JCI Insight, 3*(9), 1–12.
Du, J., Fu, C., & Sretavan, D. W. (2007). Eph/ephrin signaling as a potential therapeutic target after central nervous system injury. *Current Pharmaceutical Design, 13*(24), 2507–2518.
Du, Z., & Lovly, C. M. (2018). Mechanisms of receptor tyrosine kinase activation in cancer. *Molecular Cancer, 17*(1), 58.
Esposito, I., Kleeff, J., Bischoff, S. C., Fischer, L., Collecchi, P., Iorio, M., et al. (2002). The stem cell factor-c-kit system and mast cells in human pancreatic cancer. *Laboratory Investigation, 82*(11), 1481–1492.
Ferrara, F., D'Angelo, S., Gaiotto, T., Naranjo, L., Tian, H., Gräslund, S., et al. (2015). Recombinant renewable polyclonal antibodies. *MAbs, 7*(1), 32–41.
Fu, D. Y., Wang, Z. M., Wang, B. L., Chen, L., Yang, W. T., Shen, Z. Z., et al. (2010). Frequent epigenetic inactivation of the receptor tyrosine kinase EphA5 by promoter methylation in human breast cancer. *Human Pathology, 41*(1), 48–58.
Genander, M., & Frisen, J. (2010). Ephrins and Eph receptors in stem cells and cancer. *Current Opinion in Cell Biology, 22*(5), 611–616.
Heldin, C. H. (1995). Dimerization of cell surface receptors in signal transduction. *Cell, 80*(2), 213–223.
Herst, P. M., & Berridge, M. V. (2013). Cell hierarchy, metabolic flexibility and systems approaches to cancer treatment. *Current Pharmaceutical Biotechnology, 14*(3), 289–299.
Hoogenboom, H. R., & Chames, P. (2000). Natural and designer binding sites made by phage display technology. *Immunology Today, 21*(8), 371–378.
Hubbard, S. R. (1999). Structural analysis of receptor tyrosine kinases. *Progress in Biophysics and Molecular Biology, 71*(3–4), 343–358.
Hubbard, S. R., & Miller, W. T. (2007). Receptor tyrosine kinases: Mechanisms of activation and signaling. *Current Opinion in Cell Biology, 19*(2), 117–123.
Hubbard, S. R., & Till, J. H. (2000). Protein tyrosine kinase structure and function. *Annual Review of Biochemistry, 69*, 373–398.
Kentsis, A., Reed, C., Rice, K. L., Sanda, T., Rodig, S. J., Tholouli, E., et al. (2012). Autocrine activation of the MET receptor tyrosine kinase in acute myeloid leukemia. *Nature Medicine, 18*(7), 1118–1122.

Kohler, G., & Milstein, C. (1975). Continuous cultures of fused cells secreting antibody of predefined specificity. *Nature*, *256*(5517), 495–497.

Larsen, A. B., Pedersen, M. W., Stockhausen, M. T., Grandal, M. V., van Deurs, B., & Poulsen, H. S. (2007). Activation of the EGFR gene target EphA2 inhibits epidermal growth factor-induced cancer cell motility. *Molecular Cancer Research*, *5*(3), 283–293.

Lemmon, M. A., & Schlessinger, J. (1994). Regulation of signal transduction and signal diversity by receptor oligomerization. *Trends in Biochemical Sciences*, *19*(11), 459–463.

Lemmon, M. A., & Schlessinger, J. (2010). Cell signaling by receptor tyrosine kinases. *Cell*, *141*(7), 1117–1134.

Li, E., & Hristova, K. (2010). Receptor tyrosine kinase transmembrane domains: Function, dimer structure and dimerization energetics. *Cell Adhesion & Migration*, *4*(2), 249–254.

Liu, J. K. (2014). The history of monoclonal antibody development - progress, remaining challenges and future innovations. *Annals of Medicine and Surgery*, *3*(4), 113–116.

Longati, P., Comoglio, P. M., & Bardelli, A. (2001). Receptor tyrosine kinases as therapeutic targets: The model of the MET oncogene. *Current Drug Targets*, *2*(1), 41–55.

Manning, G., Whyte, D. B., Martinez, R., Hunter, T., & Sudarsanam, S. (2002). The protein kinase complement of the human genome. *Science*, *298*(5600), 1912–1934.

Marks, J. D., Hoogenboom, H. R., Bonnert, T. P., McCafferty, J., Griffiths, A. D., & Winter, G. (1991). By-passing immunization. Human antibodies from V-gene libraries displayed on phage. *Journal of Molecular Biology*, *222*(3), 581–597.

McDonell, L. M., Kernohan, K. D., Boycott, K. M., & Sawyer, S. L. (2015). Receptor tyrosine kinase mutations in developmental syndromes and cancer: Two sides of the same coin. *Human Molecular Genetics*, *24*(R1), R60–R66.

Pasquale, E. B. (2005). Eph receptor signalling casts a wide net on cell behaviour. *Nature Reviews. Molecular Cell Biology*, *6*(6), 462–475.

Pasquale, E. B. (2008). Eph-ephrin bidirectional signaling in physiology and disease. *Cell*, *133*(1), 38–52.

Pasquale, E. B. (2010). Eph receptors and ephrins in cancer: Bidirectional signalling and beyond. *Nature Reviews. Cancer*, *10*(3), 165–180.

Pasqualini, R., & Arap, W. (2004). Hybridoma-free generation of monoclonal antibodies. *Proceedings of the National Academy of Sciences of the United States of America*, *101*(1), 257–259.

Pawson, T., Gish, G. D., & Nash, P. (2001). SH2 domains, interaction modules and cellular wiring. *Trends in Cell Biology*, *11*(12), 504–511.

Pawson, T., & Nash, P. (2003). Assembly of cell regulatory systems through protein interaction domains. *Science*, *300*(5618), 445–452.

Peterson, A., & Seed, B. (1987). Monoclonal antibody and ligand binding sites of the T cell erythrocyte receptor (CD2). *Nature*, *329*(6142), 842–846.

Robinson, D. R., Wu, Y. M., & Lin, S. F. (2000). The protein tyrosine kinase family of the human genome. *Oncogene*, *19*(49), 5548–5557.

Sblattero, D., & Bradbury, A. (2000). Exploiting recombination in single bacteria to make large phage antibody libraries. *Nature Biotechnology*, *18*(1), 75–80.

Silva, A. P., Coelho, P. V., Anazetti, M., & Simioni, P. U. (2017). Targeted therapies for the treatment of non-small-cell lung cancer: Monoclonal antibodies and biological inhibitors. *Human Vaccines & Immunotherapeutics*, *13*(4), 843–853.

Smith, G. P. (1985). Filamentous fusion phage: Novel expression vectors that display cloned antigens on the virion surface. *Science*, *228*(4705), 1315–1317.

Staquicini, F. I., Qian, M. D., Salameh, A., Dobroff, A. S., Edwards, J. K., Cimino, D. F., et al. (2015). Receptor tyrosine kinase EphA5 is a functional molecular target in human lung cancer. *The Journal of Biological Chemistry*, *290*(12), 7345–7359.

Volinsky, N., & Kholodenko, B. N. (2013). Complexity of receptor tyrosine kinase signal processing. *Cold Spring Harbor Perspectives in Biology*, *5*(8), a009043.

Wang, T. H., Chang, J. L., Ho, J. Y., Wu, H. C., & Chen, T. C. (2012). Ephrin A5 suppresses colon cancer development by negatively regulating epidermal growth factor receptor stability. *The FEBS Journal*, *279*(2), 251–263.

Wang, T. H., Yeh, C. T., Ho, J. Y., Ng, K. F., & Chen, T. C. (2016). OncomiR miR-96 and miR-182 promote cell proliferation and invasion through targeting ephrinA5 in hepatocellular carcinoma. *Molecular Carcinogenesis*, *55*(4), 366–375.

Wang, X., Yennawar, N., & Hankey, P. A. (2014). Autoinhibition of the Ron receptor tyrosine kinase by the juxtamembrane domain. *Cell Communication and Signaling: CCS*, *12*, 28.

Wehrman, T., He, X., Raab, B., Dukipatti, A., Blau, H., & Garcia, K. C. (2007). Structural and mechanistic insights into nerve growth factor interactions with the TrkA and p75 receptors. *Neuron*, *53*(1), 25–38.

Winter, G. P. (2019). Harnessing evolution to make medicines (Nobel lecture). *Angewandte Chemie (International Ed. in English)*, *58*(41), 14438–14445.

Wu, Z., Ashlin, T. G., Xu, Q., & Wilkinson, D. G. (2019). Role of forward and reverse signaling in Eph receptor and ephrin mediated cell segregation. *Experimental Cell Research*, *381*(1), 57–65.

Yamaguchi, Y., & Pasquale, E. B. (2004). Eph receptors in the adult brain. *Current Opinion in Neurobiology*, *14*(3), 288–296.

Yamaoka, T., Kusumoto, S., Ando, K., Ohba, M., & Ohmori, T. (2018). Receptor tyrosine kinase-targeted cancer therapy. *International Journal of Molecular Sciences*, *19*(11), 3491.

Yuzawa, S., Opatowsky, Y., Zhang, Z., Mandiyan, V., Lax, I., & Schlessinger, J. (2007). Structural basis for activation of the receptor tyrosine kinase KIT by stem cell factor. *Cell*, *130*(2), 323–334.

Zhang, X., Gureasko, J., Shen, K., Cole, P. A., & Kuriyan, J. (2006). An allosteric mechanism for activation of the kinase domain of epidermal growth factor receptor. *Cell*, *125*(6), 1137–1149.

Zhang, W., Wei, X., Guo, S., Wang, J., Liu, J., & Wang, H. (2019). Differential expression of EphA5 protein in gastric carcinoma and its clinical significance. *Oncology Letters*, *17*(6), 5147–5153.

CHAPTER NINE

RTKs in pathobiology of head and neck cancers

Revathy Nadhan, Priya Srinivas*, M. Radhakrishna Pillai
Cancer Research Program, Rajiv Gandhi Centre for Biotechnology, Trivandrum, Kerala, India
*Corresponding author: e-mail address: priyasrinivas@rgcb.res.in

Contents

Abstract

Non-communicable diseases contribute to 71% of the deaths worldwide, of which cancers rank second after cardiovascular diseases. Among all the cancers, head and neck cancers (HNC) are consequential in augmenting the global cancer incidence as well as mortality. Receptor tyrosine kinases (RTKs) are emphatic for the matter that they serve as biomarkers aiding the analysis of tumor progression and metastasis as well as diagnosis, prognosis and therapeutic progression in the patients. The extensive researches on HNC have made significant furtherance in numerous targeted therapies, but for the escalating therapeutic resistance. This review explicates RTKs in HNC, their signaling pathways involved in tumorigenesis, metastasis and stemness induction, the association of non-coding RNAs with RTKs, an overview of RTK based therapy and associated

Advances in Cancer Research, Volume 147
ISSN 0065-230X
https://doi.org/10.1016/bs.acr.2020.04.008
© 2020 Elsevier Inc.
All rights reserved.

resistance in HNC, as well as a sneak peek into the HPV positive HNC and its therapy. The review extrapolates the cardinal role of RTKs and RTK based therapy as superior to other existing therapeutic interventions for HNC.

1. Introduction

Cancer remains to be the cardinal facet behind the accentuating escalation in the incidence and mortality resultant from non-communicable diseases. Amidst of the significant advancements in early diagnosis and therapeutic strategies, head and neck cancers (HNC) remains to be a prominent class contributing to the cancer incidence and mortality worldwide. Receptor Tyrosine Kinases (RTKs) are a significant class of biomarkers that have significant role in tumor progression of HNC and are extensively under research as a diagnostic and prognostic marker as well as an ideal target for treatment strategies. In this review, we provide an overview of RTKs and its subsets, unleashing the possibilities of attenuating RTK induced tumorigenesis in HNC and unraveling novel therapeutic interventions with RTKs as the central spotlight.

2. Insights into the incidence and etiology of HNC

Globally, HNC accounts for more than 850,000 cases and 450,000 deaths annually. HNC is higher in males, accounting for almost 75% of total cases and about 75% of total mortality (Cramer, Burtness, Le, & Ferris, 2019). Looking keenly into the most incident types of HNC, the cancers of the lip and oral cavity reigns the subtypes in the incidence as well as mortality regimes; further, the cancers of the larynx, nasopharynx, oropharynx, hypopharynx, para-sinuses, tongue and salivary glands add on to the list (Fig. 1). Though the incidence among the anatomical sites varies between the genders, the cancers of the lip and oral cavity contribute the highest in both the genders. While the male to female ratio for incidence varies from 6.9:1 for laryngeal cancers to 1.24:1 for cancers of the salivary glands, the mortality varies from 6.3:1 to 1.5:1 respectively. HNC incidence tends to be higher between 50 and 70 years age group, but for the HPV associated HNC, which is around 30 years. The geographical distribution of HNC varies diversely across the world. When nasopharyngeal cancers are more prevalent in areas of Hong Kong and the mouth and tongue cancers in the Indian subcontinent, the pharyngeal/laryngeal cancers are spread across

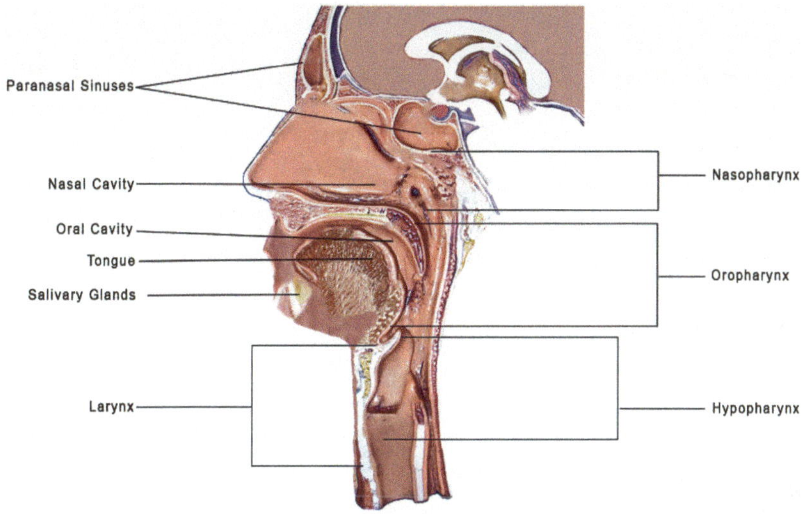

Fig. 1 An illustration of various regions relevant for head and neck cancers.

Table 1 Data on number of new cases and deaths of various head and neck cancer subtypes as per GLOBOCAN 2018.

S. no.	Cancer type	No. of new cases	No. of deaths
1.	Lip and Oral Cavity	354864 (2.0)	177384 (1.9)
2.	Larynx	177422 (1.0)	94771(1.0)
3.	Nasopharynx	129079 (0.7)	72987 (0.8)
4.	Oropharynx	92887 (0.5)	51005 (0.5)
5.	Hypopharynx	80608 (0.4)	34984 (0.4)
6.	Salivary glands	52799 (0.3)	22176 (0.2)

other parts, together attributing much of the HNC incidence to the Asian sub-continent (Bray et al., 2018).

According to GLOBOCAN 2018, Asia holds half of the new cancer cases and over half of the total cancer deaths across the globe, with global cancer deaths of 57.3% far ahead of global cancer incidence shares of 48.4%, owing to the abnormal increase in heterogeneous cancer types and fatality rates. On a global scale, the various HNC subtypes with their number of new cases and deaths are enlisted in Table 1. Among the HNC, the cancers of the lip and oral cavity rank 18th, among all the cancers, based on the number of new cases. A higher incidence cluster around Southern Asia has been exhibited by the lip and oral cavity cancers. The Age Standardized Rates (World) [ASR (W)] for the incidence of cancers in the lip and oral cavity is 12.9 in males and 4.5 in females per 100,000 individuals in South Central Asia, which is the second most incident area after Melanesia. Though the ASR (W) for cancer incidence in males in the lip and oral cavity is low in very high/high Human Development Index (HDI) areas, it ranks third in the transitioning countries especially due to the high disease burden in India, which holds 36% of the population among the lower HDI countries (Bray et al., 2018).

The Indian factsheet from the GLOBOCAN 2018 revealed that among the HNC, the lip and oral cavity cancers, ranked second in incidence among all sexes with a prevalence of 119,992 new cases (10.4% of all sites and Cum. Risk of 1.02), which is an increase of 114% as compared with that of 56,000 new cases, as depicted in GLOBOCAN 2012 (RESEARCH ICOM, 2018). Also, this HNC subtype remains to be selective cancer with the most number of new cases in men (92,011 new cases with 16.1% of all sites) and the fourth most prevalent cancers with new cases in women (27,981 new cases with 4.8% of all sites). The same subtype ranks second, after breast cancers, while assessing the mortality with 72,616 cases (9.3% of total deaths and Cum. Risk of 0.64) and a 5-year prevalence among all ages with 265,255 cases. The GLOBOCAN 2018 has also included extensive data on the other HNC subtypes like that of the larynx, nasopharynx, oropharynx, hypopharynx and salivary glands, which are depicted in Table 2. The ASR (W) incidence rates per sex for India per 100,000 mounts to 13.9 and 4.3 in males and females respectively, for the cancers in the lip and oral cavity, which is second in the ranking of all cancers; while that of the larynx being at the 9th ranking with 4.2 and 0.40 per 100,000 in males and females respectively. However, the ASR (W) for India, combining both sexes, stratifies the cancers of the lip and oral cavity to the third position, with an incidence rate of 9.1 and a mortality rate of 5.6 per 100,000 (Organization TGCO-IAfRoC-WH, 2019). As per the reports by the National Institute of Cancer Prevention and Research (NICPR), India diagnoses 0.47 million HNC cases per year; this portrays greater than 60% of the advanced disease and attributes to more than 8% of the HNC global cancer burden and deaths. The HNC incidence being 22 per 100,000 Indian populations, accounts for about 30% of all the Indian cancers, mostly within the age group of 50–60 years (Singh, Eisenberg, & Hoffman, 2018). The cancer mortality rate per 100,000 Indian women (across all ages) for lip and oral cavity is 3.3, salivary glands are 0.4, larynx is 0.3, oropharynx is 0.3, hypopharynx is 0.3 and nasopharynx is 0.2 (Singh et al., 2018).

Table 2 Data on incidence and mortality of head and neck cancers in both sexes, as per GLOBOCAN 2018.

		Incidence						Mortality					
		Males			Females			Males			Females		
S. no.	Cancer type	Cases	ASR (W)	Cumulative risk, ages birth to 74 years, %	Cases	ASR (W)	Cumulative risk, ages birth to 74 years, %	Deaths	ASR (W)	Cumulative risk, ages birth to 74 Years, %	Deaths	ASR (W)	Cumulative risk, ages birth to 74 years, %
1.	Lip and Oral Cavity	246,420	5.8	0.66	108,444	2.3	0.26	119,693	2.8	0.32	57,691	1.2	0.14
2.	Larynx	154,977	3.6	0.45	22,445	0.5	0.06	81,806	1.9	0.23	12,965	0.3	0.03
3.	Nasopharynx	93,416	2.2	0.24	35,663	0.8	0.09	54,280	1.3	0.15	18,707	0.4	0.05
4.	Oropharynx	74,472	1.8	0.21	18,415	0.4	0.05	42,116	1.0	0.12	8889	0.2	0.02
5.	Hypopharynx	67,496	1.6	0.19	13,112	0.3	0.03	29,415	0.7	0.08	5569	0.1	0.01
6.	Salivary glands	29,256	0.7	0.07	23,543	0.5	0.05	13,440	0.3	0.03	8736	0.2	0.02

The key oncogenic drivers for HNC include the long term use of tobacco and alcohol, exposure to radiations and infectious agents like the Human Papilloma Virus (HPV) and Epstein Barr Virus (EBV). The environmental and lifestyle factors contribute a greater share to the HNC incidence, wherein, the risk factors like tobacco smoking and chewing, chewing of betel quid and areca nut, especially in Asian regions and the habit of alcoholism are the risk factors of prominence. Tobacco use is the single largest risk factor for HNC which involves both direct tobacco use and second-hand smoking. However, tobacco and alcohol use together contributes to 75% of the HNC in comparison with their independent contributions, mostly in the cancers of the oral cavity, larynx, oropharynx and hypopharynx. Besides, occupational exposures to carcinogens, poor oral health and higher use of preserved or salted food, also contribute to HNC incidence. Further, when cigarette smoking poses an enhanced risk to the oropharynx, the use of smokeless tobacco increases the risk for oral cavity cancers. Alcohol consumption as an independent factor elevates the threat for cancers of larynx and pharynx (Alfouzan, 2019).

The changes in sexual habits result in high risk HPV infection and associated HNC. HPV-16 is a high-risk subtype, whose infection increases the HNC risks, though with wide geographical variations. HPV is more strongly associated with oropharyngeal cancers, especially the head and neck squamous cell carcinoma (HNSCC) and oropharyngeal squamous cell carcinoma (OPSCC), than that of the oral cavity or larynx. About 25% of HNSCC are attributed to HPV infections. HPV positivity/high tobacco and alcohol has been associated with a high risk of oral cavity cancers while HPV negativity/high tobacco and alcohol poses a higher risk of oropharyngeal cancers. Though HPV positive and negative HNC vary in their genetic makeup, the positive cases are more sensitive to therapeutic strategies and hence, positive patients have a higher survival rate. HPV-positive cases of HNSCC are prevalent among the younger males, who do not smoke or consume alcohol and mostly possess a higher socio-economic status. Also, these patients have a lower incidence of second primary cancers. The lifestyle changes including a higher number of sexual partners during lifetime and engagement in oral sex contribute to HPV positive HNC, especially the oropharyngeal cancers (Kobayashi et al., 2018; Murthy et al., 2017; Smith, Rubenstein, Haugen, Pawlita, & Turek, 2012).

3. Biomarkers for diagnosis and prognosis of HNC

Tissue biomarkers analyzed for diagnosis and prognosis of HNC involve the crucial markers assessed for classical cancer properties like cell adhesion, cell proliferation, migration, invasion, epithelial to mesenchymal transition (EMT) and so on. Demographic elements along with genetic and environmental factors should be considered for accurate diagnosis, therapeutic interventions as well as prognosis. Phenotype driven precision oncology models have been promisingly evolving in deriving accurate treatments and prognosis. The promising tissue biomarkers include EGFR, PD-L1, CCND1, PI3K, HIF-1α, PI3KCA, CDKN2A, TP53, TERT promoter, NOTCH, AKT, Ki67, AgNORs, VRK1, K-RAS, NFκB, AURK-A, ERCC1, XRCC1, VEGF and FGFR, whose mutations and gene alterations vary with HNC subtypes and can be exploited for novel therapeutic strategies (Budach & Tinhofer, 2019; Economopoulou, de Bree, Kotsantis, & Psyrri, 2019; Hsieh et al., 2019; Lampri et al., 2015). Various epigenetic modifications have also been associated with HNC including changes in gene methylation levels, variations in histone modifiers and chromatin remodelers as well as the role of ncRNAs, piRNAs, miRNAs, snoRNAs, tiRNAs, circular RNAs and lncRNAs in HNC tumorigenesis (Shilpi, Prabhat, Jayant, Harsimrut, & Bhudev, 2018).

Chemokine receptors are altered in HNC, especially the CXCR2 over-expressed in laryngeal squamous cell carcinoma, CXCR4 in nasopharyngeal carcinoma and CCR7 in numerous HNC; all of which predicts metastasis and prognosis in the patients (Bektas-Kayhan, Unur, Boy-Metin, & Cakmakoglu, 2012; Wang et al., 2004, 2005). Microsatellite instability has been analyzed in HNC, like that of LOH of 9p21, which is considered as an initial event in HNSCC (Garcia Martinez et al., 2012). CK-6.16/17, IL-6/8, MMP1/3/8/9/10 are also extensively over-expressed in HNC. miR-106b-25 cluster, miR-375, miR-451, miR-125a, miR-200a, miR-205, miR-138 and miR-38 has been reported to be deregulated in HNC (Dahiya & Dhankhar, 2016). Another biomarker cluster includes the MAGEA1-6, whose expression has been associated with worse prognosis and poor survival in HNC patients (Song et al., 2008). Centrosomal abnormalities along with over-expression of spindle checkpoint protein CDC20 as well as p53 mutations have also been reported in HNC (Yang et al., 2018).

One of the other prominent and emerging biomarkers is the PD-L1, which is expressed in tumor cells as well as tumor microenvironment (TME) (Hartmann, Bhola, & Grandis, 2016; Tan et al., 2017; Tong, Kao, & Sikora, 2012). PDL-1 has emerged as a significant immunotherapy biomarker used to select the suitable therapeutic regime in HNC patients, whose higher expression correlates with worse prognosis, poor overall survival (OS) and treatment outcomes independent of tumor origin. In HNC, high inflammation associated PD-L1 expression in tumor cells is associated with poor prognosis while enhanced expression in immune cells is associated with good prognosis (Economopoulou et al., 2019). Over-expression of Epidermal Growth Factor Receptor (EGFR) (associated with activation of mitogenic signaling) and CCND1 (which acts as a regulatory subunit of CDK4 and CDK6 for translating mitogenic signals to cell cycle progression) has been associated with HNC, which was stimulated by the use of CDK4 and CDK6 inhibitors. The resistance associated with these inhibitor drugs can be overcome by biomarker-driven patient selection. The immunoscore assessed by the presence of higher densities of CD8+ and CD3+ tumor-infiltrating lymphocytes as well as lower densities of T-regulatory cells in the TME has been studied to be indicators of better prognosis and OS in HNC patients, even in HPV negative cases (Economopoulou et al., 2019).

Minimally invasive liquid biopsy markers have been popular in prognosis, especially the hematological parameters, most prominently the neutrophil to lymphocyte ratio which is elevated in the worst prognosis for cancers in the nasopharynx, oropharynx, hypopharynx and oral cavity (Budach & Tinhofer, 2019). Circulating tumor cells (CTCs) have been extensively studied as a diagnostic biomarker for analyzing the stage of the disease, prognostic biomarker for analyzing the survival as well as therapeutic outcomes and characterization of the HNC (Payne et al., 2019). The higher neutrophil ratio creates a neutrophil extracellular trap, which sequesters CTCs to initiate metastasis to lungs and liver by modifying the extracellular matrix. Thus, therapies targeting these modified parameters prevent metastasis and recurrence. The presence of CTCs has been independently associated with increased local or distant metastases risk as well as progression-free survival (PFS) and OS. Enrichment of CTC population for analysis as well as an assessment of markers like EGFR transcripts or PD-L1 expression in CTCs would help predict accurate therapeutic strategies (Budach & Tinhofer, 2019). Differences in the expression pattern of numerous genes in single CTCs and CTC clusters have to be studied further, as the clusters have greater expression of stemness characters as well as metastatic potentialities (Payne et al., 2019).

HPV-16 is the most common subtype associated with HPV positive HNC either with the expression of its oncoprotein E6 or E7 DNA in the host cell or the presence of an extra chromosomal virus. HPV positive HNC, especially the HNSCC, uses E6 or E7 mRNA/DNA as a biomarker for HPV positivity, which along with the p16 (CDKN2A) immunohistochemistry, clarifies the prediction of recurrence-free survival, though both of them serves as independent predictors of disease progression. Further, the serum E6 or E7 protein levels or HPV-16 DNA levels in salivary oral rinses (though not much into clinical practice) can also be used as biomarkers in predicting the treatment outcome and recurrence in the HPV positive oropharyngeal cancers (Budach & Tinhofer, 2019). Patients younger than 60 years of age, who are in T stage >3 with the above median expression of CD8 positive tumor-infiltrating lymphocytes and positivity for viral protein E2 antigen, exhibit better PFS and lower death risks (Economopoulou et al., 2019).

Bio-imaging is being often correlated with biomarker expression in tumors and this has been well standardized in aiding diagnosis as well as prognosis. TNM (Tumor, Node, Metastasis) staging remains as excellent prognostic factor especially with the CD206 receptor-targeted radiopharmaceutical Technetium 99m, which accurately predicts the lymphatic neck node involvement so that neck dissection and associated morbidities can be avoided by replacement with invasive sentinel node procedure (Budach & Tinhofer, 2019). Patients with a high signature score, when a combination of radio-diagnostic signature and p16 status were analyzed, had a better response to chemotherapy (Budach & Tinhofer, 2019). Positron Emission Tomography (PET) and Single Positron Emission Computed Tomography (SPECT) employs tracer uptake correlated with the hypoxic markers like HIF-1α, CA-IX, GLUT-1 and pO2, since hypoxia results in enhanced resistance to radiotherapy in HNC (Economopoulou et al., 2019). The correlation was made not only with the hypoxic parameters but also with proliferation kinetics specific markers to analyze and predict tumor proliferation and response to therapy, like that of using the EGFR inhibitors. Magnetic Resonance Imaging (MRI) advancements have also been extensively exploited in analyzing the key biomarkers for hypoxia, proliferation, stemness etc. in HNC. Also, Cancer Stem Cell (CSC) identification by employing PET and MRI are under research wherein the evaluation of the tracer uptake could be correlated with CSC markers like CD44, CD133 and ALDH (Marcu, Reid, & Bezak, 2018).

4. RTKs as a biomarker for tumorigenesis in HNC

RTKs have extensively been analyzed as biomarkers for diagnosis, prognosis and assessing the therapeutic outcomes in cancers. Numerous classes of RTKs are present in humans. Of the 59 RTKs known so far, about 30 are known to be associated with tumorigenic signaling (Elferink & Resto, 2011). The prominent RTK families deregulated in cancers are Human Epidermal Growth Factor Receptor (HER/ERBB), Fibroblast Growth Factor Receptor (FGFR), Insulin Receptor (INSR), Insulin-like Growth Factor Receptor (IGFR), Ephrin (Eph), Platelet-Derived Growth Factor Receptor (PDGFR), Vascular Endothelial Growth Factor Receptor (VEGFR/Flt), Retinoic Acid Receptor-related Orphan Receptor (ROR), Nerve Growth Factor Receptor (NGFR), Tropomyosin Receptor Kinase (TRK), Axl, Tie, Hepatocyte Growth Factor Receptor (HGFR) and Muscle Specific Kinase (MuSK) (Choura & Rebai, 2011).

RTKs have an extracellular ligand-binding domain, a single pass transmembrane protein domain and an intracellular kinase domain. Except for the IGF receptor family, all other RTKs are monomeric in existence and upon ligand binding in the extracellular domain, it dimerizes/oligomerizes resulting in trans-autophosphorylation of the tyrosine residues in the cytoplasmic domain, releasing the *cis*-autoinhibitions and subsequent signaling cascade activation, wherein proteins with Src Homology (SH-2) or Phospho-tyrosine binding (PTB) domains bind to the phosphorylated tyrosine residues and activates pathways like MEK, RAS/MAPK, PI-3K/AKT, mTOR and JAK2/STAT signaling (Choura & Rebai, 2011; Du & Lovly, 2018). In normal conditions, the signaling is well balanced within the cells by the action of tyrosine phosphatases. However, the aberrant activation leading to oncogenic signaling is acquired by mechanisms like the gain of function mutations (like in EGFR, FGFR3, KIT etc.), over-expression and genomic amplification (EGFR, ErbB2/4, MET (c-Met), PDGFRA, Kit, Flt3 etc.), chromosomal rearrangements (BCR/ABL, NPL/ALK, EGFR-RAD51, EML4/ALK etc.) including the kinase domain duplication (EGFR-KDD, BRAF-KDD etc.) and autocrine activation. The role of miRNAs in regulating the RTKs and vice versa, along with the RTKs in the TME and the negative regulators of the RTKs, together adds on to the regulation of the RTK signaling and associated cellular processes.

Their understanding is the key to unravel potential therapeutic targets for various cancers (Du & Lovly, 2018).

In HNC, EGF binding activates the signaling cascade including MAPK, PI3K/AKT and STAT pathways. Altered EGFR activities have been reported with 58–90% of HNSCC and are mostly associated with EGFR over-expression rather than mutations or amplifications (Byeon, Ku, & Yang, 2019). EGFR amplification leading to enhanced copy numbers is correlated with poor survival, resistance to radiotherapy and loco-regional failure. Anti-EGFR antibodies, as well as Tyrosine Kinase Inhibitors (TKIs), are exploited as therapeutic strategies. EGFR inhibitor resistance has also been widely studied due to intriguing factors like EMT, the role of p53, Axl, HER3, SFK and HGF/MET (Byeon et al., 2019). Another RTK significantly activated in HNC is the IGF-1R, to which IGF-1/2 binds and activates Ras-Raf, MAPK, ERK and PI3K and activates VEGF production too, all contributing to HNC tumorigenesis. About 94% of HNSCC have altered IGF-1R and IGF-1R-EGFR cross-talk has also been well explored currently to enhance their therapeutic possibilities for HNC. MET signaling is another prominent RTK in HNC, which on binding with HGF ligand, dimerizes and autophosphorylates catalytic residues, thereby recruiting cytosolic effectors, which are common with EGFR and IGF-1R, resulting in activation of MAPK, JNK and JAK/STAT signaling pathways. MET over-expression has been observed in 85% of HNSCC, with just 10% resulting from gene amplification, suggesting a paracrine mode of over-activation in associated tumorigenesis. The common abnormalities associated with MET expression in HNC include its over-expression, missense mutations resulting in constitutive kinase activity, failure of MET down-regulatory pathways and altered interactions with other cellular molecules like EGFR, all correlating with lymph node metastasis and tumor recurrence. Anti-MET and HGF antibodies are being explored as therapeutic targets in HNC (Elferink & Resto, 2011). Axl belongs to TAM-RTKs which includes TYRO3, MER, MERTK, along with Axl. Higher Axl expression has been associated with poor PFS and OS, lymph node metastases, local recurrences and advanced tumor stages in HNC (von Massenhausen et al., 2016). ROS1, ALK, FGFR3, DDR2, RET, Axl and MERTK are also reported to be mutated in HNSCC (Elferink & Resto, 2011). Other prominent RTKs over-expressed in HNC include FGFR, VEGFR, PDGFR and RET. Various RTKs and their aberrations in head and neck cancer subtypes are enlisted in Table 3 as analyzed from c-BioPortal for Cancer Genomics.

Table 3 RTKs and their associated aberrations in head and neck cancers, as analyzed from c-BioPortal for cancer genomics (analyzed on 26/11/2019).

Gene ID	Gene alteration frequency	Mutation frequency	Amplification frequency	Multiple alterations	Deep deletion	mRNA high	Protein high
Head and Neck Squamous Cell Carcinoma (TCGA Provisional—504 patients/samples)							
EGFR (ERBB)	20.56%	2.42%	3.02%	7.86%	0.2%	6.94%	–
ERBB2	3.77%	1.79%	1.79%	0.2%	–	3.37%	–
ERBB3	8.53%	2.78%	0.2%	0.4%	–	–	1.98%
ERBB4	9.52%	4.17%	0.6%	–	1.59%	3.17%	–
FGFR1	10.32%	0.2%	6.55%	0.2%	0.99%	2.38%	–
FGFR2	5.95%	0.6%	–	0.4%	–	4.96%	–
FGFR3	6.94%	1.79%	0.4%	0.4%	1.39%	2.98%	–
FGFR4	8.73%	0.79%	–	–	–	7.94%	–
PDGFRA	7.74%	2.18%	0.4%	0.4%	–	4.76%	–
PDGFRB	6.35%	1.79%	–	–	–	4.56%	–
FLT1 (VEGFR1)	8.33%	2.98%	0.2%	–	0.4%	4.76%	–
FLT3	4.56%	1.9%	0.4%	–	0.2%	2.78%	–
FLT4	5.95%	2.58%	0.2%	–	–	3.17%	–
IGF1R	9.72%	1.59%	0.2%	0.99%	0.2%	6.75%	–
IGF2R	5.75%	1.79%	–	0.2%	0.4%	2.38%	–

INSR	4.96%	0.4%	–	0.2%	0.79%	3.57%	–
AXL	6.55%	0.99%	0.4%	1.19%	–	3.37%	0.4%
MET	8.13%	0.79%	0.79%	–	–	6.55%	–
MERTK	8.53%	0.99%	0.99%	0.6%	–	5.95%	–
RET	8.13%	2.38%	0.79%	0.2%	0.2%	4.56%	–
MST1R	4.56%	1.79%	–	–	0.79%	1.98%	–
MUSK	5.75%	1.19%	1.59%	0.2%	–	2.78%	–
TYRO3	5.75%	0.79%	–	–	0.2%	4.76%	–
NGFR	4.96%	0.2%	–	–	–	4.76%	–
NTRK1	2.18%	0.6%	0.99%	0.2%	–	0.4%	–
NTRK2	7.54%	0.6%	0.6%	–	0.2%	6.15%	–
NTRK3	3.57%	2.18%	0.6%	–	–	0.79%	–
TIE1	6.94%	1.59%	0.79%	–	–	4.56%	–
TEK	7.34%	0.99%	0.6%	0.2%	2.38%	3.17%	–
EPHA1	8.13%	1.59%	0.4%	–	0.6%	5.56%	–
EPHA2	9.33%	4.17%	0.2%	0.4%	0.4%	4.17%	–
EPHA3	6.94%	3.17%	0.2%	–	1.59%	1.98%	–
EPHA4	6.55%	1.98%	–	–	1.39%	3.17%	–
EPHA5	6.15%	4.56%	0.4%	–	–	1.19%	–

Continued

Table 3 RTKs and their associated aberrations in head and neck cancers, as analyzed from c-BioPortal for cancer genomics (analyzed on 26/11/2019).—cont'd

Gene ID	Gene alteration frequency	Mutation frequency	Amplification frequency	Multiple alterations	Deep deletion	mRNA high	Protein high
EPHA6	12.9%	2.58%	4.56%	0.79%	0.2%	4.76%	–
EPHA7	8.73%	3.97%	0.4%	0.4%	0.2%	3.77%	–
EPHA8	4.37%	1.19%	0.2%	–	–	2.98%	–
EPHA10	6.15%	1.79%	0.79%	–	–	3.57%	–
EPHB1	16.47%	2.38%	5.95%	1.79%	–	6.35%	–
EPHB2	3.77%	0.6%	0.2%	–	–	2.98%	–
EPHB3	26.39%	1.59%	15.48%	4.17%	–	5.16%	–
EPHB4	14.68%	0.99%	1.79%	3.77%	–	7.94%	–
EPHB6	6.75%	1.19%	0.2%	0.4%	0.6%	4.37%	–
Nasopharyngeal Carcinoma (Singapore, Nat Genet 2014–56 patients/samples)							
ERBB2	1.79%	1.79%	–	–	–	–	–
ERBB3	3.57%	3.57%	–	–	–	–	–
FGFR2	3.57%	3.57%	–	–	–	–	–
INSR	1.79%	1.79%	–	–	–	–	–
MET	1.79%	1.79%	–	–	–	–	–
NTRK2	1.79%	1.79%	–	–	–	–	–
TEK	1.79%	1.79%	–	–	–	–	–

Oral Squamous Cell Carcinoma (M D Anderson, Cancer Discov 2013–40 patients/samples)							
EGFR (ERBB)	5%	5%	–	–	–	–	–
ERBB4	2.5%	2.5%	–	–	–	–	–
FGFR2	2.5%	2.5%	–	–	–	–	–
PDGFRB	2.5%	2.5%	–	–	–	–	–
EPHA1	2.5%	2.5%	–	–	–	–	–
EPHA2	2.5%	2.5%	–	–	–	–	–
EPHA6	2.5%	2.5%	–	–	–	–	–
EPHA7	7.5%	7.5%	–	–	–	–	–
EPHB1	2.5%	2.5%	–	–	–	–	–
Salivary Adenoid Cystic Carcinoma (MD Anderson, Clin cancer res 2015–65 patients/samples)							
FGFR2	1.54%	1.54%	–	–	–	–	–
IGF1R	1.54%	1.54%	–	–	–	–	–

5. RTK induced de-differentiation of cancer cells in HNC and its link with stemness

Numerous CSC markers have been studied in HNSCC, which includes CD44, CD133, CD24, c-Met, ALDH, side population cells and sphere-forming ability. The key signaling pathways involved in stem cell self-renewal and differentiation processes are SHH, Wnt/β-catenin, EGFR and Notch signaling, which activates the final effectors like BM1, OCT4, SOX2, NANOG, KLF4, LIN28/Let7 to induce stemness in HNC. CD44 is a type I transmembrane glycoprotein which binds to its ligands like MMPs, growth factors, cytokines and hyaluronan and subsequently activates RTKs through pathways like Src/FAK, PI3K/AKT and HGF/c-Met to trigger cell proliferation and survival signals for inducing stemness. CD44 expression in HNSCC tissues has been correlated with malignancy, lymph node metastasis, loco-regional recurrence, drug resistance and poor outcome. In the EGFR pathway, EGF or TGF-α binds to EGFR and activates signaling resulting in poor therapeutic outcome and resistance. Higher EGFR expression has been observed in ALDH high CSCs in the HPV positive HNC cells. Also, CD44 interacts with EGFR in inducing stemness in HNC and EGFR overexpression has been associated with enhanced levels of CD44, Oct-4, Nanog and Sdf-1 (Chen et al., 2019; Chen & Wang, 2019; Jeon & Lee, 2017). EGFRvIII mutant form is present in 42% of the HNSCC (Chen et al., 2019; Chen & Wang, 2019). c-Met, which is an RTK, serves as a marker for CSCs which binds to its ligand HGF and activates the signaling with its crosstalk with c-Src, PI3K, Akt and MAPK, inducing stemness in HNC. HNC, which expresses higher levels of MET, has greater potential to induce stemness. Also, it has been reported that in HNC, the clonal selection of MET mutations has increased its percentage from 2% in primary tumors to 50% in metastatic tumors (Chen & Wang, 2019). MET signaling has also been extensively explored with stemness induction in HNC, mostly via the activation of c-Myc and STAT3 as downstream effectors. c-Met/HGF signaling regulated by HIF-1α has been reported in cancers of the salivary gland. Also, MET signaling activates the Wnt/β-catenin in instating the survival, maintenance and functional roles of CSCs in HNSCC (Rankin & Giaccia, 2016). The Cancer Associated Fibroblasts (CAFs) in the TME secrete VEGF, EGF, CCL5, cytokines and growth factors, which act via RTKs in initiating tumorigenesis. HGF secreted by CAFs act by c-Met in aiding the progression of HNC. The HPV positive HNSCC has high CD8+, high CD4+ and low CD4+/CD8+ ratio in comparison with the HPV negative cases (Miekus, 2017). PIK3R1 and PIK3CA are over-expressed in HNSCC and activated Ephs, TRKs and c-Kit pathways act as compensatory signaling mechanisms to induce CSC population independent of the PI3K pathway in HNSCC (Wang, Chan, & Cho, 2019). Also, it has been reported that periostin secreted by CAFs, binds to PTK-7 and activates Wnt/β-catenin signaling and triggers the CSC population in HNSCC (Abhold et al., 2012). Axl signaling has also been reported to be involved in HNC tumorigenesis and CSC induction (Yu et al., 2018). Conclusively, aberrant expression and activation of EGFR, c-Met and Axl have been associated with CSC induction in HNC.

6. HPV-RTK link in HNC

HPV infection is oncogenic due to the presence of two of its oncoproteins namely E6 and E7, which promotes the degradation of tumor suppressor genes p53 and pRb. About 75% of the OPSCC are HPV positive cases. In HPV positive HNC, EGFR is down-regulated owing to genetic aberrations but shows an increased expression for HER2 and HER3. HER3 over-expression in pharyngeal cancer patients was associated with lower survival rates. HPV E6 and E7 proteins inhibit the ubiquitination of HER3 by sequestering the E3 ubiquitin ligases Nrdp1 or NEDD4. In HNSCC cell lines, the levels of HER2, HER3, HER2-HER3, HER3-PI3K protein clusters were found to be higher in HPV positive rather than HPV negative cases. The over-expressed HER3 dimerizes with other RTKs and activates the signaling cascade via the PI3K pathway (Brand et al., 2017). The Axl signatures have not been significantly differentiated between HPV positive and negative cases to date. VEGF expression has been reported to be induced by the HPV E6 oncoprotein independent of the tumor suppressor p53 (Kramer et al., 2016). EGFR (HER1) exhibits a lower expression corresponding to a lower copy number in HPV positive OPSCC in comparison with the HPV negative ones. Further, mutations associated with EGFR in HPV positive HNC are a rare event. 16% of HPV negative HNSCC harbor mutations/amplification in EGFR, 4% for IGF1R and 14% for FGFR1/2/3 (Saada-Bouzid & Le Tourneau, 2019). Thus, cases with p16 positive/HPV positive HNC have in general better prognosis, lower loco-regional failures and higher survival rate (Mirghani & Blanchard, 2018). Concerning FGFR, 10% of HPV negative HNSCC harbor FGFR1 mutations or amplifications while 11% of HPV positive HNSCC harbor FGFR3 mutations or fusions (Eze, Lo, & Burtness, 2017). High IGF1R expression has been seen in HPV negative HNSCC (Kumar, Rayala, & Venkatraman, 2018). Numerous de-regulated RTKs associated with HPV positive HNC are enlisted in Fig. 2. Thus, the over-expression of RTKs, especially HER2/3, might be a cardinal factor contributing to the superior therapeutic responses in HPV positive HNC patients.

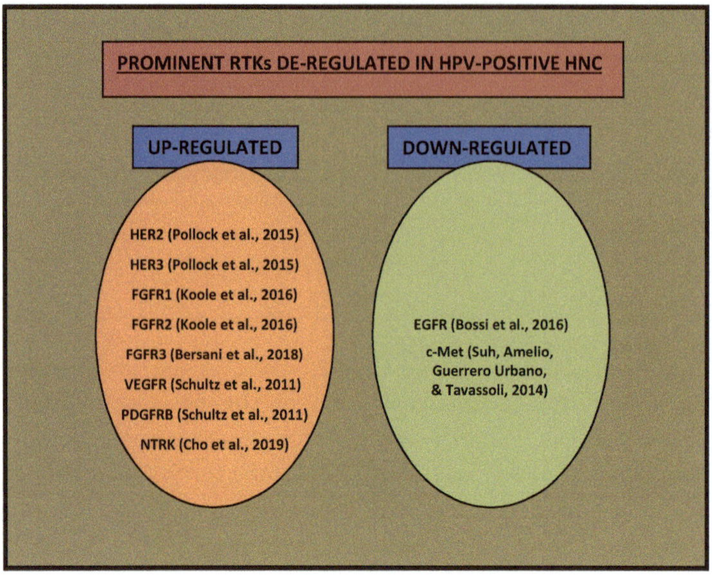

Fig. 2 Various RTKs which are up-regulated and down-regulated in HPV positive head and neck cancers (Bersani et al., 2018; Bossi et al., 2016; Cho et al., 2019; Koole et al., 2016; Pollock et al., 2015; Schultz et al., 2011; Suh, Amelio, Guerrero Urbano, & Tavassoli, 2014).

7. Global miRNAs regulating RTKs in HNC

miRNAs are small non-coding RNAs that act as either oncogenes or tumor suppressor genes in modulating the physiological processes by regulating the protein degradation either by degrading or repressing the mRNA and has been associated with various levels in the HNC tumorigenesis. Over-expression of miR-21, miR-34c, miR-31, miR-38, miR-106b, miR-375, miR-451, miR-205, miR-138, miR-155, miR-181b, miR-184 and miR-345 and down-regulation of miR-125a and miR-200a has been associated with HNSCC tumorigenesis and metastasis (Dahiya & Dhankhar, 2016; Shilpi et al., 2018). miR-34b, miR-137, miR-193a and miR-203 has been reported of being hypermethylated in HNC owing to its placement near the CpG islands. The up-regulation of miR-663 owing to hypomethylation has been reported to be contributing to resistance to chemotherapy in HNC. Also, polymorphisms in miR-146, miR-149, miR-196, miR-499 has been shown to increase the risk of HPV positive HNC in non-smokers (Shilpi et al., 2018). miR-9, miR-134 and miR-191 have also been evolving as critical biomarkers in diagnosing HNSCC (Hsieh et al., 2019). The miR-106b-25/miR17-92 cluster has been reported to initiate anti-apoptotic effects via cell cycle checkpoint deregulation affecting the PI3K/PTEN/AKT pathway (Pai & Westra, 2009). In HPV negative HNSCC, the down-regulated miR-let-7c-5p and miR-100-5p correlate with altered cell cycle control contributing to tumorigenesis (Leemans, Snijders, & Brakenhoff, 2018). Various miRNAs are associated with radioresistance in HNC, like the up-regulated miR-150/196a/210/371a-5p/34-c-5p/1323/185-3p/381/21/451 and down-regulated miR-324-3p/93-3p/4501 (Ahmad et al., 2017).

miRNAs are proven to be possessing a significant role in regulating RTKs. The miR-99 family reciprocally regulates IGF-1R in oral squamous cell carcinoma (OSCC) (Kovacova, Poprach, Buchler, Cho, & Slaby, 2018). Owing to the increased nicotine content, Snail is up-regulated which transcriptionally represses miR-493, which in turn activates and up-regulates the IGF-1R, IRS-1 and MAPK-1 and subsequent pathways in HNC (Kumar et al., 2017). miR-34a is down-regulated in HNSCC and its overexpression reduces the EGFR levels, thereby acting as a tumor suppressor in inhibiting HNC metastasis (Zhang et al., 2015). miR-27a-3p and miR-27a-5p are down-regulated in HNC, both of which targets EGFR and the latter also specifically targets mTOR and AKT in the EGFR signaling pathways. Also, the effect of EGFR TKI, erlotinib, is increased in HNC post overexpression of miR-7 (Kedmi, Sas-Chen, & Yarden, 2015). Further, miR-27a down-regulates EGFR in HNC (Le, Huang, Chen, & Lai, 2015). It has been reported that miR-7 down-regulates EGFR levels and AKT activity in inhibiting tumorigenesis in HNC by binding to the 3′UTR of EGFR mRNA. miR-7 also inhibits the IGF-1R levels in HNC (Kalinowski et al., 2012). miR-31 which is up-regulated in HNSCC and OSCC, targets 3′UTR of FIH and inhibits the HIF pathway and triggers VEGF synthesis under normoxic conditions, thus, triggering tumorigenic signals (Yu, Ma, Wu, Shu, & Gao, 2018). In OSCC and oral tongue squamous cell carcinoma (OTSCC), miR-486-3p is down-regulated and results in the up-regulation of DDR1 as the former binds to 3′UTR of DDR1 and down-regulates it (Chou et al., 2019). miR-23b and miR-27b are down-regulated in OSCC and this cluster inhibits MET as well as associated signaling in inhibiting tumorigenesis and metastasis (Fukumoto et al., 2016). Further, miR-34a inhibits the Axl expression and subsequent oncogenic signaling, as evident by its down-regulation and associated Axl over-expression and drug resistance in HNC (Giles et al., 2013). The altered HGF/c-Met signaling down-regulated miR-200c and miR-27b in HNSCC and contributes to EMT via HGF auto-activation, extracellular matrix degradation and activation of ZEB1/*E*-Cadherin pathway (Susuki et al., 2011). miR-1 and miR-206 are downregulated in HNSCC and contribute to up-regulated EGFR and c-Met in these cancers resulting in tumorigenesis (Koshizuka et al., 2017). Various miRNAs in HNC and their target RTKs analyzed by miRDB (micro RNA target prediction database) are enlisted in Table 4. Though miRNAs have been extensively studied in

Table 4 miRNAs and their target RTKs associated with head and neck cancers as analyzed from miRDB—micro RNA target prediction database.

	EGFR	ERBB2	ERBB3	ERBB4	FGFR1	FGFR2	FGFR3	FGFR4	PDGFRA	PDGFRB	VEGFR-1	VEGFR-3	IGF-1R	IGF-2R	INSR	AXL	MET
miR-21-3p														■			
miR-106b-5p			■							■							
miR-451b																	
miR-205-3p	■	■		■				■									
miR-205-5p			■	■											■		
miR-138-1-3p				■													
miR-138-2-3p							■										
miR-138-5p																	
miR-34a-3p															■		
miR-34a-5p									■							■	■
miR-34b-3p																	
miR-34b-5p																	■
miR-34c-5p									■							■	■
miR-155-3p																	
miR-181a-2-3p				■													
miR-181-a-5p									■								
miR-181b-2-3p						■											
miR-181b-3p						■											
miR-181b-5p									■								
miR-181c-5p									■								
miR-181d-5p																	
miR-345-3p						■											
miR-125a-5p			■			■											
miR-125b-2-3p				■													
miR-125b-5p				■		■											
miR-200a-3p	■								■								
miR-200a-5p																	■
miR-200b-3p																	
miR-200b-5p																■	
miR-200c-3p																	
miR-200c-5p																	
miR-137-3p				■					■								
miR-193a-3p				■													
miR-193a-5p		■															
miR-193b-3p				■													

MERTK	NGFR	NTRK1	NTRK2	NTRK3	RET	EPHA2	EPHA3	EPHA4	EPHA5	EPHA6	EPHA7	EPHA8	EPHA10	EPHB1	EPHB2	EPHB3	EPHB4	EPHB6	TIE1	TEK
			■	■				■	■		■						■			
														■						
								■			■									
								■												
								■			■									
												■								
			■																	
								■												
											■									
								■												
								■												
															■					
			■					■	■											
														■						
														■						
			■					■	■											
								■	■											
								■	■											
								■												
								■			■				■					
							■													
								■			■									
						■					■									
			■						■											
			■		■				■											
			■						■											
			■		■															
			■																	
								■	■		■	■								
									■											
			■																	
									■											

Continued

Table 4 miRNAs and their target RTKs associated with head and neck cancers as analyzed from miRDB—micro RNA target prediction database.—cont'd

	EGFR	ERBB2	ERBB3	ERBB4	FGFR1	FGFR2	FGFR3	FGFR4	PDGFRA	PDGFRB	VEGFR-1	VEGFR-3	IGF-1R	IGF-2R	INSR	AXL	MET
miR-203a-3p																	
miR-203a-5p				■									■				
miR-203b-3p																	
miR-203b-5p																	
miR-146a-3p											■			■	■		
miR-146a-5p				■					■								
miR-146b-3p																	
miR-146b-5p				■					■								
miR-149-3p										■		■				■	
miR-149-5p	■		■						■		■						
miR-196a-1-3p				■													
miR-196a-3p																	
miR-196a-5p									■								
miR-196b-5p									■								
miR-499a-3p																■	
miR-499a-5p										■							
miR-499b-3p																■	
miR-499b-5p																	
miR-9-3p																	
miR-9-5p										■				■			
miR-134-3p																	
miR-134-5p	■																
miR-17-3p																	
miR-17-5p			■						■	■	■						
miR-92a-1-5p																	
miR-92a-2-5p																	
miR-92a-3p																■	
miR-92b-3p																■	
miR-92b-5p					■												
miR-let-7c-3p									■					■			
miR-let-7c-5p													■		■		
miR-100-5p							■										
miR-27a-3p	■								■								■
miR-27b-3p	■								■						■		■
miR-150-3p										■							

MERTK	NGFR	NTRK1	NTRK2	NTRK3	RET	EPHA2	EPHA3	EPHA4	EPHA5	EPHA6	EPHA7	EPHA8	EPHA10	EPHB1	EPHB2	EPHB3	EPHB4	EPHB6	TIE1	TEK
							■		■											
							■													
											■									
														■						
			■												■					
							■								■	■				
	■			■								■	■			■				
			■													■				
■									■											
				■			■													
					■		■				■									
					■		■				■									
			■																	
									■											
			■																	
			■																	
			■																	
											■				■		■			
													■							
										■					■					
								■	■		■						■			
														■				■		
	■		■												■					
												■								
												■								
							■		■											
	■			■			■	■						■						
	■		■					■							■					
	■		■					■							■				■	
							■													

Continued

Table 4 miRNAs and their target RTKs associated with head and neck cancers as analyzed from miRDB—micro RNA target prediction database.—cont'd

	EGFR	ERBB2	ERBB3	ERBB4	FGFR1	FGFR2	FGFR3	FGFR4	PDGFRA	PDGFRB	VEGFR-1	VEGFR-3	IGF-1R	IGF-2R	INSR	AXL	MET
miR-150-5p																	
miR-371a-5p														■			
miR-371b-5p														■			■
miR-1-3p																■	■
miR-1323						■											
miR-185-3p																	
miR-185-5p													■	■			
miR-381-3p						■							■			■	
miR-21-3p														■			
miR-21-5p																	
miR-451b																	
miR-324-3p													■				
miR-324-5p																	
miR-93-3p											■						
miR-93-5p			■						■	■	■						
miR-4501																	
miR-99a-5p							■										
miR-99b-5p							■										
miR-493-3p																	
miR-493-5p									■								■
miR-7-1-3p				■									■				
miR-7-2-3p				■									■				
miR-7-5p	■			■													
miR-486-3p					■					■							
miR-486-5p																	
miR-206																■	
miR-222-3p				■													
miR-222-5p																	
miR-23a-3p				■													■
miR-23a-5p																	
miR-23b-3p				■													■
miR-23b-5p																	
miR-23c				■													■

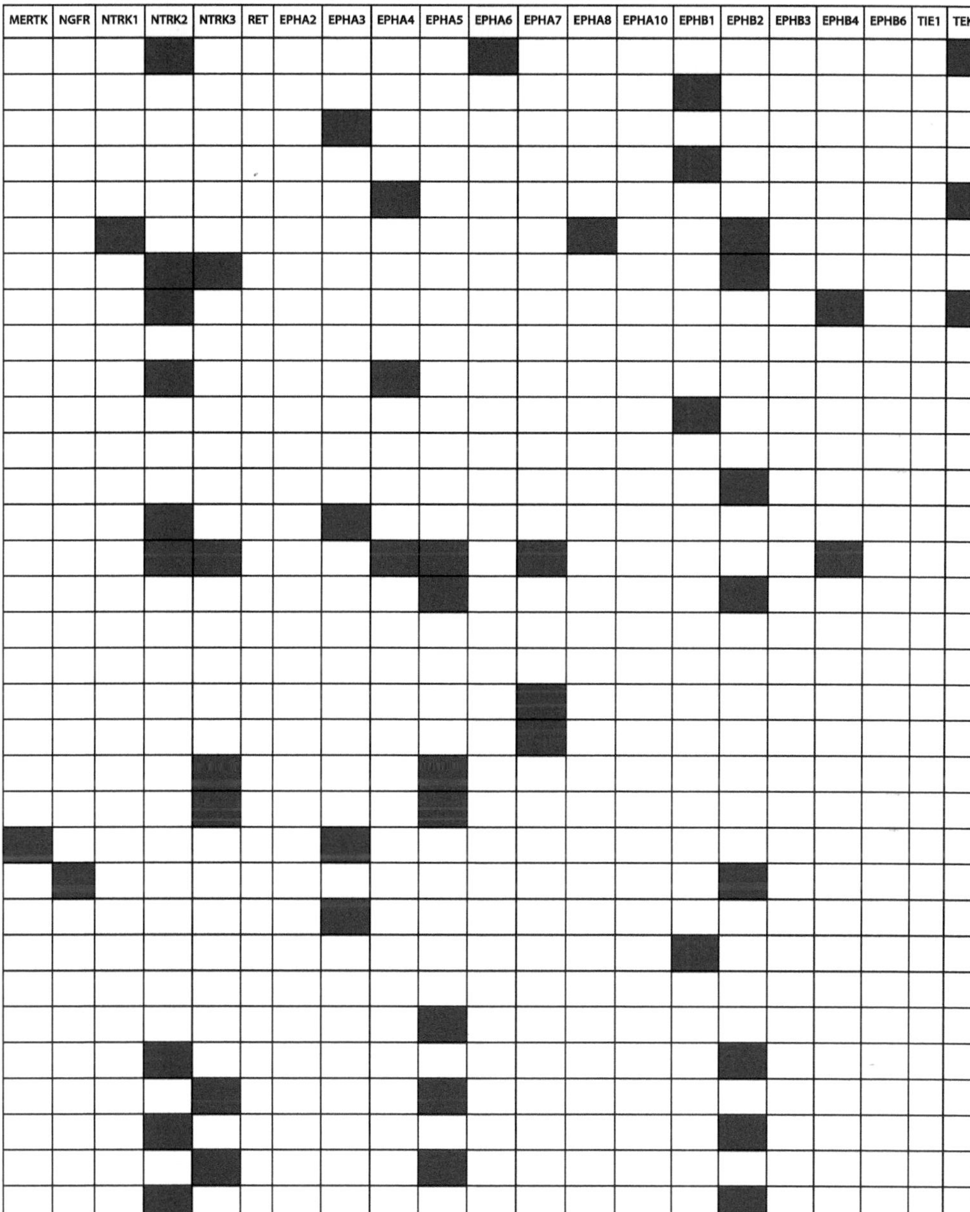
MERTK
NGFR
NTRK1
NTRK2
NTRK3
RET
EPHA2
EPHA3
EPHA4
EPHA5
EPHA6
EPHA7
EPHA8
EPHA10
EPHB1
EPHB2
EPHB3
EPHB4
EPHB6
TIE1
TEK

regulating RTKs, the regulatory role of RTKs over miRNAs are less explored. EGF has been shown to up-regulate the levels of miR31/181b/222 in oral cancers via EGFR/AKT and C/EBPβ signaling (Kedmi et al., 2015). Since recent advancements in unleashing novel therapeutic strategies employ miRNAs as critical targets, as they are components of the intricate regulatory network, the miRNAs associated with aberrant RTK signaling in HNC, could be of high potentiality.

8. Long non-coding RNA (lncRNA) and RTKs in HNC

Not just miRNAs, other non-coding RNAs like lncRNAs could also potentially aid the oncogenic transformation of normal cells by modulating RTK signaling in HNC. LncRNAs are short mRNA transcripts longer than 200 nt length, which doesn't code for proteins and actively takes part in transcriptional regulation, post-transcriptional modification, epigenetic regulation and chromosome remodeling (Garajova et al., 2017). LncRNAs namely HOXA11-AS, LINC00964, MALAT1 were found to be up-regulated in the plasma of HNSCC patients, thereby acting as circulatory biomarkers. Also, HOTAIR, GAS5 and UCA-1, are the lncRNAs which are altered in HNC and act as markers for poor prognosis. Of these, the circulating lncRNA, GAS5, serves as biomarkers to predict the treatment response to chemotherapy in HNSCC patients. In tongue squamous cell carcinoma (TSCC), MEG3 lncRNA is down-regulated and serves as a marker for poor prognosis while LINC00152 acts as a biomarker for early diagnosis and prognosis. The panel with three lncRNAs, namely KTN1-AS1, LINC00460 and RP5-894A10.6, has been reported to be prognosis biomarker panel for HNSCC (Yao et al., 2018). UCA1, MALAT1 and ADAMTS9-AS2 are associated with metastasis in TSCC while FOXCUT is associated with tumor cell proliferation and migration in OSCC. In nasopharyngeal carcinoma, HNF-1A promotes EMT, AFAP1-AS1 promotes metastasis, ROR promotes chemoresistance and LET inhibits proliferation while FIRRE, H19 and LOC541471 are up-regulated in HNSCC (Cossu et al., 2019). CASC9 lncRNA has been reported to be up-regulated in OSCC, correlated with poor survival and induce autophagic apoptosis by inhibiting AKT/mTOR pathway (Yang, Chen, Liu, & Yang, 2019).

Though numerous lncRNAs are associated with HNC, their regulatory role over RTKs or vice versa has been less analyzed. It has been reported that in HNSCC, that silent mutation destabilizes lncRNA EGFR-AS1, which shifts EGFR to the D isoform and non-canonical EGFR addiction, leading to altered sensitivity to the TKI gefitinib (Garajova et al., 2017; Tan et al., 2017). The G/G genotype exhibits higher levels of phosphorylated EGFR, AKT ad ERK and enhanced resistance to therapy (Peng, Fan, & Mo, 2018). HOTAIR is known to promote tumorigenesis, metastasis and angiogenesis, with poor prognosis by transcriptionally activating VEGFA and GRP78 and indirectly by GRP78 mediated up-regulation of Ang2 and VEGFA (Cossu et al., 2019). The rapid emergence of novel therapeutic interventions employing lncRNA gene-editing tools, in situ, could be exploited to target RTKs like EGFR and HER2, which are altered in HNC.

9. TME and RTK activation in HNC

TME constitutes the stroma surrounding the tumor cells composing of the mesenchymal cells and the extracellular matrix, which actively secretes various factors promoting tumorigenesis, angiogenesis and metastasis. TME is composed of numerous cell types, of which CAFs are the major component, with the active secretion of VEGF, EGF and HGF and subsequent activation of their RTK signaling pathways in promoting extracellular matrix degradation, angiogenesis and metastasis (Peltanova, Raudenska, & Masarik, 2019).

In HNC, under hypoxic conditions, CAFs secretes HGF, which act by paracrine mode and activate HGF/c-Met signaling, which promotes tumor cell invasion and angiogenesis by affecting the tumor cells as well as TME and also contributes to chemo/radioresistance (Byeon et al., 2019; Elferink & Resto, 2011; Hartmann et al., 2016). Under hypoxic conditions, cancer cells as well as stromal cells, especially CAFs, secrete VEGF, either by induction of HIF-1α or through activating the expression of EGF/EGFR and other factors like HGF, which promote survival, migration and differentiation of endothelial cells via PI3K, MAPK and FAK signaling, enhancing the microvasculature permeability enabling metastasis. CAFs also secrete MMPs which remodel extracellular matrix by activating EGFR signaling enabling metastasis (Pai & Westra, 2009). The cell–cell interaction between tumor vasculature and stroma is mediated by Eph receptor family members. Also, Axl is highly expressed in the TME macrophages which promote immunosuppression and pre-neoplasia. In addition to VEGFR, EGFR and PDGFR are also over-expressed within the TME in HNSCC (Du & Lovly, 2018). GAS6 lncRNA-Axl axis contributes to EMT in HNSCC by tumor cell–TME interactions (Schoumacher & Burbridge, 2017). In malignant oral leukoplakia, there is an over-expression of MMPs, FGF and receptors, FGFR-2 and FGFR-3 which is an indication of progression to OSCC. CAFs have also been reported to secrete EGF containing EFEMP1, PDGF-D, IBP5/7, which induce CSCs in HNSCC. Tumor-associated macrophages (TAMs) in HNSCC and TSCC have higher levels of EGF and TGF-β, which activate EMT by EGFR signaling. The Myeloid-Derived Suppressor Cells (MDSCs) in the TME of HNSCC induce EMT by higher expression of EGF and bFGF. Also, the primary cancer cells secrete VEGF and PlGF and this VEGF triggers CXCL1 production by TAMs, resulting in MDSC recruitment aiding metastasis. The long term co-stimulation of EGF/TGF-β promotes invasiveness in OSCC. In OSCC, paracrine activation of EGFR mediated by ADAM12 triggered secretion of HB-EGF associated with cell contact-mediated hypoxia and subsequent Notch signaling triggering invasiveness. In HNSCC, CAF derived HGF activates the c-Met/Akt pathway in anoikis. Also, VEGF secreted by TAMs and cancer cells promote trans-endothelial migration, vascular permeability and is associated with poor prognosis (Peltanova et al., 2019). To reverse MDSC mediated immunosuppression, TKIs have been employed which inhibits VEGFR1/2/3, PDGFR, c-Kit, FT3 and RET in HNSCC (Tong et al., 2012). In HNSCC, the myofibroblasts secrete HGF upon induction by TGF-β1 to promote invasion and tumorigenesis. Also, in cases of OSCC, where EMMPRIN is low, FGFR2 plays a cardinal role in tumorigenesis (Koontongkaew, 2013). Since most HNC subtypes have pre-cancerous to cancerous transformation, the influence of TME on this oncogenic transition is cardinal, especially aided with aberrant RTK expression and associated signaling. Therefore, modulation of TME could attenuate tumor progression to a larger extend.

10. Autocrine action of growth factors and chemokines on RTKs in HNC

The prominent RTKs namely EGFR, FGFR, IGF-1R, PDGFR, c-Met etc. has been over-expressed in HNC and act via autocrine as well as paracrine cascades in aiding tumorigenesis and metastasis. The autocrine action of the growth factors and chemokines through RTKs has been experimentally validated in vitro. FGFs and FGFRs are co-expressed at higher levels and FGF-2 and FGFR exhibit autocrine signaling pathways in HNSCC either independently or in association with EGFR to promote cell growth, proliferation, invasion, metastasis, EMT and angiogenesis

(English et al., 2016; Marshall et al., 2011; Tiash & Chowdhury, 2015). The over-expression of FGFRs or EGFRs might be correlated with the somatic mutations resulting in gene amplification in these cancers (Marshall et al., 2011). Autocrine stimulation of EGFR increases radioresistance in HNC (Grana, Sartor, & Cox, 2003). EGFR autocrine loop has been reported to be activated not only by its classic ligand EGF but also by TGF-α and amphiregulin secreted by HNSCC cells. Also, transactivations of EGFRs are being carried out by GPCRs in response to ligands like PGE2 and BK in HNSCC (Harari, Wheeler, & Grandis, 2009). In Abl negative HNSCC, HB-EGF exerts autocrine activation of EGFR via EGFR-Src-cortactin signaling cascade (Hayes et al., 2013). Also, c-erbB/EGFR signaling axis induces MMPs which in turn enhances the expression of other ligands like EGF, HB-EGF and TGF-1α, creating an autocrine loop stimulating oncogenesis in HNSCC (O-Charoenrat, Rhys-Evans, & Eccles, 2002). Similarly, VEGF and VEGFR-2 exhibit an autocrine loop which is known to increase the tumorigenic signals in HNSCC and Oral-laryngeal squamous cell carcinomas (Kyzas, Stefanou, Batistatou, & Agnantis, 2005). VEGF-C, secreted by HNSCC and OSCC cells, binds to Flt-4 (VEGFR-3) ad initiates autocrine signaling loop resulting in enhanced proliferation, migration and neck metastasis via up-regulation of VEGF-C and CNTN-1 (Kyzas et al., 2005; Matsuura et al., 2009). Also, higher HGF secretion have been shown by CAFs and c-Met levels has been observed in HNSCC, mostly paracrine signaling has been shown in these cancers enhancing cell proliferation, metastasis and recurrence and associated oncogenic signals (Tiash & Chowdhury, 2015). However, HGF/c-Met axis has been reported of being hyperactivated by autocrine signaling, contributing to oncogenic transformation in other cancers like acute myeloid leukemia and melanomas (Arnold, Enders, & Thomas, 2017). Considering the lower levels of IGFBP-2/3 in HNSCC cells, but higher levels of IGF-1/2 and IGF-1R, autocrine signaling has been expected in these cells contributing to tumorigenesis. Also, IGF-1R signaling activates the autocrine VEGF cascade in HNSCC (Slomiany et al., 2007; Slomiany, Black, Kibbey, Day, & Rosenzweig, 2006). Concerning PDGF-PDGFRs, though the autocrine loop activation has been reported in ovarian cancers, mammary tumors, glioblastoma and others, the autocrine signaling in HNC has not been elucidated till date (Jechlinger et al., 2006; Lokker, Sullivan, Hollenbach, Israel, & Giese, 2002; Matei et al., 2006). However, PDGF and PDGFRs are over-expressed in a variety of HNC. A mechanism of enhanced tumorigenicity was explored in HNSCC, wherein IGF-1 activated VEGF autocrine signaling cascade via HIF-1α and VEGF expression was also reported (Slomiany et al., 2006, 2007). Though the cytokines and growth factors are mostly expressed by the cancer cells themselves, targeted inhibition of these factors could act as a double-edged sword, which would also control the expression of these factors by stromal cells resulting in effective inhibition of tumor progression.

11. Tobacco associated carcinogens and RTK activation in HNC

Cigarette smoking consists of a huge number of compounds of which almost 60 of them are well-studied carcinogens. Notable among these include poly-aromatic hydrocarbons (PAHs), alkylated PAHs, aromatic amines, volatile organic hydrocarbons, N-nitrosamines, aldehydes, metals, oxidants, ethylating agents, free radicals etc.; however, the well-explored carcinogens concerning human cancers includes the PAHs, aromatic amines, N-nitrosamines and volatile organic agents. The balance between metabolic activation and detoxification processes of these carcinogens within the individuals predict their risk to cancers. The metabolites result in genetic changes like mutations, DNA adduct formation, methylation changes etc., which enhances the tumorigenic signaling in the tissues (Centers for Disease Control and Prevention (US), 2010). Various forms of intake of these carcinogens include cigarette, smokeless tobacco, pipes, cigars, bidis, hookahs and kreteks. Tobacco smoking increases HNSCC risk by 6- to 13-fold (Jethwa & Khariwala, 2017).

Nicotine in the tobacco smoke binds to the nicotinic acetylcholine receptors and induce p-EGFR translocation to the nucleus and subsequent activation of Akt pathway resulting in lymph node metastasis and drug resistance in HNSCC (Shimizu et al., 2019). Mutations in EGFR have been reported to be associated with tobacco intake in HNSCC patients (Jethwa & Khariwala, 2017). Among the tobacco components, long term exposure to NNK component as well as arecoline results in EGFR/EGF over-expression and activation of EGFR/Akt signaling to result in signals for anti-apoptosis, stemness induction and chemoresistance in HNSCC (Yang et al., 2018). Also, HNC patients who were smokers or who smoke during radiation therapy exhibit high levels of radioresistance (Browman et al., 1993). Nicotine and smokeless tobacco induces VEGF through STAT-3 activation and signaling for accelerating neovasculature in HNSCC, especially in the oral cancer cells (Macha, Matta, Chauhan, Siu, & Ralhan, 2011). In cases of HNC patients with smoking, FGFR-1 is amplified (Xie et al., 2016).

In nasopharyngeal carcinoma, nicotine-induced VEGF and HIF-1α and inhibited PEDF to promote proliferation and associated oncogenic signaling (Bredell et al., 2016). Also, in a study analyzing the expression of PDGFRA, it has been found to be higher in tobacco smokers in both normal cohorts as well as HNC group when compared with the non-smokers and also within the HNC patient cohort in comparison with the normal cohort (Ong et al., 2017). In HNSCC, nicotine induces Snail expression which downregulates miR-493, which in turn over-expresses the IGF-1R and subsequent signaling in promoting tumorigenesis (IGF-1R nicotine induction). Cigarette smoking induces c-Met and HGF expression in OSCC, aiding proliferation, metastasis, invasion, angiogenesis and poor prognosis (Sun et al., 2018). Considering the fact that tobacco use is the single largest etiological factor enhancing the HNC risk and has a prominent role in activating RTKs in inducing tumorigenesis, numerous studies have been exploring the possibilities of controlling RTKs in tobacco users, not only as an effective therapeutic strategy but also as a preventive tool.

12. Targeting RTKs in HNC

RTKs are one of the major targets for therapeutic interventions, with numerous antibodies and inhibitors being in clinical trials, as the deregulation of RTKs is frequently observed in HNC. EGFR inhibitor, cetuximab (erbitux), which is an FDA approved drug, has been used in therapy for HNC. However, despite initial results, it has been in controversy since the action of cetuximab has been doubted if through EGFR inhibition or via immunomodulatory mechanisms (Agarwal, Subash, Nayar, & Rao, 2019). Other EGFR mAb in clinical trials includes panitumumab, zalutumumab and nimotuzumab which are tried in combination with chemo-radiotherapy. Duligotuzumab (dual mAb against HER3 and EGFR) and patritumab (anti-HER3) are also being in trials for HNC (Saada-Bouzid & Le Tourneau, 2019). Other than the mAbs, EGFR TKIs are also in clinical trials which include erlotinib, gefitinib and lapatinib, of which the former two binds to the active state of EGFR, while the latter binds to the inactive state. In comparison with erlotinib and gefitinib, which are reversible acting TKIs, lapatinib (reversible dual-target TKI against EGFR and HER2), afatinib and dacomitinib (both irreversible TKIs against EGFR, HER2 and HER4) seem to be more effective against HNC. Also, EGFR antisense clinical trials have been conducted in HNC, which showed promising results, but in a smaller cohort (Egloff & Grandis, 2008).

About 14% of HPV negative HNSCC have FGFR1/2/3 mutations/amplifications and FGFR3 fusion genes have been seen in HPV positive HNC. FGFR inhibitor PD173074 showed effect with cells with fusion genes, but not with activating mutations on FGFR; however, several clinical studies are being undergone currently (Saada-Bouzid & Le Tourneau, 2019). Also, studies have been carried out in vitro in analyzing the effects of FGFR inhibitor, AZD4547 and the resistant cells were treated successfully with its combination with gefitinib (Koole et al., 2016). Other FGFR inhibitors explored for therapy in HNC are lenvatinib (dual inhibitor of VEGFR and FGFR), BGJ398 and DEBIO1347 (Dumbrava, Alfattal, Miller, & Tsimberidou, 2018). Non-selective FGFR inhibitors like ponatinib and selective ones, namely BAY1163877, are under clinical trials against HNSCC. Of all the FGFR inhibitors the pan-FGFR inhibitors including derazantinib (ARQ-087), infigratinib (BGJ398), AZD4547, erdafitinib (JNJ-42756493), pemigatinib (INCB054828) and TAS-120 seems to be promising (Dumbrava et al., 2018).

The inhibitors namely pazopanib, dovitinib and nintedanib are active against FGFR and VEGFR; however, the study with these inhibitors in HNSCC reported that dovitinib and nintedanib, but not pazopanib was active against the cells possibly due to absence of VEGFR2 (Brands et al., 2017). Various other anti-VEGFR TKIs under study against HNC include axitinib, sunitinib, sorafenib, lapatinib and vandetanib and mAbs include bevacizumab; but, due to their limited efficacies and high toxicity independent trials couldn't succeed but are progressing with combinatorial clinical trials (Saada-Bouzid & Le Tourneau, 2019). Another VEGFR mAb under study is the ranibizumab. Other VEGFR inhibitors under trial are vatalanib and cediranib (AZD2171) (Carla et al., 2012). VEGFR inhibitors improve tumor blood flow altering the TME and increase the sensitivity to radiotherapy. Hence, a combination of VEGFR inhibitors with radiotherapy as well as with EGFR inhibitors or immunomodulators is also being explored by clinical trials (Chau & Haddad, 2016).

Though there are controversial results concerning IGF-1R expression in HNC and 4% of HPV negative HNSCC harbor IGF-1R mutations/amplifications, anti-IGF-1R mAbs including figitumumab, cixutumumab (IMC-A12) were designed but did not show any efficacy in HNSCC (Koole, Brunen, et al., 2016; Saada-Bouzid & Le Tourneau, 2019). Another IGF-1R mAb is ganitumab (AMG-479) which interferes with IGF-1 and IGF-2, showed much efficacy (Boeckx et al., 2013). Also, a novel bi-specific antibody MM-141 has been developed which co-targets IGF-1R and HER3, preventing binding of their ligands to these receptors as well as inhibiting their phosphorylations and bypass of PI3K/Akt pathways. These have successfully been effective in HNSCC cells and have to be translated for clinical trial approvals (Limesand, Chibly, & Fribley, 2013). Another IGF-1R inhibitor under study was the BMS-754807 (Limesand et al., 2013). IGF-1R inhibitor, CP-751,871, has shown to sensitize HNC cells to radiation by inhibiting DNA damage repair and promoting apoptosis (Matsumoto et al., 2012). In nasopharyngeal carcinoma, IGF-1R inhibitor, linsitinib acted by suppressing ERK/Akt pathway, thereby inhibiting cell proliferation and induced radiosensitivity in combination with ionizing radiation (IR) (Wang et al., 2019).

Owing to enhanced HGF/c-Met axis in HNSCC, c-Met inhibitors were designed namely foretinib and tovantinib, but did show effective response in further clinical trials (Byeon et al., 2019). c-Met inhibitor, PF-2341066, has been reported to down-regulate β-catenin and inhibit CSCs in HNSCC (Arnold et al., 2017). Since EGFR and c-Met has similar downstream signaling, that includes MAPK and PI3K/Akt/mTOR, most of the c-Met inhibitor studies have been done in combination with EGFR inhibitors like that of SU11274 (c-Met inhibitor) and erlotinib (EGFR inhibitor), foretinib and erlotinib or lapatinib (EGFR/human epidermal growth factor (HER) 2 inhibitor) and crizotinib (c-Met inhibitor) and gefitinib. Crizotinib (PF2341066), capmatinib (INC-280), golvatinib (E7050), foretinib, tivantinib, cabozantinib, SU11274, PFO4217903 and ficlatuzumab are the key c-Met antagonists in clinical trials against HNSCC (Arnold et al., 2017; Hartmann et al., 2016; Jiang, Ye, Dong, Hu, & Xiao, 2019).

Imatinib is a partial inhibitor of PDGFR, which also inhibits VEGFR, hence used as a treatment strategy in HNSCC. Axitinib, sunitinib, lenvatinib, ponatinib, pazopanib, dovitinib (DK128), nintedanib (BIBF1120) and sorafenib are multi-target inhibitors which include PDGFRs implemented in studies for the treatment of HNC (Bran, Bran, Hormann, & Riedel, 2009).

Several inhibitors of Axl exist for HNSCC including R428, licensed as BGB324, which has entered clinical trials and neutralizing anti-Axl monoclonal antibodies namely YW327.6S2 and MAb173, which are in preclinical stages (Brand et al., 2015). Also, Axl inhibitors, TP-0903 and SLC-0211, are still at preclinical stages. UNC1062 is a MERTK inhibitor studied for the treatment of HNSCC (Schoumacher & Burbridge, 2017). Various RTK antagonists and their targets are enlisted in Table 5.

The RTK antagonists would be a promising therapeutic strategy for the matter that it would inhibit not only RTKs but also other molecules that are deregulated in HNC; it could also be possible that HNC inhibitors could be used as monotherapeutic agents.

Table 5 RTK antagonists and their targets.

S. no.	Target RTK	Type of antagonist	Drug name	References
1.	EGFR	mAb	*Cetuximab*	Taberna, Oliva, and Mesia (2019)
2.	EGFR	mAb	*Panitumumab*	Siano et al. (2017)
3.	EGFR	mAb	*Zalutumumab*	Schick, Gujral, Richards, Harrington, and Nutting (2012)
4.	EGFR	mAb	*Nimotuzumab*	Patil et al. (2018)
5.	EGFR, HER3	mAb	*Duligotuzumab*	Fayette et al. (2016)
6.	HER3	mAb	*Patritumab*	Dillon et al. (2019)
7.	EGFR	TKI	*Erlotinib*	Thinn and Hsueh (2019)
8.	EGFR	TKI	*Gefitinib*	Tang et al. (2019)
9.	EGFR, HER2,	TKI	*Lapatinib*	Weiss et al. (2016)
10.	EGFR, (HER2, HER4)	TKI	*Afatinib*	De Pauw et al. (2018)
11.	EGFR, HER2, HER4	TKI	*Dacomitinib*	Ather et al. (2013)
12.	FGFR1	TKI	*PD173074*	Sweeny et al. (2012)
13.	FGFR1/2/3	TKI	*AZD4547*	Baschnagel et al. (2017)
14.	FGFR1/2/3/4, VEGFR1/2/3, PDGFRA, RET	TKI	*Lenvatinib*	Taylor et al. (2018)
15.	FGFR1/2/3	TKI	*BAY1163877*	Héroult et al. (2015)
16.	FGFR1/2/3	TKI	*DEBIO1347*	Hyman et al. (2019)
17.	FGFR, PDGFR, VEGFR, RET	TKI	*Ponatinib*	Sweeny et al. (2012)
18.	FGFR1/2/3	TKI	*Derazantinib (ARQ-087)*	Hall et al. (2016)
19.	FGFR1/2/3	TKI	*Infigratinib (BGJ398)*	Guagnano et al. (2012)

Table 5 RTK antagonists and their targets.—cont'd

S. no.	Target RTK	Type of antagonist	Drug name	References
20.	FGFR1/2/3	TKI	*Erdafitinib (JNJ-42756493*	Bahleda et al. (2019)
21.	FGFR1/2/3	TKI	*Pemigatinib (INCB054828)*	Hollebecque et al. (2018)
22.	FGFR1/2/3/4	TKI	*TAS-120*	Kalyukina et al. (2019)
23.	VEGFR, FGFR, PDGFR	TKI	*Pazopanib*	Adkins et al. (2018)
24.	VEGFR, FGFR1/2/3, TrkA, PDGFR	TKI	*Dovitinib (DK128)*	Dillon et al. (2017)
25.	VEGFR1/2/3, FGFR1/2, PDGFR	TKI	*Nintedanib (BIBF1120)*	Kim et al. (2017)
26.	VEGFR1/2/3, PDGFR	TKI	*Axitinib*	Swiecicki, Spector, and Worden (2016)
27.	VEGFR, PDGFR	TKI	*Sunitinib*	Machiels et al. (2010)
28.	VEGFR, PDGFR	TKI	*Sorafenib*	Mockelmann et al. (2016)
29.	VEGFR, EGFR, RET	TKI	*Vandetanib*	Papadimitrakopoulou et al. (2016)
30.	VEGFR	mAb	*Bevacizumab*	Heydar et al. (2018)
31.	VEGFR	mAb	*Ranibizumab*	Burgos-Tiburcio, Santos, Arango, and Raez (2011)
32.	VEGFR	TKI	*Vatalanib*	Gauler et al. (2011)
33.	VEGFR1/2/3	TKI	*Cediranib (AZD2171)*	Wedge et al. (2005)
34.	IGF-1R	mAb	*Figitumumab*	Schmitz et al. (2012)
35.	IGF-1R	mAb	*Cixutumumab (IMC-A12)*	Ferrarotto et al. (2018)
36.	IGF-1R	mAb	*Ganitumab (AMG-479)*	Chung et al. (2011)
37.	IGF-1R, HER3	mAb	*MM-141*	Lugovskoy et al. (2011)

Continued

Table 5 RTK antagonists and their targets.—cont'd

S. no.	Target RTK	Type of antagonist	Drug name	References
38.	IGF-1R, INSR	TKI	*BMS754807*	Axelrod et al. (2015)
39.	IGF-1R	mAb	*CP751,871*	Iwasa et al. (2009)
40.	IGF-1R, INSR	TKI	*Linsitinib*	Macaulay et al. (2016)
41.	c-Met	TKI	*Foretinib*	Seiwert et al. (2013)
42.	c-Met	TKI	*Tivantinib*	Vokes et al. (2015)
43.	c-Met	TKI	*SU11274*	Seiwert et al. (2005)
44.	c-Met	TKI	*Crizotinib (PF2341066)*	Baschnagel et al. (2015)
45.	c-Met	TKI	*Capmatinib (INC280)*	Alsahafi et al. (2019)
46.	c-Met, VEGFR2	TKI	*Golvatinib (E7050)*	Arnold et al. (2017)
47.	c-Met, VEGFR2, Axl, RET	TKI	*Cabozantinib*	Kurzrock et al. (2011)
48.	c-Met	TKI	*PF 4217903*	Zou et al. (2012)
49.	HGF (ligand of c-Met)	mAb	*Ficlatuzumab*	Bauman (2018)
50.	PDGFR, VEGFR	TKI	*Imatinib*	Bruce, Slevin, Homer, McGown, and Ward (2005)
51.	Axl	TKI	*R428 (BGB324)*	Badarni et al. (2019)
52.	Axl	mAb	*YW327.6S2*	Antony and Huang (2017)
53.	Axl	mAb	*MAb173*	Brand et al. (2015)
54.	Axl	TKI	*TP0903*	Mangelson et al. (2019)
55.	Axl	TKI	*SLC-0211*	Schoumacher and Burbridge 2017)
56.	MERTK	TKI	*UNC1062*	von Massenhausen et al. (2016)

13. Therapeutic strategies for HPV positive HNC: Emphasis on RTKs

HPV attributes to a 22% risk of OSCC and 47% of tonsillar squamous cell carcinoma. HPV positive tumors have a better prognosis, lower recurrence and respond well to radio/chemotherapy. About 47% of tonsillar crypt infections and 3.9% of oral cavity infections of HPV progress to cancers. The oncoproteins E6 and E7 induce tumorigenesis by inactivating p53 and pRb respectively; also, E5 HPV protein activates EGFR signaling cascade, together which contributes to the oncogenic transformation of the host cell post HPV infection (Alsahafi et al., 2019). About 90% of HPV positive OPSCC are due to HPV-16 and most of them are in men, who are younger in age (Pan, Issaeva, & Yarbrough, 2018). Currently, early disease is treated with monotherapy either with radiotherapy or with surgery. In advanced cases, radiotherapy with concomitant chemotherapy or cetuximab or surgery followed by adjuvant radio/chemotherapy is employed (Deschuymer, Mehanna, & Nuyts, 2018).

Various clinical trials are carried out to analyze (a) replacement of cetuximab with cisplatin [De-ESCALaTE (Ireland, Netherlands, UK), TROG 12.01 (Australia, New Zealand), RTOG 1016 (USA, Canada), NCT01663259 (USA)], (b) radiotherapy dose reduction when given in combination with chemotherapy, (c) reduction of radiation dose with or without cisplatin and (d) reduction in adjuvant chemo/radiotherapy following surgery (Deschuymer et al., 2018; Taberna et al., 2017). Various concerns currently rose against the de-escalation therapeutic strategies against HPV positive HNC such that if all HPV positive patients would be potential candidates, the questionable extend of the degree of de-escalation and the question on the reliability of HPV testing would fall significant (Pan et al., 2018). Phase II clinical trial was performed with carboplatin and paclitaxel, followed by radiotherapy and paclitaxel, where effectiveness was shown in terms of reduction of radiation doses (Kobayashi et al., 2018). Other emerging therapies under study include the various immunotherapy clinical trials as well as proton therapy for the treatment of HPV positive HNSCC (Deschuymer et al., 2018). Also, the use of a DNA demethylating agent, 5-aza cytidine is being studied as a therapeutic agent against HPV positive HNC (Pan et al., 2018). Anti-PD-1 agents like pembrolizumab have also been studied in treating HPV positive HNC in combination with chemo/radiotherapies (Taberna et al., 2017). Numerous therapeutic vaccines are either under study or under clinical trials for HPV positive HNC, which includes subunit vaccines (peptide/protein/plant-peptide or protein-based), live-vector-based vaccines (bacterial/viral vector), DNA/RNA based vaccines and cell-based vaccines (dendritic cell/adoptive cell transfer-based). Two of these namely the DNA vaccine, VGX-3100, as well as the bacterial vector vaccine, ADXS11-001, are in phase III clinical trials which are quite promising vaccination strategies for HPV positive HNC in the future (Chabeda et al., 2018).

Cetuximab, which is a monoclonal antibody (mAb) against EGFR, is the targeted drug for both HPV positive and negative HNC but exhibits high limitations and resistance. In both HPV positive and negative cases, STAT3 signaling is aberrantly activated in response to several factors including enhanced EGFR and VEGFR signaling cascades and contributes to immunosuppression preventing cancer cells from lysis and confers resistance to EGFR directed or chemo/radiotherapies owing to aberrations in numerous other HER family members (Alsahafi et al., 2019). E6 and E7 have been reported to up-regulate HER3 and associated signaling in HPV positive HNSCC. Further, inhibiting HER3 by siRNA and anti-HER3 mAb KTN3379 abrogated the tumorigenic properties, which calls for anti-HER3 therapy for these cancers (Brand et al., 2017). Studies are also being carried out to look into the effectiveness of other EGFR inhibitors like panitumumab, afatinib or cetuximab in treating HPV positive HNSCC (De Felice et al., 2019). Two de-escalation studies were carried out to look into if cetuximab can replace cisplatin to reduce the toxic effects of cisplatin, namely De-ESCALaTE and RTOG 1016; however, the results showed inferior OS and PFS for cetuximab with radiotherapy than cisplatin with radiotherapy, though independent of these studies cetuximab showed lesser toxicities in comparison with cisplatin (De Felice et al., 2019; Deschuymer et al., 2018). Though HPV activates RTKs indirectly, targeting RTKs like HER3 proves to enhance the efficacy of therapeutic interventions for HPV positive HNC.

14. Drug resistance and RTKs in HNC

Drug resistance to inhibitors of over-expressed factors in cancer remains to be a major challenge in cancer therapy. Since RTKs are aberrantly activated in HNC and numerous RTK inhibitors are in clinical trials, drug resistance to these antagonists remains to confront in HNC therapy too. Major mechanisms of resistance to EGFR mAbs include over-expression of EGFR/ligands, deregulation of EGFR internalization and degradation by ubiquitination, MDG1 binding, EGFR nuclear translocation, increased SFK-mediated signaling, EGFRvIII mutated isoform, KRAS mutation, PTEN loss, enhanced heterodimerization of EGFR with HER2/3, cross-talks with VEGF/VEGFR or HGF/c-Met axis or IGF-1R, dual-specificity phosphatases, PTPRS interaction, STAT protein activation, activation of aurora kinases, p53 or cyclinD1, hypoxia induction and EMT processes in cells. Also, there exist numerous mechanisms by which HNC exhibit resistance to TKIs which include EGFR mutations, EGFR mutant isoform EGFRvIII which constitutively activates EGFR signaling, loss/mutation of PTEN, KRAS mutations, cross-talks with HGF/c-Met, HER2/3, Axl axis, VEGF/VEGFR, IGFR-1R activations and EMT processes (Carla et al., 2012).

FGFR signaling confers resistance to CSCs in HNSCC and hence a combination of drugs like cisplatin along with FGFR inhibitor would help overcome this acquired resistance. Since FGFR signaling happens via PI3K/Akt/mTOR and MAPK pathways, resistance to FGFR inhibitors might be due to mutations in these downstream signaling resulting in their constitutive/hyperactivation. Various mechanisms are exhibited within HNC for the resistance to FGFR inhibitors like activation of bypass signaling during inhibition of FGFR via c-Met, Eph3B, HER2/3 or EGFR, bypassing by activation of intracellular signaling pathways like PI3K/Akt/mTOR, MAPK, STAT3 or GSK3β, active EMT signaling in these cells and incidence of secondary gatekeeper mutations within FGFRs which prevent the binding of inhibitors to FGFRs. These can be overcome by more efficient inhibitor development or exploring better combinatorial treatment strategies involving FGFR inhibitors (Boeckx et al., 2013).

The resistance of VEGFR inhibitors or mAbs has been elucidated in HNC by various mechanisms. One of them has been explained by the hypoxia induction of these agents and this hypoxic condition selectively proliferates the tumor cells, which have the inherent mechanisms to evade these hypoxic conditions. Up-regulation of pro-angiogenic stromal cells, accumulation of endothelial progenitor cells, macrophages, neutrophils, vascular progenitor cells, pro-angiogenic monocytic cells, all of which activates angiogenic signaling in the absence of VEGFR signaling, confers VEGFR resistance. Also, VEGFR inhibitors have the potential to induce inflammatory cytokines resulting in activating angiogenic signaling and metastasis in the HNC cells (Carla et al., 2012).

One major reason for resistance against IGF-1R inhibitors by HNSCC is the existence of homologous INSR or hybrid INSR/IGF-1R receptors, which can bypass the signaling mechanisms. Growth hormone (GH) also induces the production of hepatic IGF-1 and GHR bypasses IGF-1R signaling contributing to the resistance to their inhibitors in HNC. IRS-1 also contributes to the IGF-1R resistance (Beckwith & Yee, 2015). Another cardinal player in IGF-1R resistance is the EGFR and associated signaling as both these receptors have a tendency to heterodimerize owing to the binding of IGF ligand. Also, IGF-1R can contribute to EGFR resistance (Jameson et al., 2011). Hence a better therapeutic strategy in such conditions would be a combination therapy of the IGF-1R inhibitors with the key inhibitors to the bypassing signaling pathways.

PIK3CA mutations like the hotspots E545K and H1047R confers resistance to c-Met inhibitors in HNSCC. Thus, activated PI3K/Akt signaling confers c-Met inhibitor resistance in HNC. c-Met contributes to EGFR and VEGFR resistance in HNSCC (Jiang et al., 2019). TME derived HGF and associated HGF/c-Met signaling increases radio/chemoresistance in HNSCC (Hartmann et al., 2016).

Axl mediates EGFR, VEGFR, FGFR and PI3Kα resistance in HNC (Brand et al., 2015; von Massenhausen, Sanders, et al., 2016). MERTK over-expression contributes to resistance toward Axl inhibitors in HNSCC. Axl can dimerize with EGFR, MET and HER2, thus contributing to the resistance to their inhibitors during treatment of HNC. Axl and MER axis contributes to therapeutic resistance in HNC by several mechanisms including promoting EMT, contributing to an innate immune response in TME and promoting angiogenesis (Schoumacher & Burbridge, 2017). A better strategy for overcoming the drug resistance against inhibitors would even be possible reducing the treatment duration with the administration of a higher drug dosage; however, the dosage adjustment should be carefully managed so that it would result in the least toxicity.

15. RTK signaling during radiotherapy and immunotherapy

Deregulation of RTKs has been reported to contribute to radioresistance. Infrared Radiations (IR) results in cytoprotective action in the tumor cells by stimulating ROS induced inactivation of phosphotyrosine phosphatases or homo/heterodimerization of RTKs, thereby phosphorylating RTKs like EGFR, IGF-1R, c-Met etc. along with over-expression of RTK ligands too, resulting in aberrant RTK signaling and radioresistance.

Hence, if RTK inhibitors are combined with IR for therapy, these aberrant RTK signaling will be inhibited and simultaneous induction of DNA damage by IR will sensitize cells to apoptosis as the damages will remain unrepaired, contributing to radiosensitization. Also, IR induces changes with the fluidity of the plasma membrane which results in either defective conformation of RTKs or decreased RTK expression or defective/non-binding of the ligand with RTKs, which affects the signaling. IR has been reported to localize the RTKs to either cytoplasm (endocytosis) or nucleus or even mitochondria, as in the case of EGFR in HNC and hence, the use of RTK inhibitors would prevent this translocation and signal the cells to apoptosis owing to IR induced damage in the tumor cells. The RTK (EGFR, IGF-1R, c-Met) translocation to the nucleus, prominently the EGFR, has been reported to activate the DNA DSB repair by non-homologous end joining, mismatch repair, base excision repair and homologous recombination repair pathways contributing to radioresistance, which again calls for RTK inhibitors along with radiotherapy, which results in radiosensitization. Cetuximab (EGFR mAb) in combination with radiotherapy or chemo-radiotherapy has been in clinical trials for HNSCC and have come up with better OS and PFS, except for few recent ones which didn't show it as superior in comparison with cisplatin; however, the combination has proved better than radiotherapy or chemoradiotherapy alone with better outcomes in the patients. Also, RTK TKIs erlotinib/gefitinib has also been studied in combination with radiotherapy or chemoradiotherapy with better results but came up with drug resistance and hence, recently the broad spectrum RTK TKI, afatinib, has been under clinical trials as a part of combination therapy. Further advancements have been made in targeted deliveries of RTK antagonists using liposomal or nanoparticle deliveries of the combinatorial agents inclusive of the RTK antagonists (Bhattacharya, Shetake, Pandey, & Kumar, 2018).

There have been several correlations between the RTKs and the immunotherapy against HNC. The EGFR mAb, cetuximab, not only binds to EGFR but also the CD16 receptor on dendritic cells and natural killer cells, augmenting innate and adaptive immunity including the mechanisms of ADCC and T-cell priming. The EGFR TKI, afatinib, down-regulates PDL-1 levels by inhibiting NF-κB, but blocks the immune escape by enhancing NKG2D ligand expression on tumor cells and NKG2D on the natural killer cells. Another TKI, ibrutinib, inhibits IL-2 inducible T-cell kinase to augment anti-tumor properties. The multi-kinase VEGFR1-3/FGFR1-4/PDGFR-α/RET/KIT inhibitor, lenvatinib, enhances the ratio of memory T cells and decreases the number of TAMs (Wang, Chan, & Cho, 2019).

Recent advancement in cancer immunotherapy is the adoptive immunotherapy, where modified T cells are administered which targets specific antigen on the tumor cell surface like the (CAR)-T cell therapy. Pan-erBb targeted CAR-T cell therapy is currently under trial for HNSCC treatment (Alsahafi et al., 2019). Extensive research has to be carried out with immune boosters to prevent immune resistance associated with immunotherapy in HNC.

16. Can RTK targeting be promising than current therapeutic strategies for HNC?

RTKs are a diverse class of receptors, prominent of which are the EGFR, HER, FGFR, PDGFR, IGF-1R, c-Met, VEGFR, Axl, MERTK and so on. These class of molecules is involved in a greater portion of the normal cellular physiological functions as well as the oncogenic signaling and associated properties ranging from cell proliferation, survival, metastasis, invasion, angiogenesis, EMT, stemness etc., not just restricted to the tumor cells but also the surrounding tumor milieu-TME. HNC are a heterogeneous class of molecular disorders, which makes it difficult to conclusive optimized personalized therapeutic strategies, though personalized medicine has emerged as a novel treatment tool. Innumerous RTK inhibitors, mAbs as well as silencing agents have been well explored and still being studied on at the pre-clinical stages, in vitro as well as in vivo, many have reached the clinical trials and few are FDA approved drugs (like Cetuximab, which is an EGFR mAb).

Analyses of accurate and precise biomarkers are required to analyze the progress of treatment as well as mechanisms of innate and acquired resistance while using these RTK based therapeutic agents. Nevertheless, the pre-clinical and clinical RTK based drug studies are showing great advancements, there exists a lack of predictability about the individual patient response to these drugs and the side effects associated with their long term use. Each of the RTK signaling axes contributes not only to disease progression and tumorigenesis, but also highly intertwined cross-talk mechanisms within the cancerous cells, which are a prime concern with regard to the enhanced resistance shown by these HNC toward RTK based therapies. Resistance to RTK based drugs is attributed to numerous genomic alterations, signaling cross-talks, hyperactivation of RTKs or ligands or signal components etc., which has developed various second and third-generation RTK inhibitors. Analyses of various emerging mechanisms resulting in the acquired drug resistance in these HNC to RTK based drugs, contributing to recurrence, reveal that rather than the mono-therapeutic targeting, multi-RTK targeted therapy or combinatorial approach with another conventional chemo/radio/immunotherapeutic agents would translate these strategies in an impressive manner, with higher efficacy, modulating both the tumor as well as the TME.

In short, HNC is a class of tumors that depend mostly on RTKs for survival and growth and RTKs or their ligands or subsequent signaling is either over-expressed or hyper-activated in HNC. Even with varying groups of HNC based on HPV status, RTK signatures are unique to each group. Also, RTKs span a wholesome of cellular and oncogenic functions through its signaling pathways, with a greater amount of similarities between ligands, receptors and signaling molecules, along with cross-talks between other RTKs and non-RTKs. RTKs can also serve as predictive biomarkers for therapeutic progression analysis and RTK antagonists confer radiosensitization properties to HNC cells. A better-integrated approach that affixes genetic, cellular and biochemical arms are needed for a better understanding of this complex RTK network. Thus, inhibiting RTKs in a multi-target approach or combinatorial manner with conventional therapies would efficiently inhibit multiple oncogenic properties attributed by them and serve as efficient therapeutic agents for HNC.

17. Conclusion

A better understanding of the complex RTK network and associated functionalities in normal as well as HNC cells and upgraded stratification of HNC groups based on RTK status would help bring about novel RTK based drugs against HNC and enhanced effectiveness of these drugs on HNC patients with respect to drug efficacy, reduced side effects, a lower rate of recurrence and better OS. Combinatorial therapeutic approaches, that too with multi-target based RTK drugs, rather than with monotherapeutic ones, in combination with conventional chemo/radio/immunomodulatory therapies would help bring about better therapeutic outcomes in HNC patients.

Acknowledgments

P.S. acknowledges the support from Department of Science and Technology, Government of India (EMR/2017/002222). R.N. would like to acknowledge the intra mural funding from Rajiv Gandhi Centre for Biotechnology, Thiruvananthapuram, Kerala, India, for her fellowship.

References

Abhold, E. L., Kiang, A., Rahimy, E., Kuo, S. Z., Wang-Rodriguez, J., Lopez, J. P., et al. (2012). EGFR kinase promotes acquisition of stem cell-like properties: A potential therapeutic target in head and neck squamous cell carcinoma stem cells. *PLoS One*, 7(2) e32459. (Epub 2012/03/03).

Adkins, D., Mehan, P., Ley, J., Siegel, M. J., Siegel, B. A., Dehdashti, F., et al. (2018). Pazopanib plus cetuximab in recurrent or metastatic head and neck squamous cell carcinoma: An open-label, phase 1b and expansion study. *The Lancet Oncology*, *19*(8), 1082–1093. (Epub 2018/07/14).

Agarwal, V., Subash, A., Nayar, R. C., & Rao, V. (2019). Is EGFR really a therapeutic target in head and neck cancers? *Journal of Surgical Oncology*, *119*(6), 685–686. (Epub 2019/02/01).

Ahmad, P., Sana, J., Slavik, M., Slampa, P., Smilek, P., & Slaby, O. (2017). MicroRNAs involvement in radioresistance of head and neck cancer. *Disease Markers*, *2017*, 8245345. (Epub 2017/03/23).

Alfouzan, A. (2019). Head and neck cancer pathology: Old world versus new world disease. *Nigerian Journal of Clinical Practice*, *22*(1), 1–8.

Alsahafi, E., Begg, K., Amelio, I., Raulf, N., Lucarelli, P., Sauter, T., et al. (2019). Clinical update on head and neck cancer: Molecular biology and ongoing challenges. *Cell Death & Disease*, *10*(8), 540. (Epub 2019/07/17).

Antony, J., & Huang, R. Y.-J. (2017). AXL-driven EMT state as a targetable conduit in cancer. *Cancer Research*, 77(14), 3725–3732.

Arnold, L., Enders, J., & Thomas, S. M. (2017). Activated HGF-c-met axis in head and neck cancer. *Cancers*, *9*(12), 169. (Epub 2017/12/13).

Ather, F., Hamidi, H., Fejzo, M. S., Letrent, S., Finn, R. S., Kabbinavar, F., et al. (2013). Dacomitinib, an irreversible Pan-ErbB inhibitor significantly abrogates growth in head and neck cancer models that exhibit low response to cetuximab. *PLoS One*, *8*(2), e56112 (Epub 2013/02/14).

Axelrod, M. J., Mendez, R. E., Khalil, A., Leimgruber, S. S., Sharlow, E. R., Capaldo, B., et al. (2015). Synergistic apoptosis in head and neck squamous cell carcinoma cells by co-inhibition of insulin-like growth factor-1 receptor signaling and compensatory signaling pathways. *Head & Neck*, *37*(12), 1722–1732. (Epub 2014/07/06).

Badarni, M., Prasad, M., Balaban, N., Zorea, J., Yegodayev, K. M., Joshua, B.-Z., et al. (2019). Repression of AXL expression by AP-1/JNK blockage overcomes resistance to PI3Ka therapy. *JCI Insight*, *4*(8), e125341.

Bahleda, R., Italiano, A., Hierro, C., Mita, A., Cervantes, A., Chan, N., et al. (2019). Multicenter phase I study of erdafitinib (JNJ-42756493), oral pan-fibroblast growth factor receptor inhibitor, in patients with advanced or refractory solid tumors. *Clinical Cancer Research: An Official Journal of the American Association for Cancer Research*, *25*(16), 4888–4897. (Epub 2019/05/16).

Baschnagel, A. M., Galoforo, S., Thibodeau, B. J., Ahmed, S., Nirmal, S., Akervall, J., et al. (2015). Crizotinib fails to enhance the effect of radiation in head and neck squamous cell carcinoma xenografts. *Anticancer Research*, *35*(11), 5973–5982.

Baschnagel, A. M., Li, C., Morgan, A. M., Brennan, S. R., Russo, K. A., & Harari, P. M. (2017). Abstract 1094: Combining the pan-FGFR inhibitor AZD4547 with radiation in lung and head and neck squamous cell carcinoma. *Cancer Research*, 77(Suppl. 13), 1094.

Bauman, J. E. (2018). 1124TiPRandomized, phase II study of ficlatuzumab with or without cetuximab in patients (pts) with cetuximab-resistant, recurrent/metastatic (R/M) head and neck squamous cell carcinoma (HNSCC). *Annals of Oncology*, *29*(Suppl. 8), viii372–viii399.

Beckwith, H., & Yee, D. (2015). Minireview: Were the IGF signaling inhibitors all bad? *Molecular Endocrinology*, *29*(11), 1549–1557. (Epub 2015/09/15).

Bektas-Kayhan, K., Unur, M., Boy-Metin, Z., & Cakmakoglu, B. (2012). MCP-1 and CCR2 gene variants in oral squamous cell carcinoma. *Oral Diseases*, *18*(1), 55–59. (Epub 2011/09/03).

Bersani, C., Haeggblom, L., Ursu, R. G., Giusca, S. E., Marklund, L., Ramqvist, T., et al. (2018). Overexpression of FGFR3 in HPV-positive tonsillar and base of tongue cancer is correlated to outcome. *Anticancer Research*, *38*(8), 4683–4690. (Epub 2018/08/01).

Bhattacharya, P., Shetake, N. G., Pandey, B. N., & Kumar, A. (2018). Receptor tyrosine kinase signaling in cancer radiotherapy and its targeting for tumor radiosensitization. *International Journal of Radiation Biology*, *94*(7), 628–644. (Epub 2018/05/19).

Boeckx, C., Baay, M., Wouters, A., Specenier, P., Vermorken, J. B., Peeters, M., et al. (2013). Anti-epidermal growth factor receptor therapy in head and neck squamous cell carcinoma: Focus on potential molecular mechanisms of drug resistance. *The Oncologist*, *18*(7), 850–864. (Epub 2013/07/04).

Bossi, P., Resteghini, C., Paielli, N., Licitra, L., Pilotti, S., & Perrone, F. (2016). Prognostic and predictive value of EGFR in head and neck squamous cell carcinoma. *Oncotarget*, 7(45), 74362–74379. (Epub 2016/08/25).

Bran, B., Bran, G., Hormann, K., & Riedel, F. (2009). The platelet-derived growth factor receptor as a target for vascular endothelial growth factor-mediated anti-angiogenetic therapy in head and neck cancer. *International Journal of Oncology*, *34*(1), 255–261. (Epub 2008/12/17).

Brand, T. M., Hartmann, S., Bhola, N. E., Peyser, N. D., Li, H., Zeng, Y., et al. (2017). Human papillomavirus regulates HER3 expression in head and neck cancer: Implications for targeted HER3 therapy in HPV(+) patients. *Clinical Cancer Research: An Official Journal of the American Association for Cancer Research*, *23*(12), 3072–3083. (Epub 2016/12/18).

Brand, T. M., Iida, M., Stein, A. P., Corrigan, K. L., Braverman, C. M., Coan, J. P., et al. (2015). AXL is a logical molecular target in head and neck squamous cell carcinoma. *Clinical Cancer Research: An Official Journal of the American Association for Cancer Research*, *21*(11), 2601–2612. (Epub 2015/03/15).

Brands, R. C., Knierim, L. M., De Donno, F., Steinacker, V., Hartmann, S., Seher, A., et al. (2017). Targeting VEGFR and FGFR in head and neck squamous cell carcinoma in vitro. *Oncology Reports*, *38*(3), 1877–1885. (Epub 2017/07/18).

Bray, F., Ferlay, J., Soerjomataram, I., Siegel, R. L., Torre, L. A., & Jemal, A. (2018). Global cancer statistics 2018: GLOBOCAN estimates of incidence and mortality worldwide for 36 cancers in 185 countries. *CA: A Cancer Journal for Clinicians*, *68*(6), 394–424. (Epub 2018/09/13).

Bredell, M. G., Ernst, J., El-Kochairi, I., Dahlem, Y., Ikenberg, K., & Schumann, D. M. (2016). Current relevance of hypoxia in head and neck cancer. *Oncotarget*, 7(31), 50781–50804. (Epub 2016/07/20).

Browman, G. P., Wong, G., Hodson, I., Sathya, J., Russell, R., McAlpine, L., et al. (1993). Influence of cigarette smoking on the efficacy of radiation therapy in head and neck cancer. *The New England Journal of Medicine*, *328*(3), 159–163. (Epub 1993/01/21).

Bruce, I. A., Slevin, N. J., Homer, J. J., McGown, A. T., & Ward, T. H. (2005). Synergistic effects of imatinib (STI 571) in combination with chemotherapeutic drugs in head and neck cancer. *Anti-Cancer Drugs*, *16*(7), 719–726. (Epub 2005/07/20).

Budach, V., & Tinhofer, I. (2019). Novel prognostic clinical factors and biomarkers for outcome prediction in head and neck cancer: A systematic review. *The Lancet Oncology*, *20*(6), e313–e326. (Epub 2019/06/05).

Burgos-Tiburcio, A., Santos, E. S., Arango, B. A., & Raez, L. E. (2011). Development of targeted therapy for squamous cell carcinomas of the head and neck. *Expert Review of Anticancer Therapy*, *11*(3), 373–386.

Byeon, H. K., Ku, M., & Yang, J. (2019). Beyond EGFR inhibition: Multilateral combat strategies to stop the progression of head and neck cancer. *Experimental & Molecular Medicine*, *51*(1), 8. (Epub 2019/02/01).

Carla, C., Daris, F., Cecilia, B., Francesca, B., Francesca, C., & Paolo, F. (2012). Angiogenesis in head and neck cancer: A review of the literature. *Journal of Oncology*, *2012*, 358472. (Epub 2011/12/02).

Centers for Disease Control and Prevention (US). (2010). *How tobacco smoke causes disease: The biology and behavioral basis for smoking-attributable disease: A report of the surgeon general*. Atlanta (GA): Centers for Disease Control and Prevention (US).

Chabeda, A., Yanez, R. J. R., Lamprecht, R., Meyers, A. E., Rybicki, E. P., & Hitzeroth, I. I. (2018). Therapeutic vaccines for high-risk HPV-associated diseases. *Papillomavirus Research*, *5*, 46–58. (Epub 2017/12/27).

Chau, N. G., & Haddad, R. I. (2016). Antiangiogenic agents in head and neck squamous cell carcinoma: Tired of going solo. *Cancer*, *122*(23), 3592–3595. (Epub 2016/09/21).

Chen, X., Cao, Y., Sedhom, W., Lu, L., Liu, Y., Wang, H., et al. (2019). Distinct roles of PIK3CA in the enrichment and maintenance of cancer stem cells in head and neck squamous cell carcinoma. *Molecular Oncology*, *14*, 139–158. (Epub 2019/10/11).

Chen, D., & Wang, C. Y. (2019). Targeting cancer stem cells in squamous cell carcinoma. *Precision Clinical Medicin*, *2*(3), 152–165. (Epub 2019/10/11).

Cho, Y. A., Chung, J. M., Ryu, H., Kim, E. K., Cho, B. C., & Yoon, S. O. (2019). Investigating Trk protein expression between oropharyngeal and non-oropharyngeal squamous cell carcinoma: Clinical implications and possible roles of human papillomavirus infection. *Cancer Research and Treatment*, *51*(3), 1052–1063.

Chou, S. T., Peng, H. Y., Mo, K. C., Hsu, Y. M., Wu, G. H., Hsiao, J. R., et al. (2019). MicroRNA-486-3p functions as a tumor suppressor in oral cancer by targeting DDR1. *Journal of Experimental & Clinical Cancer Research: CR*, *38*(1), 281. (Epub 2019/06/30).

Choura, M., & Rebai, A. (2011). Receptor tyrosine kinases: From biology to pathology. *Journal of Receptor and Signal Transduction Research*, *31*(6), 387–394. (Epub 2011/11/02).

Chung, C. H., Pohlmann, P. R., Rothenberg, M. L., Burkey, B. B., Parker, J., Palka, K., et al. (2011). Insulin-like growth factor-1 receptor inhibitor, AMG-479, in cetuximab-refractory head and neck squamous cell carcinoma. *Head & Neck*, *33*(12), 1804–1808. (Epub 2010/07/24).

Cossu, A. M., Mosca, L., Zappavigna, S., Misso, G., Bocchetti, M., De Micco, F., et al. (2019). Long non-coding RNAs as important biomarkers in laryngeal cancer and other head and neck tumours. *International Journal of Molecular Sciences*, *20*(14), 3444. (Epub 2019/07/25).

Cramer, J. D., Burtness, B., Le, Q. T., & Ferris, R. L. (2019). The changing therapeutic landscape of head and neck cancer. *Nature Reviews Clinical Oncology*, *16*(11), 669–683. (Epub 2019/06/14).

Dahiya, K., & Dhankhar, R. (2016). Updated overview of current biomarkers in head and neck carcinoma. *World Journal of Methodology*, *6*(1), 77–86. (Epub 2016/03/29).

De Felice, F., Tombolini, V., Valentini, V., de Vincentiis, M., Mezi, S., Brugnoletti, O., et al. (2019). Advances in the management of HPV-related oropharyngeal cancer. *Journal of Oncology*, *2019*, 9173729. (Epub 2019/05/18).

De Pauw, I., Lardon, F., Van den Bossche, J., Baysal, H., Fransen, E., Deschoolmeester, V., et al. (2018). Simultaneous targeting of EGFR, HER2, and HER4 by afatinib overcomes intrinsic and acquired cetuximab resistance in head and neck squamous cell carcinoma cell lines. *Molecular Oncology*, *12*(6), 830–854. (Epub 2018/04/01).

Deschuymer, S., Mehanna, H., & Nuyts, S. (2018). Toxicity reduction in the treatment of HPV positive oropharyngeal cancer: Emerging combined modality approaches. *Frontiers in Oncology*, *8*, 439. (Epub 2018/10/26).

Dillon, M. T., Grove, L., Newbold, K. L., Shaw, H., Brown, N. F., Mendell, J., et al. (2019). Patritumab with cetuximab plus platinum-containing therapy in recurrent or metastatic squamous cell carcinoma of the head and neck: An open-label, phase Ib study. *Clinical Cancer Research*, *25*(2), 487–495.

Dillon, P. M., Petroni, G. R., Horton, B. J., Moskaluk, C. A., Fracasso, P. M., Douvas, M. G., et al. (2017). A phase II study of dovitinib in patients with recurrent or metastatic adenoid cystic carcinoma. *Clinical Cancer Research: An Official Journal of the American Association for Cancer Research*, *23*(15), 4138–4145. (Epub 2017/04/06).

Du, Z., & Lovly, C. M. (2018). Mechanisms of receptor tyrosine kinase activation in cancer. *Molecular Cancer*, *17*(1), 58. (Epub 2018/02/20).

Dumbrava, E. I., Alfattal, R., Miller, V. A., & Tsimberidou, A. M. (2018). Complete response to a fibroblast growth factor receptor inhibitor in a patient with head and neck squamous cell carcinoma harboring FGF amplifications. *JCO Precision Oncology*, *2*(Epub 2018/01/01).

Economopoulou, P., de Bree, R., Kotsantis, I., & Psyrri, A. (2019). Diagnostic tumor markers in head and neck squamous cell carcinoma (HNSCC) in the clinical setting. *Frontiers in Oncology*, *9*, 827. (Epub 2019/09/27).

Egloff, A. M., & Grandis, J. R. (2008). Targeting epidermal growth factor receptor and SRC pathways in head and neck cancer. *Seminars in Oncology*, *35*(3), 286–297. (Epub 2008/06/12).

Elferink, L. A., & Resto, V. A. (2011). Receptor-tyrosine-kinase-targeted therapies for head and neck cancer. *Journal of Signal Transduction*, *2011*, 982879. (Epub 2011/07/22).

English, I. A., Martinez, J., El Rassi, E., Schmidt, M., Langer, E., Bornstein, S., et al. (2016). Abstract 1631: FGF2 activation of FGFR1 in head and neck squamous cell carcinoma is associated with more invasive disease and can be attenuated by FGFR inhibition. *Cancer Research*, *76*(Suppl. 14), 1631.

Eze, N., Lo, Y. C., & Burtness, B. (2017). Biomarker driven treatment of head and neck squamous cell cancer. *Cancers of the Head & Neck*, *2*, 6. (Epub 2017/08/29).

Fayette, J., Wirth, L., Oprean, C., Udrea, A., Jimeno, A., Rischin, D., et al. (2016). Randomized phase II study of duligotuzumab (MEHD7945A) vs. cetuximab in squamous cell carcinoma of the head and neck (MEHGAN Study). *Frontiers in Oncology*, *6*, 232. (Epub 2016/11/16).

Ferrarotto, R., William, W. N., Jr., Tseng, J. E., Marur, S., Shin, D. M., Murphy, B., et al. (2018). Randomized phase II trial of cixutumumab alone or with cetuximab for refractory recurrent/metastatic head and neck squamous cell carcinoma. *Oral Oncology*, *82*, 83–90. (Epub 2018/06/19).

Fukumoto, I., Koshizuka, K., Hanazawa, T., Kikkawa, N., Matsushita, R., Kurozumi, A., et al. (2016). The tumor-suppressive microRNA-23b/27b cluster regulates the MET oncogene in oral squamous cell carcinoma. *International Journal of Oncology*, *49*(3), 1119–1129. (Epub 2016/08/31).

Garajova, I., Ferracin, M., Porcellini, E., Palloni, A., Abbati, F., Biasco, G., et al. (2017). Non-coding RNAs as predictive biomarkers to current treatment in metastatic colorectal cancer. *International Journal of Molecular Sciences*, *18*(7), 1547. (Epub 2017/07/18).

Garcia Martinez, J., Perez-Escuredo, J., Garcia-Carracedo, D., Alonso-Guervos, M., Suarez-Nieto, C., Llorente-Pendas, J. L., et al. (2012). Analysis of microsatellite instability in laryngeal squamous cell carcinoma. *Acta Otorrinolaringológica Española*, *63*(2), 79–84. Epub 2011/11/22. Analisis de la inestabilidad de microsatelites en el carcinoma escamoso de laringe.

Gauler, T. C., Besse, B., Mauguen, A., Meric, J. B., Gounant, V., Fischer, B., et al. (2011). Phase II trial of PTK787/ZK 222584 (vatalanib) administered orally once-daily or in two divided daily doses as second-line monotherapy in relapsed or progressing patients with stage IIIB/IV non-small-cell lung cancer (NSCLC). *Annals of Oncology*, *23*(3), 678–687.

Giles, K. M., Kalinowski, F. C., Candy, P. A., Epis, M. R., Zhang, P. M., Redfern, A. D., et al. (2013). Axl mediates acquired resistance of head and neck cancer cells to the epidermal growth factor receptor inhibitor erlotinib. *Molecular Cancer Therapeutics*, *12*(11), 2541–2558. (Epub 2013/09/13).

Grana, T. M., Sartor, C. I., & Cox, A. D. (2003). Epidermal growth factor receptor autocrine signaling in RIE-1 cells transformed by the Ras oncogene enhances radiation resistance. *Cancer Research*, *63*(22), 7807–7814. (Epub 2003/11/25).

Guagnano, V., Kauffmann, A., Wohrle, S., Stamm, C., Ito, M., Barys, L., et al. (2012). FGFR genetic alterations predict for sensitivity to NVP-BGJ398, a selective pan-FGFR inhibitor. *Cancer Discovery*, *2*(12), 1118–1133. (Epub 2012/09/25).

Hall, T. G., Yu, Y., Eathiraj, S., Wang, Y., Savage, R. E., Lapierre, J. M., et al. (2016). Preclinical activity of ARQ 087, a novel inhibitor targeting FGFR dysregulation. *PLoS One*, *11*(9), e0162594. (Epub 2016/09/15).

Harari, P. M., Wheeler, D. L., & Grandis, J. R. (2009). Molecular target approaches in head and neck cancer: Epidermal growth factor receptor and beyond. *Seminars in Radiation Oncology*, *19*(1), 63–68. (Epub 2008/11/26).

Hartmann, S., Bhola, N. E., & Grandis, J. R. (2016). HGF/met signaling in head and neck cancer: Impact on the tumor microenvironment. *Clinical Cancer Research: An Official Journal of the American Association for Cancer Research*, *22*(16), 4005–4013. (Epub 2016/07/03).

Hayes, K. E., Walk, E. L., Ammer, A. G., Kelley, L. C., Martin, K. H., & Weed, S. A. (2013). Ableson kinases negatively regulate invadopodia function and invasion in head and neck squamous cell carcinoma by inhibiting an HB-EGF autocrine loop. *Oncogene*, *32*(40), 4766–4777. (Epub 2012/11/14).

Héroult, M., Ocker, M., Kopitz, C., Zopf, D., Hägebarth, A., Ziegelbauer, K., et al. (2015). Abstract 772: Anti-tumor efficacy of the selective pan-FGFR Inhibitor BAY 1163877 in preclinical squamous cell carcinoma models of different origin. *Cancer Research*, *75*(Suppl. 15), 772.

Heydar, H., Mansouri, K., Norooznezhad, M., Norooznezhad, F., Mohamadnia, A., & Bahrami, N. (2018). Bevacizumab inhibits angiogenic cytokines in head and neck squamous cell carcinoma: From gene to the protein. *International Journal of Hematology-Oncology and Stem Cell Research*, *12*(2), 136–141. (Epub 2018/09/21).

Hollebecque, A., Borad, M., Sahai, V., Catenacci, D. V. T., Murphy, A., Vaccaro, G., et al. (2018). 756PInterim results of fight-202, a phase II, open-label, multicenter study of INCB054828 in patients (pts) with previously treated advanced/metastatic or surgically unresectable cholangiocarcinoma (CCA) with/without fibroblast growth factor (FGF)/FGF receptor (FGFR) genetic alterations. *Annals of Oncology*, *29* (Suppl. 8), viii205–viii270. https://doi.org/10.1093/annonc/mdy282.

Hsieh, J. C., Wang, H. M., Wu, M. H., Chang, K. P., Chang, P. H., Liao, C. T., et al. (2019). Review of emerging biomarkers in head and neck squamous cell carcinoma in the era of immunotherapy and targeted therapy. *Head & Neck*, *41*(Suppl. 1), 19–45. (Epub 2019/10/02).

Hyman, D. M., Goyal, L., Grivas, P., Meric-Bernstam, F., Tabernero, J., Hu, Y., et al. (2019). FUZE clinical trial: A phase 2 study of Debio 1347 in FGFR fusion-positive advanced solid tumors irrespectively of tumor histology. *Journal of Clinical Oncology*, *37*(Suppl. 15)TPS3157-TPS.

Iwasa, T., Okamoto, I., Suzuki, M., Hatashita, E., Yamada, Y., Fukuoka, M., et al. (2009). Inhibition of insulin-like growth factor 1 receptor by CP-751,871 radiosensitizes non–small cell lung Cancer cells. *Clinical Cancer Research*, *15*(16), 5117–5125.

Jameson, M. J., Beckler, A. D., Taniguchi, L. E., Allak, A., Vanwagner, L. B., Lee, N. G., et al. (2011). Activation of the insulin-like growth factor-1 receptor induces resistance to epidermal growth factor receptor antagonism in head and neck squamous carcinoma cells. *Molecular Cancer Therapeutics*, *10*(11), 2124–2134. (Epub 2011/09/01).

Jechlinger, M., Sommer, A., Moriggl, R., Seither, P., Kraut, N., Capodiecci, P., et al. (2006). Autocrine PDGFR signaling promotes mammary cancer metastasis. *The Journal of Clinical Investigation, 116*(6), 1561–1570. (Epub 2006/06/03).

Jeon, H. M., & Lee, J. (2017). MET: Roles in epithelial-mesenchymal transition and cancer stemness. *Annals of Translational Medicine, 5*(1), 5. (Epub 2017/02/07).

Jethwa, A. R., & Khariwala, S. S. (2017). Tobacco-related carcinogenesis in head and neck cancer. *Cancer Metastasis Reviews, 36*(3), 411–423. (Epub 2017/08/13).

Jiang, X., Ye, J., Dong, Z., Hu, S., & Xiao, M. (2019). Novel genetic alterations and their impact on target therapy response in head and neck squamous cell carcinoma. *Cancer Management and Research, 11*, 1321–1336. (Epub 2019/02/26).

Kalinowski, F. C., Giles, K. M., Candy, P. A., Ali, A., Ganda, C., Epis, M. R., et al. (2012). Regulation of epidermal growth factor receptor signaling and erlotinib sensitivity in head and neck cancer cells by miR-7. *PLoS One*, 7(10), e47067. (Epub 2012/11/02).

Kalyukina, M., Yosaatmadja, Y., Middleditch, M. J., Patterson, A. V., Smaill, J. B., & Squire, C. J. (2019). TAS-120 Cancer target binding: Defining reactivity and revealing the first fibroblast growth factor receptor 1 (FGFR1) irreversible structure. *ChemMedChem, 14*(4), 494–500. (Epub 2019/01/03).

Kedmi, M., Sas-Chen, A., & Yarden, Y. (2015). MicroRNAs and growth factors: An alliance propelling tumor progression. *Journal of Clinical Medicine, 4*(8), 1578–1599. (Epub 2015/08/20).

Kim, Y., Lee, S. J., Lee, J. Y., Lee, S. H., Sun, J. M., Park, K., et al. (2017). Clinical trial of nintedanib in patients with recurrent or metastatic salivary gland cancer of the head and neck: A multicenter phase 2 study (Korean cancer Study Group HN14-01). *Cancer, 123*(11), 1958–1964. (Epub 2017/01/20).

Kobayashi, K., Hisamatsu, K., Suzui, N., Hara, A., Tomita, H., & Miyazaki, T. (2018). A review of HPV-related head and neck cancer. *Journal of Clinical Medicine*, 7(9), 241. (Epub 2018/08/29).

Koole, K., Brunen, D., van Kempen, P. M., Noorlag, R., de Bree, R., Lieftink, C., et al. (2016). FGFR1 is a potential prognostic biomarker and therapeutic target in head and neck squamous cell carcinoma. *Clinical Cancer Research: An Official Journal of the American Association for Cancer Research, 22*(15), 3884–3893. (Epub 2016/03/05).

Koole, K., Clausen, M. J., van Es, R. J., van Kempen, P. M., Melchers, L. J., Koole, R., et al. (2016). FGFR family members protein expression as prognostic markers in oral cavity and oropharyngeal squamous cell carcinoma. *Molecular Diagnosis & Therapy, 20*(4), 363–374. (Epub 2016/06/10).

Koontongkaew, S. (2013). The tumor microenvironment contribution to development, growth, invasion and metastasis of head and neck squamous cell carcinomas. *Journal of Cancer, 4*(1), 66–83. (Epub 2013/02/07).

Koshizuka, K., Hanazawa, T., Fukumoto, I., Kikkawa, N., Matsushita, R., Mataki, H., et al. (2017). Dual-receptor (EGFR and c-MET) inhibition by tumor-suppressive miR-1 and miR-206 in head and neck squamous cell carcinoma. *Journal of Human Genetics, 62*(1), 113–121. (Epub 2016/05/14).

Kovacova, J., Poprach, A., Buchler, T., Cho, W. C., & Slaby, O. (2018). MicroRNAs as predictive biomarkers of response to tyrosine kinase inhibitor therapy in metastatic renal cell carcinoma. *Clinical Chemistry and Laboratory Medicine, 56*(9), 1426–1431. (Epub 2018/02/17).

Kramer, B., Hock, C., Birk, R., Sauter, A., Stuck, B. A., Hörmann, K., et al. (2016). Targeted therapies in HPV-positive and -negative HNSCC—Alteration of EGFR and VEGFR-2 expression in vitro. *Anticancer Research, 36*(6), 2799–2807.

Kumar, A. S., Jagadeeshan, S., Pitani, R. S., Ramshankar, V., Venkitasamy, K., Venkatraman, G., et al. (2017). Snail-Modulated microRNA 493 forms a negative feedback loop with the insulin-like growth factor 1 receptor pathway and blocks tumorigenesis. *Molecular and Cellular Biology, 37*(6), e00510–e00516. (Epub 2016/12/14).

Kumar, A. S., Rayala, S. K., & Venkatraman, G. (2018). Targeting IGF1R pathway in cancer with microRNAs: How close are we? *RNA Biology*, *15*(3), 320–326. (Epub 2017/06/15).

Kurzrock, R., Sherman, S. I., Ball, D. W., Forastiere, A. A., Cohen, R. B., Mehra, R., et al. (2011). Activity of XL184 (Cabozantinib), an oral tyrosine kinase inhibitor, in patients with medullary thyroid Cancer. *Journal of Clinical Oncology*, *29*(19), 2660–2666.

Kyzas, P. A., Stefanou, D., Batistatou, A., & Agnantis, N. J. (2005). Potential autocrine function of vascular endothelial growth factor in head and neck cancer via vascular endothelial growth factor receptor-2. *Modern Pathology: An Official Journal of the United States and Canadian Academy of Pathology, Inc*, *18*(4), 485–494. (Epub 2004/10/12).

Lampri, E. S., Chondrogiannis, G., Ioachim, E., Varouktsi, A., Mitselou, A., Galani, A., et al. (2015). Biomarkers of head and neck cancer, tools or a Gordian knot? *International Journal of Clinical and Experimental Medicine*, *8*(7), 10340–10357. (Epub 2015/09/18).

Le, X., Huang, A. T., Chen, Y., & Lai, S. Y. (2015). Regulation of receptor tyrosine kinases by miRNA: Overexpression of miRNA using lentiviral inducible expression vectors. *Methods in Molecular Biology*, *1233*, 135–147. (Epub 2014/10/17).

Leemans, C. R., Snijders, P. J. F., & Brakenhoff, R. H. (2018). The molecular landscape of head and neck cancer. *Nature Reviews Cancer*, *18*(5), 269–282. (Epub 2018/03/03).

Limesand, K. H., Chibly, A. M., & Fribley, A. (2013). Impact of targeting insulin-like growth factor signaling in head and neck cancers. *Growth Hormone & IGF Research: Official Journal of the Growth Hormone Research Society and the International IGF Research Society*, *23*(5), 135–140. (Epub 2013/07/03).

Lokker, N. A., Sullivan, C. M., Hollenbach, S. J., Israel, M. A., & Giese, N. A. (2002). Platelet-derived growth factor (PDGF) autocrine signaling regulates survival and mitogenic pathways in glioblastoma cells: Evidence that the novel PDGF-C and PDGF-D ligands may play a role in the development of brain tumors. *Cancer Research*, *62*(13), 3729–3735. (Epub 2002/07/05).

Lugovskoy, A., Johnson, B., Baum, J., Adams, S., Tang, J., Kohli, N., et al. (2011). Abstract B205: Therapeutically targeting redundant, growth factor-induced prosurvival signaling with MM-141, a novel bispecific antibody targeting IGF-1R and ErbB3. In *AACR*.

Macaulay, V. M., Middleton, M. R., Eckhardt, S. G., Rudin, C. M., Juergens, R. A., Gedrich, R., et al. (2016). Phase I dose-escalation study of linsitinib (OSI-906) and erlotinib in patients with advanced solid tumors. *Clinical Cancer Research*, *22*(12), 2897–2907.

Macha, M. A., Matta, A., Chauhan, S. S., Siu, K. W., & Ralhan, R. (2011). Guggulsterone (GS) inhibits smokeless tobacco and nicotine-induced NF-kappaB and STAT3 pathways in head and neck cancer cells. *Carcinogenesis*, *32*(3), 368–380. (Epub 2010/12/24).

Machiels, J. P., Henry, S., Zanetta, S., Kaminsky, M. C., Michoux, N., Rommel, D., et al. (2010). Phase II study of sunitinib in recurrent or metastatic squamous cell carcinoma of the head and neck: GORTEC 2006-01. *Journal of Clinical Oncology: Official Journal of the American Society of Clinical Oncology*, *28*(1), 21–28. (Epub 2009/11/18).

Mangelson, R., Peterson, P., Foulks, J. M., Matsumura, Y., Mouritsen, L., Whatcott, C. J., et al. (2019). Abstract 2197: The AXL kinase inhibitor, TP-0903, demonstrates efficacy in preclinical models of colorectal cancer independent of KRAS mutation status. *Cancer Research*, *79*(Suppl. 13), 2197.

Marcu, L. G., Reid, P., & Bezak, E. (2018). The promise of novel biomarkers for head and neck cancer from an imaging perspective. *International Journal of Molecular Sciences*, *19*(9), 2511. (Epub 2018/08/29).

Marshall, M. E., Hinz, T. K., Kono, S. A., Singleton, K. R., Bichon, B., Ware, K. E., et al. (2011). Fibroblast growth factor receptors are components of autocrine signaling networks in head and neck squamous cell carcinoma cells. *Clinical Cancer Research: An Official Journal of the American Association for Cancer Research*, *17*(15), 5016–5025. (Epub 2011/06/16).

Matei, D., Emerson, R. E., Lai, Y. C., Baldridge, L. A., Rao, J., Yiannoutsos, C., et al. (2006). Autocrine activation of PDGFRalpha promotes the progression of ovarian cancer. *Oncogene, 25*(14), 2060–2069. (Epub 2005/12/07).

Matsumoto, F., Valdecanas, D. N., Mason, K. A., Milas, L., Ang, K. K., & Raju, U. (2012). The impact of timing of EGFR and IGF-1R inhibition for sensitizing head and neck cancer to radiation. *Anticancer Research, 32*(8), 3029–3035. (Epub 2012/07/31).

Matsuura, M., Onimaru, M., Yonemitsu, Y., Suzuki, H., Nakano, T., Ishibashi, H., et al. (2009). Autocrine loop between vascular endothelial growth factor (VEGF)-C and VEGF receptor-3 positively regulates tumor-associated lymphangiogenesis in oral squamoid cancer cells. *The American Journal of Pathology, 175*(4), 1709–1721. (Epub 2009/09/26).

Miekus, K. (2017). The met tyrosine kinase receptor as a therapeutic target and a potential cancer stem cell factor responsible for therapy resistance (review). *Oncology Reports, 37*(2), 647–656. (Epub 2016/12/14).

Mirghani, H., & Blanchard, P. (2018). Treatment de-escalation for HPV-driven oropharyngeal cancer: Where do we stand? *Clinical and Translational Radiation Oncology, 8*, 4–11. (Epub 2018/03/30).

Mockelmann, N., Rieckmann, T., Busch, C. J., Becker, B., Gleissner, L., Hoffer, K., et al. (2016). Effect of sorafenib on cisplatin-based chemoradiation in head and neck cancer cells. *Oncotarget, 7*(17), 23542–23551. (Epub 2016/03/26).

Murthy, V., Calcuttawala, A., Chadha, K., d'Cruz, A., Krishnamurthy, A., Mallick, I., et al. (2017). Human papillomavirus in head and neck cancer in India: Current status and consensus recommendations. *South Asian Journal of Cancer, 6*(3), 93–98. (Epub 2017/10/05).

O-Charoenrat, P., Rhys-Evans, P., & Eccles, S. (2002). A synthetic matrix metalloproteinase inhibitor prevents squamous carcinoma cell proliferation by interfering with epidermal growth factor receptor autocrine loops. *International Journal of Cancer, 100*(5), 527–533. (Epub 2002/07/19).

Ong, H. S., Gokavarapu, S., Tian, Z., Li, J., Xu, Q., Cao, W., et al. (2017). PDGFRA mRNA is overexpressed in oral cancer patients as compared to normal subjects with a significant trend of overexpression among tobacco users. *Journal of Oral Pathology & Medicine: Official Publication of the International Association of Oral Pathologists and the American Academy of Oral Pathology, 46*(8), 591–597. (Epub 2017/03/28).

Organization TGCO-IAfRoC-WH. (2019). https://gco.iarc.fr/data/factsheets/populations/356-india-fact-sheets.

Pai, S. I., & Westra, W. H. (2009). Molecular pathology of head and neck cancer: Implications for diagnosis, prognosis, and treatment. *Annual Review of Pathology, 4*, 49–70. (Epub 2008/08/30).

Pan, C., Issaeva, N., & Yarbrough, W. G. (2018). HPV-driven oropharyngeal cancer: Current knowledge of molecular biology and mechanisms of carcinogenesis. *Cancers of the Head & Neck, 3*, 12. (Epub 2019/05/17).

Papadimitrakopoulou, V. A., Frank, S. J., Cohen, E. W., Hirsch, F. R., Myers, J. N., Heymach, J. V., et al. (2016). Phase I study of vandetanib with radiation therapy with or without cisplatin in locally advanced head and neck squamous cell carcinoma. *Head & Neck, 38*(3), 439–447. (Epub 2014/10/30).

Patil, V. M., Noronha, V., Joshi, A., Agarwal, J., Laskar, S. G., Budrukkar, A., et al. (2018). Results of a randomized phase III study of nimotuzumab in combination with concurrent radiotherapy and cisplatin versus radiotherapy and cisplatin alone, in locally advanced squamous cell carcinoma of the head and neck. *Journal of Clinical Oncology, 36*(Suppl. 15), 6000.

Payne, K., Brooks, J., Spruce, R., Batis, N., Taylor, G., Nankivell, P., et al. (2019). Circulating tumour cell biomarkers in head and neck cancer: Current progress and future prospects. *Cancers, 11*(8), 1115. (Epub 2019/08/08).

Peltanova, B., Raudenska, M., & Masarik, M. (2019). Effect of tumor microenvironment on pathogenesis of the head and neck squamous cell carcinoma: A systematic review. *Molecular Cancer, 18*(1), 63. (Epub 2019/04/01).

Peng, W. X., Fan, L. T., & Mo, Y. Y. (2018). A link between a synonymous SNP and the clinical response to tyrosine kinase inhibitors. *Non-Coding RNA Investigation, 2*, 6. (Epub 2018/02/01).

Pollock, N. I., Wang, L., Wallweber, G., Gooding, W. E., Huang, W., Chenna, A., et al. (2015). Increased expression of HER2, HER3, and HER2:HER3 heterodimers in HPV-positive HNSCC using a novel proximity-based assay: Implications for targeted therapies. *Clinical Cancer Research: An Official Journal of the American Association for Cancer Research, 21*(20), 4597–4606. (Epub 2015/07/04).

Rankin, E. B., & Giaccia, A. J. (2016). The receptor tyrosine kinase AXL in cancer progression. *Cancers, 8*(11), 103. (Epub 2016/11/12).

RESEARCH ICOM. (2018). *Welfare DoHRMHF, 2018 GoI-MRN-N.* https://icmr.nic.in/sites/default/files/.

Saada-Bouzid, E., & Le Tourneau, C. (2019). Beyond EGFR targeting in SCCHN: Angiogenesis, PI3K, and other molecular targets. *Frontiers in Oncology, 9*, 74. (Epub 2019/03/01).

Schick, U., Gujral, D. M., Richards, T. M., Harrington, K. J., & Nutting, C. M. (2012). Zalutumumab in head and neck cancer. *Expert Opinion on Biological Therapy, 12*(1), 119–125. (Epub 2011/12/17).

Schmitz, S., Kaminsky-Forrett, M.-C., Henry, S., Zanetta, S., Geoffrois, L., Bompas, E., et al. (2012). Phase II study of figitumumab in patients with recurrent and/or metastatic squamous cell carcinoma of the head and neck: Clinical activity and molecular response (GORTEC 2008-02). *Annals of Oncology, 23*(8), 2153–2161.

Schoumacher, M., & Burbridge, M. (2017). Key roles of AXL and MER receptor tyrosine kinases in resistance to multiple anticancer therapies. *Current Oncology Reports, 19*(3), 19. (Epub 2017/03/03).

Schultz, J. D., Muhlheim, K., Erben, P., Hofheinz, R. D., Faber, A., Thorn, C., et al. (2011). Chemotherapeutic alteration of VEGF-/PDGF- and PDGF-Ralpha/beta expression by imatinib in HPV-transformed squamous cell carcinoma compared to HPV-negative HNSCC in vitro. *Oncology Reports, 26*(5), 1099–1109. (Epub 2011/08/02).

Seiwert, T. Y., Jagadeeswaran, R., Tretiakova, M., Husain, A. N., Martin, L., Lingen, M. W., et al. (2005). Inhibition of c-Met signaling by SU11274 in head and neck squamous cell carcinoma (HNSCC). *Cancer Research, 65* (Suppl. 9), 1405–1406.

Seiwert, T., Sarantopoulos, J., Kallender, H., McCallum, S., Keer, H. N., & Blumenschein, G., Jr. (2013). Phase II trial of single-agent foretinib (GSK1363089) in patients with recurrent or metastatic squamous cell carcinoma of the head and neck. *Investigational New Drugs, 31*(2), 417–424. (Epub 2012/08/25).

Shilpi, G., Prabhat, K., Jayant, M., Harsimrut, K., & Bhudev, C. D. (2018). Epigenetic biomarkers in head and neck cancer. *Journal of Cancer Genetics and Biomarkers, 1*(2), 41–50.

Shimizu, R., Ibaragi, S., Eguchi, T., Kuwajima, D., Kodama, S., Nishioka, T., et al. (2019). Nicotine promotes lymph node metastasis and cetuximab resistance in head and neck squamous cell carcinoma. *International Journal of Oncology, 54*(1), 283–294. (Epub 2018/11/16).

Siano, M., Molinari, F., Martin, V., Mach, N., Fruh, M., Freguia, S., et al. (2017). Multicenter phase II study of panitumumab in platinum pretreated, advanced head and neck squamous cell cancer. *The Oncologist, 22*(7), 782–e70. (Epub 2017/06/09).

Singh, S. P., Eisenberg, R., & Hoffman, G. (2018). An overview and comparative evaluation of head and neck Cancer risk factors in India and Australia. *International Journal of Otolaryngology and Head & Neck Surgery*, 7(5), 254.

Slomiany, M. G., Black, L. A., Kibbey, M. M., Day, T. A., & Rosenzweig, S. A. (2006). IGF-1 induced vascular endothelial growth factor secretion in head and neck squamous cell carcinoma. *Biochemical and Biophysical Research Communications*, *342*(3), 851–858. (Epub 2006/02/28).

Slomiany, M. G., Black, L. A., Kibbey, M. M., Tingler, M. A., Day, T. A., & Rosenzweig, S. A. (2007). Insulin-like growth factor-1 receptor and ligand targeting in head and neck squamous cell carcinoma. *Cancer Letters*, *248*(2), 269–279. (Epub 2006/09/26).

Smith, E. M., Rubenstein, L. M., Haugen, T. H., Pawlita, M., & Turek, L. P. (2012). Complex etiology underlies risk and survival in head and neck cancer human papillomavirus, tobacco, and alcohol: A case for multifactor disease. *Journal of Oncology*, *2012*, 571862. (Epub 2012/02/09).

Song, D. W., Shin, S. J., Kim, D. E., Jung, S. G., Park, J. W., & Lee, K. D. (2008). Detection of MAGE and SSX gene expressions by RT-nested PCR using common primers in head and neck cancer. *Clinical and Experimental Otorhinolaryngology*, *1*(2), 97–102. (Epub 2009/05/13).

Suh, Y., Amelio, I., Guerrero Urbano, T., & Tavassoli, M. (2014). Clinical update on cancer: Molecular oncology of head and neck cancer. *Cell Death & Disease*, *5*(1), e1018-e.

Sun, Z., Liu, Q., Ye, D., Ye, K., Yang, Z., & Li, D. (2018). Role of c-Met in the progression of human oral squamous cell carcinoma and its potential as a therapeutic target. *Oncology Reports*, *39*(1), 209–216. (Epub 2017/11/09).

Susuki, D., Kimura, S., Naganuma, S., Tsuchiyama, K., Tanaka, T., Kitamura, N., et al. (2011). Regulation of microRNA expression by hepatocyte growth factor in human head and neck squamous cell carcinoma. *Cancer Science*, *102*(12), 2164–2171. (Epub 2011/09/09).

Sweeny, L., Liu, Z., Lancaster, W., Hart, J., Hartman, Y. E., & Rosenthal, E. L. (2012). Inhibition of fibroblasts reduced head and neck cancer growth by targeting fibroblast growth factor receptor. *The Laryngoscope*, *122*(7), 1539–1544. (Epub 2012/03/31).

Swiecicki, P. L., Spector, M., & Worden, F. P. (2016). Axitinib in the treatment of head and neck malignancies. *Current Clinical Pharmacology*, *11*(2), 72–76. (Epub 2016/05/18).

Taberna, M., Mena, M., Pavon, M. A., Alemany, L., Gillison, M. L., & Mesia, R. (2017). Human papillomavirus-related oropharyngeal cancer. *Annals of Oncology: Official Journal of the European Society for Medical Oncology*, *28*(10), 2386–2398. (Epub 2017/06/22).

Taberna, M., Oliva, M., & Mesia, R. (2019). Cetuximab-containing combinations in locally advanced and recurrent or metastatic head and neck squamous cell carcinoma. *Frontiers in Oncology*, *9*, 383. (Epub 2019/06/06).

Tan, D. S. W., Chong, F. T., Leong, H. S., Toh, S. Y., Lau, D. P., Kwang, X. L., et al. (2017). Long noncoding RNA EGFR-AS1 mediates epidermal growth factor receptor addiction and modulates treatment response in squamous cell carcinoma. *Nature Medicine*, *23*(10), 1167–1175. (Epub 2017/09/19).

Tang, X., He, J., Li, B., Zheng, Y., Li, K., Zou, S., et al. (2019). Efficacy and safety of gefitinib in patients with advanced head and neck squamous cell carcinoma: A meta-analysis of randomized controlled trials. *Journal of Oncology*, *2019*, 9.

Taylor, M. H., Rasco, D. W., Brose, M. S., Vogelzang, N. J., Richey, S. L., Cohn, A. L., et al. (2018). A phase 1b/2 trial of lenvatinib plus pembrolizumab in patients with squamous cell carcinoma of the head and neck. *Journal of Clinical Oncology*, *36*(Suppl. 15), 6016.

Thinn, M. M., & Hsueh, C. T. (2019). Sustained complete response to erlotinib in squamous cell carcinoma of the head and neck: A case report. *World Journal of Clinical Cases*, 7(5), 616–622. (Epub 2019/03/14).

Tiash, S., & Chowdhury, E. H. (2015). Growth factor receptors: Promising drug targets in cancer. *Journal of Cancer Metastasis and Treatment*, *1*, 190–200.

Tong, C. C., Kao, J., & Sikora, A. G. (2012). Recognizing and reversing the immunosuppressive tumor microenvironment of head and neck cancer. *Immunologic Research*, *54*(1–3), 266–274. (Epub 2012/03/29).

Vokes, E. E., Worden, F. P., Adkins, D., Bauman, J. E., Lim, D., Sukari, A., et al. (2015). A randomized phase II trial of the MET inhibitor tivantinib + cetuximab versus cetuximab alone in patients with recurrent/metastatic head and neck cancer. *Journal of Clinical Oncology*, *33*(Suppl. 15), 6060.

von Massenhausen, A., Bragelmann, J., Billig, H., Thewes, B., Queisser, A., Vogel, W., et al. (2016). Implication of the receptor tyrosine kinase AXL in head and neck cancer progression. *International Journal of Molecular Sciences*, *18*(1), 7. (Epub 2016/12/28).

von Massenhausen, A., Sanders, C., Thewes, B., Deng, M., Queisser, A., Vogel, W., et al. (2016). MERTK as a novel therapeutic target in head and neck cancer. *Oncotarget*, 7(22), 32678–32694. (Epub 2016/04/16).

Wang, H. C., Chan, L. P., & Cho, S. F. (2019). Targeting the immune microenvironment in the treatment of head and neck squamous cell carcinoma. *Frontiers in Oncology*, *9*, 1084. (Epub 2019/11/05).

Wang, Z., Liu, G., Mao, J., Xie, M., Zhao, M., Guo, X., et al. (2019). IGF-1R inhibition suppresses cell proliferation and increases radiosensitivity in nasopharyngeal carcinoma cells. *Mediators of Inflammation*, *2019*, 5497467. (Epub 2019/08/31).

Wang, N., Wu, Q. L., Fang, Y., Mai, H. Q., Zeng, M. S., Shen, G. P., et al. (2005). Expression of chemokine receptor CXCR4 in nasopharyngeal carcinoma: Pattern of expression and correlation with clinical outcome. *Journal of Translational Medicine*, *3*, 26. (Epub 2005/06/28).

Wang, J., Xi, L., Hunt, J. L., Gooding, W., Whiteside, T. L., Chen, Z., et al. (2004). Expression pattern of chemokine receptor 6 (CCR6) and CCR7 in squamous cell carcinoma of the head and neck identifies a novel metastatic phenotype. *Cancer Research*, *64*(5), 1861–1866. (Epub 2004/03/05).

Wedge, S. R., Kendrew, J., Hennequin, L. F., Valentine, P. J., Barry, S. T., Brave, S. R., et al. (2005). AZD2171: A highly potent, orally bioavailable, vascular endothelial growth factor receptor-2 tyrosine kinase inhibitor for the treatment of cancer. *Cancer Research*, *65*(10), 4389–4400. (Epub 2005/05/19).

Weiss, J. M., Bagley, S., Hwang, W. T., Bauml, J., Olson, J. G., Cohen, R. B., et al. (2016). Capecitabine and lapatinib for the first-line treatment of metastatic/recurrent head and neck squamous cell carcinoma. *Cancer*, *122*(15), 2350–2355. (Epub 2016/05/20).

Xie, X., Wang, Z., Chen, F., Yuan, Y., Wang, J., Liu, R., et al. (2016). Roles of FGFR in oral carcinogenesis. *Cell Proliferation*, *49*(3), 261–269. (Epub 2016/05/25).

Yang, Y., Chen, D., Liu, H., & Yang, K. (2019). Increased expression of lncRNA CASC9 promotes tumor progression by suppressing autophagy-mediated cell apoptosis via the AKT/mTOR pathway in oral squamous cell carcinoma. *Cell Death & Disease*, *10*(2), 41. (Epub 2019/01/25).

Yang, S. H., Lee, T. Y., Ho, C. A., Yang, C. Y., Huang, W. Y., Lin, Y. C., et al. (2018). Exposure to nicotine-derived nitrosamine ketone and arecoline synergistically facilitates tumor aggressiveness via overexpression of epidermal growth factor receptor and its downstream signaling in head and neck squamous cell carcinoma. *PLoS One*, *13*(8), e0201267. (Epub 2018/08/28).

Yao, Y., Chen, X., Lu, S., Zhou, C., Xu, G., Yan, Z., et al. (2018). Circulating long noncoding RNAs as biomarkers for predicting head and neck squamous cell carcinoma. *Cellular Physiology and Biochemistry: International Journal of Experimental Cellular Physiology, Biochemistry, and Pharmacology*, *50*(4), 1429–1440. (Epub 2018/10/26).

Yu, T., Ma, P., Wu, D., Shu, Y., & Gao, W. (2018). Functions and mechanisms of microRNA-31 in human cancers. *Biomedicine & Pharmacotherapy*, *108*, 1162–1169. (Epub 2018/10/31).

Yu, B., Wu, K., Wang, X., Zhang, J., Wang, L., Jiang, Y., et al. (2018). Periostin secreted by cancer-associated fibroblasts promotes cancer stemness in head and neck cancer by activating protein tyrosine kinase 7. *Cell Death & Disease*, *9*(11), 1082. (Epub 2018/10/24).

Zhang, J., Wang, Y., Chen, X., Zhou, Y., Jiang, F., Chen, J., et al. (2015). MiR-34a suppresses amphiregulin and tumor metastatic potential of head and neck squamous cell carcinoma (HNSCC). *Oncotarget*, *6*(10), 7454–7469. (Epub 2015/03/13).

Zou, H. Y., Li, Q., Lee, J. H., Arango, M. E., Burgess, K., Qiu, M., et al. (2012). Sensitivity of selected human tumor models to PF-04217903, a novel selective c-Met kinase inhibitor. *Molecular Cancer Therapeutics*, *11*(4), 1036–1047. (Epub 2012/03/06).

CHAPTER TEN

Advances in EGFR/HER2-directed clinical research on breast cancer

Louis W.C. Chow[a,b,*], Erich F. Lie[b], Masakazu Toi[b,c]
[a]UNIMED Medical Institute, Hong Kong, China
[b]Organisation for Oncology and Translational Research, Hong Kong, China
[c]Department of Breast Surgery, Graduate School of Medicine, Kyoto University, Kyoto, Japan
[*]Corresponding author: e-mail address: lwcchow@ootr.org

Contents

Abstract

The discovery of the HER2 molecules has embarked a series of investigations on the efficacy and safety of different types of anti-HER2 therapies for treating breast cancer, with the clinical pathway requiring a more detailed, more precise, and more dynamics therapeutic approaches due to the heterogeneity of the disease. As the "do more" and "do less" approaches are becoming more important to personalize treatment for early HER2-positive breast cancer, recent advances aim at tackling the advanced stage of the disease by using novel therapeutic agents and combination strategies. There are also important points of consideration on prognosis and choice of therapies, including

Advances in Cancer Research, Volume 147
ISSN 0065-230X
https://doi.org/10.1016/bs.acr.2020.04.009
© 2020 Elsevier Inc.
All rights reserved.

HER2 gene copy number, HER2 heterogeneity, tissue biomarkers, blood-based biomarkers, and HER2 mutation and its treatment. Altogether, these could potentially play a vital role in the journey of HER2-positive breast cancer patient to achieve greater survival benefit and potentially a cure for the disease.

1. Introduction

Human epidermal growth factor receptor 2 (HER2)-positive breast cancer, a specific form of the disease that derives its name from its association with one of the members (HER2) of the epidermal growth factor receptor (EGFR/ErbB) family, remains a significant global health care burden despite recent improvements in the management of the disease (Harbeck et al., 2019; Hayes, 2019).

EGFR family members are transmembrane receptor tyrosine kinases that activate signal transduction pathways that regulate cell proliferation and survival. These receptors consist of multiple extracellular domains for ligand binding. The intracellular domain consists of a short juxtamembrane region and a tyrosine kinase domain (TKD) carrying the amino lobe, which is attached to a carboxyl tail that contains at least five tyrosine autophosphorylation sites (Lemmon, Schlessinger, & Ferguson, 2014). Ligand binding promotes strong homodimerization. It is also reported that different EGFR members heterodimerize too. HER2 is a non-ligand-binding member and activates when there is homodimerization or heterodimerization with other EGFR family members. Dimerization activates the carboxyl tail leading to phosphorylation of the tyrosine sites. The carboxy lobe of the "activator" kinase in an EGFR TKD dimer is thought to induce allosteric changes in the amino lobe of the "receiver" kinase and thus activate it. The phosphorylated tyrosines recruit further downstream signaling proteins, which may affect multiple cellular processes, including proliferation, survival, and differentiation, depending on receptor subtype and cellular context.

HER2 can also be activated by proteolytic cleavage of its extracellular domain (ECD) (Perrier, Gligorov, Lefèvre, & Boissan, 2018). The release of ECD into the blood stream is called ECD shedding. The shedding is increased during metastasis. The serum HER2 ECD (sHER2ECD) levels were elevated in 15% of primary breast cancer patients and reached to 45% in patients with metastatic disease. The truncated form of HER2 (p95HER2) has a 10- to 100-folds higher oncogenic potential than the full-length protein. The activation of kinase activity could be related to the release of the inhibitory domain contained within the shed ECD.

The discovery of this pathway unfolds the natural history of the disease. HER2 amplification, which occurs in approximately 20% of newly diagnosed breast cancers, is associated with a significantly shorter survival (Hayes, 2019). The relapse rate was highest among breast cancer expressing HER2. This was most evident before 1992 when anti-HER2 therapies were still undergoing investigation. However, there was significant improvement in outcomes in subsequent years as clinical trials had identified the benefits of adding anti-HER2 therapies. Despite this achievement, the current relapse rate of HER2-positive cancers still remains high especially in the early years after diagnosis (Cossetti, Tyldesley, Speers, Zheng, & Gelmon, 2015).

One of the crucial keywords of breast cancer is heterogeneity. The clinical pathway requires more detailed, more precise, and more dynamics therapeutic approaches. Research has improved our understanding that other elements play important roles in cancer management too. These include immune surveillance, DNA repair, the tumor microenvironment as well as endocrine and neural environment. HER2-positive cells differ from normal ductal cells due to the fact that HER2 amplification may be many times higher. This may be the key reason driving the behavior of the cells toward proliferation and migration (Bates, Derakhshandeh, Jones, & Webb, 2018; Wolf-Yadlin et al., 2006).

The discovery of the HER2 molecules has embarked a series of investigations on the efficacy and safety of different types of anti-HER2 therapies. They include monoclonal antibodies trastuzumab and pertuzumab, small molecules tyrosine kinase inhibitors (TKIs) like lapatinib, neratinib, and pyrotinib, as well as trastuzumab-emtansine (T-DM1). T-DM1 is an antibody-drug conjugate (ADC) covalently linked trastuzumab to a cytotoxic drug called mertansine or DM1 (called emtansine in some forms). They become the bullets at hand that clinicians can use to treat patients with different stages of HER2-positive disease.

Multiple clinical trials such as the HERA, NSABP-B31, NSABP-N9831, and BCIRG-006 showed that adjuvant trastuzumab, whether used concurrently or sequentially with chemotherapy, was associated with an improved outcome. The Cochrane Review in 2012 showed that there were 40% fewer cancer recurrences and 34% fewer deaths after adjuvant trastuzumab was introduced (Moja et al., 2012). These pivotal trials led to the approval of trastuzumab and licensed its use in the adjuvant setting for up to 12 months.

2. Treatment of early HER2-positive breast cancer

The discovery of HER2 as a proto-oncogene led to an explosion of enthusiasm toward targeting the molecule through clinical trials (Table 1) and many have been conducted in the past two decades.

A recent systemic review and meta-analysis of major adjuvant clinical trials on targeting HER2 cancers with trastuzumab shows that there is a good correlation of disease-free survival (DFS) with breast cancer-specific survival (Saad et al., 2019). The surrogate threshold effect, which is the maximum hazard ratio (HR) for DFS that statistically predicts an HR for overall survival (OS) less than 1.00 in a future trial, was found to be ranging from 0.56 to 0.8. This indicates that it is appropriate to continue to use DFS as a surrogate for OS in patients with HER2-positive early breast cancer in trials of adjuvant trastuzumab for up to 1 year.

2.1 The "do more" approach

2.1.1 Extending duration of anti-HER2 treatment

The adjuvant trastuzumab trial (HERA) was an open-label phase III trial randomizing 5102 patients with HER2-positive breast cancers, after completion of all primary therapy (including, surgery, chemotherapy, and radiotherapy), to receive trastuzumab for 1 year (once at 8 mg/kg of bodyweight intravenously, then 6 mg/kg once every 3 weeks) or for 2 years (with the same dose schedule), or to the observation group (Cameron et al., 2017). It was one of the first attempts to adopt the "do more" approach to see whether this will improve the outcome.

After a median follow-up of 11 years, patients assigned to 1 year of trastuzumab had a significantly reduced risk of a DFS event (HR = 0.76; 95% CI, 0.68–0.86) and death (HR = 0.74; 95% CI, 0.64–0.86) compared with observation. Those assigned to 2 years of adjuvant trastuzumab did not have DFS improvement when compared to patients assigned to the 1-year group. Estimates of 10-year DFS rates were 63% for observation, 69% for 1 year of trastuzumab, and 69% for 2 years of trastuzumab. The incidence of cardiac toxicity remained low in all groups and occurred mostly during the treatment phase; 7.3% in the 2-year trastuzumab group, 4.4% in the 1-year trastuzumab group, and 0.9% in the observation group.

The ExteNET study followed the approach of the HERA trial design, but instead of extending the duration of trastuzumab, patients who had completed adjuvant trastuzumab within the previous 2 years or who had residual

Table 1 Selected clinical trials on the "do more" and "do less" approach for early HER2-positive breast cancer.

"Do more"

Extending duration of anti-HER2 treatment	Treating with dual HER2 blockade	Achieving higher pCR rates	Improving treatment for residual invasive disease
• Cameron et al. (2017) (HERA; updated results) • Martin et al. (2017) (ExteNET) • Ejlertsen et al. (2018) (ExteNET; post-hoc)	• von Minckwitz et al. (2017) (APHINITY) • Piccart et al. (2019) (APHINITY; updated results)	• Untch et al. (2010) (GeparQuattro) • Gianni et al. (2012) (NEOSPHERE) • Schneeweiss et al. (2013) (TRYPHAENA) • Van Ramshorst et al. (2017) (TRAIN-2) • Baselga et al. (2012) (NeoALTTO) • Carey et al. (2013) (CALGB-40601) • Park et al. (2016) (I-SPY2) • Masuda et al. (2020) (JBCRG-20/ Neopeaks)	• von Minckwitz et al. (2019) (KATHERINE)

"Do less"

Treating with chemotherapy-sparing regimen	Shortening duration of anti-HER2 treatment	De-escalating chemotherapy in small or less aggressive cancers
• Llombart-Cussac et al. (2017) (PAMELA) • Hurvitz et al. (2019) (KRISTINE) • Bergh et al. (2019) (PREDIX HER2) • Masuda et al. (2020) (JBCRG-20/ Neopeaks)	• Earl et al. (2019) (PERSEPHONE)	• Slamon et al. (2015) (BCIRG 006) • Tolaney et al. (2015) (APT) • Tolaney, Guo, et al. (2019) (APT; updated results) • Tolaney, Trippa, et al. (2019) (ATEMPT)

disease following neoadjuvant treatment received either neratinib or placebo for an additional 1 year (Martin et al., 2017). The primary endpoint was the invasive DFS (iDFS) rate at 2 years. The trial included 2840 patients who were at higher risk of relapse, that is, T2 or greater or N1 or greater. The eligibility criterion was later amended to restrict enrollment to node-positive patients who completed trastuzumab within a year before randomization. At the 5-year follow-up, there was significant improvement in iDFS rate at 2.5% absolute difference (90.2% versus 87.7% in the placebo group; HR = 0.73; 95% CI, 0.57–0.92, $P = 0.0083$).

Unlike other previous trials on dual anti-HER2 therapy, the ExteNET study revealed that the benefit of extended therapy with neratinib was significantly stronger in the hormone receptor-positive subset of patients, with a 4.4% improvement in iDFS rate (HR = 0.60, 95% CI, 0.43–0.83; $P = 0.002$). By contrast, the improvement was negligible for patients with hormone receptor-negative breast cancer. However, in a post-hoc subset analysis, there was iDFS benefit (though statistically insignificant) if patients with hormone receptor-negative breast cancer started neratinib treatment within 6 months of completing trastuzumab therapy (Ejlertsen et al., 2018). The rate of iDFS at 5 years was 88.9% with neratinib versus 86.1% with placebo when the treatment interval was 6 months or less (HR = 0.73; 95% CI, 0.57–0.92; $P = 0.008$). The rate of iDFS at 5 years was 88.7% with neratinib versus 92.7% with placebo when the treatment interval was >6 months (HR = 1.52; 95% CI, 0.82–2.88; $P = 0.181$). These data suggested that the risk of relapse in patients with HER2-positive but hormone receptor-negative disease was greatest during the first 6 months after completing trastuzumab. Whether neratinib may help reduce this risk remains to be further elucidated.

2.1.2 Treating with dual HER2 blockade

Another "do more" approach adopted by the APHINITY trial, which randomly assigned patients with node-positive or high-risk node-negative HER2-positive, operable breast cancer to receive either pertuzumab or placebo added to standard adjuvant chemotherapy plus 1 year of treatment with trastuzumab, showed that pertuzumab significantly improved the rates of iDFS when it was added to trastuzumab and chemotherapy (von Minckwitz et al., 2017). A pre-planned subgroup analysis suggested that patients who were hormone receptor-negative seemed to benefit more than patients who were hormone receptor-positive. The patients with node-positive disease at the start of adjuvant therapy reaped a greater iDFS benefit compared with those who were node-negative. In the cohort of patients with node-positive disease, the 3-year rate of iDFS was significant at 92.0% in the pertuzumab group versus 90.2% in the placebo group, with a HR of 0.77 (95% CI, 0.62–0.96).

An updated analysis of the trial showed that there was still a continual benefit of pertuzumab on iDFS in patients with node-positive disease at a median follow-up of 74.1 months (87.9% versus 83.4% in the placebo group; HR = 0.72; 95% CI: 0.59–0.87) (Piccart et al., 2019). The achievement was remarkable with a possible huge difference on survival outcome 15 years later, with the addition of pertuzumab for the node-positive cohort. They merely highlighted the benefits of the "do more" approach for patients with aggressive HER2-positive cancers.

The ExteNET trial was completed before the APHINITY trial, so none of the patients in the ExteNET trial had received pertuzumab as part of their adjuvant therapy. Thus, future clinical trial could be carried out to evaluate the benefit of additional therapy with neratinib for patients who have had adjuvant therapy with trastuzumab and pertuzumab. Neratinib acts through a completely different mechanism than the HER2-targeted antibodies, it may provide benefit in this setting. Outside of a clinical trial, the optimal treatment approach should be discussed with the patients, taking into consideration of the risks and benefits of each treatment approach. This will enable them to understand their options and decide on their treatment plan.

2.1.3 Achieving higher pCR rates

Clinical researchers also tested the "do more" approach via neoadjuvant anti-HER2 therapeutic trials. It is generally regarded that neoadjuvant therapy should be considered for high tumor volume-to-breast ratio, lymph node-positive disease, patients with high-risk features (high grade and Ki67 values), or young age at diagnosis. The pathologic complete response (pCR) rate after neoadjuvant therapy is regarded as an important surrogate marker of event-free survival (EFS) (Cortazar et al., 2014).

The earlier trials such as the NOAH or TECHNO trials have already identified the impact of adding trastuzumab to chemotherapy on pCR (Gianni et al., 2010; Untch et al., 2011). It was reported that the addition of trastuzumab improved the pCR by a HR of 3.2 ($P < 0.001$) against those without (Untch & von Minckwitz, 2011).

Subsequent trials aimed at increasing the pCR rate by adding more cytotoxic agents to trastuzumab. One such example is the GeparQuattro study when capecitabine (X) was added to a backbone of epirubicin/

cyclophosphamide (EC) and docetaxel (T) (Untch et al., 2010). Though the combination of TX did not improve pCR, the sequential treatment of EC followed by T and then X improved pCR over the backbone arm of EC followed by T. pCR was observed in 32.9%, 31.3%, and 34.6% of patients who were treated with EC-T, EC-TX, and EC-T-X, respectively.

Another "do more" approach to boost pCR rate was to add pertuzumab to trastuzumab. NEOSPHERE study demonstrated that adding pertuzumab to the standard of docetaxel plus trastuzumab increased the total pCR in both breast and node from 21.5% to 39.3% ($P=0.0063$) (Gianni et al., 2012). Of special note is that this trial involves an arm of no chemotherapy with only trastuzumab and pertuzumab, with a pCR rate of 17.7%. Similarly the TRYPHAENA study also demonstrated that dual targeted therapy was better, especially in hormone receptor-negative cohort (Schneeweiss et al., 2013).

The question of whether anthracycline is necessary for neoadjuvant treatment of HER2-positive breast cancer was tested in the TRAIN-2 (BOOG 2012-03) trial (Van Ramshorst et al., 2017). The design compared a carboplatin arm with an epirubicin arm while all patients received both trastuzumab and pertuzumab. There was no significant difference in pCR, but hormone-receptor negative patients had a higher response rate.

Apart from using pertuzumab as a part of the dual targeted therapy, the neoALTTO study tested the efficacy of a tyrosine kinase inhibitor (TKI), lapatinib, with and without combining trastuzumab (Baselga et al., 2012). Lapatinib is a small molecule targeting the intracellular tyrosine kinase domains of the HER2 receptor. There was significant increase in pCR rate for the combination group of paclitaxel, trastuzumab, and lapatinib with an absolute difference of 21.2% ($P=0.0001$). However, such combination did not improve EFS and more patients receiving the combined treatment complained of toxicity, especially diarrhea and hepatotoxicity.

The CALGB-40601 (ALLIANCE) trial had a similar design to neoALTTO (Carey et al., 2013). A total of 305 patients with stage II-III HER2-positive breast cancers were assigned to receive before surgery weekly paclitaxel plus trastuzumab and lapatinib (THL), weekly paclitaxel plus trastuzumab (TH), and weekly paclitaxel and lapatinib (TL). Breast pCR rates were 51%, 40%, and 32%, respectively, but the difference between THL and TH was not significant. The grade or higher toxicities (including neutropenia, rash, and diarrhea) were higher among patients receiving lapatinib. These results were similar to that of the neoALTTO study. It remains controversial whether the additional of lapatinib will improve outcome.

Another TKI, neratinib, was tested via the I-SPY2 approach, which is an adaptive phase II neoadjuvant trial on high risk clinical stage II/III breast cancers (Park et al., 2016). Neratinib was added to standard chemotherapy (paclitaxel for 12 weekly cycles) to evaluate its ability to improve pCR over the same cytotoxic drug plus trastuzumab. All patients will have four cycles of 2- or 3-weekly doxorubicin/cyclophosphamide before surgery. The pCR rate for neratinib arm was 39% versus 23% for the trastuzumab arm. This was especially evident in hormone receptor-negative cancers which were 56% and 33%, respectively.

The most recent evidence in the attempt to boost pCR rate came from the JBCRG-20/Neopeaks trial (Masuda et al., 2020). This was a randomized phase II study that enrolled 204 patients with HER2-positive primary breast cancer, and assigned them in a 1:1:2 ratio to receive six cycles of docetaxel, carboplatin, trastuzumab, and pertuzumab (TCHP; arm A), four cycles of TCHP followed by four cycles of T-DM1 plus pertuzumab (KP; arm B), and four cycles of KP (arm C). Response-guided regimen was implemented for arm C, with responders receiving an additional two cycles of KP (arm C1) and non-responders receiving four cycles of 5-fluorouracil, epirubicin, and cyclophosphamide (FEC; arm C2). Estrogen receptor (ER)-positive patients received concurrent hormone therapy along with KP.

pCR rate was not found to be significantly different between the treatment arms, with arm B achieving numerically higher pCR rate (71%) than arms A (57%) and C (57%). For arm C, the pCR rate was higher in responders (63%) than non-responders who were switched to FEC (38%). Those with IHC3+ in group C (87 out of 101 patients) had a significantly higher pCR rate than those with milder HER2 expression (IHC2+/CISH+; $P<0.01$). Exploratory analysis showed that for ER-positive patients, the pCR rate was significantly higher with arm B (69%) than arms A (43.3%; $P=0.047$) and C (50.8%; $P=0.013$). In ER-negative patients, the pCR rate was comparable between the three arms. These results demonstrated that ER status might be used as an indicator in personalizing dual HER2-targeted therapy. In this regard, four cycles of TCHP followed by four cycles of KP and concurrent hormonal therapy can be considered for ER-positive patients.

2.1.4 Improving treatment for residual invasive disease

Despite the improved pCR rates with the different approaches, a substantial proportion of patients was found to have residual disease at surgery. The impact of the residual cancer burden (RCB) on relapse-free survival can be estimated by a formula suggested by Symmans et al. (2017). The formula takes into account of the proportion of active cancer cells in the residual tumor bed as well as the number and size of the lymph node metastasis, and stratified RCB into three classes. The continuous RCB index scores, which was based on predefined cut points of 1.36 and 3.28, categorized RCB into minimal RCB (RCB-I), moderate RCB (RCB-2), and extensive RCB (RCB-3). The formula confirmed that pCR (RCB = 0) patients had the best outcome whereas RCB-3 was associated with the worst survival. For HER2-positive cancers, the outcome was not good even for RCB-1 cases. In the HER2-positive cohort who received concurrent trastuzumab with sequential paclitaxel and fluorouracil, epirubicin, and cyclophosphamide, estimates of 10-year relapse-free survival rates were 95%, 77%, 47%, and 21% for pCR, RCB-1, RCB-2, and RCB-3, respectively.

CREATE-X study for HER2-negative cancers demonstrated that the addition of capecitabine after surgery for RCB cases improved DFS (Masuda et al., 2017). The KATHERINE trial, a randomized open-label phase III study, was similarly designed to study whether T-DM1 was superior to trastuzumab as adjuvant therapy for HER2-positive cancers (von Minckwitz et al., 2019). A total of 1486 patients with HER2-positive cancers (cT1-4/N0-3/M0) who had residual invasive disease in breast or axillary nodes after neoadjuvant chemotherapy plus HER2-targeted therapy at surgery were randomized to receive 14 cycles of either adjuvant T-DM1 or trastuzumab. Roughly about one-fifth of patients of either group had residual cancer of less than 1 cm and negative axillary node. Regional lymph node metastasis was found in over 50% of patients. The 3-year iDFS was 88.3% in the T-DM1 arm versus 77.0% in the trastuzumab arm, with an 11.3% absolute improvement. The HR of 0.50 (95% CI, 0.39–0.64; $P < 0.001$) implies that the risk of having an event was reduced by half when T-DM1 replaced trastuzumab as adjuvant post-operative therapy. The locoregional recurrence, contralateral breast cancer, and death without previous event were all higher for the trastuzumab arm. There is a clear advantage for patients in the T-DM1 arm compared with the trastuzumab arm in terms of risk of distant recurrence, though the data for OS were not matured at time of analysis. Distant recurrences occurred 50% more frequently in the trastuzumab arm than the T-DM1 arm. Central nervous system (CNS) events, however, were not different: 5.9% in the T-DM1 arm and 4.3% in the trastuzumab arm. There were no unexpected safety issues. All subgroups benefited from T-DM1 compared with trastuzumab, except for when patient numbers were very small (aged 65 years or older, preoperative dual HER2-directed therapy, significant residual disease at surgery). The investigators concluded that adjuvant T-DM1 represents a new standard of care in this population (who have access to this drug) and the results of KATHERINE trial could be practice changing.

2.2 The "do less" approach

2.2.1 Treating with chemotherapy-sparing regimen

The "do less" approach is becoming more important in the clinical practice. The phase II single cohort PAMELA trial adopted this approach by enrolling patients with HER2-enriched cancer and treating them before surgery with trastuzumab and lapatinib without chemotherapy (Llombart-Cussac et al., 2017). If the cancer subtypes were hormone receptor-positive, hormonal treatment with tamoxifen or letrozole was added. Intrinsic tumor subtypes were tested at baseline and day 14 by PAM50. A total of 151 patients were recruited. At the time of surgery, 30% of patients achieved pCR and this increased to 41% for the HER2-enriched subtype. The pCR rate was higher than the no chemotherapy arm of the NEOSPHERE study. This could be related to the better selection of HER2 enriched cancers in PAMELA than NEOSPHERE, as samples were tested by PAM50. However, there may still be bias as this was a phase II study with no comparator arm.

The advent of T-DM1 and its approved indication for metastatic cancers provides an opportunity to test this compound in neoadjuvant trials. The most notable example is the phase III KRISTINE study (Hurvitz et al., 2016). It was designed with an attempt to "do less" by replacing the more toxic chemotherapy with T-DM1. Four hundred and forty-four patients with operable, locally advanced or inflammatory, HER2-positive breast cancer greater than 2 cm were randomized to receive before surgery six cycles of either TCHP (control arm) or KP (study arm). The patients continued to receive 12 cycles of adjuvant anti-HER2 therapy: trastuzumab and pertuzumab in the control arm versus T-DM1 and pertuzumab in the study arm. Adjuvant chemotherapy was permitted and recommended in the study arm for patients with residual disease in the breast (>1 cm) or with lymph node metastasis.

While treatment with KP led to pCR in 44.4% of women, TCHP yielded a significantly higher pCR rate of 55.7% (absolute difference of 11.3 percentage points; 95% CI, −20.5 to −2.0; $P=0.016$). It is especially evident in tumors with IHC2+ staining (21% and 7%, respectively) and IHC3+ staining (61% and 50%, respectively) (de Haas et al., 2016). However, KP had a notably better safety profile than TCHP during the neoadjuvant treatment period, with a lower incidence of grade 3 or more adverse events (13% and 64%, respectively) and lower rate of treatment discontinuation (3% and 8%, respectively). As a result, the quality of life was better and physical functioning could be maintained longer. The 3-year EFS rate was also better for TCHP (94.2%) over KP (85.3%) with a HR of 2.61 (95% CI, 1.36–4.98).

An updated report in 2019 showed that the 3-year iDFS survival curves of TCHP and KP overlapped (92% and 93%, respectively; HR = 1.11; 95% CI, 0.52–2.4) (Hurvitz et al., 2019). Subgroup analysis showed that the curves of both treatment groups continued to overlap in both pCR (97.5% and 96.7%, respectively) and non-pCR patients (84.2% and 89.4%, respectively). The results indicated again that pCR is still a good surrogate marker for survival.

The reason of the overlap could be ascribed to the design of the trial that patients with residual disease were recommended to receive additional adjuvant chemotherapy. Twenty-five percent of the KP groups receive this kind of "rescue" type of chemotherapy. The definition of EFS is different from iDFS as the former includes disease progression and in neoadjuvant trials, it refers to disease progression while on treatment before surgery. Taking this into consideration, analysis was focused on the 15 (out of 223) patients with locoregional progression while being treated by KP. It was found that of the 14 evaluable patients with progression tested for HER2 mRNA expression, 14 (100%) was below the median. In addition, the proportion of IHC2+ and IHC3+ staining for these patients with progression were 66.7% and 33.3%, respectively. In contrast, the proportion of IHC3+ staining in patients without progression was substantially higher (91.3%). Analysis of HER2 homogeneity showed that homogenous staining was found in 20% for those with progression versus 85.1% for those without.

The PREDIX HER2 trial adopted a similar strategy as KRISTINE study on a similar patient group for attempting to identify a new potential "do less" strategy (Bergh et al., 2019). Patients were treated either by docetaxel, trastuzumab subcutaneous, and pertuzumab (arm A, 103 patients) or T-DM1 (arm B, 99 patients) for six cycles. All patients, whether arm A or B, were assessed by PET-CT at baseline and after two and six cycles of treatment. In case of no change or progressive disease, the patients were offered switch of treatment arms. The pCR rates were not different between the two arms (47% for arm A and 45% for arm B). Fourteen out of 18 patients of arm A switched to arm B, mainly because of toxicity whereas 7/9 of arm B switched because of no change or progression. These rates were not very different from those of the KRISTINE study. When analysis was performed on the decrease in SUVmax at 5 weeks after start of treatment (two cycles), patients with more than 81% decrease had a significantly higher pCR rate (74%) than those with no pCR (26%). Those with less than 45% decrease showed a pCR rate of 19% only. Thus, early evaluation during the course of neoadjuvant therapy should be considered and clinicians should motivate to switch therapy to improve outcome for those patients with lack of response to treatment.

As discussed earlier in Section 2.1.3, the JBCRG-20/Neopeaks trial found a significantly higher pCR rate in ER- and HER2-positive patients who received four cycles of TCHP followed by four cycles of KP (Masuda et al., 2020). However, the incidence of adverse events was the lowest in the KP backbone only arm, especially in arm C1 for alopecia and febrile neutropenia (5%; 0%) as compared to arms A (94.1%; 21.6%), B (86.5%; 15.4%); and C2 (81%; 33.3%). Taking into consideration the similar rate of pCR in ER-negative patients (76.2% for arm A, 73.9% for arm B, and 72.2% for arm C1), then this suggested that six cycles of KP, in patients who responded to it, seems to be a reasonable treatment for ER-negative, HER2-positive patients.

These trials therefore provide information for clinicians to consider the "do less" approach in the clinic and systemic chemotherapy may be unnecessary for some patients. Based on the findings from KRISTINE, selected patients could be treated by neoadjuvant KP with an estimated iDFS of about 97% if pCR is achieved. Likewise, patients receiving KP had a 3-year DFS of 94% in the JBCRG-20/Neopeaks trial. These patients can be spared the untoward adverse events of the toxic chemotherapy if the tumors express high levels of HER2 and greater HER2 homogeneity.

2.2.2 Shortening duration of anti-HER2 treatment

The question of 1-year or shorter duration of adjuvant anti-HER2 therapy as standard regimen was partially addressed previously (Trapani, Rajasekar, & Mathew, 2019). The SOLD and the SHORT-HER trials were of similar design. Although there was difference in choice and regimen of chemotherapy, both evaluated the concept of 9 weeks versus up to 1 year of trastuzumab. Other studies such as the PHARE, HORG, and PERSEPHONE trials evaluated 12 months versus 6 months of adjuvant trastuzumab. Except PERSEPHONE, all trials failed to prove non-inferiority of a shorter regimen of adjuvant trastuzumab.

The PERSEPHONE study was a randomized phase III trial with a non-inferiority design to challenge the empirical basis of 12 months of trastuzumab with a 6 months treatment (Earl et al., 2018, 2019). It included 4089 patients and trastuzumab was given either concurrently or sequentially with chemotherapy. Eighty-two percent of the 12-month arm and 90% of the 6-month arm received the correct number of doses. At a median follow-up of 5.4 years, the 4-years disease free survival (DFS) rate was 89.8% for the 12-month arm and 89.4% for the 6-month arm, with a HR of 1.07 (90% CI, 0.93–1.24; non-inferiority $P=0.011$). The HR was still 1.07 even after adjusting for all stratification factors (90% CI, 0.93–1.24; non-inferiority $P=0.0097$). Overall survival (OS) rate at 4 years was 94.8% and 93.8%, respectively (HR = 1.14; 90% CI, 0.95–1.37; non-inferiority $P=0.0010$). Both DFS and OS reached statistically significance for non-inferiority. The pre-defined subgroup analysis showed that 12 months of adjuvant trastuzumab may be better if the cancers were estrogen receptor-negative, when the chemotherapy type was taxane-based, when the treatment was given before surgery, or when trastuzumab was used concurrently with chemotherapy. Regarding cardiotoxicity, this was significantly less in the 6-month arm (4% versus 8%). These exciting results mark the first step to the reduction of treatment duration for many patients and particularly for those with both estrogen receptor- and HER2-positive breast cancers. There would be potential cost and resource savings especially in resource limited countries.

2.2.3 De-escalating chemotherapy in small or less aggressive cancers

De-escalation of chemotherapy is another attempt to adopt the "do less" approach. BCIRG-006 study was designed to test whether fewer cycles of chemotherapy when combined with adjuvant trastuzumab could achieve a similar survival outcome (Slamon et al., 2009). The design was also to compare the arm of doxorubicin/cyclophosphamide followed by docetaxel (AC-TH) with a novel arm of docetaxel/carboplatin (TCH). The study involved 3222 patients and at a median follow-up of 10.3 years, the DFS rate was 74.6% for the AC-TH arm and 73% for the TCH arm, whereas the OS rate was 85.9% and 83.3%, respectively (Slamon et al., 2015).

The compatibility of TCH to AC-TH in these outcomes posed the question whether higher risk HER2-positive cancers should require anthracycline-based therapy. In a subset analysis of node positive patients, the DFS rate was 69.6% for AC-TH group versus 68.4% for the TCH group. Even in patients with more than four metastatic nodes, the DFS rate was comparable between groups at 62.9% and 62.8%, respectively. These findings suggest that the benefit from adding anthracycline to the treatment regimen maybe minimal, even among high-risk patients.

The concern of having anthracycline in the treatment regimen is the incidence of adverse events related to its use. TCH-treated patients had lower incidence of arthralgia, hand-foot syndrome, stomatitis and vomiting as well as lower incidence of grade 3/4 myalgia, sensory/motor neuropathy, and nail changes. Of hematological toxicities, patients on TCH may have a higher chance of grade 3/4 anemia and thrombocytopenia, but they have lower incidence of neutropenia and leukopenia than AC-TH. In addition, cardiac events and acute leukemia occurred more frequently with AC-TH. A total of 21 patients developed grades 3/4 congestive heart failure with anthracyclines versus only four patients with TCH. Also AC-TH patients, though recovered to some degree, had a higher sustained loss of left ventricular ejection fraction (LVEF) greater than 10% from baseline; TCH patients recovered almost completely to baseline. Regarding treatment-related cases of acute leukemia, seven were associated with AC-TH versus one with TCH.

If there is no survival advantage of AC-TH over TCH, even for higher risk patients, the notion of de-escalation of chemotherapy should be examined in patients with smaller, node-negative or less aggressive type of HER2-positive cancers. The larger adjuvant trial on trastuzumab-based therapy did not include that many patients with small or less aggressive cancers. There is practically no standard of care on how to manage such patients. A few research reports, mostly retrospective studies on the outcome of patients with these small HER2-positive breast cancers, suggested that the benefit of polychemotherapy in combination with trastuzumab seemed to be marginal (Barroso-Sousa, Exman, & Tolaney, 2018).

In order to address this issue, the APT trial was designed to evaluate 12 weekly cycles of paclitaxel and trastuzumab followed by 9 months of trastuzumab monotherapy on 410 patients with cancers sized 3 cm or smaller (Tolaney et al., 2015). This was a single arm study with no comparative arm. The 3-year DFS was 98.7%. The 7-year follow up analysis showed five locoregional recurrences, six new primary contralateral breast cancers, four distant recurrences and eight deaths, resulting in a DFS rate of 93.3%, an OS rate of 95% and a recurrence-free interval (RFI) of 97.5% (Tolaney, Guo, et al., 2019). The hormone receptor-positive patients (94.6%) had a better DFS than hormone receptor-negative patients (90.7%), with a HR of 0.61 (95% CI, 0.27–1.4). The biological heterogenous nature of the disease was assessed by PAM50 gene expression analyses to characterize whether the smaller tumors were different from the larger ones. The results on 278 tumor samples showed that the different intrinsic molecular subtypes were not different from the larger cancers (66% HER2-enriched cancers, 14% luminal A cancers, 13% luminal A cancers, and 8% basal-like cancers).

The excellent long-term outcomes of APT trial provides information at the clinic that patients with small HER2 cancers, especially when hormone receptor-positive, can be considered candidates for this "do less" strategy. However this trial has limitation that this is a phase II single arm study. A randomized trial would have been ideal, but there was no standard of treatment that existed at the time of the trial to be taken as the most appropriate control arm. Nonetheless, there will always be patients especially patients with pT1a and some pT1b who may not need chemotherapy besides trastuzumab for their treatment. The balance between survival outcomes and possible toxicities from chemotherapy should be considered.

The ATEMPT trial (TBCRC-033) was designed to build further upon the results of the APT trial, with the concept of total chemotherapy free for the small and less aggressive HER2-positive cancers (Tolaney, Trippa, et al., 2019). This phase II trial included 497 patients with stage 1 HER2-positive breast cancers, N0 or microscopic N1 disease, and left ventricular ejection fraction (LVEF) of 50% or greater. They were randomized on a 3:1 basis within 90 days after surgery to receive either T-DM1 every 3 weeks for 17 cycles of 12 weekly cycles of paclitaxel plus trastuzumab (TH) followed by another 13 cycles of 3-weekly trastuzumab. The co-primary end-points were 3-year DFS in T-DM1 and comparison of incidences of clinically relevant toxicities between T-DM1 and TH. The study was not powered to assess efficacy of TH or to compare efficacy of T-DM1 to TH. It is important to note that within the trial population, 73% of the patients were IHC3+ which is an important issue for consideration of the use of T-DM1.

Regarding the hormone receptor status, 75% of patients were positive, which was higher than the usual 50% among the population of other phase II HER2 trials. This could be related to the inclusion of small tumors in this trial, as 43% of cancers were 1 cm or smaller and 57% were between 1 and 2 cm.

The 3-year DFS rate was 97.7% with T-DM1. There were 10 DFS events including two locoregional recurrences, three new contralateral breast cancers, two distant recurrences, and three non-breast cancer related death. The DFS rate was 97.5% for hormone receptor-positive patients,

98.5% for hormone receptor-negative patient, 98.5% when cancers were 1 cm or less, and 97.1% for cancers between 1 and 2 cm. The RFI was an impressive 99.1%.

The 3-year DFS rate for the TH arm was 92.8%, which was lower than then 98.7% for the same treatment regimen in the APT trial. There were seven DFS events including four locoregional recurrences, one new contralateral breast cancers, and two distant recurrences. There was no death reported in this group.

The clinically relevant toxicities (CRT) were comparable between treatment arms at a rate of 46% for both. The events included grade 4 or higher hematological toxicities, grade 3 or higher non-hematological toxicities and cardiac toxicities. Despite the comparable overall incidence rates, special note should be made that the profiles of toxicity were different between the two groups. Patients treated with T-DM1 were more likely than those on TH to have grade 2 or higher toxicities on thrombocytopenia (11% versus 1%), ALT increase (9% versus 4%), and bilirubin increase (5% versus 1%). TH had a higher neurotoxicity rate of 23% versus 11% for T-DM1. Patients were more likely to discontinue T-DM1 than TH because of toxicity (17% versus 6%). The probability of discontinuation of T-DM1 was 8.2% within 6 months and 10.7% between 6 and 12 months.

The results of the trial provide insight for clinicians to choose treatment regimen for this group of stage 1 HER2-positive cancers. Although T-DM1 was not associated with lower CRT than TH, the low DFS event rates maybe important points to consider to use T-DM1 as an alternative for patients who were afraid of TH toxicities, especially those not captured in the CRT endpoint such as alopecia and other patient-reported outcomes (PRO), which favored T-DM1.

3. Treatment of advanced HER2-positive breast cancer

HER2-positive disease is an important cancer. Patients still present when the cancer is at the advanced stage or when metastasis has occurred. One of the conditions that are still difficult to treat is primary local recurrent HER2-positive disease which is resistant to systemic chemotherapy and all known anti-HER2 therapies. The patients could survive for a long period of time despite the continual growth of cancer locally, which remains a challenge to clinicians.

Recent advances aim at tackling this condition and multiple clinical trials have been conducted aiming to treat the cancer and hopefully to identify a strategy that can improve the outcome. The results are encouraging and various guidelines have been proposed. Most trials have studied the targeted effects on the HER2 molecule by either monoclonal antibodies or TKI. However, the BOLERO-1 and BOLERO-3 studies evaluate mTOR inhibition on the activated PI3K/AKT/mTOR pathway to overcome trastuzumab resistance. The 719 HER2-positive patients recruited in BOLERO-1 study were treated as first-line treatment, with either everolimus or placebo as addition to trastuzumab plus paclitaxel (Hurvitz et al., 2015). The progression-free survival (PFS) was not found to be significantly different between the two groups, but there was 7.2 months prolongation of survival (HR = 0.66; 95% CI, 0.48–0.91; $P = 0.0049$ [protocol-specified threshold of 0.0044 was not crossed]) with the addition of everolimus in the HR-negative population.

The BOLERO-3 tested whether everolimus was effective to restore trastuzumab sensitivity from resistant cases (André et al., 2014). Patients who developed resistance to prior treatment with trastuzumab and taxane were randomly assigned to receive either everolimus or placebo. All patients received additional vinorelbine and trastuzumab. There were 569 HER2-positive patients tested and the median follow up time was 20.2 months at cut-off. The median PFS of the two arms were 7 months with everolimus and 5.78 months with placebo (HR = 0.78; 95% CI, 0.65–0.95; $P = 0.0067$).

3.1 The pivotal trials and the guidelines

The CLEOPATRA trial is a phase III study on 808 patients to receive first-line treatment with trastuzumab and docetaxel, with or without pertuzumab (Swain et al., 2015). The study showed that there was an absolute increase in OS by 15.7 months after a median follow-up of 50 months. The median OS was 56.5 months for the pertuzumab group versus 40.8 months for the control group, with a 32% reduction in mortality (HR = 0.68; 95% CI, 0.56–0.84; $P < 0.001$). The results were regarded as unprecedented.

The results of the trial were updated in June 2019 when the end-of-study analysis was reported (Swain et al., 2019). The median follow-up was 99 months with a maximum of 120 months. Fifty patients crossed over from the control group to receive pertuzumab since July 2012 and they were still counted in the control arm. The median OS was 57.1 months in the pertuzumab/trastuzumab arm in contrast to 40.8 months in the control arm, an absolute reduction of 16.3 months. The 8-year landmark OS rates were

37% and 23%, respectively. The pre-defined subgroup analysis showed that there was an OS benefit with the addition of pertuzumab in patients previously treated with trastuzumab in the adjuvant or neoadjuvant setting. The reduction in mortality was 14%.

MARIANNE study tested T-DM1 alone or in combination with pertuzumab against the standard of taxane plus trastuzumab on HER2-positive locally advanced or metastatic breast cancer patients who received no prior treatment before. The participants must have at least 6 months of disease-free period after prior adjuvant or neoadjuvant systemic treatment. The results showed that neither T-DM1 nor T-DM1 plus pertuzumab was superior to the standard of taxane plus trastuzumab (Ellis et al., 2015).

As per CLEOPATRA trial results, trastuzumab and docetaxel plus pertuzumab was better than trastuzumab and docetaxel alone, and as per MARIANNE trial, T-DM1 was not better than trastuzumab and docetaxel. Therefore, the combination of trastuzumab, docetaxel, and pertuzumab remains the preferred choice of first-line therapy for HER-positive metastatic breast cancer (Modi, 2015).

Such findings were taken into consideration that in July 2018, ASCO published its guidelines on the systemic treatment of HER2-positive advanced breast cancer (Giordano et al., 2018). The recommended first-line therapy is combination of trastuzumab, pertuzumab, and a taxane, unless the patient has a contraindication to taxanes. The second-line treatment is T-DM1. Regarding the third line and beyond treatment, clinicians should offer T-DM1 or pertuzumab if the patient has not received either before. However, for patients who have been exposed to T-DM1 and pertuzumab, there remains an option of lapatinib and capecitabine or other combinations of chemotherapy and trastuzumab, lapatinib, and trastuzumab. There is insufficient evidence to recommend one regimen over another.

Likewise, the ESO-ESMO 4th international consensus guidelines for advanced breast cancer (ABC 4) (Cardoso et al., 2018), which was released in July 2018, recommended dual blockade of trastuzumab and pertuzumab plus chemotherapy (either docetaxel or paclitaxel, with vinorelbine, nab-paclitaxel, or capecitabine as possible options) as standard first-line therapy. Trastuzumab alone or dual blockade with trastuzumab plus pertuzumab or trastuzumab plus lapatinib can be considered in patients who are not suitable for chemotherapy, with long disease-free interval, or with minimal disease burden. If the disease progressed through at least one line of trastuzumab-based therapy, T-DM1 is preferred as second-line therapy. For some patients who progressed to later lines, the combination of trastuzumab and lapatinib is a reasonable option, but the guidelines have insufficient data to recommend this combination after progression to pertuzumab or T-DM1. Trastuzumab in combination with an unused chemotherapy agent can be used as appropriate in later lines of treatment.

NCCN most recently updated its clinical practice guidelines in January 2020 (NCCN, 2020). The combination of pertuzumab, trastuzumab, and either docetaxel or paclitaxel remains the preferred regimen for recurrent or stage IV HER2-positive disease. In the absence of pertuzumab in previous regimen that includes chemotherapy plus trastuzumab, the use of dual blockade with pertuzumab and trastuzumab, with or without cytotoxic therapy, may be considered. Other recommended regimens, which can be considered as the next lines of treatment, include T-DM1, trastuzumab plus chemotherapy (paclitaxel [with or without carboplatin], docetaxel, vinorelbine, or capecitabine), trastuzumab plus lapatinib, lapatinib plus capecitabine, neratinib plus capecitabine, or trastuzumab plus other agents.

3.2 Recent clinical trials on novel therapeutic agents

The newly developed agents for treatment of late-stage HER2-positive disease, especially after THP (the combination of trastuzumab and docetaxel plus pertuzumab) and T-DM1, include TKIs (neratinib, tucatinib, and pyrotinib), new anti-HER2 antibodies (margetuximab, BTRC4017A, MCLA-128, MM-111, ZW25, and PRS343) and new ADCs (deruxtecan, SYD985, ZW49, XMT-1522, AI66, ALT-P7, MEDi1426, ARX788, and RC48).

3.2.1 Tyrosine kinase inhibitors

TKIs are small molecule inhibiting the kinase domain and blocking the downstream signal of the HER2 pathway. The possibility that they can penetrate blood-brain barrier aroused great interest to use them to treat more advanced HER2-positive cancers. The NALA study was one of such trials and it was a pivotal phase III randomized trial on neratinib, a pan-HER inhibitor (Saura et al., 2019). It involved patients who were treated with at least two lines of anti-HER2 therapies for metastatic breast cancer. Only over 30% of patients in both groups were treated with prior trastuzumab, pertuzumab, and T-DM1; and this was lower than expected for third-line conventional treatment. A total of 621 patients were randomized to receive either neratinib and capecitabine or lapatinib and capecitabine (the dose of capecitabine in the neratinib arm was lower). As a precaution against diarrhea, patients were given loperamide with first dose of neratinib and thereafter as needed.

The trial showed a PFS advantage of 2.2 months (8.8 months versus 6.6 months, $P=0.0003$) in favor of neratinib. There was also an OS benefit of 24 months versus 22 months, but the difference was statistically insignificant. The secondary end-points were also in favor of neratinib with an objective response rate (ORR) of 33% versus 27% ($P=0.12$) and a median duration of response (DOR) of 8.5 months versus 5.6 months ($P=0.0004$).

One important point to highlight regarding the NALA data was the time to intervention for CNS metastasis. There was a lower overall cumulative incidence of CNS disease with neratinib (22.8% versus 29.2%; $P=0.043$). Moreover, fewer patients on neratinib required CNS radiation (11% versus 15%). Despite these advantages, the rate of grade 3/4 diarrhea was higher in the neratinib arm (24%) than the lapatinib arm without loperamide prophylaxis (13%). However, treatment discontinuation due to treatment-emergent adverse events (TEAEs) was lower for neratinib (10.9%) than lapatinib (14.5%).

Pyrotinib is another irreversible pan-HER TKI for metastatic breast cancer patients. A recent report on a phase II open-label randomized trial to test its combination with capecitabine showed a better survival outcome than the conventional treatment with lapatinib and capecitabine (Ma et al., 2019). The report included 128 patients who had had prior chemotherapy (taxanes or anthracyclines) and about 54% of them received trastuzumab before (either in the [neo]adjuvant or metastatic setting). The study population included young patients with a median age of 48 years (range from 25 to 70 years). Over 75% of patients had visceral metastasis and about 62% of cancers were hormone receptor-positive disease. Patients in the neratinib arm received more treatment cycle than the lapatinib arm with a median of 20 cycles (range of 2–30) versus 10 cycles (range of 2–28), respectively.

The ORR of the pyrotinib arm was significantly higher than the lapatinib arm (78.5% versus 57.1%; treatment difference of 21.3%; 95% CI, 4.0–38.7%; $P=0.01$). The median PFS of the study group was also significantly longer the control group (18.1 months versus 7.0 months; adjusted HR = 0.36; 95% CI, 0.23–0.58; $P<0.001$).

Despite the survival advantage and the first trial to ORR and PFS benefits, more patients on pyrotinib developed grade 3/4 adverse events (61.3% versus 47.6%). The most common ones include hand-foot syndrome (24.6% versus 20.6%), diarrhea (15.4% versus 4.8%) and neutropenia (9.2% versus 3.2%). In order to validate these results, a phase III trial (PHOEBE) is currently being conducted to evaluate the treatment differences of pyrotinib plus capecitabine versus lapatinib plus capecitabine.

Tucatinib is another TKI, but unlike the previous two, it is a selective HER2 TKI. Because it specifically spares HER1, it is associated with less gastrointestinal and skin toxicities. In a phase Ib study on patients previously treated with different anti-HER therapies for their metastatic disease (56% of patients had CNS metastasis), the combination of tucatinib, capecitabine, and trastuzumab provided an ORR of 61% (42% in CNS metastasis) and a PFS of 7.8 months (6.7 months in CNS metastasis) (Murthy et al., 2018). There were only 33% of patients who developed grade 1/2 diarrhea and there was no grade 3/4 diarrhea.

A phase II randomized trial (HER2CLIMB) was conducted on to evaluate the efficacy of tucatinib (Murthy et al., 2020). It involved patients who had had prior treatment with trastuzumab, pertuzumab, and T-DM1. A total of 612 patients were randomized to receive the combination of tucatinib, trastuzumab, and capecitabine or trastuzumab and capecitabine alone. About half of the patients (48%) in both groups had CNS metastasis, either treated stable disease (about 60%) or untreated/treated but progressing disease (about 40%).

The addition of tucatinib significantly improved the ORR from 22.8% to 40.6% ($P<0.001$). There was also improvement of the median PFS from 5.6 months to 7.8 months (HR $=0.54$; 95% CI, 0.42–0.71; $P<0.001$) and median OS from 17.4 months to 21.9 months (HR $=0.66$; 95% CI, 0.5–0.88; $P<0.005$). The median PFS for patients with CNS metastasis improved from 5.4 months to 7.6 months (HR $=0.48$; 95% CI, 0.34–0.69; $P<0.001$). At 1 year, 25% of the tucatinib arm did not have progression whereas all patients of the control arm had progression. The most common grade 3/4 adverse event was diarrhea (13% versus 9%). The other TEAE included a higher incidence of deranged liver function test and hand-foot syndrome. Basing on these results, the phase III randomized trial called HER2CLIMB-02 has been designed to test the same types of patients with tucatinib plus T-DM1 versus T-DM1 alone.

3.2.2 New anti-HER2 monoclonal antibody

Margetuximab is the most notable example of this new class of anti-HER2 monoclonal antibody. It has similar Fab fragment with similar specificity and affinity as trastuzumab to bind to the HER2 molecule. Similarly it could also bind and activate immune cells. However, the Fc fragment of margetuximab is modified and engineered to stimulate effectors function. It has four- to fivefolds greater affinity than trastuzumab to bind to the activating Fc receptors (CD16A) of the natural killer cells but it has about sevenfolds lower affinity to bind to the inhibitory Fc receptors (CD32B) of the antigen-presenting cells, thus mounting a greater immune response (both innate and adaptive) against cancer cells. The increased CD16A engagement leads to more potent antibody-dependent cellular cytotoxicity stimulation, resulting in release of granzymes and perforins for cancer cell destruction. The reduced CD32B binding enhances APC and T-cells activities, resulting in increased immune activation, with more HER2 specific lymphocyte proliferation and antibodies production. There are F and V variants of the CD16A receptors, with similar global distribution but different affinity for the IgG Fc fragment. A greater proportion of the population is associated with either F/F (47–54%) or F/V (35–44%) genotypes compared to the V/V (9–11%) genotype. The V allotype has a higher affinity for IgG1 Fc fragment of trastuzumab and thus the V/V NK cells bind more IgG1 than F/F NK cells, leading to a better ADCC activation. This difference in CD16A biology was shown in the NSABP-B31 trial that it impacts trastuzumab outcome. The V/V genotype had a better DFS (HR = 0.118) than the F/V (HR = 0.336) and F/F (HR = 0.713) genotypes.

The efficacy of margetuximab was tested in the SOPHIA trial (Rugo, Im, Cardoso, et al., 2019; Rugo, Im, Wright, et al., 2019). This was a randomized phase III study involving 536 patients who underwent at least two lines of prior anti-HER2 therapies including pertuzumab and one to three prior treatment lines in the metastatic setting. Prior CNS metastasis was allowed provided it was treated and stable. The patients received 3-weekly cycle of margetuximab plus standard chemotherapy or 3-weekly cycle of trastuzumab plus chemotherapy. The chemotherapy was of investigator's choice and included capecitabine, eribulin, gemcitabine or vinorelbine. Over 65% of patients were treated with at least two lines of anti-HER2 therapies. All of them received trastuzumab and pertuzumab; over 90% were treated with T-DM1.

The primary PFS by central blinded analysis at the October 2018 cutoff showed that there was an improvement from 4.9 months in the trastuzumab group to 5.8 months in the margetuximab group (HR = 0.76; 95% CI, 0.59–0.98; $P = 0.033$). An update of the investigator-assessed PFS at September 2019 cutoff showed that it was similar to the primary PFS analysis, with continual significant improvement seen in the magetuximab group (HR = 0.71; 95% CI, 0.58–0.86; $P = 0.0006$), though the PFS curves tended to plateau off with time. The updated analysis also showed significantly higher ORR and clinical benefit rate (CBR) in the margetuximab group (ORR of 25.2% versus 13.7%, respectively; $P = 0.0006$ and CBR of 48.1% versus 35.6%, respectively; $P=0.0025$). The first and second interim OS analyses in October 18 and September 19, respectively, showed similar insignificant trend of benefit in the margetuximab group. The results of the final analysis will be available when over 385 OS events occur. Subgroup analysis of the second interim OS data on the F/F and F/V genotypes, at a median follow up of 15.6 months, showed a trend favoring mragetuximab with a median difference of 4.3 months (HR $=0.79$; $P=0.087$).

The V/V homozygotes did not appear to benefit from margetuximab compared to trastuzumab. However, the results should be interpreted with caution as the sample size of the V/V subgroup was small and the clinical characteristics were imbalanced between the two treatment groups; the margetuximab group consisted of older patients with more visceral or CNS metastases. The incidence of adverse events was similar between the two groups with more infusion related reactions of all grades among the margetuximab patients (13.3% versus 3.4%). The SOPHIA trial demonstrated that margetuximab is a safe monoclonal anti-HER2 antibody, which has potential for patients with late staged HER2-positive metastatic breast cancers refractory to trastuzumab, especially when they carry the F/F and F/V genotypes.

3.2.3 New antibody-drug conjugates

ADC comprises of a monoclonal antibody (for antigen specificity) and a payload (for potent cytotoxicity) connected via a linker in a single molecule. The success of T-DM1 on HER2 cancers has sparked development of new linker technologies and novel payloads. ADC delivers anticancer drug efficiently as the monoclonal antibody provides an efficient targeted delivery of the drug. When the complex is internalized, it is degraded by lysosomes or the linker is cleaved, releasing the active cytotoxic component to kill cancer cell. The new ADC can also produce the by-stander effect as the released payloads can be membrane permeable and therefore can penetrate and kill neighboring cancer cells that do not carry the antigen (Trail, Dubowchik, & Lowinger, 2018).

One of such novel ADC, known as DS-8201 or trastuzumab deruxtecan, has been tested in laboratory and early phase clinical trials. It consists of three major components: a humanized anti-HER2 IgG monoclonal antibody with the same amino acid sequence as trastuzumab, a topoisomerase I inhibitor payload which is an exatecan derivative, and a tetrapeptide-based enzymatic cleavable linker. This ADC has a high drug-to-antibody ratio of eight and a stable linker which can be cleaved by enzymes selectively expressed in cancer cells. The payload being a topoisomerase I inhibitor is a highly potent novel agent for breast cancers. It is membrane permeable and thus can exert the by-stander effect. It has a short half-life and thus can minimize the toxicity on normal cells (Nakada, Sugihara, Jikoh, Abe, & Agatsuma, 2019; Ogitani et al., 2016; Trail et al., 2018).

When DS8201 was tested in patient-derived xenograft models of breast and gastric cancers, it was found to have anti-tumor activity in HER2 low expressing cancers, which were IHC1+ or IHC2+ cancers (Ogitani et al., 2016). These tumors are usually regarded as HER2-negative and usually will not be treated by anti-HER2 therapies as they are not effective for these patients. In a phase I study, 115 HER2-positive breast cancer patients previously treated with T-DM1 received trastuzumab deruxtecan (Tamura et al., 2019). There was a confirmed ORR of 59.5% and the most common TEAEs were low grade gastrointestinal and hematologic events. The most important identified risk was interstitial lung disease (ILD) which occurred in 10% of patients. In another phase I study, trastuzumab deruxtecan was also found to have activity on HER2 low cancers (Modi et al., 2019).

A phase II study, the DESTINY-Breast01, was conducted to confirm the observed outcomes in the phase I study and to justify the recommended dose (Modi et al., 2020). The design was similar to the phase I study and tested patients with prior T-DM1 treatment. Patients with stable, treated brain metastasis were allowed but patients with history of significant ILD were excluded. The first part of the study included the pharmacokinetic stage and dose-finding stage. This part of the study identified 5.4 mg/kg as the recommended dose.

The second part of the study, the continuation stage, used this established dose on 130 patients with cancers resistant/refractory to T-DM1. Four additional patients who were intolerant to T-DM1 also participated in this stage of the study. The protocol specified analysis included all part I and part II patients ($n = 184$) treated with trastuzumab deruxtecan at dose of 5.4 mg/kg. Among these patients, 52.7% were hormone receptor-positive, 15.2% were IHC2+, 91.8% had visceral disease, and 13% has brain metastases.

The ORR by independent review was 60.9%, with 6% achieving complete response. The CBR at 6 months was 76.1% with a median DOR of 14.8 months. The median time to response was 1.6 months. The disease control rate was 97% with less than 2% progression. The PFS was 16.4 months at a median follow up of 11.1 months. The median OS was not reached at time for cut-off analysis.

The most common TEAEs were nausea, vomiting and alopecia and they were mostly low grades. However, grade 3 or higher TEAE occurred in 57.1% of patients, TEAE associated with discontinuation occurred in 25.2% of patients and TEAS associated with death occurred in 4.9% of patients. Twenty-five (13.6%) patients developed ILD at a median time to onset of 193 days and four of them died. There was no reported event of cardiac failure with decrease in left ventricular ejection fraction. These data suggested that trastuzumab deruxtecan could become a potential new standard of care for patients previously treated with T-DM1. Several clinical trials are currently underway to test the efficacy of this compound against T-DM1 in HER2-positive cancers (DESTINY-Breast02 and DESTINY-Breast03) and against chemotherapy in HER2 low cancers (DESTINY-Breast04).

Another investigational ADC called RC48 has been developed for HER2-positive cancers. It also consists of a humanized anti-HER2 antibody and a cleavable linker but the active conjugate is monomethyl auristatin E (MMAE). In a phase II study of 43 patients, a clinically meaningful ORR of 60.5% was demonstrated on metastatic urothelial cancers, including those refractory to immunotherapy (Sheng et al., 2019). The ADC was also evaluated for patients with metastatic HER2-positive breast cancers (Xu et al., 2019). Two phase I studies (C001 CANCER and C003 CANCER) showed that RC48 was well tolerated in patients with breast cancer. Pooled analysis of the data of these trials ($n = 70$) showed that this ADC was effective with a confirmed ORR of 31.4% and CBR of 38.6%. The median PFS was 5.8 months and the 6-month PFS rate was 47.5%. Further studies are designed to evaluate the efficacy of RC48 against the combination of lapatinib and capecitabine in HER2-positive metastatic breast cancers and a phase III study is being planned to test for low HER2 expression cancers.

3.3 Treatment of hormone receptor- and HER2-positive cancers

There is no standard of treatment for hormone receptor- and HER2 receptor-positive disease. While it is believed that most patients will receive chemotherapy plus anti-HER2 therapy, there remains a group of patients that will be treated with hormone therapy and/or anti-HER2 therapy. These patients are usually of advanced age with co-morbidities, relatively early stage 4 disease with minimal visceral metastasis. A recent study on the real world treatment patterns and outcome analyzed data of National

Cancer Database patients treated between 2010 and 2015 (Statler et al., 2019). Of the 6234 patients analyzed, 60.5% received hormone therapy and the rest received chemotherapy. Less than half of all patients (42.4%) received anti-HER2 therapy. The median OS for all patients was 44.4 months and the 5-year OS rate was 40%. Patients treated with combination therapy (chemotherapy or hormone therapy plus anti-HER2 therapy) had lower death rates than monotherapy. The 5-year OS rate was 47.5% for hormone therapy plus anti-HER2 therapy, 39.8% for chemotherapy plus anti-HER2 therapy, 38.5% for hormone therapy and 36.3% for chemotherapy ($P<0.001$). Multivariate analysis confirmed the superior OS rate of hormone plus anti-HER2 therapies over chemotherapy plus anti-HER2 therapy (HR = 0.74; 95% CI, 0.61–0.91; $P=0.004$). This study suggested that hormone therapy plus anti-HER2 therapy should be the preferred treatment of choice for hormone receptor- and HER2-receptor positive cancers.

Since chemotherapy may not be beneficial for hormone receptor- and HER2 receptor-positive patients, de-escalation strategies were explored in the ADAPT (Adjuvant Dynamic Marker-Adjusted Personalized Therapy) trial of the West German Study Group. It was a phase II prospective randomized neoadjuvant trial trying to compare pCR rates between T-DM1 and trastuzumab, with or without endocrine therapy (Harbeck et al., 2017).

The trial involved 375 patients who were randomly assigned to receive 12 weeks of T-DM1 alone or T-DM1 with endocrine therapy (tamoxifen for premenopausal women and aromatase inhibitor for postmenopausal women) or to receive trastuzumab with endocrine therapy. Besides pCR as the primary endpoint, the trial also included assessment of early response (proliferation decrease in Ki67 level by 30% or more or cellularity response).

Rates of near pCR as defined by ypT1a or ypT0/is were 52.9%, 52.9%, and 19.3%, respectively, for the three groups. Rates of total pCR as defined by ypT0ypN0 were 32.5%, 34.1% and 10.2%, respectively. Exploratory analyses of menopausal status with pCR rates showed that postmenopausal women had higher though insignificant pCR rates. The postmenopausal versus premenopausal pCR rates were 44.1% versus 37.9% for T-DM1 alone, 45.0% versus 38.1% for T-DM1 plus endocrine therapy, and 16.7% versus 13.6% for trastuzumab plus endocrine therapy. Early response was achieved in 67.1% of patients. Among these early responders, 35.7% achieved pCR ($P=0.005$), whereas among the non-responders, only 19.8% achieved pCR. The responders of the T-DM1 plus endocrine therapy group had the highest pCR rate of 47.4%. The pCR rate was lowest among non-responders treated by trastuzumab and endocrine therapy. The study results suggest that T-DM1 with or without endocrine therapy may be an option for hormone receptor- and HER2 receptor-positive patients, especially when they are menopaused.

Moreover, as recently shown by the JBCRG-20/Neopeaks trial (Masuda et al., 2020), a remarkably high pCR rate of 69% was achieved by four cycles TCHP followed by four cycles of KP with concurrent hormonal therapy. The authors hypothesized that HER2-positive cancer cells or clones that persisted after completion of the TCHP regimen could have responded to the subsequent treatment with KP. Further investigation is required to confirm this, but this treatment approach could perhaps be considered in selected group of patients.

In-vitro studies suggested that cyclin-dependent kinase inhibitor (CDKi) could be effective against HER2-positive cell lines, suggesting that there might be cross talks between the HER2 pathway and CDK4/6 signaling pathway.

The PATRICIA (SOLTI-1303) phase II trial attempted to address this issue by evaluating the combination of palbociclib and trastuzumab in HER2-positive metastatic breast cancers (Ciruelos et al., 2019). A total of 45 patients were recruited including 15 patients in each cohort of ER-negative (cohort A), ER-positive (cohort B1), and ER-positive with letrozole (cohort B2). The PFS rates at 6 months for cohorts A, B1, and B2 were 33.3%, 40%, and 53.3%, respectively. Luminal disease as defined by PAM50 had a higher median PFS compared to non-luminal disease (12.4 months versus 4.1 months, HR=0.37, $P=0.052$). The corresponding CBRs were 73% and 31%, respectively ($P=0.031$). The most common grade 3–4 toxicities were neutropenia (80%) and thrombocytopenia (17%). Dose reductions were required in 60% of patients. The PATRICIA trial shows that the combination of palbociclib and trastuzumab is worth further evaluation of treatment efficacy in luminal disease in a larger cohort of patients.

Pre-clinical xenograft studies using trastuzumab-sensitive and trastuzumab-acquired resistant cell lines showed that abemaciclib in combination with trastuzumab and tamoxifen had significant reduction in cancer growth (O'Brien et al., 2018). The monarcHER trial was a prospective study designed to address clinically whether this combination with abemaciclib worked (Tolaney, Wardley, et al., 2019). It was a phase II randomized trial on 237 patients treated with two or more prior lines of anti-HER2 therapies, of which prior treatment with T-DM1 and taxane was required. The patients were randomized to receive abemaciclib plus trastuzumab and fulvestrant (arm A), abemaciclib plus fulvestrant (arm B), and trastuzumab plus investigator's choice of chemotherapy (arm C). About 50% of these heavily pre-treated patients in each arm had visceral disease. The median PFS was 8.32, 5.65, and 5.69 months, respectively, for arms A, B, and C. There was significant improvement between arms A and C (HR = 0.673, $P = 0.0506$), but there was no PFS benefit observed between arms B and C. The confirmed best ORRs were 32.9%, 13.9% and 13.9% for the corresponding arms of A, B, and C. The median DOR was 12.5 months and 9.5 months for arms A and B, respectively, but it was not reached for arm C. This study suggested that the triplet therapy was better than chemotherapy plus trastuzumab and patients could be spared the untoward TEAEs related to chemotherapy.

There are other studies ongoing to further explore the values of combining CDKi with anti-HER2 therapy. The PATRICIA 2 trial is designed to be a phase II trial randomizing patients to receive the palbociclib, trastuzumab and hormone therapy combination or trastuzumab plus chemotherapy or T-DM1. The PATINA trial is a phase III study designed to test the efficacy of combination of palbociclib in a backbone of trastuzumab/pertuzumab plus aromatase inhibitor or fulvestrant. New trial has yet to be designed to test the combination of CDKi with ADC.

Other exploratory study with another combination of hormonal therapy and anti-HER2 therapy had been reported. Alpelisib, a PI3K-alpha specific inhibitor, was combined with T-DM1 in a phase I study on metastatic HER2-positive cancers resistant to trastuzumab-based therapy (Jain et al., 2018). The ORR was 43%, the CBR was 71%, and the median PFS was 8.1 months. The concept of combining anti-hormonal and anti-HER2 therapies for hormone receptor- and HER2 receptor-positive breast cancers is worth further exploration to improve outcome and quality of health.

3.4 New exploratory strategy of combining anti-HER2 therapy with immunotherapy

The PANACEA trial involved the use of pembrolizumab in trastuzumab-resistant breast cancers (Loi et al., 2019). It was a single arm phase Ib-2 trial to address the issue of immune mechanism mediating trastuzumab resistance. The treatment involved adding pembrolizumab, a programmed cell death protein I (PD-1) inhibitor, to trastuzumab. In the phase 2 part of the study, 52 patients with both programmed cell death 1 ligand 1 (PD-L1)-positive and -negative were recruited. They received a flat dose of pembrolizumab of 200 mg every 3 weeks plus standard trastuzumab. The median follow up was 13.6 months for PD-L1-positive patients and 12.2 months for PD-L1-negative patients. The ORRs were 15% and 0% for the PD-L1-positive and -negative groups, respectively. The median PFS was not significantly different between the two groups (2.7 months versus and 2.5 months, respectively), but there was a durable plateau for the PD-L1-positive group. Median OS was not reached for the PD-L1-positive group, whereas it was 7.0 months for the PD-L1-negative group. The most commonly occurring serious adverse events were dyspnoea (5%), pneumonitis (5%), pericardial effusion (3%), and upper respiratory infection (3%). There was one treatment-related death due to Lambert-Eaton syndrome in a PD-L1-negative patient.

The KATE2 trial is another attempt to evaluate immunotherapy on HER2-positive breast cancers (Emens et al., 2019a). The treatment targeted against the programmed death ligand 1 (PD-L1) which was often expressed on tumor-infiltrating immune cells. This trial was the first phase II study randomizing patients in a 2:1 fashion to receive T-DM1 plus atezolizumab (a PD-L1 inhibitor) or T-DM1 alone. The recruited patients were treated with prior trastuzumab and a taxane and progressed while on treatment for their metastatic disease or within 6 months of adjuvant therapy. Over 45% in both groups received prior pertuzumab. Although PFS and ORR were numerically higher with the addition of atezolizumab to T-DM1, the combination resulted in a higher incidence of TEAEs and did not provide any clinically significant PFS benefit (8.2 months for T-DM1 plus atezolizumab versus 6.8 months for T-DM1 alone; HR = 0.82; 95% CI, 0.55–1.23).

An updated report on OS and safety data showed that there might be a possible OS benefit by adding atezolizumab to T-DM1 in PD-L1-positive patients (Emens et al., 2019b). The 1-year OS was 89% in both groups. However, in the PD-L1-positive subgroup, the 1-year OS was higher among patients receiving the combination treatment (94.3% versus 87.5%; HR = 0.55; 95% CI, 0.43–1.8). On the other hand, the 1-year OS rates in the PD-L1 subgroup were similar (85.1% and 89.7%, respectively, HR = 0.88). In terms of safety profile, any adverse events occurred in 29% of patients on atezolizumab versus 15% on placebo. Thrombocytopenia, deranged liver enzymes, anemia, and fever were the more common adverse events seen among patients treated with atezolizumab.

3.5 Does surgery have a role to play?

Metastatic breast cancer patients live longer with improvement in systemic treatment and radiotherapy/tomotherapy for metastatic sites. The survival rate at 5 years has increased from 10% in 1970 to about 40% in women treated after 1995 (Giordano et al., 2004). Clinicians have to address the issue whether or not to remove the primary cancer in the breast at some stage, especially for those who have responded favorably to treatment previously usually regarded as palliative.

Two randomized controlled trials (RCTs) attempted to provide answer to the questions. The Indian study included women with metastatic breast cancer with objective clinical response of 50% or more to first line chemotherapy (Badwe et al., 2015). The patients were stratified by sites of metastases (visceral only, bone only, or visceral plus bone), number of metastases ($\leq$3 or $>$3), and hormone receptor expression status (estrogen/progesterone receptor-positive or both receptors negative), and then randomized to receive breast surgery plus further systemic treatment or systemic treatment alone. A total of 716 patients were recruited and of the 440 patients who responded to systemic treatment, 173 were randomized to receive breast surgery and 177 were not. The median OS was 19.2 months in the surgical treatment group and 20.5 months in the non-surgical treatment group (HR = 1.04; 95% CI, 0.81–1.34) and the 2-year OS was 41.9% and 43%, respectively. There was only one adverse event, wound infection, in the surgical treatment group. The study concluded that surgical treatment did not improve OS in patients with metastatic breast cancers and the procedure should not be offered as part of the routine clinical practice.

The Turkish study was another RCT to address this issue (Soran et al., 2018). Of the 274 patients randomized, 138 received surgery plus systemic therapy and 136 received systemic treatment alone. The median OS was 46 months for the combination group and 37 months for the group with systemic treatment alone (HR = 0.66; 95% CI, 0.49–0.88; P = 0.005). The 5 year survival rate was 41.6% and 24.4%, respectively. The survival advantage was more evident among patients with hormone receptor positive cancers (HR = 0.64; 95% CI, 0.46–0.91; P = 0.01), HER2 negative cancers (HR = 0.64; 95% CI, 0.45–0.91; 0.01), patients younger than 55 years (HR = 0.57; 95% CI, 0.38–0.86; P = 0.007) or patients with solitary bone-only metastases (HR = 0.47; 95% CI, 0.23–0.98; P = 0.04).

The contradicting results of the two RCTs could be related to the differences in the design of the studies. The Indian study assigned patients who responded favorably after systemic treatment to surgery or none, whereas patients of the Turkish study received upfront surgery followed by chemotherapy or upfront chemotherapy. About 95% of the patients of the Indian study versus 80% of the Turkish study received anthracycline. For HER2-positive cancers, trastuzumab was given only to less than 10% of patients of the Indian study but all patients of the Turkish study received trastuzumab. Whether surgery could improve survival outcome remained to be tested.

A recent retrospective study attempted to address this question by reviewing the records of 3231 patients in the National Cancer Database from 2010 (when mandatory HER2 reporting began) to 2012 (Mudgway et al., 2019). Surgical removal of the primary tumor and Cox proportional overall mortality HR were assessed. Propensity scores were matched to diminish the selection bias and adjusted for demographic, tumor, and treatment variables.

Twenty-five percent of patients had bone metastasis alone. The systemic treatment included chemotherapy/targeted therapy in 89.4%, endocrine therapy in 37.7% and radiation in 31.8%. Primary site surgery was performed for 1130 (35%) patients and the other 2101 (65%) patients did not receive any form of surgery. The medial follow-up was 21.2 months (range 0–52).

Propensity score analysis showed that surgery was associated with a 44% increased chance of survival, assuming that all patients also received systemic therapy. The median overall survival increased from 18 months to 25 months. The overall mortality HR was significantly associated with insurance (Medicare/other government versus none/Medicaid; HR = 0.36; $P < 0.0001$), chemotherapy/targeted therapy (HR = 0.76; $P = 0.008$), radiation therapy (HR = 1.33; $P = 0.0009$), ethnicity (hispanic/black versus white; HR = 1.39; $P = 0.002$) and income (lowest versus highest; HR = 1.36; $P = 0.01$). Although surgery might provide a better survival outcome for this group of patients, this was a retrospective study and potential bias toward an improved outcome could occur as doctors might be more willing to offer surgery to women who were healthier, younger, and with good response to systemic therapy.

4. Points of consideration on prognosis and choice of therapies

4.1 HER2 gene copy number

The impact of HER2 copy number has been shown to be an important factor on outcome and trastuzumab efficiency in the adjuvant setting (Borley et al., 2014). This was a retrospective study to analyze survival data of patients diagnosed with HER2-positive early breast cancer. The data on immunohistochemistry (IHC) and fluorescence in situ hybridization (FISH) testing on HER2 status were extracted from a cancer network database. HER2-positive was defined as IHC3+, or IHC2+, and

FISH-amplified with a HER2/CEP17 ratio > 2.0. High HER2 copy number (>12) appeared to be associated with consistently better response compared to patients with intermediate HER2 copy numbers (6–12). The authors concluded that HER2 gene copy number value should be included as an additional indicator for stratifying both the management and the follow-up of breast cancer patients, especially when emerging data are showing its sensitivity to trastuzumab therapy.

Another study to test the interplay between HER2 copy number and its influence on survival was carried out on 117 patients who underwent adjuvant chemotherapy and trastuzumab therapy after surgery (Adamczyk et al., 2018). Tissue samples of the formalin-fixed paraffin-embedded (FFPE) materials were tested by FISH and immunohistochemistry. The FISH test could assess HER2 gene amplification, HER2 gene copy number, and polysomy. The definition of HER2 amplification was HER2/CEP17 ratio of two or greater, that of high HER2 gene copy number was six signals per cell or more, and positive polysomy was when the average number of signals from the centromere of chromosome 17 per cell was three or more. IHC was performed on EGFR, HER3, HER4, and other proteins besides HER2.

The results showed that overexpression of HER2 protein (IHC3+) and ER negativity corresponded to higher HER2 copy number and HER2/CEP17 ratio. Tumors with polysomy of chromosome 17 were characterized by higher HER2 gene copy number but lower HER2/CEP17p ratio. Patients with tumors featuring HER3 immuno-negativity or low HER2/CEP17 ratio were characterized by 100% metastasis-free survival. The presence of both unfavorable factors (HER3 expression and high HER2/CEP17 ratio) was associated with a worse prognosis.

In the metastatic setting, HER2 (and also PTPN2) copy numbers were found to be prognostic factors as well in HER2-positive metastatic breast cancer treated with trastuzumab (Ellegård et al., 2019). The researchers used droplet digital polymerase chain reaction on genomic DNA extracted from FFPE tissue samples and immunohistochemistry for analysis on the gene copy number variation and protein expression of several components of the PI3K/AKT pathway. Of the 46 patients treated with trastuzumab for HER2-positive metastatic breast cancer, those with tumors displaying a HER2 amplification level of six or more copies had a significantly improved OS (HR = 0.4; 95% CI, 0.2–0.9) and PFS (HR = 0.3; 95% CI, 0.1–0.7) compared with patients with tumors harboring fewer HER2 copies. Copy gain ($\geq$3 copies) of the gene encoding the tyrosine phosphatase PTPN2 was associated with a shorter OS (HR = 2.0; 95% CI, 1.0–4.0) and shorter PFS (HR = 2.1; 95% CI, 1.0–4.1). The results suggested that high ERBB2 amplification level is a potential positive prognostic factor in trastuzumab-treated HER2-positive metastatic breast cancer, whereas PTPN2 gain is a potential negative prognostic factor.

One of the important highlights of this study was the association of high HER2 amplification level with the development of CNS metastases during palliative treatment. While patients demonstrating favorable extra-cranial clinical response to systemic trastuzumab, clinicians should pay attention to new development of brain metastasis, especially when headache, vomiting, and other neural related symptoms develop.

Plasma HER2 gene copy number (GCN) at baseline or after trastuzumab treatment as a prognostic factor has been retrospectively studied in HER2-positive metastatic breast cancer patients using next-generation sequencing (NGS) (Ran et al., 2018). In this study, plasma HER2 GCN of ≤2.22, >2.22–4.0, and ≥4.0 were predefined cut-off values that were used to classify 50 metastatic breast cancer patients into HER2 non-amplified, low, or high HER2 GCN group. A higher concordance of HER2 status between plasma and metastatic was reported, with a significantly higher median HER2 GCN in patients who had more metastatic diseases (≥3 sites, $P < 0.001$).

In patients who responded to trastuzumab, a more significant decrease of HER2 GCN was reported ($P = 0.037$). However, the ORR was not found to be different when assessed according to plasma HER2 status. Those with high HER2 GCN, more metastases (≥3 sites), and without anti-HER2 therapy had shorter OS (all with $P < 0.05$).

4.2 HER2 heterogeneity

Molecular heterogeneity was tested on pre-treatment and post-treatment tissue samples of CALGB-40601 patients who received neoadjuvant weekly paclitaxel plus trastuzumab with or without lapatinib (Carey et al., 2016). Gene expression profiles were extensively studied by mRNA sequencing (mRNAseq). Six-level subtype classification was generated (basal-like, luminal A, luminal B, HER2 enriched, Claudin-low and normal-like). The distribution of these six pre-treatment intrinsic subtypes was 31% HER2-enriched, 30% luminal A, 30% luminal B, 5% basal-like, 1% Claudin-low, and 2% normal-like. There were more luminal A and B subtypes (83%) when the hormone receptor was positive and more HER2-enriched subtype (51%) when the hormone receptor was negative.

The inter-tumoral heterogeneity traits affected pCR: HER2-enriched had the highest pCR rates (70%) compared with luminal A (34%) and luminal B (36%), regardless of treatment arms or hormone receptor status.

The post-treatment residual samples were also analyzed. The results were different from the pre-treatment cohort with more luminal subtypes (63%) than HER2-enriched subtype (27%). Moreover, the genomic analysis showed that the cancers switched mostly to luminal A. Twelve of 17 (71%) luminal B and six of nine (67%) HER2-enriched cancers switched to luminal A subtype after treatment and this switch was found to be related to tumor heterogeneity rather than to effect from treatment.

The KRISTINE study showed that HER2 heterogeneity is an important marker of pCR for T-DM1 in the neoadjuvant setting. As discussed in the earlier section, patients with lower HER2 mRNA expression had a higher occurrence of locoregional progression while on treatment. Moreover, homogenous staining was found in 20% of progression patients whereas it was 85% for non-progression patients. Thus, patients with less heterogenous HER2 amplified cancers responded better.

The issue of HER2 heterogeneity on pCR was further evaluated in a prospective neoadjuvant clinical trial (Filho et al., 2019). This is a phase II single arm study involving 168 patients receiving six cycles of T-DM1 plus pertuzumab. Image-guided core biopsies on two different sites within each tumor were performed for assessment of intra-tumoral heterogenous HER2 amplification. Heterogeneity was defined as either HER2 positivity by FISH in between 5% and 50% of tumor cells (as per CAP guideline) in one site or presence of HER2 negative in another. Eighty-four percent of patients had stage II disease with a median tumor size of 2.8 cm; 68% of tumors were hormone receptor-positive and 74% were IHC3+. Sixteen of 157 (10%) evaluable cases were classified as HER2 heterogenous and 13 (81%) of them were hormone receptor-positive. pCR rates were 55% in the non-heterogenous subset whereas none of the patients with HER2 heterogeneity achieved pCR. This apparently echoes the finding of the KRISTINE trial. Intra-tumoral HER2 heterogeneity may need to be considered in selection of HER2 targeted therapies without chemotherapy with a curative intent. These cancers may need additional chemotherapy besides classic anti-HER2 therapy.

4.3 Tissue biomarkers

4.3.1 Genomic markers: TOP2A, PTEN/PI3K/AKT/mTOR, MYC, and more

Biomarkers predicting chemotherapy sensitivity was looked at to see if they also have the same prediction for chemotherapy and trastuzumab combination. Topoisomerase II-alpha (TOP2A) was found to be a predictive biomarker for anthracycline chemotherapy (Press et al., 2011). It was tested on the BCIRG-006 samples and the results showed that patients with cancers having TOP2A gene amplification had a significantly better DFS and OS. There was no added benefit of trastuzumab for the TOP2A co-amplified breast cancer patients.

Other biomarkers are tested as predictors of trastuzumab response. One of these pathways is the PI3K/AKT/mTOR pathway, which is commonly involved in the evolution of breast cancer and its response to treatment (Papa & Pandolfi, 2019). The TKD of the dimerized HER2 receptor activates the MAPK pathway for cell differentiation, migration, apoptosis, and cell cycle control. TKD can also make use of the PI3K/AKT/mTOR pathway to achieve the same cellular activities but it differs from the former pathway that it can be inhibited by phosphatase and tensin homolog (PTEN). Moreover, the PI3K/AKT pathway can also produce stem transcription factors to promote the stem phenotype and lead to increased clonogenic survival. Activation of this pathway was identified in about 30% of breast cancers and included PI3K gene mutation and PTEN loss of function mutations/epigenetic changes.

Translational research has been conducted over the past decade to determine whether the PI3K-PTEN axis plays important role in trastuzumab resistance. The clinical relevance of PTEN in HER2 amplified cases was tested on 2364 samples collected in BCIRG-006 and BCIRG-005 trials (Stern et al., 2015). PTEN was analyzed by immunohistochemistry testing of FFPE tissue sections. Absence of PTEN staining occurred more often in HER2-negative than positive cancers (15.9% versus 5.4%). Though not statistically significant, complete absence of staining was associated with poorer survival outcome (both DFS and OS) in HER2-positive disease than any extent of staining (low, moderate, or high) of PTEN. Trastuzumab provided clinical benefit irrespective of PTEN status. In HER2-negative, PTEN status was not found to be associated with clinical outcome.

Researchers turn to other neoadjuvant trials to find answers whether PI3K pathway is important prognostically and whether its mutation status has any prediction of trastuzumab or lapatinib treatment on pCR. Two hundred and three biopsy samples from the neoALTTO trial were analyzed for HER2 resistance to trastuzumab/lapatinib (Shi et al., 2017). It was found that the resistance was acquired through pathway level alteration, rather than via single genes. Only PIK3CA mutation was associated with pCR (OR $=0.42$; $P=0.0185$). Mutation to RhoA pathway was associated with higher pCR to lapatinib (OR = 14.8; adjusted $P = 0.001$). Mutations in a PIK3CA-related network were associated with resistance to trastuzumab (OR = 3.0; adjusted $P = 0.09$). The combined mutation status of these two pathways could define patients who may have resistance to trastuzumab, but sensitive to lapatinib. Lapatinib could therefore be considered as substitute for trastuzumab.

DNA sequencing was successful in 181 tumors of the CALGB-40601 trial, which also tested the efficacy of trastuzumab and lapatinib combination (Carey et al., 2016). PIK3CA mutations were found in 36 (20%) of them: 25% were HER2 enriched, 7% were luminal A and 31% were luminal B tumors. Most of the mutations (93%) were in exons 9 and 20. The pCR rate was 39% for tumors with PIK3CA mutations and 47% for wild-type tumors.

Since PAMELA trial involved a similar combination of dual HER2 blockade by trastuzumab and lapatinib, its samples were analyzed together with that of the CALGB-40601 to identify genomic-based predictive biomarkers. The results of this combined analysis were reported recently (Fernandez-Martinez et al., 2019). Gene expression by mRNAseq was performed on 265 and 142 of pre-treatment tumor samples of CALGB-40601 and PAMELA trials, respectively. When tested on PAMELA samples, there was high concordance of intrinsic subtypes between nCounter PAM50 prediction test and a new HER2-positive specific gene-centering method. The new method was then applied on the CALGB-40601 samples.

The subtype distribution of the combined cohort was 62% HER2-enriched, 10% luminal A, 8% luminal B, 10% basal-like, and 10% normal-like. pCR rates were highest among the HER2 enriched than the non-HER2-enriched subtypes (48.6% versus 20.7%; $P < 0.001$). HER2-enriched subtype correlation, ERBB2 amplicon, and B-cell genomic signatures were found to be associated with pCR, whereas luminal signatures were found in non-responders. Multivariate analysis showed that HER2 subtype, ERBB2 mRNA, and IgG signature expression were independent predictors of response to paclitaxel plus trastuzumab with/without lapatinib. The 5-year EFS in the CALGB-40601 trial was better for dual anti-HER2 blockade than single (HR = 0.35; $P < 0.05$). The benefit of dual versus single anti-HER2 treatment persisted when tested for the HER2-enriched, ERBB2 high, and IgG high subpopulations.

The results from the CHER-LOB trial, a phase II study showing improved pCR rate when lapatinib was combined with chemotherapy and trastuzumab, revealed the probability of pCR was higher in PIK3CA wild-type than mutated tumors (48.4% versus 12.5%, $P=0.06$) (Guarneri et al., 2015). This study suggested PIK3CA mutation could be a potential predictive biomarker of trastuzumab/lapatinib resistance.

In order to confirm the relationship between PIK3CA mutation and pCR, the neoPHOEBE study was designed to evaluate the efficacy of buparlisib, a pan-PI3K inhibitor, on pCR (Loibl et al., 2017). The patients were stratified into two independent cohorts by PIK3CA mutation (PIK3CA mutant and PIK3CA wild-type). Within each cohort, the patients were randomized to receive either buparlisib or placebo plus trastuzumab for 6 weeks followed by buparlisib or placebo plus trastuzumab and paclitaxel. The study was terminated early because of liver toxicity. Analysis was based on the recruited 50 patients (25 patients in the buparlisib arm and the other 25 patients in the placebo arm). In each arm, 21 patients carried the wild-type PIK3CA and the other 4 patients carried the mutant PIK3CA. The pCR rate was similar between buparlisib and placebo arms (32% versus 40%, $P=0.811$). However, there was a trend toward a higher ORR in the ER-positive/HER2-positive subgroup, indicating a potential role of PI3K inhibition that warranted further evaluation with better tolerated agent.

The search for biomarkers to predict pertuzumab efficacy was carried out on samples of the CLEOPATRA trial (Baselga et al., 2014). The pooled significant prognostic markers identified were HER2 membrane protein H-score ($P=0.05$), HER2 mRNA ($P=0.008$), HER3 mRNA ($P=0.03$) and PIK3CA mutation ($P<0.001$). The HR was less than one for each indicating that high levels of these biomarkers or wild-type PIK3CA have a better prognosis.

Regarding survival outcome, superior results of PFS was demonstrated when pertuzumab was combined with trastuzumab and docetaxel, whether for wild-type (HR = 0.64; 95% CI, 0.43–0.93) or mutated (HR = 0.67; 95% CI, 0.5–0.89) PIK3CA cancers. Further analysis showed that pertuzumab patients carrying wild type PIK3CA cancers had the best outcome. The median PFS of the pertuzumab group was 21.8 months and 12.5 months for wild-type and mutated PIK3CA, respectively. The corresponding median PFS for the control group by PIK3CA status was 13.8 months and 8.6 months.

APHINITY trial was a more recent study on combining pertuzumab with trastuzumab and chemotherapy in the adjuvant setting. The iDFS benefit of adding pertuzumab was relatively small at 1.7% at 4 years, though high-risk patients (node positive or hormone receptor-negative) may have greater benefit. A nested case-control design (3 controls to match 1 distant relapse) was adopted to perform DNA (targeted) and RNA sequencing on 1023 tumor samples collected to identify biomarkers that could predict benefit of the added treatment (Krop et al., 2019). DNA analysis revealed that PI3K/PTEN/AKT alterations occurred in 37% of patients and was associated with a worse outcome (HR = 1.35; 95% CI, 1.01–1.79; $P = 0.04$). Subgroup analysis showed that there was a trend (though not statistically significant) toward decreased pertuzumab benefit when there were alterations in this pathway (HR = 1.20; 95% CI, 0.76–1.89).

The other markers that were identified to be of unfavorable prognostic significance included MYC (HR = 1.61; 95% CI, 1.16–2.23; $P = 0.00$) and ZNF703 (HR = 1.62; 95% CI, 1.07–2.47; $P = 0.02$) amplification. ZNF703 (zinc finger protein 703) is a gene that regulates that transcription of proteins essential for adhesion and proliferation of cells. This oncogene was found to be related to the poor prognosis of luminal B breast cancers. Conversely, TOP2A amplification was associated with favorable prognosis, independent of anthracycline use (HR = 0.49; 0.32–0.74; $P = 0.00$).

RNA sequencing with PAM50 showed that luminal A subtype was more favorable than the basal subtype (HR = 3.11; 95% CI, 1.44–6.6; $P = 0.003$), but it could not identify pertuzumab benefit in any of the subtypes observed. Nonetheless, HER2 copy number (6 or greater by FISH) was found to be predictive of pertuzumab benefit (HR = 0.75; 95% CI, 0.6–0.93; $P = 0.04$) and prognostic of a better outcome (HR = 0.68; 95% CI, 0.5–0.91; $P = 0.01$).

Because mTOR was a key element of the PI3K/AKT pathway, combined exploratory analysis was carried out on the tissue samples of BOLERO-1 and BOLERO-3 to evaluate whether PI3K mutation was a predictor biomarker of everolimus efficacy in HER2+ advanced breast cancer (André et al., 2016). Archival materials of 549 patients were analyzed by next generation sequencing, immunohistochemistry and Sanger sequencing. PIK3CA mutations were detected in 30% of BOLERO-1 and 32% of BOLERO-3 cases. PTEN loss was reported in 16% and 12%, respectively. PI3K pathway was hyperactive (PIK3CA mutations and/or PTEN loss and/or AKT mutation) in 47% of BOLERO-1 and 41% of BOLERO-3 samples. When both data sets were combined for analysis, PI3K pathway status was found to be predictive of PFS benefit from everolimus: HR was 0.67 (95% CI, 0.45–1.00) for PI3K mutations, 0.54 (95% CI, 0.31–0.96) for PTEN loss, and 0.67 (95% CI, 0.48–0.93) for hyperactive PI3K pathway. PIK3CA wild-type did not derive benefit from everolimus.

4.3.2 Immune markers: TILs, T-cell immune signature, and more

Besides genomic-based biomarkers, recent research showed that breast cancer could be immunogenic and could be a target for immunotherapy. Pre-treatment level of tumor-infiltrating lymphocytes (TILs) was identified to be an important predictor of pCR and prognosis (Denkert et al., 2018). Pre-treatment core biopsies of 3771 patients pooled from six German Breast Cancer Group neoadjuvant trials for assessment of stromal TILs and their relation to pCR and DFS. Data from 2560 pooled patients were available for OS analysis.

pCR rates were highest with high percentage of TILs, irrespective of intrinsic subtypes. For HER2-positive patients, the pCR rates were 32% with low TILs, 39% with intermediate TILs, and 48% with high TILs. A 10% increase in TILs was associated with a longer DFS for triple negative breast cancers (HR=0.93; 95% CI, 0.87–0.98; $P=0.011$) and HER2-positive cancers (HR=0.94; 95% CI, 0.89–0.99; $P=0.017$), but not for luminal-HER2-negative patients (HR=1.02; 95% CI, 0.96–1.09; $P=0.46$). The increase in TILs was associated with improved OS for triple negative breast cancers (HR=0.92; 95% CI, 0.86–0.99; $P=0.032$), but it had no impact on HER2-positive cancers (HR=0.94; 95% CI, 0.86–1.02; $P=0.11$) and it had negative impact on luminal-HER2-negative cancers (HR=1.10; 95% CI, 1.02–1.19; $P=0.011$), suggesting different biology of the immunological infiltrate among these subtypes.

The value of on-treatment TILs was tested on neoALTTO samples and then confirmed on PAMELA samples (Luen et al., 2019). TILs were evaluated on pre-treatment and on-treatment (week 2) core biopsy specimens to correlate with pCR and survival outcome as well as to assess its added prognostic value to clinic-pathological variables. In the neoALTTO study, 41% of the cancers were immune-enriched (with immune induction and immune persistent) and 59% immune poor (persistent low and fall in immunity). The enriched elements included dendritic cells, NK-cells and CD8+ including tissue resident memory cells. The immune enriched had a significantly higher pCR (40% versus 21%; univariate OR 2.24; 95% CI, 1.31–3.85; $P=0.003$). The on-treatment TILs added significant value to clinic-pathological parameters and pre-treatment TILs for prediction of pCR ($P=0.003$). The results were confirmed with the PAMELA samples analysis. The immune-enriched patients were also found to have significantly improved 5-year PFS (85% versus 60%; univariate HR=0.31; 95% CI, 0.18–0.54; $P<0.001$) and 5-year OS (91% versus 77%; univariate HR=0.40; 95% CI, 0.2–0.82; $P=0.012$).

The values of TILs were also demonstrated in the APHINITY adjuvant trial (Krop et al., 2019). High TIL levels (over 75 percentile) was found to be a biomarker with both prognostic and predictive values. It was associated with a better survival outcome (HR = 0.66; 95% CI, 0.49–0.88; $P = 0.005$). Patients with high TILs were also found to derive benefit from pertuzumab (HR = 0.35; 95% CI, 0.19–0.65; $P = 0.003$).

Besides TILs, a three-gene T-cell immune signature consisting of IFNG, PD-L1, and CXCL9 was also checked. High mRNA expression (over 75 percentile) of this immune signature was associated with better prognosis (HR = 0.68; 0.52–0.89; $P = 0.005$). However, it did not predict pertuzumab benefit, though high levels of IFNG (HR = 0.45; 95% CI, 0.24–0.83; $P = 0.029$) and CXCL9 (HR = 0.49; 0.27–0.89; $P = 0.054$) appeared to predict benefit of the drug.

CALGB-40601 used a different set of gene expression signatures of immune cell infiltration for assessment on response and outcome (Carey et al., 2016). It included immunoglobulin G (IgG), B-cells, T-cells, CD8, and immune cells. It was found to be significant predictor of pCR. The combined analysis of CALGB-40601 and PAMELA identified that B-cell signatures (besides ERBB2 amplicon) were associated with pCR, and IgG signature was independent predictor of response to paclitaxel plus trastuzumab with or without lapatinib.

4.4 Blood-based biomarkers

4.4.1 sHER2ECD as a biomarker

sHER2ECD is an attractive biomarker. Elevated levels (>15 ng/mL) were found in 9–23% of early breast cancer patients and in 22–73% of metastatic patients (Perrier et al., 2018). Patients with elevated sHER2ECD levels had a shorter DFS and OS. That shedding of ECD was associated with a poor survival outcome may be related to the production of the more oncogenic form of HER2 (truncated $p95^{HER}$) and the binding of trastuzumab by the shed ECD resulting in reduction of biological activity as well as decrease the delivery of anti-HER2 antibodies to active cancer cells.

sHER2ECD had also been tested to determine treatment response. Patients with pCR after neoadjuvant anti-HER2 treatment were found to have a significant decrease in sHER2ECD than those who had incomplete response. It was postulated that the binding of trastuzumab to HER2 protein blocked the protease access to the cleavage site and prevented the production of $p95^{HER}$ and sHER2ECD. Conversely, those cancers showing incomplete response might already have high levels of sHER2ECD to bind to trastuzumab to reduce its active activity.

Serum and whole-blood samples were collected from patients recruited in the CLEOPATRA study at baseline before start of treatment (Baselga et al., 2014). Serum samples were then collected at different time points during the treatment and at progression for assessment of sHER2, amphiregulin (AREG), epidermal growth factor, and transforming growth factor alpha. Low levels of sHER2 were significantly associated with better prognosis (HR > 1.0, $P = 0.04$). The other serum markers were not found to be prognostic, whether with pertuzumab or not.

The SAKK-22/99 trial was designed to test whether patients should receive sequential treatment with trastuzumab monotherapy and followed by trastuzumab-chemotherapy at progression ($n = 66$) or to receive upfront combination therapy ($n = 65$) (Eppenberger-Castori et al., 2020). Plasma $HER2_{ECD}$ (comparable with serum $HER2_{ECD}$) was measured at baseline and then serially. The baseline levels were positively correlated with HER2 gene ratio and HER2 protein expression as well as presence of visceral disease.

High baseline levels had the worst OS. At 4 years, only 21.4% of patients with baseline levels of 35 ng/mL or more were alive versus 40% of patients with lower levels. Patients treated first with trastuzumab monotherapy showed no OS difference regardless of baseline $HER2_{ECD}$ levels. However, when the time to progression rates of the two treatment arms were compared at 1 year, trastuzumab monotherapy was better than the upfront combination therapy when $HER2_{ECD}$ levels was less than 15 ng/mL (58% versus 28%). The levels of patients receiving the combination therapy dropped by more than 20% when there was response. These remained stable until progression. No such change was observed even for the responders of the trastuzumab monotherapy group, but when there was a quick drop in levels at9 weeks, there was a trend for better OS for patients treated by upfront trastuzumab monotherapy.

Despite these interesting findings, the evidence to use sHER2ECD as a biomarker and to guide therapy was not strong. ASCO does not recommend its use as predictive or prognostic marker in its current guideline.

4.4.2 Liquid biopsy: CTCs and ct-DNA

Liquid biopsy is an attractive concept to detect early relapse or monitor the treatment response of cancer. Patients can avoid invasive procedures to obtain cancer samples from deep seated or bone metastasis for analysis, especially when there are different sites of disease and tumor heterogeneity is a concern. The detection of circulating tumor cells (CTC) or circulating-DNA (ct-DNA) has been the focus of intense research. CTCs have been shown to be an important marker to predict outcome in metastatic breast cancers. It has also been shown to be an independent prognostic marker before and after adjuvant chemotherapy in 2026 patients with average to high-risk primary breast cancer (Rack et al., 2014). The prognosis was worst in patients with five or more CTCs per 30 mL of blood (HR were 4.51 [95% CI, 2.59–7.86] for DFS and 3.6 [95% CI, 1.56–8.45] for OS). The presence of persisting CTCs after chemotherapy was associated with a poorer survival outcome (HR for DFS was 1.12; 95% CI, 1.02–1.25; $P=0.02$ and HR for OS was 1.16; 95% CI, 0.99–1.37; $P=0.06$).

In another study testing CTCs at 5 years after diagnosis on a high-risk cohort of 547 patients, when a positive CTC assay was defined as the presence of at least one cell per 7.5 mL of blood, it was found that CTC was prognostic only for hormone receptor-positive cancers (Sparano et al., 2018). The recurrence rate per person year was 21.4% for CTC positive group versus 2.0% for CTC negative group (HR = 13.1; 95% CI, 4.7–36.3). Of the 23 patients with recurrence, 30.4% had a positive CTC result at a median of 2.8 years before clinical recurrence. This study provides evidence to stratify high risk patients to receive further treatment after 5 years to reduce recurrence.

A recent study carried the research further to use immunofluorescence assay to test for HER2 status of CTCs detected in patients with breast cancer (Chen et al., 2019). HER2-positive CTCs were defined as HER2 immunofluorescence intensity of 3.5 or greater times higher than the CD45 intensity. HER2-positive CTCs were detected in both HER2-positive and HER2-negative cancers, but the rate was higher in metastatic breast cancers. Though this finding is interesting, further study is required to assess whether HER2-positive CTCs can guide therapy or predict treatment response.

ct-DNA has been shown to be a powerful tool to monitor trastuzumab resistance in metastatic gastric cancer. It was also found to be prognostic biomarker in breast cancer (Garcia-Murillas et al., 2019). This study involved a cohort of 101 patients with identified somatic mutations. They received neoadjuvant chemotherapy followed by surgery or surgery followed by adjuvant chemotherapy for their breast cancer. Personalized digital polymerase chain reaction assays were performed at baseline, every 3 months for the first year of follow up, and then every 6 months. The detection of ct-DNA at diagnosis was associated with poorer relapse-free survival

(HR = 5.8; 95% CI, 1.2–27.1; $P = 0.01$). The detection of ct-DNA was associated with a high risk of future relapse (HR = 25.2; 95%CI, 6.7–95.6; $P < 0.001$). Most of the patients were tested negative at first time point and then tested positive subsequently. The median time lapsed between detection of ct-DNA and evidence of clinical relapse was 10.7 months. Almost all of the relapsed disease was extracranial.

To test whether ct-DNA was associated with response to anti-HER2 treatment, plasma DNA of 455 neoALTTO patients was collected at baseline, week 2 and before surgery for detection of PIK3CA and TP53 mutations by digital polymerase chain reaction assays (Rothé et al., 2019). The presence of ct-DNA at baseline (41%) was inversely associated with pCR (OR = 0.15; 95%CI, 0.034–0.7; $P = 0.0089$). Those patients with persistent ct-DNA detection during the course of neoadjuvant treatment had the lowest pCR.

PIK3CA and TP53 mutations were performed on tumor archival materials and ct-DNA (Ma et al., 2017). These mutations in ct-DNA (but not archival tissues) predict pyrotinib efficacy. In another study on ct-DNA to probe predictive biomarker for pyrotinib showed that multiple genetic alterations in HER2-related signaling network were associated with poorer PFS (Li et al., 2019).

The plasmaMATCH trial used ct-DNA screening to direct targeted therapies in patients with advanced breast cancer (Turner et al., 2018). The trial was designed to screen 1000 patients for ESR1, HER2, AKT1, and PIK3CA mutations as well as HER2 copy number determination. Patients with identified mutations would be treated with matched targeted therapy. In the case of HER2, patients will receive neratinib with or without fulvestrant. The trial is still undergoing recruitment but patients with mutations have been treated with matched therapies. A study with similar design, the NCI-MATCH trial, is also recruiting patients currently but it includes solid tumors and hematological malignancies.

4.5 HER2 mutation and its treatment

HER2 mutation has been identified in a small number of tumors that are not HER2-amplified as assessed by the standard immunohistochemistry or FISH assay. The mutation can occur at the extracellular, transmembrane, or TKD. They can activate HER2 signaling pathway, even when the HER2 copy number is normal. Patients with these mutations can benefit from anti-HER2 therapy.

Somatic mutation can be detected in the primary tumor, but this has not been shown to be associated with cancer metastasis and outcome. Because of tumor heterogeneity, attention has been directed toward ct-DNA to detect mutations for directing treatment. This is especially the case when the sites of the metastatic disease are within bone or difficult to access for biopsy.

The prevalence of HER2 mutation was estimated to be about 1.8% in primary tumors based on The Cancer Genome Atlas (TCGA) analysis (Cancer Genome Atlas Network, 2012). This was found to be higher (6%) in metastatic tissue specimens (Razavi et al., 2018), but the mutation rates were found to be even higher among Asian patients, being 16.7% in Koreans and 13.3% in Chinese (Lei et al., 2019). Most of the HER2 mutations occurred in HER2-negative breast cancers and they were responsible for driving metastasis (Petrelli et al., 2017). The majority of the mutations were detected primarily in exons 19–20 of the TKD (Yi et al., 2018). The most frequent mutation point was HER2 L755S (24%) in exon 19 and HER2 V777L (13.7%) in exon 20. HER2 777L mutation has been found to be related to trastuzumab resistance and reduced survival among HER2-positive (but not HER2-negative) patients.

The efficacy of anti-HER2 therapy against the different HER2 mutations has been tested. The TKD mutation remains sensitive to trastuzumab and TKI. The ECD mutations display increased sensitivity to trastuzumab (and possibly pertuzumab) and lapatinib. The transmembrane mutations are responsible for trastuzumab resistance but not TKI. Unlike HER2-amplified cancers, HER2-mutant breast cancers were not so responsive to trastuzumab but more to TKI, particularly neratinib. An ORR of 32% to neratinib was noted with responsive cancers harboring mutations mostly in the ECD and the TKD (Hyman et al., 2018).

Clinical reports on isolated cases have demonstrated tumor response when trastuzumab was combined with vinorelbine in HER2-negative metastatic breast cancer that carried HER2 777L mutation (Lei et al., 2019). Among other anti-HER2 therapy, neratinib was found to be useful in HER2-mutant cancers (Medford et al., 2019). HER2 mutations disappeared within 6 months of treatment initiation whereas other concurrent mutations remained unchanged and the patient was in prolonged remission of more than a year. The discovery of HER2 mutations and its detection in ct-DNA thus provide an opportunity for researchers to design new strategies to treat resistant HER2-negative cancers.

5. Conclusion

Ever since the discovery of the HER2 pathway and the first clinical trial using trastuzumab, the prognosis of patients with HER2-positive breast cancer has continued to improve. With the advent of the newer anti-HER2 therapies such as pertuzumab, TKIs like lapatinib, neratinib, and pyrotinib, as well as T-DM1, the "do more" and "do less" approaches are becoming more important to personalize treatment for patients with an early HER2-positive breast cancer. The goal of the "do more" approach is to achieve a better outcome, and this utilizes different strategies such as extending the duration of anti-HER2 treatment, treating with dual HER2 blockade, achieving higher pcR rate, and improving treatment for patients with residual invasive disease. The "do less" approach, on the other hand, takes into consideration the risk-benefit ratio in patients who may not need or unable to tolerate aggressive treatments. This includes strategies such as treating with chemotherapy-sparing regimen, shortening duration of anti-HER2 treatment, and de-escalating chemotherapy in small or less aggressive cancers.

Advanced HER2-positive breast cancer remains an incurable disease, and recent advances aim at tackling the disease by using novel therapeutic agents and combination strategies. The latest clinical practice guidelines recommend a first-line treatment with the combination of trastuzumab, pertuzumab, and a taxane chemotherapy, but an increasing amount of evidence from recent clinical trials have demonstrated that T-DM1 as well as TKIs, new anti-HER2 antibodies, and new ADCs could all play a role in treating the disease and improving the survival rates of patients diagnosed with the advanced stage of the disease.

There are important points to consider, including HER2 gene copy number, HER2 heterogeneity, tissue biomarkers, blood-based biomarkers, and HER2 mutation and its treatment, which could affect prognosis and choice of therapies. These factors, together with the novel therapeutic agents and combination strategies, warrant further investigations to guide clinicians in determining the best possible treatment and approach for HER2-positive breast cancer patients, with the hope of achieving a greater survival benefit and potentially a cure for the disease.

References

Adamczyk, A., Kruczak, A., Harazin-Lechowska, A., Ambicka, A., Grela-Wojewoda, A., Domagała-Haduch, M., et al. (2018). Relationship between HER2 gene status and selected potential biological features related to trastuzumab resistance and its influence on survival of breast cancer patients undergoing trastuzumab adjuvant treatment. *Oncotargets and Therapy*, *11*, 4525–4535.

André, F., Hurvitz, S., Fasolo, A., Tseng, L. M., Jerusalem, G., Wilks, S., et al. (2016). Molecular alterations and everolimus efficacy in human epidermal growth factor receptor 2-overexpressing metastatic breast cancers: Combined exploratory biomarker analysis from BOLERO-1 and BOLERO-3. *Journal of Clinical Oncology*, *34*(18), 2115–2124.

André, F., O'Regan, R., Ozguroglu, M., Toi, M., Xu, B., Jerusalem, G., et al. (2014). Everolimus for women with trastuzumab-resistant, HER2-positive, advanced breast cancer (BOLERO-3): A randomised, double-blind, placebo-controlled phase 3 trial. *The Lancet Oncology, 15*(6), 580–591.

Badwe, R., Hawaldar, R., Nair, N., Kaushik, R., Parmar, V., Siddique, S., et al. (2015). Locoregional treatment versus no treatment of the primary tumour in metastatic breast cancer: An open-label randomised controlled trial. *The Lancet Oncology, 16*(13), 1380–1388.

Barroso-Sousa, R., Exman, P., & Tolaney, S. M. (2018). De-escalating treatment in the adjuvant setting in HER2-positive breast cancer. *Future Oncology, 14*(10), 937–945.

Baselga, J., Bradbury, I., Eidtmann, H., Di Cosimo, S., de Azambuja, E., Aura, C., et al. (2012). Lapatinib with trastuzumab for HER2-positive early breast cancer (NeoALTTO): A randomised, open-label, multicentre, phase 3 trial. *Lancet, 379*(9816), 633–640.

Baselga, J., Cortés, J., Im, S. A., Clark, E., Ross, G., Kiermaier, A., et al. (2014). Biomarker analyses in CLEOPATRA: A phase III, placebo-controlled study of pertuzumab in human epidermal growth factor receptor 2-positive, first-line metastatic breast cancer. *Journal of Clinical Oncology, 32*(33), 3753–3761.

Bates, J. P., Derakhshandeh, R., Jones, L., & Webb, T. J. (2018). Mechanisms of immune evasion in breast cancer. *BMC Cancer, 18*(1), 556.

Bergh, J. C. S., Andersson, A., Bjohle, J., Bosch, A., Carlsson, L., Dreifaldt, A. C., et al. (2019). Docetaxel, trastuzumab, pertuzumab versus trastuzumab emtansine as neoadjuvant treatment of HER2-positive breast cancer: Results from the Swedish PREDIX HER2 trial identifying a new potential de-escalation standard? *Journal of Clinical Oncology, 37*(Suppl. 15), 501.

Borley, A., Mercer, T., Morgan, M., Dutton, P., Barrett-Lee, P., Brunelli, M., et al. (2014). Impact of HER2 copy number in IHC2+/FISH-amplified breast cancer on outcome of adjuvant trastuzumab treatment in a large UK cancer network. *British Journal of Cancer, 110*(8), 2139–2143.

Cameron, D., Piccart-Gebhart, M. J., Gelber, R. D., Procter, M., Goldhirsch, A., de Azambuja, E., et al. (2017). 11 years' follow-up of trastuzumab after adjuvant chemotherapy in HER2-positive early breast cancer: Final analysis of the HERceptin Adjuvant (HERA) trial. *Lancet, 389*(10075), 1195–1205.

Cancer Genome Atlas Network. (2012). Comprehensive molecular portraits of human breast tumours. *Nature, 490*(7418), 61–70.

Cardoso, F., Senkus, E., Costa, A., Papadopoulos, E., Aapro, M., André, F., et al. (2018). 4th ESO-ESMO International Consensus Guidelines for Advanced Breast Cancer (ABC 4). *Annals of Oncology, 29*(8), 1634–1657.

Carey, L. A., Berry, D. A., Cirrincione, C. T., Barry, W. T., Pitcher, B. N., Harris, L. N., et al. (2016). Molecular heterogeneity and response to neoadjuvant human epidermal growth factor receptor 2 targeting in CALGB 40601, a randomized phase III trial of paclitaxel plus trastuzumab with or without Lapatinib. *Journal of Clinical Oncology, 34*(6), 542–549.

Carey, L. A., Berry, D. A., Ollila, D., Harris, L., Krop, I. E., Weckstein, D., et al. (2013). Clinical and translational results of CALGB 40601: A neoadjuvant phase III trial of weekly paclitaxel and trastuzumab with or without lapatinib for HER2-positive breast cancer. *Journal of Clinical Oncology, 31*(Suppl. 15), 500.

Chen, W., Zhang, J., Huang, L., Chen, L., Zhou, Y., Tang, D., et al. (2019). Detection of HER2-positive circulating tumor cells using the liquidbiopsy system in breast cancer. *Clinical Breast Cancer, 19*(1), e239–e246.

Ciruelos, E., Villagrasa, P., Paré, L., Oliveira, M., Pernas, S., Cortés, J., et al. (2019). Abstract PD3-03: SOLTI-1303 PATRICIA phase II trial (STAGE 1)—Palbociclib and trastuzumab in postmenopausal patients with HER2-positive metastatic breast cancer. *Cancer Research*, *79*(4), PD3-03.

Cortazar, P., Zhang, L., Untch, M., Mehta, K., Costantino, J. P., Wolmark, N., et al. (2014). Pathological complete response and long-term clinical benefit in breast cancer: The CTNeoBC pooled analysis. *Lancet*, *384*(9938), 164–172.

Cossetti, R. J., Tyldesley, S. K., Speers, C. H., Zheng, Y., & Gelmon, K. A. (2015). Comparison of breast cancer recurrence and outcome patterns between patients treated from 1986 to 1992 and from 2004 to 2008. *Journal of Clinical Oncology*, *33*(1), 65–73.

de Haas, S. L., Hurvitz, S. A., Martin, M., Kiermaier, A., Lewis Phillips, G., Xu, J., et al. (2016). In *P6-07-09 Biomarker analysis from the neoadjuvant KRISTINE study in HER2-positive early breast cancer (EBC). Presented at the 2016 San Antonio breast cancer symposium.*

Denkert, C., von Minckwitz, G., Darb-Esfahani, S., Lederer, B., Heppner, B. I., Weber, K. E., et al. (2018). Tumour-infiltrating lymphocytes and prognosis in different subtypes of breast cancer: A pooled analysis of 3771 patients treated with neoadjuvant therapy. *The Lancet Oncology*, *19*(1), 40–50.

Earl, H. M., Hiller, L., Vallier, A. L., Loi, S., Howe, D., Higgins, H. B., et al. (2018). PERSEPHONE: 6 versus 12 months (m) of adjuvant trastuzumab in patients (pts) with HER2 positive (+) early breast cancer (EBC): Randomised phase 3 non-inferiority trial with definitive 4-year (yr) disease-free survival (DFS) results. *Journal of Clinical Oncology*, *36*(Suppl. 15), 506.

Earl, H. M., Hiller, L., Vallier, A. L., Loi, S., McAdam, K., Hughes-Davies, L., et al. (2019). 6 Versus 12 months of adjuvant trastuzumab for HER2-positive early breast cancer (PERSEPHONE): 4-year disease-free survival results of a randomised phase 3 non-inferiority trial. *Lancet*, *393*(10191), 2599–2612.

Ejlertsen, B., Barrios, C. H., Gokmen, E., Robert, N., Martinez, N., Link, J. S., et al. (2018). Timing of initiation of neratinib after trastuzumab-based adjuvant therapy in early-stage HER2+ hormone receptor (HR)-negative breast cancer: Exploratory analyses from the phase III ExteNET trial. *Journal of Clinical Oncology*, *36*(Suppl. 15), 549.

Ellegård, S., Veenstra, C., Pérez-Tenorio, G., Fagerström, V., Gårsjö, J., Gert, K., et al. (2019). ERBB2 and PTPN2 gene copy numbers as prognostic factors in HER2-positive metastatic breast cancer treated with trastuzumab. *Oncology Letters*, *17*(3), 3371–3381.

Ellis, P. A., Barrios, C. H., Eiermann, W., Toi, M., Im, Y. H., Contem, P. F., et al. (2015). Phase III, randomized study of trastuzumab emtansine (T-DM1) ± pertuzumab (P) vs trastuzumab + taxane (HT) for first-line treatment of HER2-positive MBC: Primary results from the MARIANNE study. *Journal of Clinical Oncology*, *33*(Suppl. 15), 507.

Emens, L. A., Esteva, F., Beresford, M., Saura, C., De Laurentiis, M., Kim, S. B., et al. (2019a). Abstract PD3-01: Results from KATE2, a randomized phase 2 study of atezolizumab (atezo)+trastuzumab emtansine (T-DM1) vs placebo (pbo)+T-DM1 in previously treated HER2+ advanced breast cancer (BC). *Cancer Research*, *79*(Suppl. 4), PD3-01.

Emens, L. A., Esteva, F., Beresford, M., Saura, C., De Laurentiis, M., Kim, S. B., et al. (2019b). Overall survival (OS) in kate2, a phase 2 study of programmed death ligand 1 (PD-L1) inhibitor atezolizumab (Atezo)+trastuzumab emtansine (T-DM1) vs placebo (PBO)+T-DM1 in previously treated HER2+ advanced breast cancer (BC). *Annals of Oncology*, *30*(Suppl. 5), 3050.

Eppenberger-Castori, S., Klingbiel, D., Ruhstaller, T., Dietrich, D., Rufle, D. A., Rothgiesser, K., et al. (2020). Plasma HER2ECD a promising test for patient prognosis and prediction of response in HER2 positive breast cancer: Results of a randomized study—SAKK 22/99. *BMC Cancer*, *20*(1), 114.

Fernandez-Martinez, A., Tanioka, M., Fan, C., Parker, J. S., Hoadley, K. A., Krop, I. E., et al. (2019). Genomic-based predictive biomarkers to anti-HER2 therapies:

A combined analysis of CALGB 40601 (Alliance) and PAMELA clinical trials. *Journal of Clinical Oncology*, *37*(Suppl. 15), 571.

Filho, O. M., Viale, G., Trippa, L., Li, T., Yardley, D. A., Mayer, I. A., et al. (2019). HER2 heterogeneity as a predictor of response to neoadjuvant T-DM1 plus pertuzumab: Results from a prospective clinical trial. *Journal of Clinical Oncology*, *37*(Suppl. 15), 502.

Garcia-Murillas, I., Chopra, N., Comino-Méndez, I., Beaney, M., Tovey, H., Cutts, R. J., et al. (2019). Assessment of molecular relapse detection in early-stage breast cancer. *JAMA Oncology*, *5*(10), 1473–1478. https://doi.org/10.1001/jamaoncol.2019.1838.

Gianni, L., Eiermann, W., Semiglazov, V., Manikhas, A., Lluch, A., Tjulandin, S., et al. (2010). Neoadjuvant chemotherapy with trastuzumab followed by adjuvant trastuzumab versus neoadjuvant chemotherapy alone, in patients with HER2-positive locally advanced breast cancer (the NOAH trial): A randomised controlled superiority trial with a parallel HER2-negative cohort. *Lancet*, *375*(9712), 377–384.

Gianni, L., Pienkowski, T., Im, Y. H., Roman, L., Tseng, L. M., Liu, M. C., et al. (2012). Efficacy and safety of neoadjuvant pertuzumab and trastuzumab in women with locally advanced, inflammatory, or early HER2-positive breast cancer (NeoSphere): A randomised multicentre, open-label, phase 2 trial. *The Lancet Oncology*, *13*(1), 25–32.

Giordano, S. H., Buzdar, A. U., Smith, T. L., Kau, S. W., Yang, Y., & Hortobagyi, G. N. (2004). Is breast cancer survival improving? *Cancer*, *100*(1), 44–52.

Giordano, S. H., Temin, S., Chandarlapaty, S., Crews, J. R., Esteva, F. J., Kirshner, J. J., et al. (2018). Systemic therapy for patients with advanced human epidermal growth factor receptor 2-positive breast cancer: ASCO clinical practice guideline update. *Journal of Clinical Oncology*, *36*(26), 2736–2740.

Guarneri, V., Dieci, M. V., Frassoldati, A., Maiorana, A., Ficarra, G., Bettelli, S., et al. (2015). Prospective biomarker analysis of the randomized CHER-LOB study evaluating the dual anti-her2 treatment with Trastuzumab and Lapatinib plus chemotherapy as Neoadjuvant therapy for HER2-positive breast cancer. *The Oncologist*, *20*(9), 1001–1010.

Harbeck, N., Gluz, O., Christgen, M., Kates, R. E., Braun, M., Küemmel, S., et al. (2017). De-Escalation Strategies in Human Epidermal Growth Factor Receptor 2 (HER2)-Positive Early Breast Cancer (BC): Final Analysis of the West German Study Group Adjuvant Dynamic Marker-Adjusted Personalized Therapy Trial Optimizing Risk Assessment and Therapy Response Prediction in Early BC HER2- and Hormone Receptor-Positive Phase II Randomized Trial-Efficacy, Safety, and Predictive Markers for 12 Weeks of Neoadjuvant Trastuzumab Emtansine With or Without Endocrine Therapy (ET) Versus Trastuzumab Plus ET. *Journal of Clinical Oncology*, *35*(26), 3046–3054.

Harbeck, N., Penault-Llorca, F., Cortes, J., Gnant, M., Houssami, N., Poortmans, P., et al. (2019). Breast cancer. *Nature Reviews. Disease Primers*, *5*(1), 66.

Hayes, D. F. (2019). HER2 and breast cancer—A phenomenal success story. *The New England Journal of Medicine*, *381*(13), 1284–1286.

Hurvitz, S. A., Andre, F., Jiang, Z., Shao, Z., Mano, M. S., Neciosup, S. P., et al. (2015). Combination of everolimus with trastuzumab plus paclitaxel as first-line treatment for patients with HER2-positive advanced breast cancer (BOLERO-1): A phase 3, randomised, double-blind, multicentre trial. *The Lancet Oncology*, *16*(7), 816–829.

Hurvitz, S. A., Martin, M., Jung, K. H., Huang, C. S., Harbeck, N., Valero, V., et al. (2019). Neoadjuvant trastuzumab emtansine and pertuzumab in human epidermal growth factor receptor 2-positive breast cancer: Three-year outcomes from the phase III KRISTINE study. *Journal of Clinical Oncology*, *37*(25), 2206–2216.

Hurvitz, S. A., Martin, M., Symmans, W. F., Jung, K. H., Huang, C. S., Thompson, A. M., et al. (2016). Pathologic complete response rates after neoadjuvant trastuzumab emtansine (T-DM1) + pertuzumab vs. docetaxel + carboplatin + trastuzumab +

pertuzumab (TCH + P) treatment in patients with HER2-positive (HER2 +) early breast cancer (KRISTINE/TRIO-021). *Journal of Clinical Oncology*, *34*(15), 500.

Hyman, D. M., Piha-Paul, S. A., Won, H., Rodon, J., Saura, C., Shapiro, G. I., et al. (2018). HER kinase inhibition in patients with HER2- and HER3-mutant cancers. *Nature*, *554*(7691), 189–194.

Jain, S., Shah, A. N., Santa-Maria, C. A., Siziopikou, K., Rademaker, A., Helenowski, I., et al. (2018). Phase I study of alpelisib (BYL-719) and trastuzumab emtansine (T-DM1) in HER2-positive metastatic breast cancer (MBC) after trastuzumab and taxane therapy. *Breast Cancer Research and Treatment*, *171*(2), 371–381.

Krop, I. E., Paulson, J., Campbell, C., Kiermaier, A. C., Andre, F., Fumagalli, D., et al. (2019). Genomic correlates of response to adjuvant trastuzumab (H) and pertuzumab (P) in HER2+ breast cancer (BC): Biomarker analysis of the APHINITY trial. *Journal of Clinical Oncology*, *37*(Suppl. 15), 1012.

Lei, L., Ye, W. W., Zheng, L. F., Huang, P., Shi, L., Huang, J., et al. (2019). A significant response to a combination of trastuzumab and vinorelbine in HER2-negative metastatic breast cancer with HER2 V777L mutation. *Oncotargets and Therapy*, *12*, 2931–2936.

Lemmon, M. A., Schlessinger, J., & Ferguson, K. M. (2014). The EGFR family: Not so prototypical receptor tyrosine kinases. *Cold Spring Harbor Perspectives in Biology*, *6*(4), a020768.

Li, Q., Guan, X., Chen, S., Yi, Z., Lan, B., Xing, P., et al. (2019). Safety, efficacy, and biomarker analysis of pyrotinib in combination with capecitabine in HER2-positive metastatic breast cancer patients: A phase I clinical trial. *Clinical Cancer Research*, *25*(17), 5212–5220.

Llombart-Cussac, A., Cortés, J., Paré, L., Galván, P., Bermejo, B., Martínez, N., et al. (2017). HER2-enriched subtype as a predictor of pathological complete response following trastuzumab and lapatinib without chemotherapy in early-stage HER2-positive breast cancer (PAMELA): An open-label, single-group, multicentre, phase 2 trial. *The Lancet Oncology*, *18*(4), 545–554.

Loi, S., Giobbie-Hurder, A., Gombos, A., Bachelot, T., Hui, R., Curigliano, G., et al. (2019). Pembrolizumab plus trastuzumab in trastuzumab-resistant, advanced, HER2-positive breast cancer (PANACEA): A single-arm, multicentre, phase 1b-2 trial. *The Lancet Oncology*, *20*(3), 371–382.

Loibl, S., de la Pena, L., Nekljudova, V., Zardavas, D., Michiels, S., Denkert, C., et al. (2017). Neoadjuvant buparlisib plus trastuzumab and paclitaxel for women with HER2+ primary breast cancer: A randomised, double-blind, placebo-controlled phase II trial (NeoPHOEBE). *European Journal of Cancer*, *85*, 133–145.

Luen, S. J., Griguolo, G., Nuciforo, P., Campbell, C., Fasani, R., Cortes, J., et al. (2019). On-treatment changes in tumor-infiltrating lymphocytes (TIL) during neoadjuvant HER2 therapy (NAT) and clinical outcome. *Journal of Clinical Oncology*, *37*(Suppl. 15), 574.

Ma, F., Li, Q., Chen, S., Zhu, W., Fan, Y., Wang, J., et al. (2017). Phase I Study and Biomarker Analysis of Pyrotinib, a Novel Irreversible Pan-ErbB Receptor Tyrosine Kinase Inhibitor, in Patients With Human Epidermal Growth Factor Receptor 2-Positive Metastatic Breast Cancer. *Journal of Clinical Oncology*, *35*(27), 3105–3112.

Ma, F., Ouyang, Q., Li, W., Jiang, Z., Tong, Z., Liu, Y., et al. (2019). Pyrotinib or lapatinib combined with capecitabine in HER2-positive metastatic breast cancer with prior taxanes, anthracyclines, and/or trastuzumab: A randomized, phase II study. *Journal of Clinical Oncology*, *37*(29), 2610–2619.

Martin, M., Holmes, F. A., Ejlertsen, B., Delaloge, S., Moy, B., Iwata, H., et al. (2017). Neratinib after trastuzumab-based adjuvant therapy in HER2-positive breast cancer (ExteNET): 5-year analysis of a randomised, double-blind, placebo-controlled, phase 3 trial. *The Lancet Oncology*, *18*(12), 1688–1700.

Masuda, N., Lee, S. J., Ohtani, S., Im, Y. H., Lee, E. S., Yokota, I., et al. (2017). Adjuvant capecitabine for breast cancer after preoperative chemotherapy. *The New England Journal of Medicine*, *376*(22), 2147–2159.

Masuda, N., Ohtani, S., Takano, T., Inoue, K., Suzuki, E., Nakamura, R., et al. (2020). A randomized, 3-arm, neoadjuvant, phase 2 study comparing docetaxel + carboplatin + trastuzumab + pertuzumab (TCbHP), TCbHP followed by trastuzumab emtansine and pertuzumab (T-DM1 + P), and T-DM1 + P in HER2-positive primary breast cancer. *Breast Cancer Research and Treatment*, *180*(1), 135–146. https://doi.org/10.1007/s10549-020-05524-6.

Medford, A. J., Dubash, T. D., Juric, D., Spring, L., Niemierko, A., Vidula, N., et al. (2019). Blood-based monitoring identifies acquired and targetable driver HER2 mutations in endocrine-resistant metastatic breast cancer. *NPJ Precision Oncology*, *3*, 18.

Modi, S. (2015). In *HER2-positive breast cancer: Beyond trastuzumab. Presented at the 2015 ASCO annual meeting*.

Modi, S., Ohtani, S., Lee, C. S., Wang, K., Saxena, K., & Cameron, D. A. (2019). A phase III, multicenter, randomized, open label trial of [fam-] trastuzumab deruxtecan (DS-8201a) versus investigator's choice in HER2-low breast cancer. *Journal of Clinical Oncology*, *37*(Suppl. 15), TPS1102.

Modi, S., Saura, C., Yamashita, T., Park, Y. H., Kim, S. B., Tamura, K., et al. (2020). Trastuzumab deruxtecan in previously treated HER2-positive breast cancer. *The New England Journal of Medicine*, *382*, 610–621. https://doi.org/10.1056/NEJMoa1914510.

Moja, L., Tagliabue, L., Balduzzi, S., Parmelli, E., Pistotti, V., Guarneri, V., et al. (2012). Trastuzumab containing regimens for early breast cancer. *Cochrane Database of Systematic Reviews*, *4*, CD006243.

Mudgway, R., de Paz Villanueva, C. C., Lin, A. C., Senthil, M., Garberoglio, C. A., & Lum, S. S. (2019). The impact of primary tumor surgery on survival in HER2 positive stage IV breast cancer patients in the current era of targeted therapy. *Cancer Research*, *79*(Suppl. 13), 4873.

Murthy, R., Borges, V. F., Conlin, A., Chaves, J., Chamberlain, M., Gray, T., et al. (2018). Tucatinib with capecitabine and trastuzumab in advanced HER2-positive metastatic breast cancer with and without brain metastases: A non-randomised, open-label, phase 1b study. *The Lancet Oncology*, *19*(7), 880–888.

Murthy, R. K., Loi, S., Okines, A., Paplomata, E., Hamilton, E., Hurvitz, S. A., et al. (2020). Tucatinib, trastuzumab, and capecitabine for HER2-positive metastatic breast cancer. *The New England Journal of Medicine*, *382*, 597–609. https://doi.org/10.1056/NEJMoa1914609.

Nakada, T., Sugihara, K., Jikoh, T., Abe, Y., & Agatsuma, T. (2019). The latest research and development into the antibody-drug conjugate, [fam-] trastuzumab deruxtecan (DS-8201a), for HER2 cancer therapy. *Chemical & Pharmaceutical Bulletin (Tokyo)*, *67*(3), 173–185.

NCCN Clinical Practice Guidelines in Oncology. (2020). *Breast cancer. Version 1.2020. January 15, 2020. Accessed at https://www.nccn.org/professionals/physician_gls/pdf/breast.pdf.*

O'Brien, N., Conklin, D., Beckmann, R., Luo, T., Chau, K., Thomas, J., et al. (2018). Preclinical activity of abemaciclib alone or in combination with antimitotic and targeted therapies in breast cancer. *Molecular Cancer Therapeutics*, *17*(5), 897–907.

Ogitani, Y., Aida, T., Hagihara, K., Yamaguchi, J., Ishii, C., Harada, N., et al. (2016). DS-8201a, A novel HER2-targeting ADC with a novel DNA topoisomerase I inhibitor, demonstrates a promising antitumor efficacy with differentiation from T-DM1. *Clinical Cancer Research*, *22*(20), 5097–5108.

Papa, A., & Pandolfi, P. P. (2019). The PTEN^{-}PI3K axis in cancer. *Biomolecules*, *9*(4) pii: E153.

Park, J. W., Liu, M. C., Yee, D., Yau, C., van 't Veer, L. J., Symmans, W. F., et al. (2016). Adaptive randomization of neratinib in early breast cancer. *The New England Journal of Medicine*, *375*(1), 11–22.

Perrier, A., Gligorov, J., Lefèvre, G., & Boissan, M. (2018). The extracellular domain of Her2 in serum as a biomarker of breast cancer. *Laboratory Investigation*, *98*(6), 696–707.

Petrelli, F., Tomasello, G., Barni, S., Lonati, V., Passalacqua, R., & Ghidini, M. (2017). Clinical and pathological characterization of HER2 mutations in human breast cancer: A systematic review of the literature. *Breast Cancer Research and Treatment*, *166*(2), 339–349.

Piccart, M., Procter, M., Fumagalli, D., de Azambuja, D., Clark, E., Ewer, M. S., et al. (2019). In *GS1-04. Interim overall survival analysis of APHINITY (BIG 4-11): A randomized multicenter, double-blind, placebo-controlled trial comparing chemotherapy plus trastuzumab plus pertuzumab versus chemotherapy plus trastuzumab plus placebo as adjuvant therapy in patients with operable HER2-positive early breast cancer. Presented at the 2019 San Antonio Breast Cancer Symposium.*

Press, M. F., Sauter, G., Buyse, M., Bernstein, L., Guzman, R., Santiago, A., et al. (2011). Alteration of topoisomerase II-alpha gene in human breast cancer: Association with responsiveness to anthracycline-based chemotherapy. *Journal of Clinical Oncology*, *29*(7), 859–867.

Rack, B., Schindlbeck, C., Jückstock, J., Andergassen, U., Hepp, P., Zwingers, T., et al. (2014). Circulating tumor cells predict survival in early average-to-high risk breast cancer patients. *Journal of the National Cancer Institute*, *106*(5), pii: dju066.

Ran, R., Huang, W., Lin, S., Niu, Y., Rugo, H. S., Kong, W., et al. (2018). Prognostic role of plasma HER2 gene copy number in patients with HER2 positive metastatic breast cancer. *Journal of Clinical Oncology*, *36*(Suppl. 15), e13018.

Razavi, P., Chang, M. T., Xu, G., Bandlamudi, C., Ross, D. S., Vasan, N., et al. (2018). The genomic landscape of endocrine-resistant advanced breast cancers. *Cancer Cell*, *34*(3), 427–438.

Rothé, F., Silva, M. J., Venet, D., Campbell, C., Bradburry, I., Rouas, G., et al. (2019). Circulating tumor DNA in HER2-amplified breast cancer: A translational research substudy of the NeoALTTO phase III trial. *Clinical Cancer Research*, *25*(12), 3581–3588.

Rugo, H. S., Im, S. A., Cardoso, F., Cortes, J., Curigliano, G., Pegram, M. D., et al. (2019). In *GS1-02. Phase 3 SOPHIA study of margetuximab + chemotherapy vs trastuzumab + chemotherapy in patients with HER2+ metastatic breast cancer after prior anti-HER2 therapies: Second interim overall survival analysis. Presented at the 2019 San Antonio breast cancer symposium.*

Rugo, H. S., Im, S. A., Wright, G. L. S., Escriva-de-Romani, S., DeLaurentiis, M., Cortes, J., et al. (2019). SOPHIA primary analysis: A phase 3 (P3) study of margetuximab (M) + chemotherapy (C) versus trastuzumab (T) + C in patients (pts) with HER2+ metastatic (met) breast cancer (MBC) after prior anti-HER2 therapies (Tx). *Journal of Clinical Oncology*, *37*(Suppl. 15), 1000.

Saad, E. D., Squifflet, P., Burzykowski, T., Quinaux, E., Delaloge, S., Mavroudis, D., et al. (2019). Disease-free survival as a surrogate for overall survival in patients with HER2-positive, early breast cancer in trials of adjuvant trastuzumab for up to 1 year: A systematic review and meta-analysis. *The Lancet Oncology*, *20*(3), 361–370.

Saura, C., Oliveira, M., Feng, Y. H., Dai, M. S., Hurvitz, S. A., Kim, S. B., et al. (2019). Neratinib + capecitabine versus lapatinib + capecitabine in patients with HER2+ metastatic breast cancer previously treated with $\geq$ 2 HER2-directed regimens: Findings from the multinational, randomized, phase III NALA trial. *Journal of Clinical Oncology*, *37*(Suppl. 15), 1002.

Schneeweiss, A., Chia, S., Hickish, T., Harvey, V., Eniu, A., Hegg, R., et al. (2013). Pertuzumab plus trastuzumab in combination with standard neoadjuvant anthracycline-containing and anthracycline-free chemotherapy regimens in patients with HER2-positive early breast cancer: A randomized phase II cardiac safety study (TRYPHAENA). *Annals of Oncology*, *24*(9), 2278–2284.

Sheng, X., Zhou, A. P., Yao, X., Shi, Y., Luo, H., Shi, B., et al. (2019). A phase II study of RC48-ADC in HER2-positive patients with locally advanced or metastatic urothelial carcinoma. *Journal of Clinical Oncology*, *37*(Suppl. 15), 4509.

Shi, W., Jiang, T., Nuciforo, P., Hatzis, C., Holmes, E., Harbeck, N., et al. (2017). Pathway level alterations rather than mutations in single genes predict response to HER2-targeted therapies in the neo-ALTTO trial. *Annals of Oncology*, *28*(1), 128–135.

Slamon, D. J., Eiermann, W., Robert, N. J., Giermek, J., Martin, M., Jasiowka, M., et al. (2015). In *S5-04 Ten year follow-up of the BCIRG-006 trial comparing doxorubicin plus cyclophosphamide followed by docetaxel (AC® T) with doxorubicin plus cyclophosphamide followed by docetaxel and trastuzumab (AC® TH) with docetaxel, carboplatin and trastuzumab (TCH) in HER2 + early breast cancer patients. Presented at the 2015 San Antonio breast cancer symposium.*

Slamon, D., Eiermann, W., Robert, N., Pienkowski, T., Martin, M., Rolski, J., et al. (2009). Phase III Randomized Trial Comparing Doxorubicin and Cyclophosphamide Followed by Docetaxel (AC→T) with Doxorubicin and Cyclophosphamide Followed by Docetaxel and Trastuzumab (AC→TH) with Docetaxel, Carboplatin and Trastuzumab (TCH) in Her2neu Positive Early Breast Cancer Patients: BCIRG 006 Study. *Cancer Research*, *69*(Suppl. 24), 62.

Soran, A., Ozmen, V., Ozbas, S., Karanlik, H., Muslumanoglu, M., Igci, A., et al. (2018). Randomized trial comparing resection of primary tumor with no surgery in stage IV breast cancer at presentation: Protocol MF07-01. *Annals of Surgical Oncology*, *25*(11), 3141–3149.

Sparano, J., O'Neill, A., Alpaugh, K., Wolff, A. C., Northfelt, D. W., Dang, C. T., et al. (2018). Association of circulating tumor cells with late recurrence of estrogen receptor-positive breast cancer: A secondary analysis of a randomized clinical trial. *JAMA Oncology*, *4*(12), 1700–1706.

Statler, A. B., Hobbs, B. P., Wei, W., Gupta, A., Blake, C. N., & Nahleh, Z. A. (2019). Real-world treatment patterns and outcomes in HR+/HER2+ metastatic breast cancer patients: A national cancer database analysis. *Scientific Reports*, *9*(1), 18126.

Stern, H. M., Gardner, H., Burzykowski, T., Elatre, W., O'Brien, C., Lackner, M. R., et al. (2015). PTEN loss is associated with worse outcome in HER2-Amplified breast cancer patients but is not associated with trastuzumab resistance. *Clinical Cancer Research*, *21*(9), 2065–2074.

Swain, S. M., Baselga, J., Kim, S. B., Ro, J., Semiglazov, V., Campone, M., et al. (2015). Pertuzumab, trastuzumab, and docetaxel in HER2-positive metastatic breast cancer. *The New England Journal of Medicine*, *372*(8), 724–734.

Swain, S. M., Miles, D., Kim, S. B., Im, Y. H., Im, S. A., Semiglazov, V., et al. (2019). End-of-study analysis from the phase III, randomized, double-blind, placebo (Pla)-controlled CLEOPATRA study of first-line (1L) pertuzumab (P), trastuzumab (H), and docetaxel (D) in patients (pts) with HER2-positive metastatic breast cancer (MBC). *Journal of Clinical Oncology*, *37*(Suppl. 15), 1020.

Symmans, W. F., Wei, C., Gould, R., Yu, X., Zhang, Y., Liu, M., et al. (2017). Long-term prognostic risk after neoadjuvant chemotherapy associated with residual cancer burden and breast cancer subtype. *Journal of Clinical Oncology*, *35*(10), 1049–1060.

Tamura, K., Tsurutani, J., Takahashi, S., Iwata, H., Krop, I. E., Redfern, C., et al. (2019). Trastuzumab deruxtecan (DS-8201a) in patients with advanced HER2-positive breast cancer previously treated with trastuzumab emtansine: A dose-expansion, phase 1 study. *The Lancet Oncology*, *20*(6), 816–826.

Tolaney, S. M., Barry, W. T., Dang, C. T., Yardley, D. A., Moy, B., Marcom, P. K., et al. (2015). Adjuvant paclitaxel and trastuzumab for node-negative, HER2-positive breast cancer. *The New England Journal of Medicine*, *372*(2), 134–141.

Tolaney, S. M., Guo, H., Pernas, S., Barry, W. T., Dillon, D. A., Ritterhouse, L., et al. (2019). Seven-year follow-up analysis of adjuvant paclitaxel and trastuzumab trial for node-negative, human epidermal growth factor receptor 2-positive breast cancer. *Journal of Clinical Oncology*, *37*(22), 1868–1875.

Tolaney, S. M., Trippa, L., Barry, W., Hu, J., Dang, C., Yardley, D., et al. (2019). In *TBCRC 033: A randomized phase II study of adjuvant trastuzumab emtansine (T-DM1) vs paclitaxel (T) in combination with trastuzumab (H) for stage I HER2-positive breast cancer (BC) (ATEMPT). Presented at the 2019 San Antonio breast cancer symposium*.

Tolaney, S. M., Wardley, A. M., Zambelli, S., Hilton, J., Troso-Sandoval, T., Ricci, F., et al. (2019). MonarcHER: A randomized phase II study of abemaciclib plus trastuzumab with or without fulvestrant versus trastuzumab plus standard-of-care chemotherapy in women with HR+, HER2+ advanced breast cancer. *Annals of Oncology*, *30*(Suppl. 5), LBA23.

Trail, P. A., Dubowchik, G. M., & Lowinger, T. B. (2018). Antibody drug conjugates for treatment of breast cancer: Novel targets and diverse approaches in ADC design. *Pharmacology & Therapeutics*, *181*, 126–142.

Trapani, D., Rajasekar, A. K. A., & Mathew, A. (2019). More options for adjuvant treatment of HER2-positive breast cancer: How to choose wisely? *International Journal of Cancer*, *145*(11), 2901–2906.

Turner, N., Bye, H., Kernaghan, S., Proszek, P., Fribbens, C., Moretti, L., et al. (2018). Abstract OT1-06-03: The plasmaMATCH trial: A multiple parallel cohort, open-label, multi-centre phase II clinical trial of ctDNA screening to direct targeted therapies in patients with advanced breast cancer (CRUK/15/010). *Cancer Research*, *78*(Suppl. 4), OT1-06-03.

Untch, M., Fasching, P. A., Konecny, G. E., Hasmüller, S., Lebeau, A., Kreienberg, R., et al. (2011). Pathologic complete response after neoadjuvant chemotherapy plus trastuzumab predicts favorable survival in human epidermal growth factor receptor 2-overexpressing breast cancer: Results from the TECHNO trial of the AGO and GBG study groups. *Journal of Clinical Oncology*, *29*(25), 3351–3357.

Untch, M., Rezai, M., Loibl, S., Fasching, P. A., Huober, J., Tesch, H., et al. (2010). Neoadjuvant treatment with trastuzumab in HER2-positive breast cancer: Results from the GeparQuattro study. *Journal of Clinical Oncology*, *28*(12), 2024–2031.

Untch, M., & von Minckwitz, G. (2011). Neoadjuvant chemotherapy: Early response as a guide for further treatment: Clinical, radiological, and biological. *Journal of the National Cancer Institute. Monographs*, *2011*(43), 138–141.

Van Ramshorst, M. S., Werkhoven, E. V., Mandjes, I. A., Kemper, I., Dezentjé, V. O., Oving, I. M., et al. (2017). A phase III trial of neoadjuvant chemotherapy with or without anthracyclines in the presence of dual HER2-blockade for HER2+ breast cancer: The TRAIN-2 study (BOOG 2012-03). *Journal of Clinical Oncology*, *35*(Suppl. 15), 507.

von Minckwitz, G., Huang, C. S., Mano, M. S., Loibl, S., Mamounas, E. P., Untch, M., et al. (2019). Trastuzumab emtansine for residual invasive HER2-positive breast cancer. *The New England Journal of Medicine*, *380*(7), 617–628.

von Minckwitz, G., Procter, M., de Azambuja, E., Zardavas, D., Benyunes, M., Viale, G., et al. (2017). Adjuvant pertuzumab and trastuzumab in early HER2-positive breast cancer. *The New England Journal of Medicine*, *377*(2), 122–131.

Wolf-Yadlin, A., Kumar, N., Zhang, Y., Hautaniemi, S., Zaman, M., Kim, H. D., et al. (2006). Effects of HER2 overexpression on cell signaling networks governing proliferation and migration. *Molecular Systems Biology*, *2*, 54.

Xu, B., Wang, J., Fang, J., Chen, X., Han, Y., Li, Q., et al. (2019). In *PD4-06. Early clinical development of RC48-ADC in patients with HER2 positive metastatic breast cancer. Presented at the 2019 San Antonio breast cancer symposium*.

Yi, Z., Ma, F., Guan, Y., Chang, L., Wang, W., Guan, X., et al. (2018). Impact of HER2 mutation status on personalized molecular targeted therapy in advanced breast cancers. *Journal of Clinical Oncology*, *36*(Suppl. 15), 1039.

FSC
www.fsc.org
MIX
Papier | Fördert gute Waldnutzung
FSC® C083411

Kolibri 360 GmbH
Ferdinand-Jühlke-Straße 7
99095 Erfurt, Deutschland
produktsicherheit@kolibri360.de